Brownian Motion

This eagerly awaited textbook offers a broad and deep exposition of Brownian motion. Extensively class tested, it leads the reader from the basics to the latest research in the area.

Starting with the construction of Brownian motion, the book then proceeds to sample path properties such as continuity and nowhere differentiability. Notions of fractal dimension are introduced early and are used throughout the book to describe fine properties of Brownian paths. The relation of Brownian motion and random walk is explored from several viewpoints, including a development of the theory of Brownian local times from random walk embeddings. Stochastic integration is introduced as a tool, and an accessible treatment of the potential theory of Brownian motion clears the path for an extensive treatment of intersections of Brownian paths. An investigation of exceptional points on the Brownian path and an appendix on SLE processes, by Oded Schramm and Wendelin Werner, lead directly to recent research themes.

'This splendid account of the modern theory of Brownian motion puts special emphasis on sample path properties and connections with harmonic functions and potential theory, without omitting such important topics as stochastic integration, local times or relations with random walk. The most significant properties of Brownian motion are derived via powerful and elegant methods. This book, which fills a gap in the existing literature, will be of interest both to the beginner, for the clarity of exposition and the judicious choice of topics, and to the specialist, who will find neat approaches to many classical results and to some more recent ones. This beautiful book will soon become a must for anybody who is interested in Brownian motion and its applications.'

Jean-François Le Gall, Université Paris 11 (Paris-Sud, Orsay)

'*Brownian Motion* by Mörters and Peres, a modern and attractive account of one of the central topics of probability theory, will serve both as an accessible introduction at the level of a Master's course and as a work of reference for fine properties of Brownian paths. The unique focus of the book on Brownian motion gives it a satisfying concreteness and allows a rapid approach to some deep results.

The introductory chapters, besides providing a careful account of the theory, offer some helpful points of orientation towards an intuitive and mature grasp of the subject matter. The authors have made many contributions to our understanding of path properties, fractal dimensions and potential theory for Brownian motion, and this expertise is evident in the later chapters of the book. I particularly liked the marking of the 'leaves' of the theory by stars, not only because this offers a chance to skip on, but also because these are often the high points of our present knowledge.'

James Norris, University of Cambridge

'This excellent book does a beautiful job of covering a good deal of the theory of Brownian motion in a very user-friendly fashion. The approach is hands-on which makes it an attractive book for a first course on the subject. It also contains topics not usually covered, such as the 'intersection-equivalence' approach to multiple points as well as the study of slow and fast points. Other highlights include detailed connections with random fractals and a short overview of the connections with SLE. I highly recommend it.'

Jeff Steif, Chalmers University of Technology

Brownian Motion

PETER MÖRTERS AND YUVAL PERES

with an appendix by Oded Schramm and Wendelin Werner

CAMBRIDGE
UNIVERSITY PRESS

CAMBRIDGE
UNIVERSITY PRESS

University Printing House, Cambridge CB2 8BS, United Kingdom

Cambridge University Press is part of the University of Cambridge.

It furthers the University's mission by disseminating knowledge in the pursuit of education, learning and research at the highest international levels of excellence.

www.cambridge.org
Information on this title: www.cambridge.org/9780521760188

© P. Mörters and Y. Peres 2010

First published 2010

A catalogue record for this publication is available from the British Library

ISBN 978-0-521-76018-8 Hardback

Contents

Preface

The aim of this book is to introduce Brownian motion as central object of probability theory and discuss its properties, putting particular emphasis on sample path properties. Our hope is to capture as much as possible the spirit of Paul Lévy's investigations on Brownian motion, by moving quickly to the fascinating features of the Brownian motion process, and filling in more and more details into the picture as we move along.

Inevitably, while exploring the nature of Brownian paths one encounters a great variety of other subjects: Hausdorff dimension serves from early on in the book as a tool to quantify subtle features of Brownian paths, stochastic integrals helps us to get to the core of the invariance properties of Brownian motion, and potential theory is developed to enable us to control the probability the Brownian motion hits a given set.

An important idea of this book is to make it as *interactive* as possible and therefore we have included more than 100 exercises collected at the end of each of the ten chapters. Exercises marked with the symbol §̲ have either a hint, a reference to a solution, or a full solution given at the end of the book. We have also marked some theorems with a star to indicate that the results will not be used in the remainder of the book and may be skipped on first reading. At the end of the book we have given a short list of selected open research problems dealing with the material of the book.

This book grew out of lectures given by Yuval Peres at the Statistics Department, University of California, Berkeley in Spring 1998. We are grateful to the students who attended the course and wrote the first draft of the notes: Diego Garcia, Yoram Gat, Diogo A. Gomes, Charles Holton, Frédéric Latrémolière, Wei Li, Ben Morris, Jason Schweinsberg, Bálint Virág, Ye Xia and Xiaowen Zhou. The first draft of these notes, about 80 pages in volume, was edited by Bálint Virág and Elchanan Mossel and at this stage corrections were made by Serban Nacu and Yimin Xiao. The notes were distributed via the internet and turned out to be very popular — this demand motivated us to expand these notes to a full book hopefully retaining the character of the original notes.

Peter Mörters lectured on the topics of this book in the Graduate School in Mathematical Sciences at the University of Bath in Autumn 2003, thanks are due to the audience, and in particular to Alex Cox and Pascal Vogt, for their contributions. Yuval Peres thanks

Pertti Mattila for the invitation to lecture on this material at the joint summer school in Jyväskyla, August 1999, and Peter Mörters thanks Michael Scheutzow for the invitation to lecture at the Berlin graduate school in probability in Stralsund, April 2003.

When it became clear that the new developments around the stochastic Loewner evolution would open a new chapter in the story of Brownian motion we discussed the inclusion of a chapter on this topic. Realising that doing this rigorously in detail would go beyond the scope of this book, we asked Oded Schramm to provide an appendix describing the new developments in a less formal manner. Oded agreed and immediately started designing the appendix, but his work was cut short by his tragic and premature death in 2008. We are very grateful that Wendelin Werner accepted the task of completing this appendix at very short notice.

Several people read drafts of the book at various stages, supplied us with helpful lists of corrections, and suggested or tested exercises and references. We thank Anselm Adelmann, Tonci Antunovic, Christian Bartsch, Noam Berger, Jian Ding, Uta Freiberg, Nina Gantert, Subhroshekhar Gosh, Ben Hough, Davar Khoshnevisan, Richard Kiefer, Achim Klenke, Michael Kochler, Manjunath Krishnapur, David Levin, Nathan Levy, Arjun Malhotra, Jason Miller, Asaf Nachmias, Weiyang Ning, Marcel Ortgiese, Ron Peled, Jim Pitman, Michael Scheutzow, Perla Sousi, Jeff Steif, Kamil Szczegot, Ran Tessler, Hermann Thorisson, and Brigitta Vermesi.

We also thank several people who have contributed pictures, namely Ben Hough, Marcel Ortgiese, Yelena Shvets and David Wilson. The cover shows a planar Brownian motion with points coloured according to the occupation measure of a small neighbourhood, we thank Raissa d'Souza for providing the picture.

Peter Mörters
Yuval Peres

Frequently used notation

Numbers:

$\lceil x \rceil$ the smallest integer bigger or equal to x

$\lfloor x \rfloor$ the largest integer smaller or equal to x

$\mathfrak{Re}(z), \mathfrak{Im}(z)$ the real, resp. imaginary, part of the complex number z

i   the imaginary unit

Topology of Euclidean space $\mathbb{R}^d$:

$\mathbb{R}^d$ Euclidean space consisting of all column vectors $x = (x_1, \ldots, x_d)^{\mathrm{T}}$

$|\cdot|$ Euclidean norm $|x| = \sqrt{\sum_{i=1}^{d} x_i^2}$

$\mathcal{B}(x, r)$ the open ball of radius $r > 0$ centred in $x \in \mathbb{R}^d$, i.e. $\mathcal{B}(x, r) = \{ y \in \mathbb{R}^d : |x - y| < r \}$

$\overline{U}$ closure of the set $U \subset \mathbb{R}^d$

∂U boundary of the set $U \subset \mathbb{R}^d$

$\mathfrak{B}(A)$ the collection of all Borel subsets of $A \subset \mathbb{R}^d$

Binary relations:

$a \wedge b$ the minimum of a and b

$a \vee b$ the maximum of a and b

$X \stackrel{\mathrm{d}}{=} Y$ the random variables X and Y have the same distribution

$X_n \stackrel{\mathrm{d}}{\to} X$ the random variables X_n converge to X in distribution, see Section 12.1 in the appendix

$a(n) \asymp b(n)$ the ratio of the two sides is bounded from above and below by positive constants that do not depend on n

$a(n) \sim b(n)$ the ratio of the two sides converges to one

Vectors, functions, and measures:

I_d $d \times d$ identity matrix

1_A indicator function with $1_A(x) = 1$ if $x \in A$ and 0 otherwise

δ_x	Dirac measure with mass concentrated on x, i.e. $\delta_x(A) = 1$ if $x \in A$ and 0 otherwise
f^+	the positive part of the function f, i.e. $f^+(x) = f(x) \vee 0$
f^-	the negative part of the function f, i.e. $f^-(x) = -(f(x) \wedge 0)$
$\mathcal{L}_d$ or $\mathcal{L}$	Lebesgue measure on $\mathbb{R}^d$
$\sigma_{x,r}$	$(d-1)$-dimensional surface measure on $\partial\mathcal{B}(x,r) \subset \mathbb{R}^d$ if $x = 0, r = 1$ we also write $\sigma = \sigma_{0,1}$
$\varpi_{x,r}$	uniform distribution on $\partial\mathcal{B}(x,r)$, $\varpi_{x,r} = \frac{\sigma_{x,r}}{\sigma_{x,r}(\partial\mathcal{B}(x,r))}$, if $x = 0, r = 1$ we also write $\varpi = \varpi_{0,1}$

Function spaces:

$\mathbf{C}(K)$	the topological space of all continuous functions on the compact $K \subset \mathbb{R}^d$, equipped with the supremum norm $\|f\| = \sup_{x \in K} |f(x)|$
$\mathbf{L}^p(\mu)$	the Banach space of equivalence classes of functions f with finite $\mathbf{L}^p$-norm $\|f\|_p = \left(\int f^p \, d\mu \right)^{1/p}$. If $\mu = \mathcal{L}_{|K}$ we write $\mathbf{L}^p(K)$.
$\mathbf{D}[0,1]$	the Dirichlet space consisting of functions $F \in \mathbf{C}[0,1]$ such that for some $f \in \mathbf{L}^2[0,1]$ and all $t \in [0,1]$ we have $F(t) = \int_0^t f(s) \, ds$.

Probability measures and σ-algebras:

$\mathbb{P}_x$	a probability measure on a measure space $(\Omega, \mathcal{A})$ such that the process $\{B(t) : t \geq 0\}$ is a Brownian motion started in x
$\mathbb{E}_x$	the expectation associated with $\mathbb{P}_x$
$\mathfrak{p}(t,x,y)$	the transition density of Brownian motion $\mathbb{P}_x\{B(t) \in A\} = \int_A \mathfrak{p}(t,x,y) \, dy$
$\mathcal{F}^0(t)$	the smallest σ-algebra that makes $\{B(s) : 0 \leq s \leq t\}$ measurable
$\mathcal{F}^+(t)$	the right-continuous augmentation $\mathcal{F}^+(t) = \bigcap_{s > t} \mathcal{F}^0(s)$.

Stopping times:

For any Borel sets $A_1, A_2, \ldots \subset \mathbb{R}^d$ and a Brownian motion $B : [0, \infty) \to \mathbb{R}^d$,

$$\tau(A_1) := \inf\{t \geq 0 : B(t) \in A_1\}, \qquad \text{the entry time into } A_1,$$

$$\tau(A_1, \ldots, A_n) := \begin{cases} \inf\{t \geq \tau(A_1, \ldots, A_{n-1}) : B(t) \in A_n\}, & \text{if } \tau(A_1, \ldots, A_{n-1}) < \infty, \\ \infty, & \text{otherwise.} \end{cases}$$

the time to enter A_1 and then A_2 and so on until A_n.

Systems of subsets in $\mathbb{R}^d$:

For any fixed d-dimensional unit cube Cube $= x + [0,1]^d$ we denote:

$\mathfrak{D}_k$ family of all half-open dyadic subcubes $D = x + \prod_{i=1}^d \left[k_i 2^{-k}, (k_i + 1)2^{-k} \right) \subset \mathbb{R}^d$, $k_i \in \{0, \ldots, 2^k - 1\}$, of side length 2^{-k}

$\mathfrak{D}$ all half-open dyadic cubes $\mathfrak{D} = \bigcup_{k=0}^{\infty} \mathfrak{D}_k$ in Cube

$\mathfrak{C}_k$ family of all compact dyadic subcubes $D = x + \prod_{i=1}^d \left[k_i 2^{-k}, (k_i + 1)2^{-k} \right] \subset \mathbb{R}^d$, $k_i \in \{0, \ldots, 2^k - 1\}$, of side length 2^{-k}

$\mathfrak{C}$ all compact dyadic cubes $\mathfrak{C} = \bigcup_{k=0}^{\infty} \mathfrak{C}_k$ in Cube.

Potential theory:

For a metric space (E, ρ) and mass distribution μ on E:

$\phi_\alpha(x)$ the α-potential of a point $x \in E$ defined as $\phi_\alpha(x) = \int \frac{d\mu(y)}{\rho(x,y)^\alpha}$,

$I_\alpha(\mu)$ the α-energy of the measure μ defined as $I_\alpha(\mu) = \iint \frac{d\mu(x)\, d\mu(y)}{\rho(x,y)^\alpha}$,

$\text{Cap}_\alpha(E)$ the α-capacity of E defined as $\text{Cap}_\alpha(E) = \sup\{I_\alpha(\mu)^{-1} : \mu(E) = 1\}$.

For a general kernel $K \colon E \times E \to [0, \infty]$:

$U_\mu(x)$ the potential of μ at x defined as $U_\mu(x) = \int K(x,y)\, d\mu(y)$,

$I_K(\mu)$ K-energy of μ defined as $I_K(\mu) = \iint K(x,y)\, d\mu(x)\, d\mu(y)$,

$\text{Cap}_K(E)$ K-capacity of E defined as $\text{Cap}_K(E) = \sup\{I_K(\mu)^{-1} : \mu(E) = 1\}$.

If $K(x,y) = f(\rho(x,y))$ we also write:

$I_f(\mu)$ instead of $I_K(\mu)$,

$\text{Cap}_f(E)$ instead of $\text{Cap}_K(E)$.

Sets and processes associated with Brownian motion:

For a linear Brownian motion $\{B(t) \colon t \geqslant 0\}$:

$\{M(t) \colon t \geqslant 0\}$ the maximum process defined by $M(t) = \sup_{s \leqslant t} B(s)$,

Rec   the set of record points $\{t \geqslant 0 \colon B(t) = M(t)\}$,

Zeros   the set of zeros $\{t \geqslant 0 \colon B(t) = 0\}$.

For a Brownian motion $\{B(t) \colon t \geqslant 0\}$ in $\mathbb{R}^d$ for $d \geqslant 1$:

Graph(A) the graph $\{(t, B(t)) \colon t \in A\} \subset \mathbb{R}^{d+1}$,

Range(A) the range $\{B(t) \colon t \in A\} \subset \mathbb{R}^d$.

Occasionally these notions are used for functions $f \colon [0, \infty) \to \mathbb{R}^d$ which are not necessarily Brownian sample paths, which we indicate by appending a subindex f to the notion.

Motivation

Much of probability theory is devoted to describing the *macroscopic picture* emerging in random systems defined by a host of *microscopic random effects*. Brownian motion is the macroscopic picture emerging from a particle moving randomly in d-dimensional space without making very big jumps. On the microscopic level, at any time step, the particle receives a random displacement, caused for example by other particles hitting it or by an external force, so that, if its position at time zero is S_0, its position at time n is given as $S_n = S_0 + \sum_{i=1}^{n} X_i$, where the displacements $X_1, X_2, X_3, \ldots$ are assumed to be independent, identically distributed random variables with values in $\mathbb{R}^d$. The process $\{S_n : n \geqslant 0\}$ is a random walk, the displacements represent the microscopic inputs. When we think about the macroscopic picture, what we mean is questions such as:

- Does S_n drift to infinity?
- Does S_n return to the neighbourhood of the origin infinitely often?
- What is the speed of growth of $\max\{|S_1|, \ldots, |S_n|\}$ as $n \to \infty$?
- What is the asymptotic number of windings of $\{S_n : n \geqslant 0\}$ around the origin?

It turns out that not all the features of the microscopic inputs contribute to the macroscopic picture. Indeed, if they exist, only the *mean* and *covariance* of the displacements are shaping the picture. In other words, all random walks whose displacements have the same mean and covariance matrix give rise to the same macroscopic process, and even the assumption that the displacements have to be independent and identically distributed can be substantially relaxed. This effect is called *universality*, and the macroscopic process is often called a *universal object*. It is a common approach in probability to study various phenomena through the associated universal objects.

If the jumps of a random walk are sufficiently tame to become negligible in the macroscopic picture, in particular if it has finite mean and variance, any continuous time stochastic process $\{B(t) : t \geqslant 0\}$ describing the macroscopic features of this random walk should have the following properties:

(1) for all times $0 \leqslant t_1 \leqslant t_2 \leqslant \ldots \leqslant t_n$ the random variables

$$B(t_n) - B(t_{n-1}),\ B(t_{n-1}) - B(t_{n-2}),\ \ldots,\ B(t_2) - B(t_1)$$

are independent; we say that the process has *independent increments*,

1

(2) the distribution of the increment $B(t + h) - B(t)$ does not depend on t; we say that the process has *stationary increments*,

(3) the process $\{B(t): t \geqslant 0\}$ has almost surely continuous paths.

It follows (with some work) from the central limit theorem that these features imply that there exists a vector $\mu \in \mathbb{R}^d$ and a matrix $\Sigma \in \mathbb{R}^{d \times d}$ such that

(4) for every $t \geqslant 0$ and $h \geqslant 0$ the increment $B(t + h) - B(t)$ is multivariate normally distributed with mean $h\mu$ and covariance matrix $h\Sigma\Sigma^{\mathrm{T}}$.

Hence any process with the features (1)-(3) above is characterised by just three parameters,

- the *initial distribution*, i.e. the law of $B(0)$,
- the *drift vector μ*,
- the *diffusion matrix Σ*.

The process $\{B(t): t \geqslant 0\}$ is called a *Brownian motion with drift μ and diffusion matrix Σ*. If the drift vector is zero, and the diffusion matrix is the identity we simply say the process is a *Brownian motion*. If $B(0) = 0$, i.e. the motion is started at the origin, we use the term *standard Brownian motion*.

Suppose we have a standard Brownian motion $\{B(t): t \geqslant 0\}$. If X is a random variable with values in $\mathbb{R}^d$, μ a vector in $\mathbb{R}^d$ and Σ a $d \times d$ matrix, then it is easy to check that $\{\tilde{B}(t): t \geqslant 0\}$ given by

$$\tilde{B}(t) = \tilde{B}(0) + \mu t + \Sigma B(t), \text{ for } t \geqslant 0,$$

is a process with the properties (1)-(4) with initial distribution X, drift vector μ and diffusion matrix Σ. Hence the macroscopic picture emerging from a random walk with finite variance can be fully described by a standard Brownian motion.

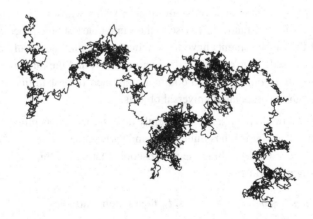

Fig. 0.1. The range of a planar Brownian motion $\{B(t): 0 \leqslant t \leqslant 1\}$.

In *Chapter 1* we start exploring Brownian motion by looking at dimension $d = 1$. Here Brownian motion is a random continuous function and we ask about its *regularity*, for example: For which parameters α is the random function $B \colon [0, 1] \to \mathbb{R}$ α-Hölder continuous? Is the random function $B \colon [0, 1] \to \mathbb{R}$ differentiable? The surprising answer to the second question was given by Paley, Wiener and Zygmund in 1933: Almost surely, the random function $B \colon [0, 1] \to \mathbb{R}$ is *nowhere* differentiable! This is particularly interesting, as it is not easy to construct a continuous, nowhere differentiable function without the help of randomness. We give a modern proof of the Paley, Wiener and Zygmund theorem, see Theorem 1.30.

In *Chapter 2* we move to general dimension d. We prove and explore the strong Markov property, which roughly says that at suitable random times Brownian motion starts afresh, see Theorem 2.16. Among the facts we derive from this property are that the set of all points visited by a Brownian motion in $d \geqslant 2$ has area zero, but the set of times when Brownian motion in $d = 1$ revisits the origin is uncountable. Besides these sample path properties, the strong Markov property is also the key to some fascinating distributional identities. It enables us to understand, for example, the process $\{M(t) \colon t \geqslant 0\}$ of the running maxima $M(t) = \max_{0 \leqslant s \leqslant t} B(s)$ of Brownian motion in $d = 1$, the process $\{T_a \colon a \geqslant 0\}$ of the first hitting times $T_a = \inf\{t \geqslant 0 \colon B(t) = a\}$ of level a of a Brownian motion in $d = 1$, and the process of the vertical first hitting positions of the lines $\{(x, y) \in \mathbb{R}^2 \colon x = a\}$ by a Brownian motion in $d = 2$, as a function of a.

In *Chapter 3* we explore the rich relations of Brownian motion to harmonic analysis. In particular we learn how Brownian motion helps solving the classical *Dirichlet problem*.

Fig. 0.2. Brownian motion and the Dirichlet problem

For its formulation in the planar case, fix a connected open set $U \subset \mathbb{R}^2$ with nice boundary, and let $\varphi \colon \partial U \to \mathbb{R}$ be continuous. The harmonic functions $f \colon U \to \mathbb{R}$ on the domain U are characterised by the differential equation

$$\frac{\partial^2 f}{\partial x_1^2}(x) + \frac{\partial^2 f}{\partial x_2^2}(x) = 0 \quad \text{for all } x \in U.$$

The Dirichlet problem is to find, for a given domain U and boundary data φ, a continuous function $f\colon U \cup \partial U \to \mathbb{R}$, which is harmonic on U and agrees with φ on ∂U. In Theorem 3.12 we show that the unique solution of this problem is given as

$$f(x) = \mathbb{E}\big[\varphi(B(T)) \mid B(0) = x\big], \text{ for } x \in \overline{U},$$

where $\{B(t)\colon t \geqslant 0\}$ is a Brownian motion and $T = \inf\{t \geqslant 0\colon B(t) \notin U\}$ is the first exit time from U. We exploit this result, for example, to show exactly in which dimensions a particle following a Brownian motion drifts to infinity, see Theorem 3.20.

In *Chapter 4* we provide one of the major tools in our study of Brownian motion, the concept of Hausdorff dimension, and show how it can be applied in the context of Brownian motion. Indeed, when describing the sample paths of a Brownian motion one frequently encounters questions of the size of a given set: How big is the set of all points visited by a Brownian motion in the plane? How big is the set of double-points of a planar Brownian motion? How big is the set of times where Brownian motion visits a given set, say a point? For an example, let $\{B(t)\colon t \geqslant 0\}$ be Brownian motion on the real line and look at Zeros $= \{t \geqslant 0\colon B(t) = 0\}$, the set of its zeros. Although $t \mapsto B(t)$ is a continuous function, Zeros is an infinite set. This set is *big*, as it is an uncountable set without isolated points. However, it is also *small* in the sense that its Lebesgue measure is zero. Indeed, Zeros is a fractal set and we show in Theorem 4.24 that its Hausdorff dimension is $1/2$.

In *Chapter 5* we explore the relationship of random walk and Brownian motion. We prove a theorem which justifies our initial point of view that Brownian motion is the macroscopic picture emerging from a large class of random walks: By *Donsker's invariance principle* one can obtain Brownian motion by taking scaled copies of a random walk and taking a limit in distribution. This result is called an invariance principle because all random walks whose increments have mean zero and finite variance essentially produce the same limit, a Brownian motion. Donsker's invariance principle is also a major tool in deriving results for random walks from those of Brownian motion, and vice versa. Both directions can be useful: In some cases the fact that Brownian motion is a continuous time process is an advantage over discrete time random walks. For example, as we discuss below, Brownian motion has scaling invariance properties, which can be a powerful tool in the study of its path properties. In other cases it is a major advantage that (simple) random walk is a discrete object and combinatorial arguments can be the right tool to derive important features. Chapter 5 offers a number of case studies for the mutually beneficial relationship between Brownian motion and random walks. Beyond Donsker's invariance principle, there is a second fascinating aspect of the relationship between random walk and Brownian motion: Given a Brownian motion in $d = 1$, we can sample from its path at certain carefully chosen times, and thus construct every random walk with mean zero and finite variance. Finding these times is called the *Skorokhod embedding problem* and we shall give two different solutions to it. The embedding problem is also the main tool in our proof of Donsker's invariance principle.

In *Chapter 6* we look again at Brownian motion in dimension $d = 1$. For a random walk on the integers running for a finite amount of time, we can define a 'local time' at a point $z \in \mathbb{Z}$ by simply counting how many times the walk visits z. Can we define an analogous quantity for Brownian motion? In Chapter 6 we show that this is possible, and offer an elegant construction of Brownian local time based on a random walk approximation. A first highlight of this chapter arises when we aim to describe the local times: If a Brownian path is started at some positive level $a > 0$ and stopped upon hitting zero, we can describe the process of local times in x as a function of x, for $0 \leqslant x \leqslant a$. The resulting process is distributed like the square of the modulus of a planar Brownian motion. This is the famous *Ray–Knight theorem*. The second highlight of this chapter is related to the nature of local time at a fixed point. The Brownian local time in x is no longer the number of visits to the point x by a Brownian motion – if x is visited at all, this number would be infinite – but we shall see that it can be described as the Hausdorff measure of the set of times at which the motion visits x.

Because Brownian motion arises as the scaling limit of a great variety of different random walks, it naturally has a number of invariance properties. One of the most important invariance properties of Brownian motion is *conformal invariance*, which we discuss in *Chapter 7*. To make this plausible think of an angle-preserving linear mapping $L \colon \mathbb{R}^d \to \mathbb{R}^d$, like a rotation followed by multiplication by a. Take a random walk started in zero with increments of mean zero and covariance matrix the identity, and look at its image under L. This image is again a random walk and its increments are distributed like LX. Appropriately rescaled as in Donsker's invariance principle, both random walks converge to a Brownian motion, the second one with a slightly different covariance matrix. This process can be identified as a time-changed Brownian motion $\{B(a^2 t) \colon t \geqslant 0\}$. This easy observation has a deeper, local counterpart for planar Brownian motion: Suppose that $\phi \colon U \to V$ is a conformal mapping of a simply connected domain $U \subset \mathbb{R}^2$ onto a domain $V \subset \mathbb{R}^2$. Conformal mappings are locally angle-preserving and the Riemann mapping theorem of complex analysis tells us that a lot of such domains and mappings exist.

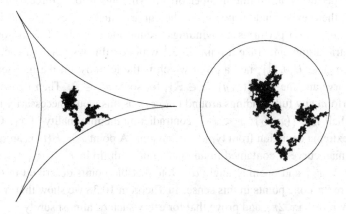

Fig. 0.3. A conformal mapping of Brownian paths

Suppose that $\{B(t) \colon t \geqslant 0\}$ is a standard Brownian motion started in some point $x \in U$ and $\tau = \inf\{t > 0 \colon B(t) \notin U\}$ is the first exit time of the path from the domain U. Then it turns out that the image process $\{\phi(B(t)) \colon 0 \leqslant t \leqslant \tau\}$ is a *time-changed* Brownian motion in the domain V, stopped when it leaves V, see Theorem 7.20. In order to prove this we have to develop a little bit of the theory of stochastic integration with respect to a Brownian motion, and we give a lot of further applications of this tool in Chapter 7.

In *Chapter 8* we develop the potential theory of Brownian motion. The problem which is the motivation behind this is, given a compact set $A \subset \mathbb{R}^d$, to find the probability that a Brownian motion $\{B(t): t \geqslant 0\}$ hits the set A, i.e. that there exists $t > 0$ with $B(t) \in A$. This problem is answered in the best possible way by Theorem 8.24, which is a modern extension of a classical result of Kakutani: The hitting probability can be approximated by the capacity of A with respect to the Martin kernel up to a factor of two.

With a wide range of tools at our hand, in *Chapter 9* we study the self-intersections of Brownian motion: For example, a point $x \in \mathbb{R}^d$ is called a double point of $\{B(t): t \geqslant 0\}$ if there exist times $0 < t_1 < t_2$ such that $B(t_1) = B(t_2) = x$. In which dimensions does Brownian motion have double points? How big is the set of double points? We show that in dimensions $d \geqslant 4$ no double points exist, in dimension $d = 3$ double points exist and the set of double points has Hausdorff dimension one, and in dimension $d = 2$ double points exist and the set of double points has Hausdorff dimension two. In dimension $d = 2$ we find a surprisingly complex situation: While every point $x \in \mathbb{R}^2$ is almost surely not visited by a Brownian motion, there exist (random) points in the plane, which are visited infinitely often, even uncountably often. This result, Theorem 9.24, is one of the highlights of this book.

Chapter 10 deals with exceptional points for Brownian motion and Hausdorff dimension spectra of families of exceptional points. To explain an example, we look at a Brownian motion in the plane run for one time unit, which is a continuous curve $\{B(t): t \in [0,1]\}$. In Chapter 7 we see that, for any point on the curve, almost surely, the Brownian motion performs an infinite number of full windings in both directions around this point. Still, there exist random points on the curve, which are exceptional in the sense that Brownian motion performs no windings around them at all. This follows from an easy geometric argument: Take a point in $\mathbb{R}^2$ with coordinates (x_1, x_2) such that $x_1 = \min\{x: (x, x_2) \in B[0,1]\}$, i.e. a point which is the leftmost on the intersection of the Brownian curve and the line $\{(z, y): z \in \mathbb{R}\}$, for some $x_2 \in \mathbb{R}$. Then Brownian motion does not perform any full windings around (x_1, x_2), as this would necessarily imply that it crosses the half-line $\{(x, x_2): x < x_2\}$, contradicting the minimality of x_1. One can ask for a more extreme deviation from typical behaviour: A point $x = B(t)$ is an α-cone point if the Brownian curve is contained in an open cone with tip in $x = (x_1, x_2)$, central axis $\{(x_1, x): x > x_2\}$ and opening angle α. Note that the points described in the previous paragraph are 2π-cone points in this sense. In Theorem 10.38 we show that α-cone points exist exactly if $\alpha \in [\pi, 2\pi]$, and prove that for every such α, almost surely,

$$\dim \left\{ x \in \mathbb{R}^2 : x \text{ is an } \alpha\text{-cone point} \right\} = 2 - \frac{2\pi}{\alpha}.$$

This is an example of a Hausdorff dimension spectrum, a topic which has been at the centre of some research activity at the beginning of the current millennium.

1

Brownian motion as a random function

In this chapter we focus on one-dimensional, or linear, Brownian motion. We start with Paul Lévy's construction of Brownian motion and discuss two fundamental sample path properties, continuity and differentiability. We then discuss the Cameron–Martin theorem, which shows that sample path properties for Brownian motion with drift can be obtained from the corresponding results for driftless Brownian motion.

1.1 Paul Lévy's construction of Brownian motion

1.1.1 Definition of Brownian motion

Brownian motion is closely linked to the normal distribution. Recall that a random variable X is normally distributed with mean μ and variance σ^2 if

$$\mathbb{P}\{X > x\} = \frac{1}{\sqrt{2\pi\sigma^2}} \int_x^\infty e^{-\frac{(u-\mu)^2}{2\sigma^2}} \, du, \qquad \text{for all } x \in \mathbb{R}.$$

Definition 1.1. A real-valued stochastic process $\{B(t) \colon t \geqslant 0\}$ is called a **(linear) Brownian motion** with start in $x \in \mathbb{R}$ if the following holds:
- $B(0) = x$,
- the process has **independent increments,** i.e. for all times $0 \leqslant t_1 \leqslant t_2 \leqslant \ldots \leqslant t_n$ the increments $B(t_n) - B(t_{n-1})$, $B(t_{n-1}) - B(t_{n-2})$, $\ldots$, $B(t_2) - B(t_1)$ are independent random variables,
- for all $t \geqslant 0$ and $h > 0$, the increments $B(t+h) - B(t)$ are normally distributed with expectation zero and variance h,
- almost surely, the function $t \mapsto B(t)$ is continuous.

We say that $\{B(t) \colon t \geqslant 0\}$ is a **standard Brownian motion** if $x = 0$. ◇

We will address the nontrivial question of the *existence* of a Brownian motion in Section 1.1.2. For the moment let us step back and look at some technical points. We have defined Brownian motion as a *stochastic process* $\{B(t) \colon t \geqslant 0\}$ which is just a family of (uncountably many) random variables $\omega \mapsto B(t, \omega)$ defined on a single probability space $(\Omega, \mathcal{A}, \mathbb{P})$. At the same time, a stochastic process can also be interpreted as a *random function* with the sample functions defined by $t \mapsto B(t, \omega)$. The *sample path properties* of a stochastic process are the properties of these random functions, and it is these properties we will be most interested in in this book.

7

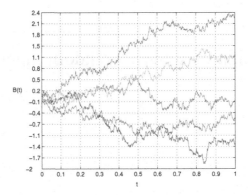

Fig. 1.1. Graphs of five sampled Brownian motions

By the **finite-dimensional distributions** of a stochastic process $\{B(t)\colon t \geqslant 0\}$ we mean the laws of all the finite dimensional random vectors

$$\big(B(t_1), B(t_2), \ldots, B(t_n)\big), \text{ for all } 0 \leqslant t_1 \leqslant t_2 \leqslant \ldots \leqslant t_n.$$

To describe these joint laws it suffices to describe the joint law of $B(0)$ and the increments

$$\big(B(t_1) - B(0), B(t_2) - B(t_1), \ldots, B(t_n) - B(t_{n-1})\big), \text{ for all } 0 \leqslant t_1 \leqslant t_2 \leqslant \ldots \leqslant t_n.$$

This is what we have done in the first three items of the definition, which specify the finite-dimensional distributions of Brownian motion. However, the last item, almost sure continuity, is also crucial, and this is information which goes beyond the finite-dimensional distributions of the process in the sense above, technically because the set $\{\omega \in \Omega\colon t \mapsto B(t,\omega) \text{ continuous}\}$ is in general not in the σ-algebra generated by the random vectors $(B(t_1), B(t_2), \ldots, B(t_n)), n \in \mathbb{N}$.

Example 1.2 Suppose that $\{B(t)\colon t \geqslant 0\}$ is a Brownian motion and U is an independent random variable, which is uniformly distributed on $[0,1]$. Then the process $\{\tilde{B}(t)\colon t \geqslant 0\}$ defined by

$$\tilde{B}(t) = \begin{cases} B(t) & \text{if } t \neq U, \\ 0 & \text{if } t = U, \end{cases}$$

has the same finite-dimensional distributions as a Brownian motion, but is discontinuous if $B(U) \neq 0$, i.e. with probability one, and hence this process is not a Brownian motion. ◇

We see that, if we are interested in the sample path properties of a stochastic process, we may need to specify more than just its finite-dimensional distributions. Suppose $\mathfrak{X}$ is a property a function might or might not have, like continuity, differentiability, etc. We say that a process $\{X(t)\colon t \geqslant 0\}$ **has property $\mathfrak{X}$ almost surely** if there exists $A \in \mathcal{A}$ such that $\mathbb{P}(A) = 1$ and $A \subset \{\omega \in \Omega\colon t \mapsto X(t,\omega) \text{ has property } \mathfrak{X}\}$. Note that the set on the right need not lie in $\mathcal{A}$.

1.1.2 Paul Lévy's construction of Brownian motion

It is a substantial issue whether the conditions imposed on the finite-dimensional distributions in the definition of Brownian motion allow the process to have continuous sample paths, or whether there is a contradiction. In this section we show that there is no contradiction and, fortunately, Brownian motion exists.

Theorem 1.3 (Wiener 1923) *Standard Brownian motion exists.*

We construct Brownian motion as a uniform limit of continuous functions, to ensure that it automatically has continuous paths. Recall that we need only construct a *standard* Brownian motion $\{B(t)\colon t \geqslant 0\}$, as $X(t) = x + B(t)$ is a Brownian motion with starting point x. The proof exploits properties of Gaussian random vectors, which are the higher-dimensional analogue of the normal distribution.

Definition 1.4. A random vector $X = (X_1, \ldots, X_n)$ is called a **Gaussian random vector** if there exists an $n \times m$ matrix A, and an n-dimensional vector b such that $X^{\mathrm{T}} = AY + b$, where Y is an m-dimensional vector with independent standard normal entries. $\diamond$

Basic facts about Gaussian random variables are collected in Appendix 12.2.

Proof of Wiener's theorem. We first construct Brownian motion on the interval $[0, 1]$ as a random element on the space $\mathbf{C}[0, 1]$ of continuous functions on $[0, 1]$. The idea is to construct the right joint distribution of Brownian motion step by step on the finite sets

$$\mathcal{D}_n = \left\{ \tfrac{k}{2^n} : 0 \leqslant k \leqslant 2^n \right\}$$

of dyadic points. We then interpolate the values on $\mathcal{D}_n$ linearly and check that the uniform limit of these continuous functions exists and is a Brownian motion.

To do this let $\mathcal{D} = \bigcup_{n=0}^{\infty} \mathcal{D}_n$ and let $(\Omega, \mathcal{A}, \mathbb{P})$ be a probability space on which a collection $\{Z_t \colon t \in \mathcal{D}\}$ of independent, standard normally distributed random variables can be defined. Let $B(0) := 0$ and $B(1) := Z_1$. For each $n \in \mathbb{N}$ we define the random variables $B(d)$, $d \in \mathcal{D}_n$ such that

(1) for all $r < s < t$ in $\mathcal{D}_n$ the random variable $B(t) - B(s)$ is normally distributed with mean zero and variance $t - s$, and is independent of $B(s) - B(r)$,

(2) the vectors $(B(d)\colon d \in \mathcal{D}_n)$ and $(Z_t \colon t \in \mathcal{D} \setminus \mathcal{D}_n)$ are independent.

Note that we have already done this for $\mathcal{D}_0 = \{0, 1\}$. Proceeding inductively we may assume that we have succeeded in doing it for some $n - 1$. We then define $B(d)$ for $d \in \mathcal{D}_n \setminus \mathcal{D}_{n-1}$ by

$$B(d) = \frac{B(d - 2^{-n}) + B(d + 2^{-n})}{2} + \frac{Z_d}{2^{(n+1)/2}}.$$

Note that the first summand is the linear interpolation of the values of B at the neighbouring points of d in $\mathcal{D}_{n-1}$. Therefore $B(d)$ is independent of $(Z_t \colon t \in \mathcal{D} \setminus \mathcal{D}_n)$ and the second property is fulfilled.

Moreover, as $\frac{1}{2}[B(d+2^{-n})-B(d-2^{-n})]$ depends only on $(Z_t : t \in \mathcal{D}_{n-1})$, it is independent of $Z_d/2^{(n+1)/2}$. By our induction assumptions both terms are normally distributed with mean zero and variance $2^{-(n+1)}$. Hence their sum $B(d) - B(d - 2^{-n})$ and their difference $B(d + 2^{-n}) - B(d)$ are independent and normally distributed with mean zero and variance 2^{-n} by Corollary 12.12.

Indeed, all increments $B(d) - B(d - 2^{-n})$, for $d \in \mathcal{D}_n \setminus \{0\}$, are independent. To see this it suffices to show that they are pairwise independent, as the vector of these increments is Gaussian. We have seen in the previous paragraph that pairs $B(d) - B(d - 2^{-n})$, $B(d + 2^{-n}) - B(d)$ with $d \in \mathcal{D}_n \setminus \mathcal{D}_{n-1}$ are independent. The other possibility is that the increments are over intervals separated by some $d \in \mathcal{D}_{n-1}$. Choose $d \in \mathcal{D}_j$ with this property and minimal j, so that the two intervals are contained in $[d - 2^{-j}, d]$, respectively $[d, d + 2^{-j}]$. By induction the increments over these two intervals of length 2^{-j} are independent, and the increments over the intervals of length 2^{-n} are constructed from the independent increments $B(d) - B(d - 2^{-j})$, respectively $B(d + 2^{-j}) - B(d)$, using a disjoint set of variables $(Z_t : t \in \mathcal{D}_n)$. Hence they are independent and this implies the first property, and completes the induction step.

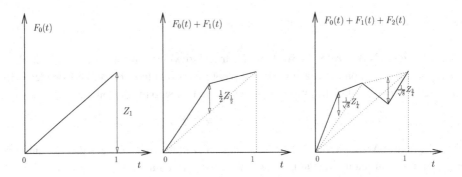

Fig. 1.2. The first three steps in the construction of Brownian motion

Having thus chosen the values of the process on all dyadic points, we interpolate between them. Formally, define

$$F_0(t) = \begin{cases} Z_1 & \text{for } t = 1, \\ 0 & \text{for } t = 0, \\ \text{linear} & \text{in between,} \end{cases}$$

and, for each $n \geqslant 1$,

$$F_n(t) = \begin{cases} 2^{-(n+1)/2} Z_t & \text{for } t \in \mathcal{D}_n \setminus \mathcal{D}_{n-1} \\ 0 & \text{for } t \in \mathcal{D}_{n-1} \\ \text{linear between consecutive points in } \mathcal{D}_n. \end{cases}$$

These functions are continuous on $[0, 1]$ and, for all n and $d \in \mathcal{D}_n$,

$$B(d) = \sum_{i=0}^{n} F_i(d) = \sum_{i=0}^{\infty} F_i(d), \tag{1.1}$$

see Figure 1.2 for an illustration. This can be seen by induction. It holds for $n = 0$. Suppose that it holds for $n - 1$. Let $d \in \mathcal{D}_n \setminus \mathcal{D}_{n-1}$. Since for $0 \leqslant i \leqslant n - 1$ the function F_i is linear on $[d - 2^{-n}, d + 2^{-n}]$, we get

$$\sum_{i=0}^{n-1} F_i(d) = \sum_{i=1}^{n-1} \frac{F_i(d - 2^{-n}) + F_i(d + 2^{-n})}{2} = \frac{B(d - 2^{-n}) + B(d + 2^{-n})}{2}.$$

Since $F_n(d) = 2^{-(n+1)/2} Z_d$, this gives (1.1).

On the other hand, we have, by definition of Z_d and by Lemma 12.9 of the appendix, for $c > 1$ and large n,

$$\mathbb{P}\{|Z_d| \geqslant c\sqrt{n}\} \leqslant \exp\left(\frac{-c^2 n}{2}\right),$$

so that the series

$$\sum_{n=0}^{\infty} \mathbb{P}\{ \text{ there exists } d \in \mathcal{D}_n \text{ with } |Z_d| \geqslant c\sqrt{n}\} \leqslant \sum_{n=0}^{\infty} \sum_{d \in \mathcal{D}_n} \mathbb{P}\{|Z_d| \geqslant c\sqrt{n}\}$$

$$\leqslant \sum_{n=0}^{\infty} (2^n + 1) \exp\left(\frac{-c^2 n}{2}\right),$$

converges as soon as $c > \sqrt{2 \log 2}$. Fix such a c. By the Borel–Cantelli lemma there exists a random (but almost surely finite) N such that for all $n \geqslant N$ and $d \in \mathcal{D}_n$ we have $|Z_d| < c\sqrt{n}$. Hence, for all $n \geqslant N$,

$$\|F_n\|_\infty < c\sqrt{n} 2^{-n/2}. \tag{1.2}$$

This upper bound implies that, almost surely, the series

$$B(t) = \sum_{n=0}^{\infty} F_n(t)$$

is uniformly convergent on $[0, 1]$. We denote the continuous limit by $\{B(t) \colon t \in [0, 1]\}$.

It remains to check that the increments of this process have the right finite-dimensional distributions. This follows directly from the properties of B on the dense set $\mathcal{D} \subset [0, 1]$ and the continuity of the paths. Indeed, suppose that $t_1 < t_2 < \cdots < t_n$ are in $[0, 1]$. We find $t_{1,k} \leqslant t_{2,k} \leqslant \cdots \leqslant t_{n,k}$ in $\mathcal{D}$ with $\lim_{k \uparrow \infty} t_{i,k} = t_i$ and infer from the continuity of B that, for $1 \leqslant i \leqslant n - 1$,

$$B(t_{i+1}) - B(t_i) = \lim_{k \uparrow \infty} B(t_{i+1,k}) - B(t_{i,k}).$$

As $\lim_{k \uparrow \infty} \mathbb{E}[B(t_{i+1,k}) - B(t_{i,k})] = 0$ and

$$\lim_{k \uparrow \infty} \mathrm{Cov}\left(B(t_{i+1,k}) - B(t_{i,k}), B(t_{j+1,k}) - B(t_{j,k})\right)$$

$$= \lim_{k \uparrow \infty} \mathbf{1}_{\{i=j\}} \left(t_{i+1,k} - t_{i,k}\right) = \mathbf{1}_{\{i=j\}} \left(t_{i+1} - t_i\right),$$

the increments $B(t_{i+1}) - B(t_i)$ are, by Proposition 12.15 of the appendix, independent Gaussian random variables with mean 0 and variance $t_{i+1} - t_i$, as required.

We have thus constructed a continuous process $B \colon [0,1] \to \mathbb{R}$ with the same finite-dimensional distributions as Brownian motion. Take a sequence $B_0, B_1, \ldots$ of independent $\mathbf{C}[0,1]$-valued random variables with the distribution of this process, and define $\{B(t) \colon t \geqslant 0\}$ by gluing together the parts, more precisely by

$$B(t) = B_{\lfloor t \rfloor}(t - \lfloor t \rfloor) + \sum_{i=0}^{\lfloor t \rfloor - 1} B_i(1), \text{ for all } t \geqslant 0.$$

This defines a continuous random function $B \colon [0, \infty) \to \mathbb{R}$ and one can see easily from what we have shown so far that it is a standard Brownian motion. ∎

Remark 1.5 If Brownian motion is constructed as a family $\{B(t) \colon t \geqslant 0\}$ of random variables on some probability space Ω, it is sometimes useful to know that the mapping $(t, \omega) \mapsto B(t, \omega)$ is measurable on the product space $[0, \infty) \times \Omega$. Exercise 1.2 shows that this can be achieved by Lévy's construction. ◇

Remark 1.6 A stochastic process $\{Y(t) \colon t \geqslant 0\}$ is called a **Gaussian process**, if for all $t_1 < t_2 < \ldots < t_n$ the vector $(Y(t_1), \ldots, Y(t_n))$ is a Gaussian random vector. It is shown in Exercise 1.3 that Brownian motion with start in $x \in \mathbb{R}$ is a Gaussian process. ◇

1.1.3 Simple invariance properties of Brownian motion

One of the themes of this book is that many natural sets that can be derived from the sample paths of Brownian motion are in some sense *random fractals*. An intuitive approach to fractals is that they are sets which have an interesting geometric structure at all scales.

A key rôle in this behaviour is played by the very simple *scaling invariance* property of Brownian motion, which we now formulate. It identifies a transformation on the space of functions, which changes the individual Brownian random functions but leaves their distribution unchanged.

Lemma 1.7 (Scaling invariance) *Suppose $\{B(t) \colon t \geqslant 0\}$ is a standard Brownian motion and let $a > 0$. Then the process $\{X(t) \colon t \geqslant 0\}$ defined by $X(t) = \frac{1}{a} B(a^2 t)$ is also a standard Brownian motion.*

Proof. Continuity of the paths, independence and stationarity of the increments remain unchanged under the scaling. It remains to observe that $X(t) - X(s) = \frac{1}{a}(B(a^2 t) - B(a^2 s))$ is normally distributed with expectation 0 and variance $(1/a^2)(a^2 t - a^2 s) = t - s$. ∎

Remark 1.8 Scaling invariance has many useful consequences. As an example, let $a < 0 < b$, and look at $T(a, b) = \inf\{t \geqslant 0 \colon B(t) = a \text{ or } B(t) = b\}$, the first exit time of a one-dimensional standard Brownian motion from the interval $[a, b]$. Then, with $X(t) = \frac{1}{a} B(a^2 t)$ we have

$$\mathbb{E}T(a, b) = a^2 \, \mathbb{E} \inf\left\{t \geqslant 0 \colon X(t) = 1 \text{ or } X(t) = b/a\right\} = a^2 \, \mathbb{E}T(1, b/a),$$

which implies that $\mathbb{E}T(-b, b)$ is a constant multiple of b^2. Also

$$\mathbb{P}\{\{B(t) \colon t \geqslant 0\} \text{ exits } [a, b] \text{ at } a\} = \mathbb{P}\{\{X(t) \colon t \geqslant 0\} \text{ exits } [1, b/a] \text{ at } 1\}$$

is only a function of the ratio b/a. The scaling invariance property will be used extensively in all the following chapters, and we shall often use the phrase that a fact holds 'by Brownian scaling' to indicate this. ◇

We shall discuss a very powerful extension of the scaling invariance property, the *conformal invariance property*, in Chapter 7 of the book. A further useful invariance property of Brownian motion, invariance under time inversion, can be identified easily. As above, the transformation on the space of functions changes the individual Brownian random functions without changing the distribution.

Theorem 1.9 (Time inversion) *Suppose $\{B(t) \colon t \geqslant 0\}$ is a standard Brownian motion. Then the process $\{X(t) \colon t \geqslant 0\}$ defined by*

$$X(t) = \begin{cases} 0 & \text{for } t = 0, \\ tB(1/t) & \text{for } t > 0, \end{cases}$$

is also a standard Brownian motion.

Proof. Recall that the finite-dimensional distributions $(B(t_1), \ldots, B(t_n))$ of Brownian motion are Gaussian random vectors and are therefore characterised by $\mathbb{E}[B(t_i)] = 0$ and $\mathrm{Cov}(B(t_i), B(t_j)) = t_i$ for $0 \leqslant t_i \leqslant t_j$.

Obviously, $\{X(t) \colon t \geqslant 0\}$ is also a Gaussian process and the Gaussian random vectors $(X(t_1), \ldots, X(t_n))$ have expectation zero. The covariances, for $t > 0$, $h \geqslant 0$, are given by

$$\mathrm{Cov}(X(t+h), X(t)) = (t+h)t\, \mathrm{Cov}(B(1/(t+h)), B(1/t))$$

$$= t(t+h)\frac{1}{t+h} = t.$$

Hence the law of all the finite-dimensional distributions

$$\big(X(t_1), X(t_2), \ldots, X(t_n)\big), \text{ for } 0 \leqslant t_1 \leqslant \cdots \leqslant t_n,$$

are the same as for Brownian motion. The paths of $t \mapsto X(t)$ are clearly continuous for all $t > 0$ and in $t = 0$ we use the following two facts: First, as the set $\mathbb{Q}$ of rationals is countable, the distribution of $\{X(t) \colon t \geqslant 0, t \in \mathbb{Q}\}$ is the same as for a Brownian motion, and hence

$$\lim_{\substack{t \downarrow 0 \\ t \in \mathbb{Q}}} X(t) = 0 \text{ almost surely.}$$

And second, $\mathbb{Q} \cap (0, \infty)$ is dense in $(0, \infty)$ and $\{X(t) \colon t \geqslant 0\}$ is almost surely continuous on $(0, \infty)$, so that

$$0 = \lim_{\substack{t \downarrow 0 \\ t \in \mathbb{Q}}} X(t) = \lim_{t \downarrow 0} X(t) \text{ almost surely.}$$

Hence $\{X(t) \colon t \geqslant 0\}$ has almost surely continuous paths, and is a Brownian motion. ∎

Remark 1.10 The symmetry inherent in the time inversion property becomes more apparent if one considers the *Ornstein–Uhlenbeck diffusion* $\{X(t)\colon t \in \mathbb{R}\}$, which is given by

$$X(t) = e^{-t}B(e^{2t}) \text{ for all } t \in \mathbb{R}.$$

This is a Markov process (this will be explained properly in Chapter 2.2.3), such that $X(t)$ is standard normally distributed for all t. It is a diffusion with a drift towards the origin proportional to the distance from the origin. Unlike Brownian motion, the Ornstein–Uhlenbeck diffusion is time reversible: The time inversion formula gives that $\{X(t)\colon t \geqslant 0\}$ and $\{X(-t)\colon t \geqslant 0\}$ have the same law. For t near $-\infty$, $X(t)$ relates to the Brownian motion near time 0, and for t near ∞, $X(t)$ relates to the Brownian motion near ∞.   ◇

Time inversion is a useful tool to relate the properties of Brownian motion in a neighbourhood of time $t = 0$ to properties at infinity. To illustrate the use of time inversion we exploit Theorem 1.9 to get an interesting statement about the long-term behaviour from an easy statement at the origin.

Corollary 1.11 (Law of large numbers) *Almost surely,* $\lim\limits_{t\to\infty} \dfrac{B(t)}{t} = 0.$

Proof.   Let $\{X(t)\colon t \geqslant 0\}$ be as defined in Theorem 1.9. Using this theorem, we see that $\lim_{t\to\infty} B(t)/t = \lim_{t\to\infty} X(1/t) = X(0) = 0$ almost surely.   ∎

In the next two chapters we discuss the two basic analytic properties of Brownian motion as a random function, its *continuity* and *differentiability* properties.

1.2 Continuity properties of Brownian motion

The definition of Brownian motion already requires that the sample functions are continuous almost surely. This implies that on the interval $[0, 1]$ (or any other compact interval) the sample functions are uniformly continuous, i.e. there exists some (random) function φ with $\lim_{h\downarrow 0} \varphi(h) = 0$ called a **modulus of continuity** of the function $B : [0, 1] \to \mathbb{R}$, such that

$$\limsup_{h\downarrow 0} \sup_{0\leqslant t\leqslant 1-h} \frac{|B(t+h) - B(t)|}{\varphi(h)} \leqslant 1. \tag{1.3}$$

Can we achieve such a bound with a deterministic function φ, i.e. is there a nonrandom modulus of continuity for the Brownian motion? The answer is yes, as the following theorem shows.

Theorem 1.12 *There exists a constant $C > 0$ such that, almost surely, for every sufficiently small $h > 0$ and all $0 \leqslant t \leqslant 1 - h$,*

$$\left|B(t+h) - B(t)\right| \leqslant C\sqrt{h\log(1/h)}.$$

Proof.   This follows quite elegantly from Lévy's construction of Brownian motion. Recall the notation introduced there and that we have represented Brownian motion as a series

$$B(t) = \sum_{n=0}^{\infty} F_n(t),$$

where each F_n is a piecewise linear function. The derivative of F_n exists almost everywhere, and by definition and (1.2), for any $c > \sqrt{2\log 2}$ there exists a (random) $N \in \mathbb{N}$ such that, for all $n > N$,

$$\|F_n'\|_\infty \leqslant \frac{2\|F_n\|_\infty}{2^{-n}} \leqslant 2c\sqrt{n}2^{n/2}.$$

Now for each $t, t + h \in [0,1]$, using the mean-value theorem,

$$|B(t+h) - B(t)| \leqslant \sum_{n=0}^{\infty} |F_n(t+h) - F_n(t)| \leqslant \sum_{n=0}^{\ell} h\|F_n'\|_\infty + \sum_{n=\ell+1}^{\infty} 2\|F_n\|_\infty.$$

Hence, using (1.2) again, we get for all $\ell > N$, that this is bounded by

$$h\sum_{n=0}^{N} \|F_n'\|_\infty + 2ch \sum_{n=N}^{\ell} \sqrt{n}2^{n/2} + 2c \sum_{n=\ell+1}^{\infty} \sqrt{n}2^{-n/2}.$$

We now suppose that h is (again random and) small enough that the first summand is smaller than $\sqrt{h\log(1/h)}$ and that ℓ defined by $2^{-\ell} < h \leqslant 2^{-\ell+1}$ exceeds N. For this choice of ℓ the second and third summands are also bounded by constant multiples of $\sqrt{h\log(1/h)}$ as both sums are dominated by their largest element. Hence we get (1.3) with a deterministic function $\varphi(h) = C\sqrt{h\log(1/h)}$.   ∎

This upper bound is pretty close to the optimal result. The following lower bound confirms that the only missing bit is the precise value of the constant.

Theorem 1.13 *For every constant $c < \sqrt{2}$, almost surely, for every $\varepsilon > 0$ there exist $0 < h < \varepsilon$ and $t \in [0, 1-h]$ with*

$$|B(t+h) - B(t)| \geqslant c\sqrt{h\log(1/h)}.$$

Proof.   Let $c < \sqrt{2}$ and define, for integers $k, n \geqslant 0$, the events

$$A_{k,n} = \left\{ B((k+1)e^{-n}) - B(ke^{-n}) > c\sqrt{n}e^{-n/2} \right\}.$$

Then, using Lemma 12.9, for any $k \geqslant 0$,

$$\mathbb{P}(A_{k,n}) = \mathbb{P}\{B(e^{-n}) > c\sqrt{n}e^{-n/2}\} = \mathbb{P}\{B(1) > c\sqrt{n}\} \geqslant \frac{c\sqrt{n}}{c^2 n + 1}\frac{1}{\sqrt{2\pi}}e^{-c^2 n/2}.$$

By our assumption on c, we have $e^n \mathbb{P}(A_{k,n}) \to \infty$ as $n \uparrow \infty$. Therefore, using $1 - x \leqslant e^{-x}$ for all x,

$$\mathbb{P}\left(\bigcap_{k=0}^{\lfloor e^n - 1\rfloor} A_{k,n}^c \right) = (1 - \mathbb{P}(A_{0,n}))^{e^n} \leqslant \exp(-e^n \mathbb{P}(A_{0,n})) \to 0.$$

By considering $h = e^{-n}$ one can now see that, for any $\varepsilon > 0$,

$$\mathbb{P}\{|B(t+h) - B(t)| \leqslant c\sqrt{h\log(1/h)} \, \forall h \in (0,\varepsilon), t \in [0, 1-h]\} = 0. \qquad \blacksquare$$

One can determine the constant c in the best possible modulus of continuity $\varphi(h) = c\sqrt{h\log(1/h)}$ precisely. Indeed, our proof of the lower bound yields a value of $c = \sqrt{2}$, which turns out to be optimal. This striking result is due to Paul Lévy.

Theorem* 1.14 (Lévy's modulus of continuity (1937)) *Almost surely,*

$$\limsup_{h\downarrow 0} \ \sup_{0\leqslant t\leqslant 1-h} \ \frac{|B(t+h) - B(t)|}{\sqrt{2h\log(1/h)}} = 1.$$

Remark 1.15 We come back to the modulus of continuity of Brownian motion in Chapter 10, where we prove a substantial extension, the spectrum of fast times of Brownian motion. We will not use Theorem 1.14 in the sequel as Theorem 1.12 is sufficient to discuss all problems where an upper bound on the increase of a Brownian motion is needed. Hence the proof of Lévy's modulus of continuity may be skipped on first reading.   ◇

In the light of Theorem 1.13, we only need to prove the upper bound. We first look at increments over a class of intervals, which is chosen to be sparse, but big enough to approximate arbitrary intervals. More precisely, given natural numbers n, m, we let $\Lambda_n(m)$ be the collection of all intervals of the form

$$\left[(k-1+b)2^{-n+a}, (k+b)2^{-n+a} \right],$$

for $k \in \{1, \ldots, 2^n\}$, $a, b \in \{0, \frac{1}{m}, \ldots, \frac{m-1}{m}\}$. We further define $\Lambda(m) := \bigcup_n \Lambda_n(m)$.

Lemma 1.16 *For any fixed m and $c > \sqrt{2}$, almost surely, there exists $n_0 \in \mathbb{N}$ such that, for any $n \geqslant n_0$,*

$$|B(t) - B(s)| \leqslant c\sqrt{(t-s)\log\tfrac{1}{(t-s)}} \qquad \text{for all } [s,t] \in \Lambda_m(n).$$

Proof. From the tail estimate for a standard normal random variable X, see Lemma 12.9, we obtain

$$\mathbb{P}\Big\{ \sup_{k\in\{1,\ldots,2^n\}} \sup_{a,b\in\{0,\frac{1}{m},\ldots,\frac{m-1}{m}\}}$$

$$\left| B\big((k-1+b)2^{-n+a}\big) - B\big((k+b)2^{-n+a}\big) \right| > c\sqrt{2^{-n+a}\log(2^{n+a})} \Big\}$$

$$\leqslant 2^n m^2 \, \mathbb{P}\{X > c\sqrt{\log(2^n)}\}$$

$$\leqslant \frac{m^2}{c\sqrt{\log(2^n)}} \frac{1}{\sqrt{2\pi}} 2^{n(1-c^2/2)},$$

and as the right hand side is summable, the result follows from the Borel–Cantelli lemma. $\blacksquare$

Lemma 1.17 *Given $\varepsilon > 0$ there exists $m \in \mathbb{N}$ such that for every interval $[s,t] \subset [0,1]$ there exists an interval $[s',t'] \in \Lambda(m)$ with $|t - t'| < \varepsilon\,(t - s)$ and $|s - s'| < \varepsilon\,(t - s)$.*

Proof. Choose m large enough to ensure that $1/m < \varepsilon/4$ and $2^{1/m} < 1 + \varepsilon/2$. Given an interval $[s,t] \subset [0,1]$, we first pick n such that $2^{-n} \leqslant t - s < 2^{-n+1}$, then $a \in \{0, 1/m, \ldots, (m-1)/m\}$ such that $2^{-n+a} \leqslant t - s < 2^{-n+a+1/m}$. Next, pick $k \in \{1, \ldots, 2^n\}$ such that $(k-1)2^{-n+a} < s \leqslant k2^{-n+a}$, and $b \in \{0, 1/m, \ldots, (m-1)/m\}$ such that $(k-1+b)2^{-n+a} < s \leqslant (k-1+b+1/m)2^{-n+a}$. Let $s' = (k-1+b)2^{-n+a}$, then

$$|s - s'| \leqslant \tfrac{1}{m}2^{-n+a} \leqslant \tfrac{\varepsilon}{4}2^{-n+1} \leqslant \tfrac{\varepsilon}{2}\,(t - s).$$

Choosing $t' = (k + b)2^{-n+a}$ ensures that $[s',t'] \in \Lambda_n(m)$ and, moreover,

$$|t - t'| \leqslant |s - s'| + |(t - s) - (t' - s')| \leqslant \tfrac{\varepsilon}{2}\,(t - s) + \left(2^{-n+a+1/m} - 2^{-n+a}\right)$$
$$\leqslant \tfrac{\varepsilon}{2}\,(t - s) + \tfrac{\varepsilon}{2}2^{-n+a} \leqslant \varepsilon\,(t - s),$$

as required. ∎

Proof of Theorem 1.14. Given $c > \sqrt{2}$, pick $0 < \varepsilon < 1$ small enough to ensure that $\tilde{c} := c - \varepsilon > \sqrt{2}$ and $m \in \mathbb{N}$ as in Lemma 1.17. Using Lemma 1.16 we choose $n_0 \in \mathbb{N}$ large enough that, for all $n \geqslant n_0$ and all intervals $[s',t'] \in \Lambda_n(m)$, almost surely,

$$|B(t') - B(s')| \leqslant \tilde{c}\sqrt{(t' - s')\log\tfrac{1}{(t'-s')}}.$$

Now let $[s,t] \subset [0,1]$ be arbitrary, with $t - s < 2^{-n_0} \wedge \varepsilon$, and pick $[s',t'] \in \Lambda(m)$ with $|t - t'| < \varepsilon\,(t - s)$ and $|s - s'| < \varepsilon\,(t - s)$. Then, recalling Theorem 1.12, we obtain

$$\big|B(t) - B(s)\big| \leqslant \big|B(t) - B(t')\big| + \big|B(t') - B(s')\big| + \big|B(s') - B(s)\big|$$
$$\leqslant C\sqrt{|t - t'|\log\tfrac{1}{|t-t'|}} + \tilde{c}\sqrt{(t' - s')\log\tfrac{1}{t'-s'}} + C\sqrt{|s - s'|\log\tfrac{1}{|s-s'|}}$$
$$\leqslant \left(4C\sqrt{\varepsilon} + \tilde{c}\sqrt{(1 + 2\varepsilon)(1 - \log(1 - 2\varepsilon))}\right)\sqrt{(t - s)\log\tfrac{1}{t-s}}.$$

By making $\varepsilon > 0$ small, the first factor on the right can be chosen arbitrarily close to c. This completes the proof of the upper bound, and hence of the theorem. ∎

Remark 1.18 The limsup in Theorem 1.14 may be replaced by a limit, see Exercise 1.7. ◇

Definition 1.19. A function $f\colon [0,\infty) \to \mathbb{R}$ is said to be **locally α-Hölder continuous at** $x \geqslant 0$, if there exists $\varepsilon > 0$ and $c > 0$ such that

$$|f(x) - f(y)| \leqslant c\,|x - y|^\alpha, \qquad \text{for all } y \geqslant 0 \text{ with } |y - x| < \varepsilon.$$

We refer to $\alpha > 0$ as the **Hölder exponent** and to $c > 0$ as the **Hölder constant**. ◇

Clearly, α-Hölder continuity gets stronger, as the exponent α gets larger. The results of this chapter so far indicate that, for Brownian motion, the transition between paths which are α-Hölder continuous and paths which are not happens at $\alpha = 1/2$.

Corollary 1.20 *If $\alpha < 1/2$, then, almost surely, Brownian motion is everywhere locally α-Hölder continuous.*

Proof.   Let $C > 0$ be as in Theorem 1.12. Applying this theorem to the Brownian motions $\{B(t) - B(k) \colon t \in [k, k+1]\}$, where k is a nonnegative integer, we see that, almost surely, for every k there exists $h(k) > 0$ such that for all $t \in [k, k+1)$ and $0 < h < (k + 1 - t) \wedge h(k)$,

$$\left| B(t + h) - B(t) \right| \leqslant C \sqrt{h \log(1/h)} \leqslant C \, h^\alpha \, .$$

Doing the same to the Brownian motions $\{\tilde{B}(t) \colon t \in [k, k+1]\}$ with $\tilde{B}(t) = B(k + 1 - t) - B(k + 1)$ gives the full result.   ∎

Remark 1.21 This result is optimal in the sense that, for $\alpha > 1/2$, almost surely, at every point, Brownian motion fails to be locally α-Hölder continuous, see Exercise 1.9. Points where Brownian motion is locally $1/2$-Hölder continuous exist almost surely, but they are very rare. We come back to this issue when discussing 'slow points' of Brownian motion in Chapter 10.   ◇

1.3 Nondifferentiability of Brownian motion

Having proved in the previous section that Brownian motion is somewhat *regular*, let us see why it is *erratic*. One manifestation is that the paths of Brownian motion have no intervals of monotonicity.

Theorem 1.22 *Almost surely, for all $0 < a < b < \infty$, Brownian motion is not monotone on the interval $[a, b]$.*

Proof.   First fix a nondegenerate interval $[a, b]$, i.e. an interval of positive length. If it is an interval of monotonicity, i.e. if $B(s) \leqslant B(t)$ for all $a \leqslant s \leqslant t \leqslant b$, then we pick numbers $a = a_1 \leqslant \ldots \leqslant a_{n+1} = b$ and divide $[a, b]$ into n sub-intervals $[a_i, a_{i+1}]$. Each increment $B(a_i) - B(a_{i+1})$ has to have the same sign. As the increments are independent, this has probability $2 \cdot 2^{-n}$, and taking $n \to \infty$ shows that the probability that $[a, b]$ is an interval of monotonicity must be zero. Taking a countable union gives that, almost surely, there is no nondegenerate interval of monotonicity with rational endpoints, but each nondegenerate interval would have a nondegenerate rational sub-interval.   ∎

In order to discuss differentiability of Brownian motion we make use of the *time inversion* trick, which allows us to relate differentiability at $t = 0$ to a long-term property. This property is a complementary result to the law of large numbers: Whereas Corollary 1.11 asserts that Brownian motion grows slower than linearly, the next proposition shows that the limsup growth of $B(t)$ is faster than $\sqrt{t}$.

Proposition 1.23 *Almost surely,*

$$\limsup_{n\to\infty} \frac{B(n)}{\sqrt{n}} = +\infty, \quad and \quad \liminf_{n\to\infty} \frac{B(n)}{\sqrt{n}} = -\infty. \qquad (1.4)$$

For the proof of Proposition 1.23 we use the Hewitt–Savage 0-1 law for exchangeable events, which we briefly recall. Readers unfamiliar with the result are invited to give a proof as Exercise 1.10.

Definition 1.24. Let $X_1, X_2, \ldots$ be a sequence of random variables on a probability space $(\Omega, \mathcal{F}, \mathbb{P})$ and consider a set A of sequences such that

$$\{X_1, X_2, \ldots \in A\} \in \mathcal{F}.$$

The event $\{X_1, X_2, \cdots \in A\}$ is called **exchangeable** if

$$\{X_1, X_2, \ldots \in A\} \subset \{X_{\sigma_1}, X_{\sigma_2}, \ldots \in A\}$$

for all finite permutations $\sigma \colon \mathbb{N} \to \mathbb{N}$. Here *finite permutation* means that σ is a bijection with $\sigma_n = n$ for all sufficiently large n. ◇

Lemma 1.25 (Hewitt–Savage 0-1 law) *If E is an exchangeable event for an independent, identically distributed sequence, then $\mathbb{P}(E)$ is 0 or 1.*

Proof of Proposition 1.23. We clearly have, by Fatou's lemma,

$$\mathbb{P}\{B(n) > c\sqrt{n} \text{ infinitely often}\} \geqslant \limsup_{n\to\infty} \mathbb{P}\{B(n) > c\sqrt{n}\}.$$

By the scaling property, the expression in the $\limsup$ equals $\mathbb{P}\{B(1) > c\}$, which is positive. Let $X_n = B(n) - B(n-1)$, and note that

$$\{B(n) > c\sqrt{n} \text{ infinitely often}\} = \Big\{\sum_{j=1}^{n} X_j > c\sqrt{n} \text{ infinitely often}\Big\}$$

is an exchangeable event. Hence the Hewitt–Savage 0-1 law gives that, with probability one, $B(n) > c\sqrt{n}$ infinitely often. Taking the intersection over all positive integers c gives the first part of the statement and the second part is proved analogously. ∎

Remark 1.26 It is natural to ask whether there exists a 'gauge' function $\varphi \colon [0, \infty) \to [0, \infty)$ such that $B(t)/\varphi(t)$ has a $\limsup$ which is greater than 0 but less than ∞. An answer will be given by the law of the iterated logarithm in the first section of Chapter 5.◇

For a function f, we define the **upper** and **lower right derivatives**

$$D^* f(t) \;=\; \limsup_{h\downarrow 0} \frac{f(t+h) - f(t)}{h},$$

and

$$D_* f(t) \;=\; \liminf_{h\downarrow 0} \frac{f(t+h) - f(t)}{h}.$$

We now show that for any fixed time t, almost surely, Brownian motion is not differentiable at t. For this we use Proposition 1.23 and the invariance under time inversion.

Theorem 1.27 *Fix $t \geqslant 0$. Then, almost surely, Brownian motion is not differentiable at t. Moreover, $D^* B(t) = +\infty$ and $D_* B(t) = -\infty$.*

Proof.   Given a standard Brownian motion B we construct a further Brownian motion X by time inversion as in Theorem 1.9. Then

$$D^* X(0) \geqslant \limsup_{n \to \infty} \frac{X(\frac{1}{n}) - X(0)}{\frac{1}{n}} \geqslant \limsup_{n \to \infty} \sqrt{n}\, X(\tfrac{1}{n}) = \limsup_{n \to \infty} \frac{B(n)}{\sqrt{n}},$$

which is infinite by Proposition 1.23. Similarly, $D_* X(0) = -\infty$, showing that X is not differentiable at 0. Now let $t > 0$ be arbitrary and $\{B(t) : t \geqslant 0\}$ a Brownian motion. Then $X(s) = B(t + s) - B(t)$ defines a standard Brownian motion and differentiability of X at zero is equivalent to differentiability of B at t.   ∎

While the previous proof shows that every t is almost surely a point of nondifferentiability for the Brownian motion, this does *not* imply that almost surely *every* t is a point of non-differentiability for the Brownian motion! The order of the quantifiers *for all t* and *almost surely* in results like Theorem 1.27 is of vital importance. Here the statement holds for all Brownian paths outside a set of probability zero, which may depend on t, and the union of all these sets of probability zero may not itself be a set of probability zero.

To illustrate this point, consider the following example: The argument in the proof of Theorem 1.27 also shows that the Brownian motion X crosses 0 for arbitrarily small values $s > 0$. Defining the level sets $Z(t) = \{s > 0 : X(s) = X(t)\}$, this shows that every t is almost surely an accumulation point from the right for $Z(t)$. But not every point $t \in [0, 1]$ is an accumulation point from the right for $Z(t)$. For example the last zero of $\{X(t) : t \geqslant 0\}$ before time 1 is, by definition, never an accumulation point from the right for $Z(t) = Z(0)$. This example illustrates that there can be random *exceptional times* at which Brownian motion exhibits atypical behaviour. These times are so rare that any fixed (i.e. nonrandom) time is almost surely not of this kind.

Remark 1.28   The behaviour of Brownian motion at a fixed time $t > 0$ reflects the behaviour at *typical times* in the following sense: Suppose $\mathfrak{X}$ is a measurable event (a set of paths) such that $\{B(t) : t \geqslant 0\} \in \mathfrak{X}$ almost surely. By stationarity of the increments this implies $\mathbb{P}\{\{B(t + s) - B(t) : s \geqslant 0\} \in \mathfrak{X}\} = 1$ for all fixed $t \geqslant 0$. Moreover, almost surely, the set of exceptional times $\{t : \{B(t + s) - B(t) : s \geqslant 0\} \notin \mathfrak{X}\}$ has Lebesgue measure zero. Indeed, using the joint measurability mentioned in Remark 1.5 and Fubini's theorem,

$$\mathbb{E} \int_0^\infty \mathbf{1}\{t \colon \{B(t + s) - B(s) : s \geqslant 0\} \notin \mathfrak{X}\}\, dt = \int_0^\infty \mathbb{P}\{\{B(s) : s \geqslant 0\} \notin \mathfrak{X}\}\, dt = 0.$$

For example, the previous result shows that, almost surely, the path of a Brownian motion is not differentiable at Lebesgue-almost every time t.   ◇

Remark 1.29 Exercise 1.11 shows that, almost surely, there exist times $t_*, t^* \in [0, 1)$ with $D^* B(t^*) \leqslant 0$ and $D_* B(t_*) \geqslant 0$. Hence the almost sure behaviour at a fixed point t, which is described in Theorem 1.27, does not hold at all points simultaneously. ◇

Theorem 1.30 (Paley, Wiener and Zygmund 1933) *Almost surely, Brownian motion is nowhere differentiable. Furthermore, almost surely, for all t,*

$$\text{either} \quad D^* B(t) = +\infty \quad \text{or} \quad D_* B(t) = -\infty \quad \text{or both.}$$

Proof. Suppose that there is a $t_0 \in [0, 1]$ such that $-\infty < D_* B(t_0) \leqslant D^* B(t_0) < \infty$. Then

$$\limsup_{h \downarrow 0} \frac{|B(t_0 + h) - B(t_0)|}{h} < \infty,$$

and, using the boundedness of Brownian motion on $[0, 2]$, this implies that for some finite constant M there exists t_0 with

$$\sup_{h \in [0,1]} \frac{|B(t_0 + h) - B(t_0)|}{h} \leqslant M.$$

It suffices to show that this event has probability zero for any M. From now on fix M. If t_0 is contained in the binary interval $[(k-1)/2^n, k/2^n]$ for $n > 2$, then for all $1 \leqslant j \leqslant 2^n - k$ the triangle inequality gives

$$\left| B\left((k+j)/2^n\right) - B\left((k+j-1)/2^n\right) \right|$$
$$\leqslant \left| B\left((k+j)/2^n\right) - B(t_0)\right| + \left|B(t_0) - B\left((k+j-1)/2^n\right)\right|$$
$$\leqslant M(2j+1)/2^n.$$

Define events

$$\Omega_{n,k} := \left\{ \left| B\left((k+j)/2^n\right) - B\left((k+j-1)/2^n\right) \right| \leqslant M(2j+1)/2^n \text{ for } j = 1, 2, 3 \right\}.$$

Then by independence of the increments and the scaling property, for $1 \leqslant k \leqslant 2^n - 3$,

$$\mathbb{P}(\Omega_{n,k}) \leqslant \prod_{j=1}^{3} \mathbb{P}\left\{ \left| B\left((k+j)/2^n\right) - B\left((k+j-1)/2^n\right) \right| \leqslant M(2j+1)/2^n \right\}$$
$$\leqslant \mathbb{P}\left\{ |B(1)| \leqslant 7M/\sqrt{2^n} \right\}^3,$$

which is at most $(7M2^{-n/2})^3$, since the normal density is bounded by $1/2$. Hence

$$\mathbb{P}\left(\bigcup_{k=1}^{2^n - 3} \Omega_{n,k} \right) \leqslant 2^n (7M2^{-n/2})^3 = (7M)^3 2^{-n/2},$$

which is summable over all n. Hence, by the Borel–Cantelli lemma,

$$\mathbb{P}\left\{ \text{ there is } t_0 \in [0, 1] \text{ with } \sup_{h \in [0,1]} \frac{|B(t_0 + h) - B(t_0)|}{h} \leqslant M \right\}$$
$$\leqslant \mathbb{P}\left(\bigcup_{k=1}^{2^n - 3} \Omega_{n,k} \text{ for infinitely many } n \right) = 0. \qquad \blacksquare$$

Remark 1.31 The proof of Theorem 1.30 can be tightened to prove that, for any $\alpha > \frac{1}{2}$, the sample paths of Brownian motion are, almost surely, nowhere locally α-Hölder continuous, see Exercise 1.9.   ◇

Remark 1.32 There is an abundance of interesting statements about the right derivatives of Brownian motion, which we state as exercises at the end of the chapter. As a taster we mention here that Lévy [Le54] asked whether, almost surely, $D^*B(t) \in \{-\infty, \infty\}$ for every $t \in [0, 1)$. Exercise 1.13 shows that this is not the case.   ◇

Another important regularity property, which Brownian motion does *not* possess is to be of bounded variation. We first define what it means for a function to be of bounded variation.

Definition 1.33. A right-continuous function $f \colon [0, t] \to \mathbb{R}$ is a function of **bounded variation** if

$$V_f^{(1)}(t) := \sup \sum_{j=1}^{k} \left| f(t_j) - f(t_{j-1}) \right| < \infty,$$

where the supremum is over all $k \in \mathbb{N}$ and partitions $0 = t_0 \leqslant t_1 \leqslant \cdots \leqslant t_{k-1} \leqslant t_k = t$. If the supremum is infinite f is said to be of **unbounded variation**.   ◇

Remark 1.34 It is not hard to show that f is of bounded variation if and only if it can be written as the difference of two increasing functions.   ◇

Theorem 1.35 *Suppose that the sequence of partitions*

$$0 = t_0^{(n)} \leqslant t_1^{(n)} \leqslant \cdots \leqslant t_{k(n)-1}^{(n)} \leqslant t_{k(n)}^{(n)} = t$$

is nested, i.e. at each step one or more partition points are added, and the mesh

$$\Delta(n) := \sup_{1 \leqslant j \leqslant k(n)} \left\{ t_j^{(n)} - t_{j-1}^{(n)} \right\}$$

converges to zero. Then, almost surely,

$$\lim_{n \to \infty} \sum_{j=1}^{k(n)} \left(B(t_j^{(n)}) - B(t_{j-1}^{(n)}) \right)^2 = t,$$

and therefore Brownian motion is of unbounded variation.

Remark 1.36 For a sequence of partitions as above, we call

$$\lim_{n \to \infty} \sum_{j=1}^{k(n)} \left(B(t_j^{(n)}) - B(t_{j-1}^{(n)}) \right)^2$$

the **quadratic variation** of Brownian motion. The fact that Brownian motion has finite quadratic variation will be of crucial importance in Chapter 7, however, the analogy to the notion of bounded variation of a function is not perfect: In Exercise 1.15 we find a sequence of partitions

$$0 = t_0^{(n)} \leqslant t_1^{(n)} \leqslant \cdots \leqslant t_{k(n)-1}^{(n)} \leqslant t_{k(n)}^{(n)} = t$$

with mesh converging to zero, such that almost surely

$$\limsup_{n \to \infty} \sum_{j=1}^{k(n)} \left(B(t_j^{(n)}) - B(t_{j-1}^{(n)}) \right)^2 = \infty.$$

In particular, the condition that the partitions in Theorem 1.35 are nested cannot be dropped entirely, though it can be replaced by other conditions, see Exercise 1.16. ◇

The proof of Theorem 1.35 is based on the following simple lemma.

Lemma 1.37 *If X, Z are independent, symmetric random variables in $\mathbf{L}^2$, then*

$$\mathbb{E}\left[(X + Z)^2 \,\big|\, X^2 + Z^2\right] = X^2 + Z^2.$$

Proof. By symmetry of Z we have

$$\mathbb{E}\left[(X + Z)^2 \,\big|\, X^2 + Z^2\right] = \mathbb{E}\left[(X - Z)^2 \,\big|\, X^2 + Z^2\right].$$

Both sides of the equation are finite, so that we can take the difference and obtain

$$\mathbb{E}\left[XZ \,\big|\, X^2 + Z^2\right] = 0,$$

and the result follows immediately. ∎

Proof of Theorem 1.35. By the Hölder property, we can find, for any $\alpha \in (0, 1/2)$, an n such that $|B(a) - B(b)| \leqslant |a - b|^\alpha$ for all $a, b \in [0, t]$ with $|a - b| \leqslant \Delta(n)$. Hence

$$\sum_{j=1}^{k(n)} \left| B(t_j^{(n)}) - B(t_{j-1}^{(n)}) \right| \geqslant \Delta(n)^{-\alpha} \sum_{j=1}^{k(n)} \left(B(t_j^{(n)}) - B(t_{j-1}^{(n)}) \right)^2.$$

Therefore, once we show that the random variables

$$X_n := \sum_{j=1}^{k(n)} \left(B(t_j^{(n)}) - B(t_{j-1}^{(n)}) \right)^2$$

converge almost surely to a positive random variable it follows immediately that Brownian motion is almost surely of unbounded variation. By inserting elements in the sequence, if necessary, we may assume that at each step exactly one point is added to the partition.

To see that $\{X_n : n \in \mathbb{N}\}$ converges we use the theory of martingales in discrete time, see Appendix 12.3 for basic facts on martingales. We denote by $\mathcal{G}_n$ the σ-algebra generated by the random variables $X_n, X_{n+1}, \ldots$. Then

$$\mathcal{G}_\infty := \bigcap_{k=1}^{\infty} \mathcal{G}_k \subset \cdots \subset \mathcal{G}_{n+1} \subset \mathcal{G}_n \subset \cdots \subset \mathcal{G}_1.$$

We show that $\{X_n : n \in \mathbb{N}\}$ is a reverse martingale, i.e. that almost surely,

$$X_n = \mathbb{E}[X_{n-1} \mid \mathcal{G}_n] \qquad \text{for all } n \geqslant 2.$$

This is easy with the help of Lemma 1.37. Indeed, if $s \in (t_1, t_2)$ is the inserted point we apply it to the symmetric, independent random variables $B(s) - B(t_1)$, $B(t_2) - B(s)$ and denote by $\mathcal{F}$ the σ-algebra generated by $(B(s) - B(t_1))^2 + (B(t_2) - B(s))^2$. Then

$$\mathbb{E}\left[(B(t_2) - B(t_1))^2 \mid \mathcal{F}\right] = (B(s) - B(t_1))^2 + (B(t_2) - B(s))^2,$$

and hence

$$\mathbb{E}\left[(B(t_2) - B(t_1))^2 - (B(s) - B(t_1))^2 - (B(t_2) - B(s))^2 \mid \mathcal{F}\right] = 0,$$

which implies that $\{X_n : n \in \mathbb{N}\}$ is a reverse martingale.

By the Lévy downward theorem, see Theorem 12.26 in the appendix,

$$\lim_{n \uparrow \infty} X_n = \mathbb{E}[X_1 \mid \mathcal{G}_\infty] \qquad \text{almost surely.}$$

The limit has expectation $\mathbb{E}[X_1] = t$ and, by Fatou's lemma, its variance is bounded by

$$\liminf_{n \uparrow \infty} \mathbb{E}\left[(X_n - \mathbb{E}X_n)^2\right] = \liminf_{n \uparrow \infty} 3 \sum_{j=1}^{k(n)} \left(t_j^{(n)} - t_{j-1}^{(n)}\right)^2 \leqslant 3t \liminf_{n \uparrow \infty} \Delta(n) = 0.$$

Hence, $\mathbb{E}[X_1 \mid \mathcal{G}_\infty] = t$ almost surely, as required. ∎

1.4 The Cameron–Martin theorem

In the previous two sections we have obtained results about the almost sure behaviour of a Brownian motion $\{B(t) : t \geqslant 0\}$ without drift. In this section we ask whether these results hold as well for a Brownian motion with drift $\{B(t) + \mu t : t \geqslant 0\}$ or, more generally, for which time-dependent drift functions F the process $\{B(t) + F(t) : t \geqslant 0\}$ has the same behaviour as a Brownian motion path. This section can be skipped on first reading.

We denote by $\mathbb{L}_0$ the law of standard Brownian motion $\{B(t) : t \in [0, 1]\}$, and for a function $F : [0, 1] \to \mathbb{R}$ write $\mathbb{L}_F$ for the law of $\{B(t) + F(t) : t \in [0, 1]\}$. We ask, for which functions F any set A with $\mathbb{L}_0(A) = 0$ also satisfies $\mathbb{L}_F(A) = 0$, in other words, for which F is $\mathbb{L}_F$ absolutely continuous with respect to $\mathbb{L}_0$?

Clearly, necessary conditions are continuity of F and $F(0) = 0$. However, these conditions are not sufficient. Denote by $\mathbf{D}[0, 1]$ the **Dirichlet space**

$$\mathbf{D}[0, 1] = \left\{ F \in \mathbf{C}[0, 1] : \text{ exists } f \in \mathbf{L}^2[0, 1] \text{ such that } F(t) = \int_0^t f(s)\, ds \; \forall t \in [0, 1] \right\}.$$

Given $F \in \mathbf{D}[0, 1]$ the associated f is uniquely determined as an element of $\mathbf{L}^2[0, 1]$, and is denoted by F', the derivative of F.

Recall that for two nonzero measures μ and ν on the same space we write $\mu \perp \nu$, and say that μ and ν are *singular* if there exists a Borel set A with $\mu(A) = 0$ and $\nu(A^c) = 0$. Otherwise, we say that they are *equivalent* if they are mutually absolutely continuous, i.e. if $\mu \ll \nu$ and $\nu \ll \mu$.

Theorem 1.38 (Cameron–Martin) *Let $F \in \mathbf{C}[0,1]$ satisfy $F(0) = 0$.*

   (1) *If $F \notin \mathbf{D}[0,1]$ then $\mathbb{L}_F \perp \mathbb{L}_0$.*
   (2) *If $F \in \mathbf{D}[0,1]$ then $\mathbb{L}_F$ and $\mathbb{L}_0$ are equivalent.*

Remark 1.39 As a consequence we see that *any* almost sure property of the Brownian motion B also holds almost surely for $B + F$, when $F \in \mathbf{D}[0,1]$. Conversely, when $F \notin \mathbf{D}[0,1]$ *some* almost sure property of Brownian motion fails for $B + F$, see also Exercise 1.18. ◇

Before proving the theorem we make some preparations. For $F \in \mathbf{C}[0,1]$ and $n > 0$, denote

$$Q_n(F) = 2^n \sum_{j=1}^{2^n} \left[F\left(\tfrac{j}{2^n}\right) - F\left(\tfrac{j-1}{2^n}\right) \right]^2 .$$

Lemma 1.40 *Let $F \in \mathbf{C}[0,1]$ satisfy $F(0) = 0$. Then $\{Q_n(F): n \geqslant 1\}$ is an increasing sequence, and*

$$F \in \mathbf{D}[0,1] \iff \sup_n Q_n(F) < \infty .$$

Moreover, if $F \in \mathbf{D}[0,1]$, then $Q_n(F) \to \|F'\|_2^2$ as $n \to \infty$.

Proof. The general inequality $(a+b)^2 \leqslant 2a^2 + 2b^2$ gives

$$\left[F\left(\tfrac{j}{2^n}\right) - F\left(\tfrac{j-1}{2^n}\right) \right]^2 \leqslant 2\left[F\left(\tfrac{2j-1}{2^{n+1}}\right) - F\left(\tfrac{j-1}{2^n}\right) \right]^2 + 2\left[F\left(\tfrac{j}{2^n}\right) - F\left(\tfrac{2j-1}{2^{n+1}}\right) \right]^2 .$$

Summing this inequality over $j \in \{1, \ldots, 2^n\}$ yields that $Q_n(F)$ is increasing in n. For $F \in \mathbf{D}[0,1]$ with $F' = f$, we can write, using Cauchy–Schwarz,

$$Q_n(F) = 2^n \sum_{j=1}^{2^n} \left(\int_{(j-1)2^{-n}}^{j2^{-n}} f \, dt \right)^2 \leqslant \sum_{j=1}^{2^n} \int_{(j-1)2^{-n}}^{j2^{-n}} f^2 \, dt = \|f\|_2^2 .$$

Assume now that $\sup_n Q_n(F) < \infty$. For any $t \in [0,1]$ that is not a dyadic rational and for each $n \geqslant 1$, there is a unique interval of the form $[\tfrac{k-1}{2^n}, \tfrac{k}{2^n}]$ (for some integer $k > 0$), to which t belongs. Denote this interval by $I_n(t) = [a_n, b_n]$ and observe that for t uniformly distributed in $[0,1]$, given $I_1(t), \ldots, I_n(t)$, the interval $I_{n+1}(t)$ is equally likely to be each of the two halves of $I_n(t)$. This implies that

$$Y_n(t) = 2^n [F(b_n) - F(a_n)] ,$$

defines a martingale with respect to the filtration $(\sigma(I_n): n = 0, 1, \ldots)$. Furthermore,

$$\mathbb{E}Y_n^2 = 2^{2n} \sum_{k=1}^{2^n} \frac{1}{2^n} \left[F\left(\tfrac{k}{2^n}\right) - F\left(\tfrac{k-1}{2^n}\right) \right]^2 = Q_n(F) .$$

Hence $\{Y_n : n = 0, 1, \dots\}$ is a martingale bounded in $\mathbf{L}^2$. By the convergence theorem for $\mathbf{L}^2$-bounded martingales, see Theorem 12.28 in the appendix, there is a random variable Y in $\mathbf{L}^2[0, 1]$ such that $Y_n \to Y$ almost surely and in $\mathbf{L}^2$. For fixed j and m we have that

$$F\left(\tfrac{j}{2^m}\right) = \int_0^{\frac{j}{2^m}} Y_n(t)\, dt \to \int_0^{\frac{j}{2^m}} Y(t)\, dt \quad \text{as } n \to \infty.$$

Let $G(x) = \int_0^x Y(t)\, dt$. Since $F\left(\tfrac{j}{2^m}\right) = G\left(\tfrac{j}{2^m}\right)$ for any j and m and F, G are continuous, we deduce that $F(x) = G(x)$ for all $x \in [0, 1]$. Therefore $F \in \mathbf{D}[0, 1]$ and $F' = Y$ almost everywhere. As $\mathbb{E} Y_n^2 \to \mathbb{E} Y^2$ we conclude that $Q_n(F) \to \|F'\|_2^2$. ∎

We use the result of Lemma 1.40 to construct a very basic stochastic integral with respect to Brownian motion.

Lemma 1.41 (Paley–Wiener stochastic integral) *Let $\{B(t) : t \geqslant 0\}$ be standard Brownian motion, and suppose $F \in \mathbf{D}[0, 1]$. Then the sequence*

$$\xi_n = 2^n \sum_{j=1}^{2^n} \left[F\left(\tfrac{j}{2^n}\right) - F\left(\tfrac{j-1}{2^n}\right)\right] \left[B\left(\tfrac{j}{2^n}\right) - B\left(\tfrac{j-1}{2^n}\right)\right]$$

converges almost surely and in $\mathbf{L}^2$. We denote the limit of ξ_n by $\int_0^1 F'\, dB$.

Proof. Recall from Lévy's construction of Brownian motion that

$$B\left(\tfrac{2j-1}{2^n}\right) = \tfrac{1}{2}\left[B\left(\tfrac{2j-2}{2^n}\right) + B\left(\tfrac{2j}{2^n}\right)\right] + \sigma_n Z\left(\tfrac{2j-1}{2^n}\right) \tag{1.5}$$

where $\sigma_n = 2^{-(n+1)/2}$ and $Z(t)$, for t binary rational, are i.i.d. standard normal random variables. Therefore

$$\xi_n - \xi_{n-1} = 2^n \sigma_n \sum_{j=1}^{2^{n-1}} \left[2F\left(\tfrac{2j-1}{2^n}\right) - F\left(\tfrac{2j-2}{2^n}\right) - F\left(\tfrac{2j}{2^n}\right)\right] Z\left(\tfrac{2j-1}{2^n}\right).$$

This implies that $\{\xi_n : n \geqslant 1\}$ is a martingale. The definition of ξ_n readily yields that $\mathbb{E} \xi_n^2 = Q_n(F)$. Since $F \in \mathbf{D}[0, 1]$, Lemma 1.40 implies that $\sup_n \mathbb{E} \xi_n^2$ is bounded, and thus the convergence theorem for $\mathbf{L}^2$-bounded martingales concludes the proof. ∎

Remark 1.42 Denote by $\mathcal{D}_n = \{j2^{-n} : j = 0, \dots, 2^n\}$ the dyadic partition of the interval $[0, 1]$. Let $\mathcal{F}_n$ be the σ-algebra in $\mathbf{C}[0, 1]$ determined by the restriction map to $\mathcal{D}_n$. Then the σ-algebras $(\mathcal{F}_n : n \geqslant 1)$ generate the Borel σ-algebra in $\mathbf{C}[0, 1]$. ◇

Proof of Theorem 1.38. For any $x \in \mathbf{C}[0, 1]$ and $n > 0$, we write

$$\nabla_j^{(n)} x = x\left(\tfrac{j}{2^n}\right) - x\left(\tfrac{j-1}{2^n}\right),$$

sometimes dropping the superindex when n is fixed. For $x \in \mathbf{C}[0, 1]$, we write

$$H_n(x) = 2^{n-1} \left[\sum_{j=1}^{2^n} (\nabla_j^{(n)} F)^2 - 2 \sum_{j=1}^{2^n} \nabla_j^{(n)} x \, \nabla_j^{(n)} F\right].$$

When we look at the finite-dimensional distributions of $\mathbb{L}_0$ and $\mathbb{L}_F$ on a finite set of times such as $\mathcal{D}_n$, the Radon–Nikodým derivative $\frac{d\mathbb{L}_F}{d\mathbb{L}_0}|_{\mathcal{D}_n}$ is the ratio of the two Lebesgue densities, provided they exist. Hence we obtain

$$\frac{d\mathbb{L}_F}{d\mathbb{L}_0}\Big|_{\mathcal{D}_n}(x) = \prod_{j=1}^{2^n} \exp\left\{-\frac{[\nabla_j x - \nabla_j F]^2}{2^{1-n}}\right\} \exp\left\{\frac{(\nabla_j x)^2}{2^{1-n}}\right\} = e^{-H_n(x)}. \tag{1.6}$$

By Theorem 12.32 (a) the process given by $\frac{d\mathbb{L}_F}{d\mathbb{L}_0}|_{\mathcal{D}_n} = e^{-H_n}$ is a nonnegative martingale with respect to $\mathbb{L}_0$. (This can also be checked directly, see Exercise 1.17.) It therefore converges $\mathbb{L}_0$-almost surely to a nonnegative finite limit, and hence H_n converges $\mathbb{L}_0$-almost surely, possibly to ∞. We have

$$\mathbb{E}_{\mathbb{L}_0} H_n = \int H_n(x)\, d\mathbb{L}_0(x) = \frac{1}{2} Q_n(F),$$

and

$$\text{Var}_{\mathbb{L}_0} H_n = Q_n(F).$$

Thus, by Chebyshev's inequality, we get

$$\mathbb{P}_{\mathbb{L}_0}\left\{H_n \leqslant \tfrac{1}{4} Q_n(F)\right\} \leqslant \frac{16}{Q_n(F)}.$$

If $F \notin \mathbf{D}[0,1]$, then Lemma 1.40 implies that $\mathbb{L}_0$-almost surely $H_n \to \infty$. By Theorem 12.32 of the appendix, we conclude that $\mathbb{L}_F \perp \mathbb{L}_0$.

For the converse, suppose that $F \in \mathbf{D}[0,1]$. By Lemma 1.41, we have

$$H_n(x) \longrightarrow \frac{1}{2}\|F'\|_2^2 - \int_0^1 F'\, dB \qquad \mathbb{L}_0\text{-almost everywhere}.$$

We conclude by (1.6) and Theorem 12.32 (iii) that $\mathbb{L}_F \ll \mathbb{L}_0$. To finish the proof of the theorem, observe that $\mathbb{L}_F \ll \mathbb{L}_0$ if and only if $\mathbb{L}_0 \ll \mathbb{L}_{-F}$. ∎

Remark 1.43 The proof of Theorem 1.38 and an easy scaling also show that, for any $t > 0$ and $F \in \mathbf{D}[0,t]$, the density of $\mathbb{L}_F$ with respect to $\mathbb{L}_0$ is given as

$$\frac{d\mathbb{L}_F}{d\mathbb{L}_0}(B) = \exp\left\{-\frac{1}{2}\int_0^t F'(s)^2\, ds + \int_0^t F'\, dB\right\} \qquad \text{for } \mathbb{L}_0\text{-almost every } B \in \mathbf{C}[0,t].$$

Choosing $F(s) = \mu s$ and applying Brownian scaling we obtain that the density of Brownian motion with drift μ with respect to a driftless Brownian motion on $\mathbf{C}[0,t]$ is

$$\frac{d\mathbb{L}_F}{d\mathbb{L}_0}(B) = \exp\left\{-\tfrac{1}{2}\mu^2 t + \mu B(t)\right\} \qquad \text{for } \mathbb{L}_0\text{-almost every } B \in \mathbf{C}[0,t]. \qquad \diamond$$

We now have a second look at the construction of Brownian motion and the Cameron–Martin theorem, now from a Hilbert space perspective.

Let $\{\varphi_n : n = 0, 1, \ldots\}$ be an orthonormal basis of $\mathbf{L}^2[0, 1]$. For example we may take the *trigonometric basis*

$$\{\varphi_n : n = 0, 1, \ldots\} = \{1\} \cup \{\sqrt{2}\cos(\pi nt) \colon n = 1, 2, \ldots\}, \qquad (1.7)$$

or the *Haar basis*

$$\{\varphi_n : n = 0, 1, \ldots\} = \{1\} \cup \{\varphi_{m,k} \; : \; m \geqslant 1 \text{ and } 1 \leqslant k \leqslant 2^{m-1}\}, \qquad (1.8)$$

where $n = 2^{m-1} - 1 + k$ and

$$\varphi_{m,k} = \sqrt{2^{m-1}}\left(1_{\left[\frac{2k-2}{2^m}, \frac{2k-1}{2^m}\right]} - 1_{\left[\frac{2k-1}{2^m}, \frac{2k}{2^m}\right]}\right), \qquad (1.9)$$

see Exercise 1.20. Consider the Dirichlet space $\mathbf{D}[0, 1]$ endowed with the inner product

$$\langle F, G \rangle_{\mathbf{D}[0,1]} = \langle F', G' \rangle_{\mathbf{L}^2[0,1]} \; .$$

Define $\{\Phi_n : n = 0, 1, \ldots\}$ by

$$\Phi_n(t) = \int_0^t \varphi_n(s)ds \, .$$

As this integration is an isometry from $\mathbf{L}^2[0, 1]$ to $\mathbf{D}[0, 1]$, we deduce that $\{\Phi_n : n = 0, 1, \ldots\}$ is an orthonormal basis for $\mathbf{D}[0, 1]$. Furthermore, by Cauchy–Schwarz,

$$\left| \int_0^t f(s)\, ds - \int_0^t g(s)\, ds \right| \leqslant \|f - g\|_2 \, ;$$

therefore, if $F_n \to F$ in $\mathbf{D}[0, 1]$ then $F_n \to F$ uniformly. Thus for any $F \in \mathbf{D}[0, 1]$, the series

$$F = \sum_{n=0}^{\infty} \langle \varphi_n, F' \rangle_{\mathbf{L}^2} \, \Phi_n = \sum_{n=0}^{\infty} \langle \Phi_n, F \rangle_{\mathbf{D}} \, \Phi_n \, ,$$

converges in $\mathbf{D}[0, 1]$ and uniformly.

Let $\{\Phi_n : n = 0, 1, \ldots\}$ be an orthonormal basis in $\mathbf{D}[0, 1]$, where $\Phi_n(t) = \int_0^t \varphi_n(s)\, ds$, and let $\{Z_n : n = 0, 1, \ldots\}$ be i.i.d. standard normal random variables. For each fixed $t \in [0, 1]$, we have

$$\sum_{n=0}^{\infty} \Phi_n^2(t) = \sum_{n=0}^{\infty} \langle 1_{[0,t]}, \varphi_n \rangle_{\mathbf{L}^2[0,1]}^2 = \|1_{[0,t]}\|_2^2 = t$$

by Parseval's identity. Therefore, for fixed t, the series

$$W(t) = \sum_{n=0}^{\infty} Z_n \Phi_n(t) \qquad (1.10)$$

converges almost surely and in $\mathbf{L}^2$, since the partial sums form an $\mathbf{L}^2$-bounded martingale. However, the series almost surely does not converge in $\mathbf{D}[0, 1]$ since $\sum_{n=0}^{\infty} Z_n^2 = \infty$ almost surely; we show below that it almost surely *does* converge uniformly in $\mathbf{C}[0, 1]$ for a suitable choice of $\{\Phi_n : n = 0, 1, \ldots\}$. Almost sure uniform convergence of (1.10) implies that the sum is a standard Brownian motion on $[0, 1]$, since it is continuous and has the correct covariance. Namely,

$$\mathrm{Cov}(W(t), W(s)) = \mathbb{E} \sum_{n=0}^{\infty} Z_n^2 \int_0^t \varphi_n(u)\, du \int_0^s \varphi_n(u)\, du$$

$$= \sum_{n=0}^{\infty} \langle 1_{[0,t]}, \varphi_n \rangle \langle 1_{[0,s]}, \varphi_n \rangle = \langle 1_{[0,t]} 1_{[0,s]} \rangle = s \wedge t\,,$$

where the convergence of (1.10) in $\mathbf{L}^2$ is used to interchange summation and integration.

Proposition 1.44 *For the Haar basis (1.8), the series (1.10) converges uniformly in* $\mathbf{C}[0, 1]$ *with probability one.*

Proof. We can write the series (1.10) more explicitly using (1.9),

$$W(t) = tZ_0 + \sum_{m=1}^{\infty} \sum_{k=1}^{2^m - 1} Z_{m,k} \Phi_{m,k}(t)\,, \tag{1.11}$$

where Z_0 and $\{Z_{m,k}\}$ are i.i.d. standard normal variables and $\Phi_{m,k} = \int_0^t \varphi_{m,k}(s)\, ds$. The tail estimate for standard normal distributions, see Lemma 12.9 in the appendix, gives

$$\sum_{k=1}^{2^m - 1} \mathbb{P}(|Z_{m,k}| \geqslant \sqrt{2m}) \leqslant 2^m e^{-m}$$

which is summable over $m \geqslant 1$. Thus, almost surely, the bound $|Z_{m,k}| \leqslant \sqrt{2m}$ holds in (1.11) with at most finitely many exceptions. Since $|\Phi_{m,k}(x)| \leqslant 2^{-m/2}$ for all $x \in [0, 1]$, the series (1.11) converges uniformly with probability one. ∎

Remark 1.45 For the Haar basis (1.8), the construction of Brownian motion via the series (1.11) coincides with Lévy's construction as given in Theorem 1.3. ◇

The construction (1.10) yields an alternative proof for the positive direction of the Cameron–Martin theorem. Given $F \in \mathbf{D}[0, 1]$, we show that $\mathbb{L}_0 \ll \mathbb{L}_F$. Write

$$F = \sum_{n=0}^{\infty} a_n \Phi_n, \quad \text{with } \sum_{n=0}^{\infty} a_n^2 < \infty\,,$$

where Φ_n is the integrated Haar basis (or any other orthonormal basis of $\mathbf{D}[0, 1]$ for which the series (1.10) converges uniformly almost surely). Then,

$$W + F = \sum_{n=0}^{\infty} (Z_n + a_n) \Phi_n\,,$$

where, as usual, $\{Z_n\}$ are i.i.d. standard normal. Proving $\mathbb{L}_0 \ll \mathbb{L}_F$ is thus equivalent to proving that the law of the vector $(Z_n : n = 0, 1, \ldots)$ is absolutely continuous to the law of $(Z_n + a_n : n = 0, 1, \ldots)$. To this end we could use Kakutani's absolute-continuity criterion for product measures, see e.g. 14.17 in [Wi91]; however it is also simple to apply Theorem 12.32 of the appendix directly.

Indeed, let $R_n(z_0, \ldots, z_n)$ denote the Radon–Nikodým derivative of the law of the shifted Gaussian vector $(Z_j + a_j \colon j = 0, 1, \ldots)$ with respect to the law of the standard Gaussian vector $(Z_j \colon j = 0, 1, \ldots)$. Then

$$R_n(z_0, \ldots, z_n) = \prod_{j=0}^{n} \frac{e^{-(z_j - a_j)^2/2}}{e^{-z_j^2/2}} = \exp\left\{ \sum_{j=0}^{n} a_j z_j - \sum_{j=0}^{n} a_j^2/2 \right\}.$$

As $\sum_{j=0}^{n} a_j Z_j$ is a martingale bounded in $\mathbf{L}^2$ and $\sum_{j=0}^{\infty} a_j^2 < \infty$, we conclude that

$$\lim_{n \to \infty} R_n(Z_0, \ldots, Z_n)$$

almost surely exists and is positive. Theorem 12.32 (iii) then implies that $\mathbb{L}_0 \ll \mathbb{L}_F$.

Exercises

Exercise 1.1. Let $\{B(t) \colon t \geqslant 0\}$ be a Brownian motion with arbitrary starting point. Show that, for all $s, t \geqslant 0$, we have $\operatorname{Cov}(B(s), B(t)) = s \wedge t$.

Exercise 1.2. ⑤ Show that, in Theorem 1.3, Brownian motion is constructed as a jointly measurable function $(\omega, t) \mapsto B(\omega, t)$ on $\Omega \times [0, \infty)$.

Exercise 1.3. ⑤ Show that Brownian motion with start in $x \in \mathbb{R}$ is a Gaussian process.

Exercise 1.4. Show that, for every point $x \in \mathbb{R}$, there exists a *two-sided Brownian motion* $\{B(t) \colon t \in \mathbb{R}\}$ with $B(0) = x$, which has continuous paths, independent increments and the property that, for all $t \in \mathbb{R}$ and $h > 0$, the increments $B(t + h) - B(t)$ are normally distributed with expectation zero and variance h.

Exercise 1.5. ⑤ Fix $x, y \in \mathbb{R}$. The *Brownian bridge* with start in x and end in y is the process $\{X(t) \colon 0 \leqslant t \leqslant 1\}$ defined by

$$X(t) = B(t) - t\big(B(1) - y\big), \qquad \text{for } 0 \leqslant t \leqslant 1,$$

where $\{B(t) \colon t \geqslant 0\}$ is a Brownian motion started in x. The Brownian bridge is an almost surely continuous process such that $X(0) = x$ and $X(1) = y$.

(a) Show that, for every bounded $f \colon \mathbb{R}^n \to \mathbb{R}$,

$$\mathbb{E}\big[f\big(X(t_1), \ldots, X(t_n)\big)\big] = \int f(x_1, \ldots, x_n) \frac{\mathfrak{p}(t_1, x, x_1)}{\mathfrak{p}(1, x, y)}$$

$$\times \prod_{i=2}^{n} \mathfrak{p}(t_i - t_{i-1}, x_i, x_{i+1}) \mathfrak{p}(1 - t_n, x_n, y)\, dx_1 \ldots dx_n,$$

for all $0 < t_1 < \cdots < t_n < 1$ where

$$\mathfrak{p}(t, x, y) = \frac{1}{\sqrt{2\pi t}} e^{-\frac{(y-x)^2}{2t}}.$$

(b) Infer that, for any $t_0 < 1$, the laws of the processes $\{X(t): 0 \leqslant t \leqslant t_0\}$ and $\{B(t): 0 \leqslant t \leqslant t_0\}$ are mutually absolutely continuous, and the Radon–Nikodým derivative evaluated at $\{\psi(t): 0 \leqslant t \leqslant t_0\}$ is a function of $\psi(t_0)$.

Exercise 1.6. $\boxed{\text{S}}$ Prove the law of large numbers in Corollary 1.11 directly.
Hint. Use the law of large numbers for sequences of independent identically distributed random variables to show that $\lim_{n \to \infty} B(n)/n = 0$. Then show that $B(t)$ does not oscillate too much between n and $n + 1$.

Exercise 1.7. $\boxed{\text{S}}$ Show the following improvement to Theorem 1.14: Almost surely,

$$\lim_{h \downarrow 0} \sup_{0 \leqslant t \leqslant 1-h} \frac{|B(t+h) - B(t)|}{\sqrt{2h \log(1/h)}} = 1.$$

Exercise 1.8. $\boxed{\text{S}}$ Let $f: [0, 1] \to \mathbb{R}$ be a continuous function with $f(0) = 0$. Then, for a standard Brownian motion $\{B(t): t \geqslant 0\}$ and $\epsilon > 0$, we have

$$\mathbb{P}\Big\{ \sup_{0 \leqslant t \leqslant 1} |B(t) - f(t)| < \varepsilon \Big\} > 0.$$

Exercise 1.9. $\boxed{\text{S}}$ Show that, if $\alpha > 1/2$, then, almost surely, at every point, Brownian motion fails to be locally α-Hölder continuous.

Exercise 1.10. $\boxed{\text{S}}$ Show that, if E is an exchangeable event for an independent, identically distributed sequence, then $\mathbb{P}(E)$ is 0 or 1.

Exercise 1.11. Show that, for a Brownian motion $\{B(t): t \geqslant 0\}$,

(a) for all $t \geqslant 0$ we have $\mathbb{P}\{t \text{ is a local maximum}\} = 0$;
(b) almost surely local maxima exist;
(c) almost surely, there exist $t_*, t^* \in [0, 1)$ with $D^* B(t^*) \leqslant 0$ and $D_* B(t_*) \geqslant 0$.

Exercise 1.12. $\boxed{\text{S}}$ Let $f \in \mathbf{C}[0, 1]$ be any fixed continuous function. Show that, almost surely, the function $\{B(t) + f(t): t \in [0, 1]\}$ is nowhere differentiable.

Exercise 1.13. $\boxed{\text{S}}$ Show that, almost surely, there exists a time t at which $D^* B(t) = 0$.

Exercise 1.14. $\boxed{\text{S}}$ Show that, almost surely,

$$D^* B(t_0) = -\infty,$$

where t_0 is uniquely determined by

$$B(t_0) = \max_{0 \leqslant t \leqslant 1} B(t).$$

Hint. Try this exercise *after* the discussion of the strong Markov property in Chapter 2.

Exercise 1.15. ⑤

(a) Show that, almost surely, there exists a family

$$0 = t_0^{(n)} \leqslant t_1^{(n)} \leqslant \cdots \leqslant t_{k(n)-1}^{(n)} \leqslant t_{k(n)}^{(n)} = t$$

of (random) partitions such that

$$\lim_{n \uparrow \infty} \sum_{j=1}^{k(n)} \left(B(t_j^{(n)}) - B(t_{j-1}^{(n)}) \right)^2 = \infty .$$

Hint. Use the construction of Brownian motion to pick a partition consisting of dyadic intervals, such that the increment of Brownian motion over any chosen interval is large relative to the square root of its length.

(b) Construct a (nonrandom) sequence of partitions

$$0 = t_0^{(n)} \leqslant t_1^{(n)} \leqslant \cdots \leqslant t_{k(n)-1}^{(n)} \leqslant t_{k(n)}^{(n)} = t$$

with mesh converging to zero, such that, almost surely,

$$\limsup_{n \to \infty} \sum_{j=1}^{k(n)} \left(B(t_j^{(n)}) - B(t_{j-1}^{(n)}) \right)^2 = \infty.$$

Exercise 1.16. ⑤ Consider a (not necessarily nested) sequence of partitions

$$0 = t_0^{(n)} \leqslant t_1^{(n)} \leqslant \cdots \leqslant t_{k(n)-1}^{(n)} \leqslant t_{k(n)}^{(n)} = t$$

with mesh converging to zero.

(a) Show that, in the sense of $\mathbf{L}^2$-convergence,

$$\lim_{n \to \infty} \sum_{j=1}^{k(n)} \left(B(t_j^{(n)}) - B(t_{j-1}^{(n)}) \right)^2 = t.$$

(b) Show that, if additionally

$$\sum_{n=1}^{\infty} \sum_{j=1}^{k(n)} \left(t_j^{(n)} - t_{j-1}^{(n)} \right)^2 < \infty,$$

then the convergence in (a) also holds almost surely.

Exercise 1.17. ⑤ Using the notation as in Remark 1.42 and below, for a fixed function $F \in \mathbf{C}[0,1]$ and a Brownian motion $B \in \mathbf{C}[0,1]$ we denote

$$H_n = 2^{n-1} \left[\sum_{j=1}^{2^n} (\nabla_j^{(n)} F)^2 - 2 \sum_{j=1}^{2^n} \left(\nabla_j^{(n)} B \right) \left(\nabla_j^{(n)} F \right) \right].$$

Show directly that $\{e^{-H_n} : n \geqslant 1\}$ is a martingale with respect to the filtration $(\mathcal{F}_n : n \geqslant 1)$.

Exercise 1.18. By the Cameron-Martin theorem for a Brownian motion B and $F \in$ $\mathbf{D}[0,1]$, the function $B + F$ has almost surely finite quadratic variation. Show that there exist continuous functions $F \notin \mathbf{D}[0,1]$ such that $B + F$ has infinite quadratic variation almost surely.

Exercise 1.19. ⑤ Let $F \in \mathbf{D}[0,1]$. The Cameron-Martin theorem together with the Hölder continuity of Brownian motion implies that F is Hölder continuous with exponent α, for all $\alpha < 1/2$. Prove directly that F is Hölder continuous with exponent $1/2$.

Exercise 1.20. Show that the Haar system $\{\varphi_n : n = 0, 1, \ldots\}$ constructed in (1.8) is complete in $\mathbf{L}^2[0,1]$.
Hint. It suffices to show that this system spans all step functions where the steps are dyadic intervals of length at least 2^{-m}. This can be verified by induction on m.

Notes and comments

The first study of the mathematical process of Brownian motion is due to Bachelier in [Ba00] in the context of modelling stock market fluctuations, see [DE06] for a modern edition. Bachelier's work was long forgotten and has only recently been rediscovered, today an international society for mathematical finance is named after him. The physical phenomenon of Brownian motion is usually attributed to Brown [Br28] and was explained by Einstein in [Ei05], see also [Ei56]. Einstein's explanation of the phenomenon was also a milestone in the establishment of the atomistic world view of physics. The first rigorous construction of mathematical Brownian motion is due to Wiener [Wi23], and in his honour Brownian motion is sometimes called the *Wiener process*. Moreover, the space of continuous function equipped with the distribution of standard Brownian motion is often called *Wiener space*. There is also a generalisation of Wiener's approach to the construction of more general Gaussian measures on separable Banach space, which is called the abstract Wiener space, see Kallianpur [Ka71].

As explained in the introduction, Brownian motion describes the macroscopic picture emerging from a random walk if its increments are sufficiently tame not to cause jumps which are visible in the macroscopic description. If this is not the case the class of *Lévy processes* and within this class the *stable processes* offer a macroscopic description. A very good book dealing with Lévy processes is Bertoin [Be96] and a recommended introductory course in the subject is Kyprianou [Ky06].

There is a variety of constructions of Brownian motion in the literature. The approach we have followed goes back to one of the great pioneers of Brownian motion, the French mathematician *Paul Lévy*, see [Le48]. Lévy's construction has the advantage that continuity properties of Brownian motion can be obtained from the construction. An alternative is to first show that a Markov process with the correct transition probabilities can be constructed, and then to use an abstract criterion, like Kolmogorov's criterion for the existence of a continuous version of the process. See, for example, Revuz and Yor [RY94], Karatzas and Shreve [KS91] and Kahane [Ka85] for further alternative constructions. For the Haar basis (1.8), the construction of Brownian motion via the series (1.11) is exactly Lévy's interpolation construction, expressed in more fancy language. Nevertheless, the Hilbert space point of view is essential in studies of more general Gaussian processes, see the excellent book by Janson [Ja97]. For a proof that the series (1.10) converges uniformly for the trigonometric basis (1.7) and more on the Hilbert space perspective, see Kahane [Ka85].

Gaussian processes, only briefly mentioned here, are one of the richest and best understood class of processes in probability theory. Some good references for this are Adler [Ad90] and Lifshits [Li95]. A lot of effort in current research is put into trying to extend our understanding of Brownian motion to more general Gaussian processes like the so-called *fractional Brownian motion*. The main difficulty is that these processes do not have the extremely useful Markov property — which we shall discuss in the next chapter, and which we will make heavy use of throughout the book.

The modulus of continuity, Theorem 1.14, goes back to Lévy [Le37]. Observe that this result describes continuity of Brownian motion near its *worst* time. By contrast, the law of the iterated logarithm in the form of Corollary 5.3 shows that at a *typical* time the continuity properties of Brownian motion are better: For every fixed time $t > 0$ and $c > \sqrt{2}$, almost surely, there exists $\varepsilon > 0$ with $|B(t) - B(t + h)| \leqslant c\sqrt{h \log\log(1/h)}$ for all $|h| < \varepsilon$. In Chapter 10 we explore for how many times $t > 0$ we are close to the worst case scenario.

The existence of points where Brownian motion is locally $1/2$-Hölder continuous is a very tricky question. Dvoretzky [Dv63] showed that, for a sufficiently small $c > 0$, almost surely no point satisfies $1/2$-local Hölder continuity with Hölder constant c. Later, Davis [Da83] and, independently, Greenwood and Perkins [GP83] identified the maximal possible Hölder constant, we will discuss their work in Chapter 10.

There is a lot of discussion about nowhere differentiable, continuous functions in the analysis literature of the early twentieth century. Examples are Weierstrass' function, see e.g. [MG84], and van der Waerden's function, see e.g. [Bi82]. Nowhere differentiability of Brownian motion was first shown by Paley, Wiener and Zygmund in [PWZ33], but the proof we give is due to Dvoretzky, Erdős and Kakutani [DEK61]. Besides the discussion of special examples of such functions, the statement that in some sense 'most' or 'almost all' continuous functions are nowhere differentiable is particularly fascinating. A topological form of this statement is that nowhere differentiability is a generic property for the space $\mathbf{C}([0, 1])$ in the sense of Baire category. A newer, measure theoretic approach based on an idea of Christensen [Ch72], which was later rediscovered by Hunt, Sauer, and Yorke [HSY92], is the notion of prevalence. A subset A of a separable Banach space X is called *prevalent* if there exists a Borel probability measure μ on X such that $\mu(x + A) = 1$ for any $x \in X$. A strengthening of the proof of Theorem 1.30, see Exercise 1.12, shows that the set of nowhere differentiable functions is prevalent.

The time t where $D^*B(t) = 0$ which we constructed in Exercise 1.13 is an exceptional time, i.e. a time where Brownian motion behaves differently from almost every other time. In Chapter 10 we enter a systematic discussion of such times, and in particular address the question how many exceptional points (in terms of Hausdorff dimension) of a certain type exist. The set of times where $D^*B(t) = 0$ has Hausdorff dimension $1/4$, see Barlow and Perkins [BP84].

The interesting fact that the 'true' quadratic variation of Brownian motion, taken as a supremum over arbitrary partitions with mesh going to zero, is infinite is a result of Lévy, see [Le40]. Finer variation properties of Brownian motion have been studied by Taylor in [Ta72]. He shows, for example, that the ψ-variation

$$V^\psi = \sup \sum_{i=1}^k \psi\big(|B(t_i) - B(t_{i-1})|\big),$$

where the supremum is taken over all partitions $0 = t_0 < \cdots < t_k = 1$, $k \in \mathbb{N}$, is finite almost surely for $\psi_1(s) = s^2/(2 \log\log(1/s))$, but is infinite for any ψ with $\psi(s)/\psi_1(s) \to \infty$ as $s \downarrow 0$.

2

Brownian motion as a strong Markov process

In this chapter we discuss the strong Markov property of Brownian motion. We also briefly discuss Markov processes in general and show that some processes, which can be derived from Brownian motion, are also Markov processes. We then exploit these facts to get finer properties of Brownian sample paths.

2.1 The Markov property and Blumenthal's 0-1 law

For the discussion of the Markov property we include higher dimensional Brownian motion, which can be defined easily by requiring the characteristics of a linear Brownian motion in every component, and independence of the components.

Definition 2.1. If $B_1, \ldots, B_d$ are independent linear Brownian motions started in $x_1, \ldots, x_d$, then the stochastic process $\{B(t): t \geqslant 0\}$ given by

$$B(t) = (B_1(t), \ldots, B_d(t))^{\mathrm{T}}$$

is called a **d-dimensional Brownian motion** started in $(x_1, \ldots, x_d)^{\mathrm{T}}$. The d-dimensional Brownian motion started in the origin is also called **standard Brownian motion**. One-dimensional Brownian motion is also called **linear**, two-dimensional Brownian motion **planar Brownian motion**. ◇

Notation 2.2. Throughout this book we write $\mathbb{P}_x$ for the probability measure which makes the d-dimensional process $\{B(t): t \geqslant 0\}$ a Brownian motion started in $x \in \mathbb{R}^d$, and $\mathbb{E}_x$ for the corresponding expectation. ◇

Suppose now that $\{X(t): t \geqslant 0\}$ is a stochastic process. Intuitively, the **Markov property** says that if we know the process $\{X(t): t \geqslant 0\}$ on the interval $[0, s]$, for the prediction of the future $\{X(t): t \geqslant s\}$ this is as useful as just knowing the endpoint $X(s)$. Moreover, a process is called a **(time-homogeneous) Markov process** if it starts afresh at any fixed time s. Slightly more precisely this means that, supposing the process can be started in any point $X(0) = x \in \mathbb{R}^d$, the time-shifted process $\{X(s+t): t \geqslant 0\}$ has the same distribution as the process started in $X(s) \in \mathbb{R}^d$. We shall formalise the notion of a Markov process later in this chapter, but start by giving a straight formulation of the facts for a Brownian motion.

36

Fig. 2.1. Brownian motion starts afresh at time s.

Note that two stochastic processes $\{X(t)\colon t \geqslant 0\}$ and $\{Y(t)\colon t \geqslant 0\}$ are called **independent**, if for any sets $t_1, \ldots, t_n \geqslant 0$ and $s_1, \ldots, s_m \geqslant 0$ of times the vectors $(X(t_1), \ldots, X(t_n))$ and $(Y(s_1), \ldots, Y(s_m))$ are independent.

Theorem 2.3 (Markov property) *Suppose that $\{B(t)\colon t \geqslant 0\}$ is a Brownian motion started in $x \in \mathbb{R}^d$. Let $s > 0$, then the process $\{B(t+s) - B(s)\colon t \geqslant 0\}$ is again a Brownian motion started in the origin and it is independent of the process $\{B(t)\colon 0 \leqslant t \leqslant s\}$.*

Proof. It is easy to check that $\{B(t+s) - B(s)\colon t \geqslant 0\}$ satisfies the definition of a d-dimensional Brownian motion. The independence statement follows directly from the independence of the increments of a Brownian motion. ∎

We now improve this result slightly and introduce some useful terminology.

Definition 2.4.

(a) A **filtration** on a probability space $(\Omega, \mathcal{F}, \mathbb{P})$ is a family $(\mathcal{F}(t)\colon t \geqslant 0)$ of σ-algebras such that $\mathcal{F}(s) \subset \mathcal{F}(t) \subset \mathcal{F}$ for all $s < t$.

(b) A probability space together with a filtration is called a **filtered probability space**.

(c) A stochastic process $\{X(t)\colon t \geqslant 0\}$ defined on a filtered probability space with filtration $(\mathcal{F}(t)\colon t \geqslant 0)$ is called **adapted** if $X(t)$ is $\mathcal{F}(t)$-measurable for any $t \geqslant 0$. ◇

Suppose we have a Brownian motion $\{B(t)\colon t \geqslant 0\}$ defined on some probability space, then we can define a filtration $(\mathcal{F}^0(t)\colon t \geqslant 0)$ by letting

$$\mathcal{F}^0(t) = \sigma\big(B(s)\colon 0 \leqslant s \leqslant t\big)$$

be the σ-algebra generated by the random variables $B(s)$, for $0 \leqslant s \leqslant t$. With this definition, the Brownian motion is obviously adapted to the filtration. Intuitively, this σ-algebra contains all the information available from observing the process up to time t.

By Theorem 2.3, the process $\{B(t+s) - B(s)\colon t \geqslant 0\}$ is independent of $\mathcal{F}^0(s)$. In a first step, we improve this and allow a slightly larger (augmented) σ-algebra $\mathcal{F}^+(s)$ defined by

$$\mathcal{F}^+(s) = \bigcap_{t>s} \mathcal{F}^0(t).$$

Clearly, the family $(\mathcal{F}^+(t)\colon t \geqslant 0)$ is again a filtration and $\mathcal{F}^+(s) \supset \mathcal{F}^0(s)$, but intuitively $\mathcal{F}^+(s)$ is a bit larger than $\mathcal{F}^0(s)$, allowing an additional infinitesimal glance into the future.

Theorem 2.5 *For every $s \geqslant 0$ the process $\{B(t+s) - B(s)\colon t \geqslant 0\}$ is independent of the σ-algebra $\mathcal{F}^+(s)$.*

Proof.   By continuity $B(t+s) - B(s) = \lim_{n\to\infty} B(s_n + t) - B(s_n)$ for a strictly decreasing sequence $\{s_n\colon n \in \mathbb{N}\}$ converging to s. By Theorem 2.3, for any $t_1, \ldots, t_m \geqslant 0$, the vector $(B(t_1 + s) - B(s), \ldots, B(t_m + s) - B(s)) = \lim_{j\uparrow\infty}(B(t_1 + s_j) - B(s_j), \ldots, B(t_m + s_j) - B(s_j))$ is independent of $\mathcal{F}^+(s)$, and so is the process $\{B(t+s) - B(s)\colon t \geqslant 0\}$.   ∎

Remark 2.6 An alternative way of stating this is that conditional on $\mathcal{F}^+(s)$ the process $\{B(t+s)\colon t \geqslant 0\}$ is a Brownian motion started in $B(s)$.   ◇

We now look at the **germ σ-algebra** $\mathcal{F}^+(0)$, which heuristically comprises all events defined in terms of Brownian motion on an infinitesimal small interval to the right of the origin.

Theorem 2.7 (Blumenthal's 0-1 law) *Let $x \in \mathbb{R}^d$ and $A \in \mathcal{F}^+(0)$. Then $\mathbb{P}_x(A) \in \{0,1\}$.*

Proof.   Using Theorem 2.5 for $s = 0$ we see that any $A \in \sigma(B(t)\colon t \geqslant 0)$ is independent of $\mathcal{F}^+(0)$. This applies in particular to $A \in \mathcal{F}^+(0)$, which therefore is independent of itself, hence has probability zero or one.   ∎

As a first application we show that a standard linear Brownian motion has positive and negative values and zeros in every small interval to the right of 0. We have studied this remarkable property of Brownian motion already by different means, in the discussion following Theorem 1.27.

Theorem 2.8 *Suppose $\{B(t)\colon t \geqslant 0\}$ is a linear Brownian motion. Define $\tau = \inf\{t > 0\colon B(t) > 0\}$ and $\sigma = \inf\{t > 0\colon B(t) = 0\}$. Then*

$$\mathbb{P}_0\{\tau = 0\} = \mathbb{P}_0\{\sigma = 0\} = 1.$$

Proof.   The event

$$\{\tau = 0\} = \bigcap_{n=1}^{\infty} \Big\{ \text{ there is } 0 < \varepsilon < 1/n \text{ such that } B(\varepsilon) > 0 \Big\}$$

is clearly in $\mathcal{F}^+(0)$. Hence we just have to show that this event has positive probability.

This follows, as $\mathbb{P}_0\{\tau \leqslant t\} \geqslant \mathbb{P}_0\{B(t) > 0\} = 1/2$ for $t > 0$. Hence $\mathbb{P}_0\{\tau = 0\} \geqslant 1/2$ and we have shown the first part. The same argument works replacing $B(t) > 0$ by $B(t) < 0$ and from these two facts $\mathbb{P}_0\{\sigma = 0\} = 1$ follows, using the intermediate value property of continuous functions.   ∎

A further application is a 0-1 law for the **tail σ-algebra** of a Brownian motion. Define $\mathcal{G}(t) = \sigma(B(s) \colon s \geqslant t)$. Let $\mathcal{T} = \bigcap_{t \,\geqslant\, 0} \mathcal{G}(t)$ be the σ-algebra of all **tail events**.

Theorem 2.9 (Zero-one law for tail events) *Let $x \in \mathbb{R}^d$ and suppose $A \in \mathcal{T}$ is a tail event. Then $\mathbb{P}_x(A) \in \{0, 1\}$.*

Proof.   It suffices to look at the case $x = 0$. Under the time inversion of Brownian motion, the tail σ-algebra is mapped on the germ σ-algebra, which contains only sets of probability zero or one, by Blumenthal's 0-1 law.   ∎

Remark 2.10 In Exercise 2.2 we shall see that, for any tail event $A \in \mathcal{T}$, the probability $\mathbb{P}_x(A)$ is independent of x. For a germ event $A \in \mathcal{F}^+(0)$, however, the probability $\mathbb{P}_x(A)$ may depend on x.   ◇

As final example of this section we now exploit the Markov property to study the local and global extrema of a linear Brownian motion.

Theorem 2.11 *For a linear Brownian motion $\{B(t) \colon 0 \leqslant t \leqslant 1\}$, almost surely,*

   (a) *every local maximum is a strict local maximum;*

   (b) *the set of times where the local maxima are attained is countable and dense;*

   (c) *the global maximum is attained at a unique time.*

Proof.   We first show that, given two nonoverlapping closed time intervals, i.e. such that their interiors are disjoint, the maxima of Brownian motion on them are different almost surely, see Figure 2.2 for an illustration. Let $[a_1, b_1]$ and $[a_2, b_2]$ be two fixed intervals with $b_1 \leqslant a_2$. Denote by m_1 and m_2, the maxima of Brownian motion on these two intervals. Note first that, by the Markov property together with Theorem 2.8, almost surely $B(a_2) < m_2$. Hence this maximum agrees with maximum in the interval $[a_2 - \frac{1}{n}, b_2]$, for some $n \in \mathbb{N}$, and we may therefore assume in the proof that $b_1 < a_2$.

Applying the Markov property at time b_1 we see that the random variable $B(a_2) - B(b_1)$ is independent of $m_1 - B(b_1)$. Using the Markov property at time a_2 we see that $m_2 - B(a_2)$ is also independent of both these variables. The event $m_1 = m_2$ can be written as

$$B(a_2) - B(b_1) = m_1 - B(b_1) - (m_2 - B(a_2)).$$

Conditioning on the values of the random variables $m_1 - B(b_1)$ and $m_2 - B(a_2)$, the left hand side is a continuous random variable and the right hand side a constant, hence this event has probability 0.

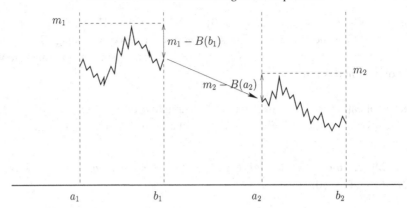

Fig. 2.2. The random variables $m_1 - B(b_1)$ and $m_2 - B(b_2)$ are independent of the increment $B(a_2) - B(b_1)$.

(a) By the statement just proved, almost surely, all nonoverlapping pairs of nondegenerate compact intervals with rational endpoints have different maxima. If Brownian motion however has a non-strict local maximum, there are two such intervals where Brownian motion has the same maximum.

(b) In particular, almost surely, the maximum over any nondegenerate compact interval with rational endpoints is not attained at an endpoint. Hence every such interval contains a local maximum, and the set of times where local maxima are attained is dense. As every local maximum is strict, this set has at most the cardinality of the collection of these intervals.

(c) Almost surely, for any rational number $q \in [0, 1]$ the maximum in $[0, q]$ and in $[q, 1]$ are different. Note that, if the global maximum is attained for two points $t_1 < t_2$ there exists a rational number $t_1 < q < t_2$ for which the maximum in $[0, q]$ and in $[q, 1]$ agree.   ∎

2.2 The strong Markov property and the reflection principle

Heuristically, the Markov property states that Brownian motion is started anew at each *deterministic time instance*. It is a crucial property of Brownian motion that this holds also for an important class of random times. These random times are called *stopping times*.

The basic idea is that a random time T is a stopping time if we can decide whether $\{T \leqslant t\}$ by just knowing the path of the stochastic process up to time t. Think of the situation that T is the first moment where some random event related to the process happens.

Definition 2.12.   A random variable T with values in $[0, \infty]$, defined on a probability space with filtration $(\mathcal{F}(t) : t \geqslant 0)$ is called a **stopping time** with respect to $(\mathcal{F}(t) : t \geqslant 0)$ if $\{T \leqslant t\} \in \mathcal{F}(t)$, for every $t \geqslant 0$.   ◇

Remark 2.13 We formulate some basic facts about stopping times in general:

- Every deterministic time $t \geqslant 0$ is a stopping time with respect to every filtration $(\mathcal{F}(t) : t \geqslant 0)$.

- If $(T_n \colon n = 1, 2, \ldots)$ is an increasing sequence of stopping times with respect to $(\mathcal{F}(t) \colon t \geqslant 0)$ and $T_n \uparrow T$, then T is also a stopping time with respect to $(\mathcal{F}(t) \colon t \geqslant 0)$. This is so because

$$\{T \leqslant t\} = \bigcap_{n=1}^{\infty} \{T_n \leqslant t\} \in \mathcal{F}(t) \, .$$

- Let T be a stopping time with respect to $(\mathcal{F}(t) \colon t \geqslant 0)$. Define times T_n by

$$T_n = (m+1)2^{-n} \text{ if } m2^{-n} \leqslant T < (m+1)2^{-n} \, .$$

In other words, we stop at the first time of the form $k2^{-n}$ after T. It is easy to see that T_n is a stopping time with respect to $(\mathcal{F}(t) \colon t \geqslant 0)$. We will use it later as a discrete approximation to T. ◇

Remark 2.14 Recall from Section 2.1 the definition of the σ-algebras $(\mathcal{F}^0(t) \colon t \geqslant 0)$ and $(\mathcal{F}^+(t) \colon t \geqslant 0)$ associated with Brownian motion.

- Every stopping time T with respect to $(\mathcal{F}^0(t) \colon t \geqslant 0)$ is also a stopping time with respect to $(\mathcal{F}^+(t) \colon t \geqslant 0)$ as $\mathcal{F}^0(t) \subset \mathcal{F}^+(t)$ for every $t \geqslant 0$.

- Suppose H is a closed set, for example a singleton. Then the first hitting time $T = \inf\{t \geqslant 0 \colon B(t) \in H\}$ of the set H is a stopping time with respect to $(\mathcal{F}^0(t) \colon t \geqslant 0)$. Indeed, we note that

$$\{T \leqslant t\} = \bigcap_{n=1}^{\infty} \bigcup_{s \in \mathbb{Q} \cap (0,t)} \bigcup_{x \in \mathbb{Q}^d \cap H} \{B(s) \in \mathcal{B}(x, \tfrac{1}{n})\} \in \mathcal{F}^0(t).$$

- Suppose $G \subset \mathbb{R}^d$ is open, then

$$T = \inf\{t \geqslant 0 \colon B(t) \in G\}$$

is a stopping time with respect to the filtration $(\mathcal{F}^+(t) \colon t \geqslant 0)$, but *not necessarily* with respect to $(\mathcal{F}^0(t) \colon t \geqslant 0)$. To see this note that, by continuity of Brownian motion,

$$\{T \leqslant t\} = \bigcap_{s > t} \{T < s\} = \bigcap_{s > t} \bigcup_{r \in \mathbb{Q} \cap (0,s)} \{B(r) \in G\} \in \mathcal{F}^+(t),$$

so that T is a stopping time with respect to $(\mathcal{F}^+(t) \colon t \geqslant 0)$. However, supposing that G is bounded and the starting point not contained in cl G, we may fix a path $\gamma \colon [0, t] \to \mathbb{R}^d$ with $\gamma(0, t) \cap \text{cl}\, G = \emptyset$ and $\gamma(t) \in \partial G$. Then the σ-algebra $\mathcal{F}^0(t)$ contains no nontrivial subset of $\{B(s) = \gamma(s) \, \forall 0 \leqslant s \leqslant t\}$, i.e. no subset other than the empty set and the set itself. If we had $\{T \leqslant t\} \in \mathcal{F}^0(t)$, the set

$$\{B(s) = \gamma(s) \text{ for all } 0 \leqslant s \leqslant t, T = t\}$$

would be in $\mathcal{F}^0(t)$ and (as indicated in Figure 2.3) a nontrivial subset of this set, which is a contradiction. ◇

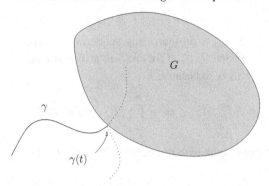

Fig. 2.3. At time t the path γ hits the boundary of G, see the arrow. The two possible dotted continuations indicate that the path may or may not satisfy $T = t$.

Because the first hitting times of open or closed sets play an important rôle, the last item in Remark 2.14 shows that when dealing with Brownian motion it is often preferable to work with stopping times with respect to the richer filtration $(\mathcal{F}^+(t): t \geqslant 0)$ instead of $(\mathcal{F}^0(t): t \geqslant 0)$. Therefore in the case of Brownian motion we make the convention that, unless stated otherwise, notions of stopping time, etc. *always refer to the filtration* $(\mathcal{F}^+(t): t \geqslant 0)$. As this filtration is larger, our choice produces more stopping times.

The crucial property which distinguishes $(\mathcal{F}^+(t): t \geqslant 0)$ from $(\mathcal{F}^0(t): t \geqslant 0)$ is **right-continuity**, which means that

$$\bigcap_{\varepsilon > 0} \mathcal{F}^+(t + \varepsilon) = \mathcal{F}^+(t).$$

To see this note that

$$\bigcap_{\varepsilon > 0} \mathcal{F}^+(t + \varepsilon) = \bigcap_{n=1}^{\infty} \bigcap_{k=1}^{\infty} \mathcal{F}^0(t + 1/n + 1/k) = \mathcal{F}^+(t).$$

The next result indicates the technical advantage of right-continuous filtrations.

Proposition 2.15 *Suppose a random variable T with values in $[0, \infty]$ satisfies $\{T < t\} \in \mathcal{F}(t)$, for every $t \geqslant 0$, and $(\mathcal{F}(t): t \geqslant 0)$ is right-continuous, then T is a stopping time with respect to $(\mathcal{F}(t): t \geqslant 0)$.*

Proof. Suppose that T satisfies the conditions of the theorem. Then

$$\{T \leqslant t\} = \bigcap_{k=1}^{\infty} \{T < t + 1/k\} \in \bigcap_{n=1}^{\infty} \mathcal{F}(t + 1/n) = \mathcal{F}(t),$$

using the right-continuity of $(\mathcal{F}(t): t \geqslant 0)$ in the last step. ∎

We define, for every stopping time T, the σ-algebra

$$\mathcal{F}^+(T) = \{A \in \mathcal{A}: A \cap \{T \leqslant t\} \in \mathcal{F}^+(t) \text{ for all } t \geqslant 0\}.$$

This means that the part of A that lies in $\{T \leqslant t\}$ should be measurable with respect to the information available at time t. Heuristically, this is the collection of events that happened before the stopping time T. In particular, it is easy to see that the random path $\{B(t): t \leqslant T\}$ is $\mathcal{F}^+(T)$-measurable. As in the proof of the last theorem we can infer that for right-continuous filtrations like our $(\mathcal{F}^+(t): t \geqslant 0)$ the event $\{T < t\}$ may replace $\{T \leqslant t\}$ without changing the definition.

We can now state and prove the *strong Markov property* for Brownian motion, which was rigorously established by Hunt [Hu56] and Dynkin [Dy57].

Theorem 2.16 (Strong Markov property) *For every almost surely finite stopping time T, the process*

$$\{B(T+t) - B(T): t \geqslant 0\}$$

is a standard Brownian motion independent of $\mathcal{F}^+(T)$.

Remark 2.17 An alternative form of the strong Markov property is that, for any bounded measurable $f: \mathbf{C}([0,\infty), \mathbb{R}^d) \to \mathbb{R}$ and $x \in \mathbb{R}^d$, we have almost surely

$$\mathbb{E}_x\left[f\left(\{B(T+t): t \geqslant 0\}\right) \mid \mathcal{F}^+(T)\right] = \mathbb{E}_{B(T)}\left[f\left(\{\tilde{B}(t): t \geqslant 0\}\right)\right],$$

where the expectation on the right is with respect to a Brownian motion $\{\tilde{B}(t): t \geqslant 0\}$ started in the fixed point $B(T)$. ◇

Proof. We first show our statement for the stopping times T_n which discretely approximate T from above, $T_n = (m+1)2^{-n}$ if $m2^{-n} \leqslant T < (m+1)2^{-n}$, see Remark 2.13. Write $B_k = \{B_k(t): t \geqslant 0\}$ for the Brownian motion defined by $B_k(t) = B(t + k/2^n) - B(k/2^n)$, and $B_* = \{B_*(t): t \geqslant 0\}$ for the process defined by $B_*(t) = B(t + T_n) - B(T_n)$. Suppose that $E \in \mathcal{F}^+(T_n)$. Then, for every event $\{B_* \in A\}$, we have

$$\mathbb{P}\left(\{B_* \in A\} \cap E\right) = \sum_{k=0}^{\infty} \mathbb{P}\left(\{B_k \in A\} \cap E \cap \{T_n = k2^{-n}\}\right)$$

$$= \sum_{k=0}^{\infty} \mathbb{P}\{B_k \in A\} \mathbb{P}\left(E \cap \{T_n = k2^{-n}\}\right),$$

using that $\{B_k \in A\}$ is independent of $E \cap \{T_n = k2^{-n}\} \in \mathcal{F}^+(k2^{-n})$ by Theorem 2.5. Now, by Theorem 2.3, $\mathbb{P}\{B_k \in A\} = \mathbb{P}\{B \in A\}$ does not depend on k, and hence we get

$$\sum_{k=0}^{\infty} \mathbb{P}\{B_k \in A\} \mathbb{P}\left(E \cap \{T_n = k2^{-n}\}\right) = \mathbb{P}\{B \in A\} \sum_{k=0}^{\infty} \mathbb{P}\left(E \cap \{T_n = k2^{-n}\}\right)$$

$$= \mathbb{P}\{B \in A\}\mathbb{P}(E),$$

which shows that B_* is a Brownian motion and independent of E, hence of $\mathcal{F}^+(T_n)$.

It remains to generalise this to general stopping times T. As $T_n \downarrow T$ we have that

$$\{B(s+T_n) - B(T_n): s \geqslant 0\}$$

is a Brownian motion independent of $\mathcal{F}^+(T_n) \supset \mathcal{F}^+(T)$. Hence the increments

$$B(s+t+T) - B(t+T) = \lim_{n \to \infty} B(s+t+T_n) - B(t+T_n)$$

of the process $\{B(r+T) - B(T): r \geqslant 0\}$ are independent and normally distributed with mean zero and variance s. As the process is obviously almost surely continuous, it is a Brownian motion. Moreover all increments, $B(s+t+T) - B(t+T) = \lim B(s+t+T_n) - B(t+T_n)$, and hence the process itself, are independent of $\mathcal{F}^+(T)$. ∎

Remark 2.18 Let $\tau = \inf\{t \geqslant 0: B(t) = \max_{0 \leqslant s \leqslant 1} B(s)\}$. It is intuitively clear that τ is *not* a stopping time. To prove it, recall that almost surely $\tau < 1$. The increment $B(\tau + t) - B(\tau)$ is negative in a small neighbourhood to the right of 0, which contradicts the strong Markov property and Theorem 2.8. ◇

2.2.1 The reflection principle

We will see many applications of the strong Markov property later, however, the next result, the reflection principle, is particularly interesting. The reflection principle states that Brownian motion reflected at some stopping time T is still a Brownian motion.

Theorem 2.19 (Reflection principle) *If T is a stopping time and $\{B(t): t \geqslant 0\}$ is a standard Brownian motion, then the process $\{B^*(t): t \geqslant 0\}$ called* **Brownian motion reflected at** *T and defined by*

$$B^*(t) = B(t)1_{\{t \leqslant T\}} + (2B(T) - B(t))1_{\{t > T\}}$$

is also a standard Brownian motion.

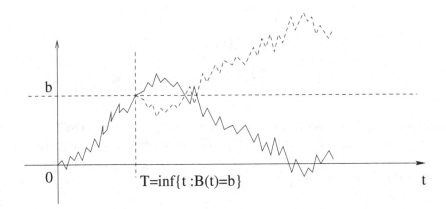

Fig. 2.4. The reflection principle in the case of the first hitting time of level b.

Proof.   If T is finite, by the strong Markov property both paths

$$\{B(t+T) - B(T): t \geqslant 0\} \text{ and } \{-(B(t+T) - B(T)): t \geqslant 0\} \tag{2.1}$$

are Brownian motions and independent of the beginning $\{B(t) \colon 0 \leqslant t \leqslant T\}$. The concatenation mapping, which takes a continuous path $\{g(t) \colon t \geqslant 0\}$ and glues it to the end point of a finite continuous path $\{f(t) \colon 0 \leqslant t \leqslant T\}$ to form a new continuous path, is measurable. Hence the process arising from glueing the first path in (2.1) to $\{B(t) \colon 0 \leqslant t \leqslant T\}$ and the process arising from glueing the second path in (2.1) to $\{B(t) \colon 0 \leqslant t \leqslant T\}$ have the same distribution. The first is just $\{B(t) \colon t \geqslant 0\}$, the second is $\{B^*(t) \colon t \geqslant 0\}$, as introduced in the statement. ∎

Remark 2.20 For a linear Brownian motion, consider

$$\tau = \inf\left\{t \geqslant 0 \colon B(t) = \max_{0 \leqslant s \leqslant 1} B(s)\right\}$$

and let $\{B^*(t) \colon t \geqslant 0\}$ be the reflection at τ defined as in Theorem 2.19. Recall from Remark 2.18 that τ is not a stopping time. Not only is the reflected process *not* Brownian motion, but its law is singular with respect to that of Brownian motion. Indeed, τ is a point of increase of the reflected process by construction, whereas we shall see in Theorem 5.14 that Brownian motion almost surely has no such point. ◇

Now we apply the reflection principle in the case of linear Brownian motion. Let $M(t) = \max_{0 \leqslant s \leqslant t} B(s)$. A priori it is not at all clear what the distribution of this random variable is, but we can determine it as a consequence of the reflection principle.

Theorem 2.21 *If $a > 0$ then $\mathbb{P}_0\{M(t) > a\} = 2\mathbb{P}_0\{B(t) > a\} = \mathbb{P}_0\{|B(t)| > a\}$.*

Proof. Let $T = \inf\{t \geqslant 0 \colon B(t) = a\}$ and let $\{B^*(t) \colon t \geqslant 0\}$ be Brownian motion reflected at the stopping time T. Then

$$\{M(t) > a\} = \{B(t) > a\} \cup \{M(t) > a, \ B(t) \leqslant a\}.$$

This is a disjoint union and the second summand coincides with event $\{B^*(t) \geqslant a\}$. Hence the statement follows from the reflection principle. ∎

Remark 2.22 Theorem 2.21 is most useful when combined with a tail estimate for the Gaussian as in Lemma 12.9 in the appendix. For example, for an upper bound we obtain, for all $a > 0$,

$$\mathbb{P}_0\{M(t) > a\} \leqslant \frac{\sqrt{2t}}{a\sqrt{\pi}} \exp\left\{-\frac{a^2}{2t}\right\}. \qquad ◇$$

2.2.2 The area of planar Brownian motion

Continuous curves in the plane can still be extremely wild. Space-filling curves, like the Peano curve, can map the time interval $[0, 1]$ continuously on sets of positive area, see for example [La98]. We now show that the range of planar Brownian motion has zero area. The Markov property and the reflection principle play an important rôle in the proof.

Suppose $\{B(t) \: : \: t \geqslant 0\}$ is planar Brownian motion. We denote the Lebesgue measure on $\mathbb{R}^d$ by $\mathcal{L}_d$, and use the symbol $f * g$ to denote the **convolution** of the functions f and g given, whenever well-defined, by

$$f * g\,(x) := \int f(y)g(x-y)\,dy.$$

For a set $A \subset \mathbb{R}^d$ and $x \in \mathbb{R}^d$ we write $A + x := \{a + x \colon a \in A\}$.

Lemma 2.23 *If $A_1, A_2 \subset \mathbb{R}^2$ are Borel sets with positive area, then*

$$\mathcal{L}_2\big(\{x \in \mathbb{R}^2 : \mathcal{L}_2(A_1 \cap (A_2 + x)) > 0\}\big) > 0.$$

Proof.   We may assume A_1 and A_2 are bounded. By Fubini's theorem,

$$\int_{\mathbb{R}^2} 1_{A_1} * 1_{-A_2}(x)\,dx = \int_{\mathbb{R}^2} \int_{\mathbb{R}^2} 1_{A_1}(w) 1_{A_2}(w-x)\,dw\,dx$$

$$= \int_{\mathbb{R}^2} 1_{A_1}(w) \left(\int_{\mathbb{R}^2} 1_{A_2}(w-x)\,dx \right) dw$$

$$= \mathcal{L}_2(A_1)\mathcal{L}_2(A_2) > 0.$$

Thus $1_{A_1} * 1_{-A_2}(x) > 0$ on a set of positive area. But

$$1_{A_1} * 1_{-A_2}(x) = \int 1_{A_1}(y)\, 1_{-A_2}(x-y)\,dy = \int 1_{A_1}(y)\, 1_{A_2+x}(y)\,dy$$

$$= \mathcal{L}_2(A_1 \cap (A_2 + x)),$$

proving the lemma.   ∎

We are now ready to prove Lévy's theorem on the area of planar Brownian motion.

Theorem 2.24 (Lévy 1940) *Almost surely, $\mathcal{L}_2(B[0,1]) = 0$.*

Proof.   Let $X = \mathcal{L}_2(B[0,1])$ denote the area of $B[0,1]$. First we check that $\mathbb{E}[X] < \infty$. Note that $X > a$ only if the Brownian motion leaves the square centred in the origin of side length $\sqrt{a}$. Hence, using Theorem 2.21 and Lemma 12.9 of the appendix,

$$\mathbb{P}\{X > a\} \leqslant 2\,\mathbb{P}\big\{ \max_{t \in [0,1]} |W(t)| > \sqrt{a}/2 \big\} = 4\,\mathbb{P}\{W(1) > \sqrt{a}/2\} \leqslant 4e^{-a/8},$$

for $a > 1$, where $\{W(t) \colon t \geqslant 0\}$ is standard one-dimensional Brownian motion. Hence,

$$\mathbb{E}[X] = \int_0^\infty \mathbb{P}\{X > a\}\,da \leqslant 4 \int_1^\infty e^{-a/8}da + 1 < \infty.$$

Note that $B(3t)$ and $\sqrt{3}B(t)$ have the same distribution, and hence

$$\mathbb{E}\mathcal{L}_2(B[0,3]) = 3\mathbb{E}\mathcal{L}_2(B[0,1]) = 3\mathbb{E}[X].$$

Note that we have $\mathcal{L}_2(B[0,3]) \leqslant \sum_{j=0}^{2} \mathcal{L}_2(B[j, j+1])$ with equality if and only if for

$0 \leqslant i < j \leqslant 2$ we have $\mathcal{L}_2(B[i, i+1] \cap B[j, j+1]) = 0$. On the other hand, for $j = 0, 1, 2$, we have $\mathbb{E}\mathcal{L}_2(B[j, j+1]) = \mathbb{E}[X]$ and

$$3\mathbb{E}[X] = \mathbb{E}\mathcal{L}_2(B[0,3]) \leqslant \sum_{j=0}^{2} \mathbb{E}\mathcal{L}_2(B[j, j+1]) = 3\mathbb{E}[X],$$

whence, almost surely, the intersection of any two of the $B[j, j+1]$ has measure zero. In particular, $\mathcal{L}_2(B[0,1] \cap B[2,3]) = 0$ almost surely.

Now we can use the Markov property to define two Brownian motions, $\{B_1(t) : t \in [0,1]\}$ by $B_1(t) = B(t)$, and $\{B_2(t) : t \in [0,1]\}$ by $B_2(t) = B(t+2) - B(2) + B(1)$. The random variable $Y := B(2) - B(1)$ is independent of both Brownian motions. For $x \in \mathbb{R}^2$, let $R(x)$ denote the area of the set $B_1[0,1] \cap (x + B_2[0,1])$, and note that $\{R(x) : x \in \mathbb{R}^2\}$ is independent of Y. Then

$$0 = \mathbb{E}[\mathcal{L}_2(B[0,1] \cap B[2,3])] = \mathbb{E}[R(Y)] = (2\pi)^{-1} \int_{\mathbb{R}^2} e^{-|x|^2/2}\, \mathbb{E}[R(x)]\, dx,$$

where we are averaging with respect to the Gaussian distribution of $B(2) - B(1)$. Thus, for $\mathcal{L}_2$-almost all x, we have $R(x) = 0$ almost surely and hence, by Fubini's theorem,

$$\mathcal{L}_2(\{x \in \mathbb{R}^2 : R(x) > 0\}) = 0, \qquad \text{almost surely.}$$

From Lemma 2.23 we get that, almost surely, $\mathcal{L}_2(B[0,1]) = 0$ or $\mathcal{L}_2(B[2,3]) = 0$. The observation that $\mathcal{L}_2(B[0,1])$ and $\mathcal{L}_2(B[2,3])$ are identically distributed and independent completes the proof that $\mathcal{L}_2(B[0,1]) = 0$ almost surely. ∎

Remark 2.25 How big is the range, or path, of Brownian motion? We have seen that the Lebesgue measure of a planar Brownian path is zero almost surely, but a more precise answer needs the concept of Hausdorff measure and dimension, which we develop in Chapter 4. ◇

Corollary 2.26 *For any points $x, y \in \mathbb{R}^d$, $d \geqslant 2$, we have $\mathbb{P}_x\{y \in B(0,1]\} = 0$.*

Proof. Observe that, by projection onto the first two coordinates, it suffices to prove this result for $d = 2$. Note that Theorem 2.24 holds for Brownian motion with arbitrary starting point $y \in \mathbb{R}^2$. By Fubini's theorem, for any fixed $y \in \mathbb{R}^2$,

$$\int_{\mathbb{R}^2} \mathbb{P}_y\{x \in B[0,1]\}\, dx = \mathbb{E}_y \mathcal{L}_2(B[0,1]) = 0.$$

Hence, for $\mathcal{L}_2$-almost every point x, we have $\mathbb{P}_y\{x \in B[0,1]\} = 0$. By symmetry of Brownian motion,

$$\mathbb{P}_y\{x \in B[0,1]\} = \mathbb{P}_0\{x - y \in B[0,1]\} = \mathbb{P}_0\{y - x \in B[0,1]\} = \mathbb{P}_x\{y \in B[0,1]\}.$$

We infer that $\mathbb{P}_x\{y \in B[0,1]\} = 0$, for $\mathcal{L}_2$-almost every point x. For any $\varepsilon > 0$ we thus have, almost surely, $\mathbb{P}_{B(\varepsilon)}\{y \in B[0,1]\} = 0$. Hence,

$$\mathbb{P}_x\{y \in B(0,1]\} = \lim_{\varepsilon \downarrow 0} \mathbb{P}_x\{y \in B[\varepsilon,1]\} = \lim_{\varepsilon \downarrow 0} \mathbb{E}_x \mathbb{P}_{B(\varepsilon)}\{y \in B[0, 1-\varepsilon]\} = 0,$$

where we have used the Markov property in the second step. ∎

Remark 2.27 Loosely speaking, planar Brownian motion almost surely does not hit singletons. Which other sets are not hit by Brownian motion? This clearly depends on the size and shape of the set in some intricate way, and a precise answer will use the notion of capacity, which we study in Chapter 8. ◇

2.2.3 The zero set of Brownian motion

As a further application of the strong Markov property we have a first look at the properties of the zero set $\{t \geqslant 0 : B(t) = 0\}$ of one-dimensional Brownian motion. We prove that this set is a closed set with no isolated points (sometimes called a **perfect** set). This is perhaps surprising since, almost surely, a Brownian motion has isolated zeros from the left, for instance the first zero after $1/2$, or from the right, like the last zero before $1/2$.

Theorem 2.28 *Let* $\{B(t) : t \geqslant 0\}$ *be a one dimensional Brownian motion and*

$$\mathsf{Zeros} = \{t \geqslant 0 : B(t) = 0\}$$

its **zero set**. *Then, almost surely,* Zeros *is a closed set with no isolated points.*

Proof. Clearly, with probability one, Zeros is closed because Brownian motion is continuous almost surely. To prove that no point of Zeros is isolated we consider the following construction: For each rational $q \in [0, \infty)$ consider the first zero after q, i.e.,

$$\tau_q = \inf\{t \geqslant q : B(t) = 0\}.$$

Note that τ_q is an almost surely finite stopping time. Since Zeros is closed, the inf is almost surely a minimum. By the strong Markov property, applied to τ_q, we have that for each q, almost surely τ_q is not an isolated zero from the right. But, since there are only countably many rationals, we conclude that almost surely, for all rational q, the zero τ_q is not isolated from the right.

Our next task is to prove that the remaining points of Zeros are not isolated from the left. So we claim that any $0 < t \in$ Zeros which is different from τ_q for all rational q is not an isolated point from the left. To see this take a sequence $q_n \uparrow t$, $q_n \in \mathbb{Q}$. Define $t_n = \tau_{q_n}$. Clearly $q_n \leqslant t_n < t$ and so $t_n \uparrow t$. Thus t is not isolated from the left. ∎

Remark 2.29 Theorem 2.28 implies that Zeros is uncountable, see Exercise 2.9. ◇

2.3 Markov processes derived from Brownian motion

In this section, we define the concept of a Markov process. Our motivation is that various processes derived from Brownian motion are Markov processes. Among the examples are the reflection of Brownian motion in zero, and the process $\{T_a : a \geqslant 0\}$ of times T_a when a Brownian motion reaches level a for the first time. We assume that the reader is familiar with the notion of conditional expectation given a σ-algebra, see [Wi91] for a reference.

Definition 2.30. A function $p \colon [0, \infty) \times \mathbb{R}^d \times \mathfrak{B} \to \mathbb{R}$, where $\mathfrak{B}$ is the Borel σ-algebra in $\mathbb{R}^d$, is a **Markov transition kernel** provided

(1) $p(\cdot, \cdot, A)$ is measurable as a function of (t, x), for each $A \in \mathfrak{B}$;

(2) $p(t, x, \cdot)$ is a Borel probability measure on $\mathbb{R}^d$ for all $t \geqslant 0$ and $x \in \mathbb{R}^d$, when integrating a function f with respect to this measure we write

$$\int f(y) \, p(t, x, dy);$$

(3) for all $A \in \mathfrak{B}$, $x \in \mathbb{R}^d$ and $t, s > 0$,

$$p(t + s, x, A) = \int_{\mathbb{R}^d} p(t, y, A) \, p(s, x, dy).$$

An adapted process $\{X(t) \colon t \geqslant 0\}$ is a **(time-homogeneous) Markov process** with transition kernel p with respect to a filtration $(\mathcal{F}(t) \colon t \geqslant 0)$, if for all $t \geqslant s$ and Borel sets $A \in \mathfrak{B}$ we have, almost surely,

$$\mathbb{P}\{X(t) \in A \mid \mathcal{F}(s)\} = p(t - s, X(s), A). \qquad \diamond$$

Observe that $p(t, x, A)$ is the probability that the process takes a value in A at time t, if it is started at the point x. Readers familiar with *Markov chains* can recognise the pattern behind this definition: The Markov transition kernel p plays the rôle of the transition matrix P in this setup. The next two examples are easy consequences of the Markov property for Brownian motion.

Example 2.31 Brownian motion is a Markov process and for its transition kernel p the distribution $p(t, x, \cdot)$ is a normal distribution with mean x and variance t. Similarly, d-dimensional Brownian motion is a Markov process and $p(t, x, \cdot)$ is a Gaussian with mean x and covariance matrix t times identity. Note that property (3) in the definition of the Markov transition kernel is just the fact that the sum of two independent Gaussian random vectors is a Gaussian random vector with the sum of the covariance matrices. $\qquad \diamond$

Notation 2.32. The transition kernel of d-dimensional Brownian motion is described by probability measures $p(t, x, \cdot)$ with densities denoted throughout this book by

$$\mathfrak{p}(t, x, y) = (2\pi t)^{-d/2} \exp\left(-\frac{|x - y|^2}{2t}\right). \qquad \diamond$$

Example 2.33 The *reflected one-dimensional Brownian motion* $\{X(t) \colon t \geqslant 0\}$ defined by $X(t) = |B(t)|$ is a Markov process. Moreover, its transition kernel $p(t, x, \cdot)$ is the law of $|Y|$ for Y normally distributed with mean x and variance t, which we call the *modulus normal distribution* with parameters x and t. $\qquad \diamond$

We now prove a famous theorem of Paul Lévy, which shows that the difference of the maximum process of a Brownian motion and the Brownian motion itself is a reflected Brownian motion. To be precise, this means that the difference of the processes has the same finite-dimensional distributions as a reflected Brownian motion, and is also almost surely continuous.

Theorem 2.34 (Lévy 1948) *Let $\{M(t)\colon t \geqslant 0\}$ be the maximum process of a linear standard Brownian motion $\{B(t)\colon t \geqslant 0\}$, i.e. the process defined by*

$$M(t) = \max_{0 \leqslant s \leqslant t} B(s).$$

Then, the process $\{Y(t)\colon t \geqslant 0\}$ defined by $Y(t) = M(t) - B(t)$ is a reflected Brownian motion.

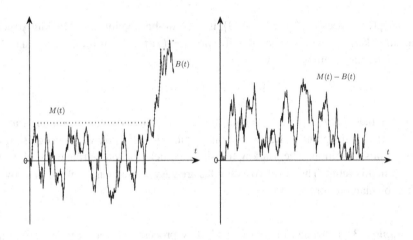

Fig. 2.5. On the left, the processes $\{B(t)\colon t \geqslant 0\}$ with associated maximum process $\{M(t)\colon t \geqslant 0\}$ indicated by the dashed curve. On the right the process $\{M(t) - B(t)\colon t \geqslant 0\}$.

Proof. The main step is to show that the process $\{Y(t)\colon t \geqslant 0\}$ is a Markov process and its Markov transition kernel $p(t, x, \cdot)$ has modulus normal distribution with parameters x and t. Once this is established, it is immediate that the finite-dimensional distributions of this process agree with those of a reflected Brownian motion. Obviously, $\{Y(t)\colon t \geqslant 0\}$ has almost surely continuous paths. For the main step, fix $s > 0$, consider the two processes $\{\hat{B}(t)\colon t \geqslant 0\}$ defined by

$$\hat{B}(t) = B(s + t) - B(s) \text{ for } t \geqslant 0,$$

and $\{\hat{M}(t)\colon t \geqslant 0\}$ defined by

$$\hat{M}(t) = \max_{0 \leqslant u \leqslant t} \hat{B}(u) \text{ for } t \geqslant 0.$$

Because $Y(s)$ is $\mathcal{F}^+(s)$-measurable, it suffices to check that conditional on $\mathcal{F}^+(s)$, for every $t \geqslant 0$, the random variable $Y(s + t)$ has the same distribution as $|Y(s) + \hat{B}(t)|$. Indeed, this directly implies that $\{Y(t)\colon t \geqslant 0\}$ is a Markov process with the same transition kernel as the reflected Brownian motion. To prove the claim fix $s, t \geq 0$ and observe that $M(s + t) = M(s) \vee (B(s) + \hat{M}(t))$, and hence

$$Y(s + t) = (M(s) \vee (B(s) + \hat{M}(t))) - (B(s) + \hat{B}(t)).$$

Using the fact that $(a \vee b) - c = (a - c) \vee (b - c)$, we have

$$Y(s + t) = (Y(s) \vee \hat{M}(t)) - \hat{B}(t).$$

To finish, it suffices to check, for every $y \geq 0$, that $y \vee \hat{M}(t) - \hat{B}(t)$ has the same distribution as $|y + \hat{B}(t)|$. For any $a \geq 0$ write

$$P_1 = \mathbb{P}\{y - \hat{B}(t) > a\}, \qquad P_2 = \mathbb{P}\{y - \hat{B}(t) \leq a \text{ and } \hat{M}(t) - \hat{B}(t) > a\}.$$

Then $\mathbb{P}\{y \vee \hat{M}(t) - \hat{B}(t) > a\} = P_1 + P_2$. Since $\{\hat{B}(t) \colon t \geq 0\}$ has the same distribution as $\{-\hat{B}(t) \colon t \geq 0\}$ we have $P_1 = \mathbb{P}\{y + \hat{B}(t) > a\}$. To study the second term it is useful to define the time reversed Brownian motion $\{W(u) \colon 0 \leq u \leq t\}$ by $W(u) := \hat{B}(t - u) - \hat{B}(t)$. Note that this process is also a Brownian motion for $0 \leq u \leq t$ since it is continuous and its finite dimensional distributions are Gaussian with the right covariances. Let $M_W(t) = \max_{0 \leq u \leq t} W(u)$. Then $M_W(t) = \hat{M}(t) - \hat{B}(t)$. Since $W(t) = -\hat{B}(t)$, we have

$$P_2 = \mathbb{P}\{y + W(t) \leq a \text{ and } M_W(t) > a\}.$$

Using the reflection principle by reflecting $\{W(u) \colon 0 \leq u \leq t\}$ at the first time it hits a, we get another Brownian motion $\{W^*(u) \colon 0 \leq u \leq t\}$. In terms of this Brownian motion we have $P_2 = \mathbb{P}\{W^*(t) \geq a + y\}$. Since it has the same distribution as $\{-\hat{B}(t) \colon t \geq 0\}$, it follows that $P_2 = \mathbb{P}\{y + \hat{B}(t) \leq -a\}$. The Brownian motion $\{\hat{B}(t) \colon t \geq 0\}$ has continuous distribution, and so, by adding P_1 and P_2, we get $\mathbb{P}\{y \vee \hat{M}(t) - \hat{B}(t) > a\} = \mathbb{P}\{|y + \hat{B}(t)| > a\}$. This proves the main step and, consequently, the theorem. ∎

While, as seen above, $\{M(t) - B(t) \colon t \geq 0\}$ is a Markov process, it is important to note that the maximum process $\{M(t) \colon t \geq 0\}$ itself is not a Markov process. However the times when new maxima are achieved form a Markov process, as the following theorem shows.

Theorem 2.35 *For any $a \geq 0$ define the stopping times*

$$T_a = \inf\{t \geq 0 \colon B(t) = a\}.$$

Then $\{T_a \colon a \geq 0\}$ is an increasing Markov process with transition kernel given by the densities

$$p(a, t, s) = \frac{a}{\sqrt{2\pi(s-t)^3}} \exp\left(-\frac{a^2}{2(s-t)}\right) \mathbb{1}\{s > t\}, \qquad \text{for } a > 0.$$

*This process is called the **stable subordinator** of index $\frac{1}{2}$.*

Remark 2.36 As the transition densities satisfy the *shift-invariance property*

$$p(a, t, s) = p(a, 0, s - t) \quad \text{for all } a \geq 0 \text{ and } s, t \geq 0,$$

the stable subordinators $\{T_a \colon a \geq 0\}$ have stationary and independent increments. ◇

Proof. Fix $a \geqslant b \geqslant 0$ and note that for all $t \geqslant 0$ we have

$$\{T_a - T_b = t\}$$
$$= \big\{ B(T_b + s) - B(T_b) < a - b, \text{ for } s < t, \text{ and } B(T_b + t) - B(T_b) = a - b \big\}.$$

By the strong Markov property of Brownian motion this event is independent of $\mathcal{F}^+(T_b)$ and therefore in particular of $\{T_d : d \leqslant b\}$. This proves the Markov property of $\{T_a : a \geqslant 0\}$. The form of the transition kernel follows from the reflection principle,

$$\mathbb{P}\{T_a - T_b \leqslant t\} = \mathbb{P}\{T_{a-b} \leqslant t\} = \mathbb{P}\Big\{ \max_{0 \leqslant s \leqslant t} B(s) \geqslant a - b \Big\}$$

$$= 2\mathbb{P}\{B(t) \geqslant a - b\} = 2 \int_{a-b}^{\infty} \frac{1}{\sqrt{2\pi t}} \exp\Big(-\frac{x^2}{2t} \Big)\, dx$$

$$= \int_0^t \frac{1}{\sqrt{2\pi s^3}} (a - b) \exp\Big(-\frac{(a-b)^2}{2s} \Big)\, ds,$$

where we used the substitution $x = \sqrt{t/s}\,(a - b)$ in the last step. ∎

In a similar way there is another important Markov process, the Cauchy process, hidden in the planar Brownian motion, see Figure 2.6.

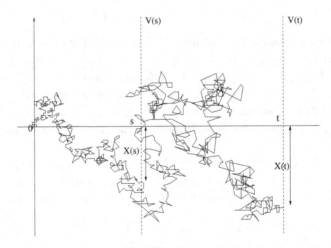

Fig. 2.6. The Cauchy process embedded in planar Brownian motion

Theorem 2.37 *Let $\{B(t): t \geqslant 0\}$ be a planar Brownian motion and denote $B(t) = (B_1(t), B_2(t))$. Define a family $(V(a): a \geqslant 0)$ of vertical lines by*

$$V(a) = \{(x, y) \in \mathbb{R}^2 : x = a\},$$

and let $T(a) = \tau(V(a))$ be the first hitting time of $V(a)$. Then the process $\{X(a): a \geqslant 0\}$ defined by $X(a) := B_2(T(a))$ is a Markov process with transition kernel given by

$$p(a, x, A) = \frac{1}{\pi} \int_A \frac{a}{a^2 + (x - y)^2}\, dy\,.$$

*This process is called the **Cauchy process**.*

Proof. The Markov property of $\{X(a)\colon a \geqslant 0\}$ is a consequence of the strong Markov property of Brownian motion for the stopping times $T(a)$, and the fact that $T(a) < T(b)$ for all $a < b$. In order to calculate the transition density recall from Theorem 2.35 that $T(a)$, which is the first time when the one-dimensional Brownian motion $\{B_1(s)\colon s \geqslant 0\}$ hits level a, has density

$$\frac{a}{\sqrt{2\pi s^3}} \exp\left(-\frac{a^2}{2s}\right).$$

$T(a)$ is independent of $\{B_2(s)\colon s \geqslant 0\}$ and therefore the density of $B_2(T(a))$ is (in the variable x)

$$\int_0^\infty \frac{1}{\sqrt{2\pi s}} \exp\left(-\frac{x^2}{2s}\right) \frac{a}{\sqrt{2\pi s^3}} \exp\left(-\frac{a^2}{2s}\right) ds = \int_0^\infty \frac{a\, e^{-\sigma}}{\pi(a^2 + x^2)} \, d\sigma = \frac{a}{\pi(a^2 + x^2)},$$

where the integral is evaluated using the substitution $\sigma = \frac{1}{2s}(a^2 + x^2)$. ∎

Remark 2.38 As in the case of stable subordinators, see Remark 2.36, one can see from the form of the transition kernel that the Cauchy process has independent, stationary increments. Alternative proofs of Theorem 2.37, avoiding the explicit evaluation of integrals will be given in Exercise 2.19 and Exercise 7.5. ◇

2.4 The martingale property of Brownian motion

In the previous section we have taken a particular feature of Brownian motion, the Markov property, and introduced an abstract class of processes, the Markov processes, which share this feature. We have seen that a number of process derived from Brownian motion are again Markov processes and this insight helped us getting new information about Brownian motion. In this section we follow a similar plan, taking a different feature of Brownian motion, the martingale property, as a starting point.

Definition 2.39. A real-valued stochastic process $\{X(t)\colon t \geqslant 0\}$ is a **martingale** with respect to a filtration $(\mathcal{F}(t)\colon t \geqslant 0)$ if it is adapted to the filtration, $\mathbb{E}|X(t)| < \infty$ for all $t \geqslant 0$ and, for any pair of times $0 \leqslant s \leqslant t$,

$$\mathbb{E}\big[X(t)\,|\,\mathcal{F}(s)\big] = X(s) \text{ almost surely.}$$

The process is called a **submartingale** if $\geqslant$ holds, and a **supermartingale** if $\leqslant$ holds in the display above. ◇

Remark 2.40 Intuitively, a martingale is a process where the current state $X(t)$ is always the best prediction for its further states. In this sense, martingales describe *fair games*. If $\{X(t)\colon t \geqslant 0\}$ is a martingale, the process $\{|X(t)|\colon t \geqslant 0\}$ need not be a martingale, but it still is a submartingale, as a simple application of the triangle inequality shows. ◇

Example 2.41 For a one-dimensional Brownian motion $\{B(t)\colon t \geqslant 0\}$ we have

$$
\begin{aligned}
\mathbb{E}\big[B(t)\,\big|\,\mathcal{F}^+(s)\big] &= \mathbb{E}\big[B(t) - B(s)\,\big|\,\mathcal{F}^+(s)\big] + B(s) \\
&= \mathbb{E}\big[B(t) - B(s)\big] + B(s) = B(s), \text{ for } 0 \leqslant s \leqslant t,
\end{aligned}
$$

using Theorem 2.5 in the second step. Hence Brownian motion is a martingale.   ◇

We now state two useful facts about martingales, which we will exploit extensively: The *optional stopping theorem* and *Doob's maximal inequality*. Both of these results are well-known in the discrete time setting and there is a reminder in Appendix 12.3. The natural extension of these results to the continuous time setting is the content of our propositions.

The optional stopping theorem provides a condition under which the defining equation for martingales can be extended from fixed times $0 \leqslant s \leqslant t$ to stopping times $0 \leqslant S \leqslant T$. We are focussing on *continuous* martingales, which means that, almost surely, their sample paths are continuous.

Proposition 2.42 (Optional stopping theorem) *Suppose $\{X(t)\colon t \geqslant 0\}$ is a continuous martingale, and $0 \leqslant S \leqslant T$ are stopping times. If the process $\{X(t \wedge T)\colon t \geqslant 0\}$ is dominated by an integrable random variable X, i.e. $|X(t \wedge T)| \leqslant X$ almost surely, for all $t \geqslant 0$, then*

$$
\mathbb{E}\big[X(T)\,\big|\,\mathcal{F}(S)\big] = X(S), \text{ almost surely.}
$$

Proof.   The best way to prove this is to prove the result first for martingales in discrete time, and then extend the result by approximation. The result for discrete time is provided in Theorem 12.27 of the appendix. Let us explain the approximation step here.
Fix $N \in \mathbb{N}$ and define a discrete time martingale by $X_n = X(T \wedge n2^{-N})$ and stopping times $S' = \lfloor 2^N S \rfloor + 1$ and $T' = \lfloor 2^N T \rfloor + 1$, with respect to the filtration $(\mathcal{G}(n)\colon n \in \mathbb{N})$ given by $\mathcal{G}(n) = \mathcal{F}(n2^{-N})$. Obviously X_n is dominated by an integrable random variable and hence the discrete time result gives $\mathbb{E}\big[X_{T'}\,\big|\,\mathcal{G}(S')\big] = X_{S'}$, which translates as $\mathbb{E}\big[X(T)\,\big|\,\mathcal{F}(S_N)\big] = X(T \wedge S_N)$, for $S_N = 2^{-N}(\lfloor 2^N S \rfloor + 1)$. Hence, for $A \in \mathcal{F}(S)$, using dominated convergence,

$$
\begin{aligned}
\int_A X(T)\,d\mathbb{P} &= \lim_{N \uparrow \infty} \int_A \mathbb{E}\big[X(T)\mid\mathcal{F}(S_N)\big]\,d\mathbb{P} = \int_A \lim_{N \uparrow \infty} X(T \wedge S_N)\,d\mathbb{P} \\
&= \int_A X(S)\,d\mathbb{P},
\end{aligned}
$$

and hence the claim follows from the definition of conditional expectation.   ∎

The following inequality will also be of great use to us.

Proposition 2.43 (Doob's maximal inequality) *Suppose $\{X(t)\colon t \geqslant 0\}$ is a continuous martingale and $p > 1$. Then, for any $t \geqslant 0$,*

$$
\mathbb{E}\Big[\big(\sup_{0 \leqslant s \leqslant t} |X(s)|\big)^p\Big] \leqslant \big(\tfrac{p}{p-1}\big)^p \mathbb{E}\big[|X(t)|^p\big].
$$

Proof. Again this is proved for martingales in discrete time in our appendix, see Theorem 12.30, and can be extended by approximation. Fix $N \in \mathbb{N}$ and define a discrete time martingale by $X_n = X(tn2^{-N})$ with respect to the filtration $(\mathcal{G}(n): n \in \mathbb{N})$ given by $\mathcal{G}(n) = \mathcal{F}(tn2^{-N})$. By the discrete version of Doob's maximal inequality,

$$\mathbb{E}\Big[\Big(\sup_{1 \leqslant k \leqslant 2^N} |X_k|\Big)^p\Big] \leqslant \big(\tfrac{p}{p-1}\big)^p \mathbb{E}\big[|X_{2^N}|^p\big] = \big(\tfrac{p}{p-1}\big)^p \mathbb{E}\big[|X(t)|^p\big].$$

Letting $N \uparrow \infty$ and using monotone convergence gives the claim. ∎

We now use the martingale property and the optional stopping theorem to prove Wald's lemmas for Brownian motion. These results identify the first and second moments of the value of Brownian motion at well-behaved stopping times.

Theorem 2.44 (Wald's lemma for Brownian motion) *Let $\{B(t): t \geqslant 0\}$ be a standard linear Brownian motion, and T be a stopping time such that either*

  (i) $\mathbb{E}[T] < \infty$, *or*
  (ii) $\{B(t \wedge T): t \geqslant 0\}$ *is dominated by an integrable random variable.*

Then we have $\mathbb{E}[B(T)] = 0$.

Remark 2.45 The proof of Wald's lemma is based on an optional stopping argument. An alternative proof of (i), which uses only the strong Markov property and the law of large numbers, is suggested in Exercise 2.7. Also, the moment condition (i) in Theorem 2.44 can be relaxed, see Theorem 2.50 for an optimal criterion. ◇

Proof. We first show that a stopping time satisfying condition (i), also satisfies condition (ii). So suppose $\mathbb{E}[T] < \infty$, and define

$$M_k = \max_{0 \leqslant t \leqslant 1} |B(t + k) - B(k)| \quad \text{and } M = \sum_{k=1}^{\lceil T \rceil} M_k.$$

Then

$$\mathbb{E}[M] = \mathbb{E}\Big[\sum_{k=1}^{\lceil T \rceil} M_k\Big] = \sum_{k=1}^{\infty} \mathbb{E}\big[1\{T > k-1\} M_k\big] = \sum_{k=1}^{\infty} \mathbb{P}\{T > k-1\} \mathbb{E}[M_k]$$
$$= \mathbb{E}[M_0] \mathbb{E}[T + 1] < \infty,$$

where, using Fubini's theorem and Remark 2.22,

$$\mathbb{E}[M_0] = \int_0^{\infty} \mathbb{P}\big\{ \max_{0 \leqslant t \leqslant 1} |B(t)| > x\big\}\, dx \leqslant 1 + \int_1^{\infty} \tfrac{2\sqrt{2}}{x\sqrt{\pi}} \exp\big\{ -\tfrac{x^2}{2}\big\}\, dx < \infty.$$

Now note that $|B(t \wedge T)| \leqslant M$, so that (ii) holds. It remains to observe that under condition (ii) we can apply the optional stopping theorem with $S = 0$, which yields that $\mathbb{E}[B(T)] = 0$. ∎

Corollary 2.46 *Let* $S \leqslant T$ *be stopping times and* $\mathbb{E}[T] < \infty$. *Then*

$$\mathbb{E}\big[(B(T))^2\big] = \mathbb{E}\big[(B(S))^2\big] + \mathbb{E}\big[(B(T) - B(S))^2\big].$$

Proof. The tower property of conditional expectation gives

$$\mathbb{E}\big[(B(T))^2\big] = \mathbb{E}\big[(B(S))^2\big] + 2\mathbb{E}\Big[B(S)\mathbb{E}[B(T) - B(S) \mid \mathcal{F}(S)]\Big]$$
$$+ \mathbb{E}\big[(B(T) - B(S))^2\big].$$

Note that $\mathbb{E}[T] < \infty$ implies $\mathbb{E}[T - S \mid \mathcal{F}(S)] < \infty$ almost surely. Hence the strong Markov property at time S together with Wald's lemma imply $\mathbb{E}[B(T) - B(S) \mid \mathcal{F}(S)] = 0$ almost surely, so that the middle term vanishes. ∎

To find the second moment of $B(T)$ and thus prove Wald's second lemma, we identify a further martingale derived from Brownian motion.

Lemma 2.47 *Suppose* $\{B(t) \colon t \geqslant 0\}$ *is a linear Brownian motion. Then the process*

$$\big\{B(t)^2 - t \colon t \geqslant 0\big\}$$

is a martingale.

Proof. The process is adapted to the natural filtration of Brownian motion and

$$\mathbb{E}\big[B(t)^2 - t \mid \mathcal{F}^+(s)\big]$$
$$= \mathbb{E}\big[(B(t) - B(s))^2 \mid \mathcal{F}^+(s)\big] + 2\,\mathbb{E}\big[B(t)B(s) \mid \mathcal{F}^+(s)\big] - B(s)^2 - t$$
$$= (t - s) + 2B(s)^2 - B(s)^2 - t = B(s)^2 - s,$$

which completes the proof. ∎

Theorem 2.48 (Wald's second lemma) *Let* T *be a stopping time for standard Brownian motion such that* $\mathbb{E}[T] < \infty$. *Then*

$$\mathbb{E}\big[B(T)^2\big] = \mathbb{E}[T].$$

Proof. Look at the martingale $\{B(t)^2 - t \colon t \geqslant 0\}$ and define stopping times

$$T_n = \inf\{t \geqslant 0 \colon |B(t)| = n\}$$

so that $\{B(t \wedge T \wedge T_n)^2 - t \wedge T \wedge T_n \colon t \geqslant 0\}$ is dominated by the integrable random variable $n^2 + T$. By the optional stopping theorem we get $\mathbb{E}[B(T \wedge T_n)^2] = \mathbb{E}[T \wedge T_n]$. By Corollary 2.46 we have $\mathbb{E}[B(T)^2] \geqslant \mathbb{E}[B(T \wedge T_n)^2]$. Hence, by monotone convergence,

$$\mathbb{E}\big[B(T)^2\big] \geqslant \lim_{n \to \infty} \mathbb{E}\big[B(T \wedge T_n)^2\big] = \lim_{n \to \infty} \mathbb{E}\big[T \wedge T_n\big] = \mathbb{E}[T].$$

Conversely, now using Fatou's lemma in the first step,

$$\mathbb{E}\big[B(T)^2\big] \leqslant \liminf_{n \to \infty} \mathbb{E}\big[B(T \wedge T_n)^2\big] = \liminf_{n \to \infty} \mathbb{E}\big[T \wedge T_n\big] \leqslant \mathbb{E}[T]. \qquad ∎$$

Wald's lemmas suffice to obtain exit probabilities and expected exit times for a linear Brownian motion. In Chapter 3 we shall explore the corresponding problem for higher-dimensional Brownian motion using harmonic functions.

Theorem 2.49 *Let $a < 0 < b$ and, for a standard linear Brownian motion $\{B(t) : t \geqslant 0\}$, define $T = \min\{t \geqslant 0 : B(t) \in \{a, b\}\}$. Then*

- $\mathbb{P}\{B(T) = a\} = \dfrac{b}{|a| + b}$ *and* $\mathbb{P}\{B(T) = b\} = \dfrac{|a|}{|a| + b}$.

- $\mathbb{E}[T] = |a|b$.

Proof. Let $T = \tau(\{a, b\})$ be the first exit time from the interval $[a, b]$. This stopping time satisfies the condition of the optional stopping theorem, as $|B(t \wedge T)| \leqslant |a| \vee b$. Hence, by Wald's first lemma,

$$0 = \mathbb{E}[B(T)] = a\mathbb{P}\{B(T) = a\} + b\mathbb{P}\{B(T) = b\}.$$

Together with the easy equation $\mathbb{P}\{B(T) = a\} + \mathbb{P}\{B(T) = b\} = 1$ one can solve this, and obtain $\mathbb{P}\{B(T) = a\} = b/(|a| + b)$, and $\mathbb{P}\{B(T) = b\} = |a|/(|a| + b)$. To use Wald's second lemma, we check that $\mathbb{E}[T] < \infty$. For this purpose note that

$$\mathbb{E}[T] = \int_0^\infty \mathbb{P}\{T > t\}\, dt = \int_0^\infty \mathbb{P}\{B(s) \in (a, b) \text{ for all } s \in [0, t]\}\, dt,$$

and that, for $t \geqslant k \in \mathbb{N}$ the integrand is bounded by the kth power of $\max_{x \in (a,b)} \mathbb{P}_x\{B(1) \in (a, b)\}$, i.e. decreases exponentially. Hence the integral is finite.

Now, by Wald's second lemma and the exit probabilities, we obtain

$$\mathbb{E}[T] = \mathbb{E}[B(T)^2] = \frac{a^2 b}{|a| + b} + \frac{b^2 |a|}{|a| + b} = |a|b. \qquad \blacksquare$$

We now discuss a strengthening of Theorem 2.44, which works with a weaker moment condition. This theorem will not be used in the remainder of the book and can be skipped on first reading. We shall see in Exercise 2.13 that the condition we give is in some sense optimal.

Theorem* 2.50 *Let $\{B(t) : t \geqslant 0\}$ be a standard linear Brownian motion and T a stopping time with $\mathbb{E}[T^{1/2}] < \infty$. Then $\mathbb{E}[B(T)] = 0$.*

Proof. Let $\{M(t) : t \geqslant 0\}$ be the maximum process of $\{B(t) : t \geqslant 0\}$ and T a stopping time with $\mathbb{E}[T^{1/2}] < \infty$. Let $\tau = \lceil \log_4 T \rceil$, so that $B(t \wedge T) \leqslant M(4^\tau)$. In order to get $\mathbb{E}[B(T)] = 0$ from the optional stopping theorem it suffices to show that the majorant is integrable, i.e. that

$$\mathbb{E} M(4^\tau) < \infty.$$

Define a discrete time stochastic process $\{X_k : k \in \mathbb{N}\}$ by $X_k = M(4^k) - 2^{k+1}$, and observe that τ is a stopping time with respect to the filtration $(\mathcal{F}^+(4^k) : k \in \mathbb{N})$. Moreover, the process $\{X_k : k \in \mathbb{N}\}$ is a supermartingale. Indeed,

$$\mathbb{E}\big[X_k \,|\, \mathcal{F}_{k-1}\big] \leqslant M(4^{k-1}) + \mathbb{E}\Big[\max_{0 \leqslant t \leqslant 4^k - 4^{k-1}} B(t)\Big] - 2^{k+1},$$

and the supermartingale property follows as

$$\mathbb{E}\Big[\max_{0 \leqslant t \leqslant 4^k - 4^{k-1}} B(t)\Big] = \sqrt{4^k - 4^{k-1}}\,\mathbb{E}\Big[\max_{0 \leqslant t \leqslant 1} B(t)\Big] \leqslant 2^k,$$

using that, by the reflection principle, Theorem 2.21, and the Cauchy–Schwarz inequality,

$$\mathbb{E}\Big[\max_{0 \leqslant t \leqslant 1} B(t)\Big] = \mathbb{E}|B(1)| \leqslant \big(\mathbb{E}[B(1)^2]\big)^{\frac{1}{2}} = 1.$$

Now let $t = 4^\ell$ and use the supermartingale property for $\tau \wedge \ell$ to get

$$\mathbb{E}\big[M(4^\tau \wedge t)\big] = \mathbb{E}\big[X_{\tau \wedge \ell}\big] + \mathbb{E}\big[2^{\tau \wedge \ell + 1}\big] \leqslant \mathbb{E}[X_0] + 2\,\mathbb{E}\big[2^\tau\big].$$

Note that $X_0 = M(1) - 2$, which has finite expectation and, by our assumption on the moments of T, we have $\mathbb{E}[2^\tau] < \infty$. Thus, by monotone convergence,

$$\mathbb{E}\big[M(4^\tau)\big] = \lim_{t \uparrow \infty} \big[M(4^\tau \wedge t)\big] < \infty,$$

which completes the proof of the theorem. ∎

Given the function $f \colon \mathbb{R} \to \mathbb{R}$, $f(x) = x^2$, we were able, in Lemma 2.47, to subtract a suitable term from $f(B(t))$ to obtain a martingale. To get a feeling for what we wish to subtract in the case of a general f, we look at the analogous problem for the simple random walk $\{S_n : n \in \mathbb{N}\}$. A straightforward calculation gives, for $f \colon \mathbb{Z} \to \mathbb{R}$,

$$\mathbb{E}\big[f(S_{n+1}) \,\big|\, \sigma(S_1, \ldots, S_n)\big] - f(S_n) = \tfrac{1}{2}\big(f(S_n + 1) - 2f(S_n) + f(S_n - 1)\big)$$
$$= \tfrac{1}{2}\,\tilde{\Delta}f(S_n),$$

where $\tilde{\Delta}$ is the second difference operator $\tilde{\Delta}f(x) := f(x+1) - 2f(x) + f(x-1)$. Hence

$$f(S_n) - \tfrac{1}{2}\sum_{k=0}^{n-1} \tilde{\Delta}f(S_k)$$

defines a (discrete time) martingale. In the Brownian motion case, one would expect a similar result with $\tilde{\Delta}f$ replaced by its continuous analogue, the Laplacian

$$\Delta f(x) = \sum_{i=1}^{d} \frac{\partial^2 f}{\partial x_i^2}.$$

Theorem 2.51 *Let $f \colon \mathbb{R}^d \to \mathbb{R}$ be twice continuously differentiable, and $\{B(t) : t \geqslant 0\}$ be a d-dimensional Brownian motion. Further suppose that, for all $t > 0$ and $x \in \mathbb{R}^d$, we have $\mathbb{E}_x|f(B(t))| < \infty$ and $\mathbb{E}_x \int_0^t |\Delta f(B(s))|\,ds < \infty$. Then the process $\{X(t) : t \geqslant 0\}$ defined by*

$$X(t) = f(B(t)) - \tfrac{1}{2}\int_0^t \Delta f(B(s))\,ds$$

is a martingale.

Proof.   For any $0 \leqslant s < t$,

$$\mathbb{E}\big[X(t) \,\big|\, \mathcal{F}(s)\big]$$

$$= \mathbb{E}_{B(s)}\big[f(B(t-s))\big] - \tfrac{1}{2} \int_0^s \Delta f(B(u))\,du - \int_0^{t-s} \mathbb{E}_{B(s)}\big[\tfrac{1}{2}\Delta f(B(u))\big]\,du.$$

Now, using integration by parts and $\tfrac{1}{2}\Delta\mathfrak{p}(t,x,y) = \tfrac{\partial}{\partial t}\mathfrak{p}(t,x,y)$, we find

$$\mathbb{E}_{B(s)}\big[\tfrac{1}{2}\Delta f(B(u))\big] = \tfrac{1}{2}\int \mathfrak{p}(u,B(s),x)\,\Delta f(x)\,dx$$

$$= \tfrac{1}{2}\int \Delta\mathfrak{p}(u,B(s),x)\,f(x)\,dx = \int \tfrac{\partial}{\partial u}\mathfrak{p}(u,B(s),x)\,f(x)\,dx,$$

and hence

$$\int_0^{t-s} \mathbb{E}_{B(s)}\big[\tfrac{1}{2}\Delta f(B(u))\big]\,du = \lim_{\varepsilon\downarrow 0}\int\Big[\int_\varepsilon^{t-s}\tfrac{\partial}{\partial u}\mathfrak{p}(u,B(s),x)\,du\Big]f(x)\,dx$$

$$= \int \mathfrak{p}(t-s,B(s),x)\,f(x)\,dx - \lim_{\varepsilon\downarrow 0}\int\mathfrak{p}(\varepsilon,B(s),x)\,f(x)\,dx$$

$$= \mathbb{E}_{B(s)}\big[f(B(t-s))\big] - f(B(s)),$$

and this confirms the martingale property.   ∎

Example 2.52 Using $f(x) = x^2$ in Theorem 2.51 yields the familiar martingale $\{B(t)^2 - t \colon t \geqslant 0\}$. Using $f(x) = x^3$ we obtain the martingale $\{B(t)^3 - 3\int_0^t B(s)\,ds \colon t \geqslant 0\}$ and not the familiar martingale $\{B(t)^3 - 3tB(t)\colon t \geqslant 0\}$. Of course, the difference $\{\int_0^t (B(t) - B(s))\,ds \colon t \geqslant 0\}$ is a martingale.   ◇

The next lemma states a fundamental principle, which we will discuss further in Chapter 7, see in particular Theorem 7.18.

Corollary 2.53 *Suppose* $f \colon \mathbb{R}^d \to \mathbb{R}$ *satisfies* $\Delta f(x) = 0$ *and* $\mathbb{E}_x|f(B(t))| < \infty$, *for every* $x \in \mathbb{R}^d$ *and* $t > 0$. *Then the process* $\{f(B(t)) \colon t \geqslant 0\}$ *is a martingale.*

Example 2.54 The function $f \colon \mathbb{R}^2 \to \mathbb{R}$ given by $f(x_1, x_2) = e^{x_1}\cos x_2$ satisfies $\Delta f(x) = 0$. Hence $X(t) = e^{B_1(t)}\cos B_2(t)$ defines a martingale, where $\{B_1(t)\colon t \geqslant 0\}$ and $\{B_2(t)\colon t \geqslant 0\}$ are independent linear Brownian motions.   ◇

Exercises

Exercise 2.1. Show that the definition of d-dimensional Brownian motion is invariant under an orthonormal change of coordinates. More precisely, if A is a $d \times d$-matrix with $AA^{\mathrm{T}} = I_d$ and $\{B(t)\colon t \geqslant 0\}$ is Brownian motion, then so is $\{AB(t)\colon t \geqslant 0\}$.

Exercise 2.2. Show that for any tail event $A \in \mathcal{T}$ the probability $\mathbb{P}_x(A)$ is independent of x, whereas for a germ event $A \in \mathcal{F}^+(0)$ the probability $\mathbb{P}_x(A)$ may depend on x.

Exercise 2.3. § Show that

(i) If $S \leqslant T$ are stopping times, then $\mathcal{F}^+(S) \subset \mathcal{F}^+(T)$.
(ii) If $T_n \downarrow T$ are stopping times, then $\mathcal{F}^+(T) = \bigcap_{n=1}^{\infty} \mathcal{F}^+(T_n)$.
(iii) If T is a stopping time, then the random variable $B(T)$ is $\mathcal{F}^+(T)$-measurable.

Exercise 2.4. Let $\{B(t) \colon -\infty < t < \infty\}$ be a two-sided Brownian motion as defined in Exercise 1.4, but including the d-dimensional case. A real valued random variable τ is

- a **stopping time** if $\{\tau \leqslant t\} \in \mathcal{F}^+(t) := \bigcap_{n=1}^{\infty} \sigma(B(s) \colon -\infty < s \leqslant t + \frac{1}{n})$,
- a **reverse stopping time** if $\{\tau \leqslant t\} \in \mathcal{G}^-(t) := \bigcap_{n=1}^{\infty} \sigma(B(s) \colon t - \frac{1}{n} \leqslant s < \infty)$.

For a stopping time τ let $\mathcal{F}^+(\tau)$ be the collection of events A with $A \cap \{\tau \leqslant t\} \in \mathcal{F}^+(t)$, for a reverse stopping time τ let $\mathcal{G}^-(\tau)$ be the collection of events A with $A \cap \{\tau \geqslant t\} \in \mathcal{G}^-(t)$. Show that

(a) $\{B(\tau + t) - B(\tau) \colon t \geqslant 0\}$ is a standard Brownian motion independent of $\mathcal{F}^+(\tau)$,

(b) $\{B(\tau - t) - B(\tau) \colon t \geqslant 0\}$ is a standard Brownian motion independent of $\mathcal{G}^-(\tau)$.

Exercise 2.5. Let $\{B(t) \colon 0 \leqslant t \leqslant 1\}$ be a linear Brownian motion and $F \in \mathbf{D}[0,1]$. Show that, almost surely, the set $\{t \in [0,1] \colon B(t) = F(t)\}$ is a perfect set.
Hint. Use the Cameron–Martin theorem, see Theorem 1.38.

Exercise 2.6. Let $\{B(t) \colon 0 \leqslant t \leqslant 1\}$ be a linear Brownian motion and

$$\tau = \sup\{t \in [0,1] \colon B(t) = 0\}.$$

Show that, almost surely, there exist times $t_n < s_n < \tau$ with $t_n \uparrow \tau$ such that

$$B(t_n) < 0 \qquad \text{and} \qquad B(s_n) > 0.$$

Exercise 2.7. § Let $\{B(t) \colon t \geqslant 0\}$ be a standard Brownian motion on the line, and T be a stopping time with $\mathbb{E}[T] < \infty$. Define an increasing sequence of stopping times by $T_1 = T$ and $T_n = T(B_n) + T_{n-1}$ where the stopping time $T(B_n)$ is the same function as T, but associated with the Brownian motion $\{B_n(t) \colon t \geqslant 0\}$ given by

$$B_n(t) = B(t + T_{n-1}) - B(T_{n-1}).$$

(a) Show that, almost surely,

$$\lim_{n \uparrow \infty} \frac{B(T_n)}{n} = 0.$$

(b) Show that $B(T)$ is integrable.

(c) Show that, almost surely,

$$\lim_{n \uparrow \infty} \frac{B(T_n)}{n} = \mathbb{E}\big[B(T)\big].$$

Combining (a) and (c) implies that $\mathbb{E}\big[B(T)\big] = 0$, which is Wald's lemma.

Exercise 2.8. Show that, for any $x > 0$ and measurable set $A \subset [0, \infty)$,

$$\mathbb{P}_x\{B(s) \geqslant 0 \text{ for all } 0 \leqslant s \leqslant t \text{ and } B(t) \in A\} = \mathbb{P}_x\{B(t) \in A\} - \mathbb{P}_{-x}\{B(t) \in A\}.$$

Exercise 2.9. ⑤ Show that any nonempty, closed set with no isolated points is uncountable. Note that this applies, in particular, to the zero set of linear Brownian motion.

Exercise 2.10. The *Ornstein–Uhlenbeck diffusion* is the process $\{X(t) \colon t \in \mathbb{R}\}$, given by

$$X(t) = e^{-t} B(e^{2t}) \text{ for all } t \in \mathbb{R},$$

see also Remark 1.10. Show that $\{X(t) \colon t \geqslant 0\}$ and $\{X(-t) \colon t \geqslant 0\}$ are Markov processes and find their Markov transition kernels.

Exercise 2.11. Let $x, y \in \mathbb{R}^d$ and $\{B(t) \colon t \geqslant 0\}$ a d-dimensional Brownian motion started in x. Define the d-dimensional *Brownian bridge* $\{X(t) \colon 0 \leqslant t \leqslant 1\}$ with start in x and end in y by

$$X(t) = B(t) - t\big(B(1) - y\big), \qquad \text{for } 0 \leqslant t \leqslant 1.$$

Show that the Brownian bridge is *not* a time-homogeneous Markov process.

Exercise 2.12. Find two stopping times $S \leqslant T$ with $\mathbb{E}[S] < \infty$ such that

$$\mathbb{E}[(B(S))^2] > \mathbb{E}[(B(T))^2].$$

Exercise 2.13. ⑤ The purpose of this exercise is to show that the moment condition in Theorem 2.50 is optimal. Let $\{B(t) \colon t \geqslant 0\}$ be a standard linear Brownian motion and define $T = \inf\{t \geqslant 0 \colon B(t) = 1\}$, so that $B(T) = 1$ almost surely. Show that

$$\mathbb{E}[T^\alpha] < \infty \qquad \text{for all } \alpha < 1/2.$$

Exercise 2.14. Let $\{B(t) \colon t \geqslant 0\}$ be a standard linear Brownian motion

(a) Show that there exists a stopping time T with $\mathbb{E}T = \infty$ but $\mathbb{E}[(B(T))^2] < \infty$.

(b) Show that, for every stopping time T with $\mathbb{E}T = \infty$ and $\mathbb{E}\sqrt{T} < \infty$, we have

$$\mathbb{E}\big[B(T)^2\big] = \infty.$$

Exercise 2.15. Let $\{B(t) \colon t \geqslant 0\}$ be a linear Brownian motion.

(a) Show that, for $\sigma > 0$, the process $\{\exp(\sigma B(t) - \frac{\sigma^2 t}{2}) \colon t \geqslant 0\}$ is a martingale.

(b) Show, by taking derivatives $\frac{\partial^n}{\partial \sigma^n}$ at 0, that the following processes are martingales.

- $\{B(t)^2 - t : t \geqslant 0\}$,
- $\{B(t)^3 - 3tB(t) : t \geqslant 0\}$, and
- $\{B(t)^4 - 6tB(t)^2 + 3t^2 : t \geqslant 0\}$.

(c) Find $\mathbb{E}[T^2]$ for $T = \min\{t \geqslant 0 : B(t) \in \{a, b\}\}$ and $a < 0 < b$.

Exercise 2.16. Ⓢ Let $\{B(t) : t \geqslant 0\}$ be a linear Brownian motion and $a, b > 0$. Show that

$$\mathbb{P}_0\{B(t) = a + bt \text{ for some } t > 0\} = e^{-2ab}.$$

Exercise 2.17. Ⓢ Let $R > 0$ and $A = \{-R, R\}$. Denote by $\tau(A)$ the first hitting time of A, and by T_x the first hitting times of the point $x \in \mathbb{R}$. Consider a linear Brownian motion started at $x \in [0, R]$, and prove that

(a) $\mathbb{E}_x[\tau(A)] = R^2 - x^2$.

(b) $\mathbb{E}_x[T_R \,|\, T_R < T_0] = \frac{R^2 - x^2}{3}$.

Hint. In (b) use one of the martingales of Exercise 2.15(b).

Exercise 2.18. Let $\{B(t) : t \geqslant 0\}$ be a linear Brownian motion.

(a) Use the optional stopping theorem for the martingale in Exercise 2.15(a) to show that, with $\tau_a = \inf\{t \geqslant 0 : B(t) = a\}$,

$$\mathbb{E}_0\left[e^{-\lambda \tau_a}\right] = e^{-a\sqrt{2\lambda}}, \quad \text{for all } \lambda, a > 0.$$

(b) Show that, with $\tau_{-a} = \inf\{t \geqslant 0 : B(t) = -a\}$, we have

$$\mathbb{E}_0\left[e^{-\lambda \tau_a}\right] = \mathbb{E}_0\left[e^{-\lambda \tau_a} \mathbf{1}\{\tau_a < \tau_{-a}\}\right] + \mathbb{E}_0\left[e^{-\lambda \tau_{-a}} \mathbf{1}\{\tau_{-a} < \tau_a\}\right] e^{-2a\sqrt{2\lambda}}.$$

(c) Deduce that $\tau = \tau_a \wedge \tau_{-a}$ satisfies

$$\mathbb{E}_0\left[e^{-\lambda \tau}\right] = \operatorname{sech}(a\sqrt{2\lambda}),$$

where $\operatorname{sech}(x) = \frac{2}{e^x + e^{-x}}$.

Exercise 2.19. In this exercise we interpret $\mathbb{R}^2$ as the complex plane. Hence a planar Brownian motion becomes a complex Brownian motion. A complex-valued stochastic process is called a martingale, if its real and imaginary parts are martingales. Let $\{B(t) : t \geqslant 0\}$ be a complex Brownian motion started in i, the imaginary unit.

(a) Show that $\{e^{i\lambda B(t)} : t \geqslant 0\}$ is a martingale, for any $\lambda \in \mathbb{R}$.

(b) Let T be the first time when $\{B(t) : t \geqslant 0\}$ hits the real axis. Using the optional stopping theorem at T, show that

$$\mathbb{E}\left[e^{i\lambda B(T)}\right] = e^{-\lambda}.$$

Inverting the Fourier transform, the statement of (b) means that $B(T)$ is Cauchy distributed, a fact we already know from an explicit calculation, see Theorem 2.37.

Exercise 2.20. § Let $f \colon \mathbb{R}^d \to \mathbb{R}$ be twice continuously differentiable, $\{B(t) \colon t \geqslant 0\}$ a d-dimensional Brownian motion such that $\mathbb{E}_x \int_0^t e^{-\lambda s} |f(B(s))|\, ds < \infty$ and $\mathbb{E}_x \int_0^t e^{-\lambda s} |\Delta f(B(s))|\, ds < \infty$, for any $x \in \mathbb{R}^d$ and $t > 0$.

(a) Show that the process $\{X(t) \colon t \geqslant 0\}$ defined by

$$X(t) = e^{-\lambda t} f(B(t)) - \int_0^t e^{-\lambda s} \left(\tfrac{1}{2} \Delta f(B(s)) - \lambda f(B(s)) \right) ds$$

is a martingale.

(b) Suppose U is a bounded open set, $\lambda \geqslant 0$, and $u \colon U \to \mathbb{R}$ is a bounded solution of

$$\tfrac{1}{2} \Delta u(x) = \lambda u(x), \qquad \text{for } x \in U,$$

and $\lim_{x \to x_0} u(x) = f(x_0)$ for all $x_0 \in \partial U$. Show that,

$$u(x) = \mathbb{E}_x \left[f(B(\tau)) e^{-\lambda \tau} \right],$$

where $\tau = \inf\{t \geqslant 0 \colon B(t) \notin U\}$.

Notes and comments

The Markov property is central to any discussion of Brownian motion. The discussion of this chapter is only a small fraction of what has to be said, and the Markov property will be omnipresent in the rest of the book. The name goes back to Markov's paper [Ma06] where the Markovian dependence structure was introduced and a law of large numbers for dependent random variables was proved. The strong Markov property had been used for special stopping times, like hitting times of a point, since the 1930s. Hunt [Hu56] formalised the idea and gave rigorous proofs, and so did, independently, Dynkin [Dy57].

Zero-one laws are classics in probability theory. We have already encountered the powerful Hewitt–Savage law and there are more to come. Blumenthal's zero-one law was first proved in [Bl57]. It holds well beyond the setting of Brownian motion, for a class of Markov processes called *Feller processes*, which includes all processes with stationary, independent increments.

The reflection principle is usually attributed to D. André [An87], who stated a variant for random walks. His concern was the ballot problem: if two candidates in a ballot receive a, respectively b votes, with $a > b$, what is the probability that the first candidate was always in the lead during the counting of the votes? See the classical text of Feller [Fe68] for more on this problem. A formulation of the reflection principle for Brownian motion was given by Lévy [Le39], though apparently not based on the rigorous foundation of the strong Markov property. We shall later use a higher-dimensional version of the reflection principle, where a Brownian motion in $\mathbb{R}^d$ is reflected in a hyperplane.

The class of Markov processes, defined in this chapter, has a rich and fascinating theory of its own, and some aspects are discussed in the books Rogers and Williams [RW00a, RW00b] and Chung [Ch82]. A typical feature of this theory is its strong connection to analysis and potential theory, which stems from the key rôle played by the transition semigroup in their definition. This aspect is emphasised in different ways in the books by Blumenthal and Getoor [BG68] and Bass [Ba98]. Many of the important examples of Markov processes can be derived from Brownian motion in one way or the other, and this is an excellent motivation for further study of the theory. Amongst them are stable Lévy processes, like the Cauchy process or stable subordinators, the Bessel processes, and diffusions.

The intriguing relationship uncovered in Theorem 2.34 has found numerous extensions and complementary results, among them Pitman's $2M - B$ theorem, which we will discuss in Section 5.5, which describes the process $\{2M(t) - B(t): t \geqslant 0\}$ as a 3-dimensional Bessel process or, equivalently, a Brownian motion conditioned to stay positive.

The concept of martingales is due to Doob, see [Do53]. They are an important class of stochastic processes in their own right and one of the gems of modern probability theory. A gentle introduction, mostly in discrete time, is Williams [Wi91], while Revuz and Yor [RY94] discuss continuous martingales and the rich relations to Brownian motion. A fascinating fact, due to Dambis [Da65], Dubins, and Schwarz [DS65], is that for every continuous martingale $\{M(t): t \geqslant 0\}$ with unbounded quadratic variation there exists a time-change, i.e. a reparametrisation $t \mapsto T_t$ such that $T_t, t \geqslant 0$ are stopping times, such that $t \mapsto M(T_t)$ is a Brownian motion.

The martingale featuring in Exercise 2.15 (a) plays an important rôle in the context of the Cameron–Martin theorem. It represents the density of the law of a Brownian motion with constant drift, with respect to the law of a driftless Brownian motion on the space $C[0, t]$, see Remark 1.43. See also Freedman [Fr83] for a nice treatment of this connection. Girsanov's theorem offers a more systematic approach to mutual densities, which is best understood in the language of semimartingales, see for example Revuz and Yor [RY94]. Theorem 2.50 establishes a special case of an important result in martingale theory, the Burkholder–Davis–Gundy inequalities, see [BDG72] for the original paper and Theorem 3.28 of Karatzas and Shreve [KS91] or Rogers and Williams [RW00b] for a textbook treatment. A presentation closer to ours is in Proposition VII-2-3(b) of Neveu [Ne75]. Exercise 2.17 appears in similar form in Stern [St75].

3

Harmonic functions, transience and recurrence

In this chapter we explore the relation of harmonic functions and Brownian motion. This approach will be particularly useful for d-dimensional Brownian motion for $d > 1$. It allows us to study the fundamental questions of transience and recurrence of Brownian motion, investigate the classical Dirichlet problem of electrostatics, and provide the background for the deeper investigations of probabilistic potential theory, which will follow in Chapter 8.

3.1 Harmonic functions and the Dirichlet problem

Let U be a **domain**, i.e. a connected open set $U \subset \mathbb{R}^d$, and ∂U be its boundary. Suppose that its closure $\overline{U}$ is a homogeneous body and its boundary is electrically charged, the charge given by some continuous function $\varphi \colon \partial U \to \mathbb{R}$. The *Dirichlet problem* asks for the voltage $u(x)$ at some point $x \in U$. Kirchhoff's laws state that u must be a *harmonic function* in U. We therefore start by discussing the basic features of harmonic functions.

Definition 3.1. Let $U \subset \mathbb{R}^d$ be a domain. A function $u \colon U \to \mathbb{R}$ is **harmonic** (on U) if it is twice continuously differentiable and, for any $x \in U$,

$$\Delta u(x) := \sum_{j=1}^{d} \frac{\partial^2 u}{\partial x_j^2}(x) = 0.$$

If instead of the last condition only $\Delta u(x) \geqslant 0$, then the function u is **subharmonic**. ◇

To begin with we give two useful reformulations of the harmonicity condition, called the **mean value properties**, which do not make explicit reference to differentiability.

Theorem 3.2 *Let $U \subset \mathbb{R}^d$ be a domain and $u \colon U \to \mathbb{R}$ measurable and locally bounded. The following conditions are equivalent:*

  (i) *u is harmonic;*
  (ii) *for any ball $\mathcal{B}(x,r) \subset U$, we have*

$$u(x) = \frac{1}{\mathcal{L}(\mathcal{B}(x,r))} \int_{\mathcal{B}(x,r)} u(y)\, dy;$$

(iii) *for any ball* $\mathcal{B}(x,r) \subset U$,

$$u(x) = \frac{1}{\sigma_{x,r}(\partial\mathcal{B}(x,r))} \int_{\partial\mathcal{B}(x,r)} u(y)\,d\sigma_{x,r}(y),$$

where $\sigma_{x,r}$ *is the surface measure on* $\partial\mathcal{B}(x,r)$.

Remark 3.3 We use the following version of Green's identity,

$$\int_{\partial\mathcal{B}(x,r)} \frac{\partial u}{\partial n}(y)\,d\sigma_{x,r}(y) = \int_{\mathcal{B}(x,r)} \Delta u(y)\,dy, \tag{3.1}$$

where $n(y)$ is the outward normal vector of the ball at y, see [Ba95]. One can avoid the use of this identity and prove the result by purely probabilistic means, see Exercise 8.1. ⋄

Proof. **(ii)** ⇒ **(iii)** Assume u has the mean value property (ii). Define $\psi\colon (0,\infty) \to \mathbb{R}$ by

$$\psi(r) = r^{1-d} \int_{\partial\mathcal{B}(x,r)} u(y)\,d\sigma_{x,r}(y).$$

Then, for any $r > 0$, we have

$$r^d\,\mathcal{L}(\mathcal{B}(x,1))\,u(x) = \mathcal{L}(\mathcal{B}(x,r))\,u(x) = \int_{\mathcal{B}(x,r)} u(y)\,dy = \int_0^r \psi(s)\,s^{d-1}\,ds.$$

Differentiating with respect to r gives $\psi(r) = d\,\mathcal{L}(\mathcal{B}(x,1))u(x)$ for almost all $r \in (0,\infty)$. As $dr^{d-1}\mathcal{L}(\mathcal{B}(x,1)) = \sigma_{x,r}(\partial\mathcal{B}(x,r))$ we infer that

$$u(x) = \frac{1}{\sigma_{x,r}(\partial\mathcal{B}(x,r))} \int_{\partial\mathcal{B}(x,r)} u(y)\,d\sigma_{x,r}(y), \text{ for almost all } r \in (0,\infty). \tag{3.2}$$

Suppose $g\colon [0,\infty) \to [0,\infty)$ is a smooth function with compact support in $[0,\varepsilon)$ and $\int g(|x|)\,dx = 1$. Integrating (3.2) one obtains

$$u(x) = \int u(y)g(|x-y|)\,dy$$

for all $x \in U$ and sufficiently small $\varepsilon > 0$. As convolution of a smooth function with a bounded function produces a smooth function, we observe that u is infinitely often differentiable in U. In particular, this implies that (3.2) holds indeed for all $r > 0$, proving (iii).

(iii) ⇒ **(ii)** Fix $s > 0$, multiply (iii) by $\sigma_{x,r}(\partial\mathcal{B}(x,r))$ and integrate over all radii $0 < r < s$.

(iii) ⇒ **(i)** We have seen above that (iii) implies that u is infinitely often differentiable in U. Now suppose that $\Delta u \neq 0$, so that there exists a small ball $\mathcal{B}(x,\varepsilon) \subset U$ such that either $\Delta u(x) > 0$ on $\mathcal{B}(x,\varepsilon)$, or $\Delta u(x) < 0$ on $\mathcal{B}(x,\varepsilon)$. With the notation from above,

$$0 = \psi'(r) = r^{1-d} \int_{\partial\mathcal{B}(x,r)} \frac{\partial u}{\partial n}(y)\,d\sigma_{x,r}(y) = r^{1-d} \int_{\mathcal{B}(x,r)} \Delta u(y)\,dy,$$

using (3.1). This is a contradiction.

(i) ⇒ **(iii)** Suppose that u is harmonic and $\mathcal{B}(x,r) \subset U$. Using (3.1), we obtain that

$$\psi'(r) = r^{1-d} \int_{\partial\mathcal{B}(x,r)} \frac{\partial u}{\partial n}(y)\,d\sigma_{x,r}(y) = r^{1-d} \int_{\mathcal{B}(x,r)} \Delta u(y)\,dy = 0.$$

Hence ψ is constant, and as $\lim_{r\downarrow 0} \psi(r) = \sigma_{0,1}(\mathcal{B}(0,1))\,u(x)$, we obtain (iii). ∎

Remark 3.4 A twice differentiable function $u \colon U \to \mathbb{R}$ is subharmonic if and only if

$$u(x) \leqslant \frac{1}{\mathcal{L}(\mathcal{B}(x,r))} \int_{\mathcal{B}(x,r)} u(y) \, dy \qquad \text{for any ball } \mathcal{B}(x,r) \subset U. \qquad (3.3)$$

This can be obtained in a way very similar to Theorem 3.2, see also Exercise 3.1. ◇

An important property satisfied by harmonic, and in fact subharmonic, functions is the maximum principle. This is one of the key principles of analysis.

Theorem 3.5 (Maximum principle) *Suppose $u \colon \mathbb{R}^d \to \mathbb{R}$ is a function, which is subharmonic on an open connected set $U \subset \mathbb{R}^d$.*

(i) *If u attains its maximum in U, then u is a constant.*

(ii) *If u is continuous on $\bar{U}$ and U is bounded, then*

$$\max_{x \in \bar{U}} u(x) = \max_{x \in \partial U} u(x).$$

Remark 3.6 If u is harmonic, the theorem may be applied to both u and $-u$. Hence the conclusions of the theorem also hold with 'maximum' replaced by 'minimum'. ◇

Proof. (i) Let M be the maximum. Note that $V = \{x \in U : u(x) = M\}$ is relatively closed in U. Since U is open, for any $x \in V$, there is a ball $\mathcal{B}(x,r) \subset U$. By the mean-value property of u, see Remark 3.4,

$$M = u(x) \leqslant \frac{1}{\mathcal{L}(\mathcal{B}(x,r))} \int_{\mathcal{B}(x,r)} u(y) \, dy \leqslant M.$$

Equality holds everywhere, and as $u(y) \leqslant M$ for all $y \in \mathcal{B}(x,r)$, we infer that $u(y) = M$ almost everywhere on $\mathcal{B}(x,r)$. By continuity this implies $\mathcal{B}(x,r) \subset V$. Hence V is also open, and by assumption nonempty. Since U is connected we get that $V = U$. Therefore, u is constant on U.

(ii) Since u is continuous and $\bar{U}$ is closed and bounded, u attains a maximum on $\bar{U}$. By (i) the maximum has to be attained on ∂U. ∎

Corollary 3.7 *Suppose $u_1, u_2 \colon \mathbb{R}^d \to \mathbb{R}$ are functions, which are harmonic on a bounded domain $U \subset \mathbb{R}^d$ and continuous on $\bar{U}$. If u_1 and u_2 agree on ∂U, then they are identical.*

Proof. By Theorem 3.5(ii) applied to $u_1 - u_2$ we obtain that

$$\sup_{x \in \bar{U}} \big\{ u_1(x) - u_2(x) \big\} = \sup_{x \in \partial U} \big\{ u_1(x) - u_2(x) \big\} = 0.$$

Hence $u_1(x) \leqslant u_2(x)$ for all $x \in \bar{U}$. Applying the same argument to $u_2 - u_1$, one sees that $\sup_{x \in \bar{U}} \{ u_2(x) - u_1(x) \} = 0$. Hence $u_1(x) = u_2(x)$ for all $x \in \bar{U}$. ∎

We can now formulate the basic fact on which the relationship of Brownian motion and harmonic functions rests.

Theorem 3.8 *Suppose U is a domain, $\{B(t): t \geqslant 0\}$ a Brownian motion started inside U and $\tau = \tau(\partial U) = \min\{t \geqslant 0: B(t) \in \partial U\}$ the first hitting time of its boundary. Let $\varphi: \partial U \to \mathbb{R}$ be measurable, and such that the function $u: U \to \mathbb{R}$ with*

$$u(x) = \mathbb{E}_x\left[\varphi(B(\tau))\,\mathbf{1}\{\tau < \infty\}\right], \quad \text{for every } x \in U, \tag{3.4}$$

is locally bounded. Then u is a harmonic function.

Proof.   The proof uses only the strong Markov property of Brownian motion and the mean value characterisation of harmonic functions. For a ball $\mathcal{B}(x, \delta) \subset U$ let $\tilde\tau = \inf\{t > 0: B(t) \notin \mathcal{B}(x, \delta)\}$, then the strong Markov property implies that

$$u(x) = \mathbb{E}_x\left[\mathbb{E}_x\left[\varphi(B(\tau))\,\mathbf{1}\{\tau < \infty\} \,\big|\, \mathcal{F}^+(\tilde\tau)\right]\right] = \mathbb{E}_x\left[u\big(B(\tilde\tau)\big)\right]$$
$$= \int_{\partial\mathcal{B}(x,\delta)} u(y)\,\varpi_{x,\delta}(dy),$$

where $\varpi_{x,\delta}$ is the uniform distribution on the sphere $\partial\mathcal{B}(x, \delta)$. Therefore, u has the mean value property and, as it is also locally bounded, it is harmonic on U by Theorem 3.2. ∎

Definition 3.9.   Let U be a domain in $\mathbb{R}^d$ and let ∂U be its boundary. Suppose $\varphi: \partial U \to \mathbb{R}$ is a continuous function on its boundary. A continuous function $v: \overline{U} \to \mathbb{R}$ is a **solution to the Dirichlet problem** with boundary value φ, if it is harmonic on U and $v(x) = \varphi(x)$ for $x \in \partial U$.   ◇

The Dirichlet problem was posed by Gauss in 1840. In fact Gauss thought he showed that there is always a solution, but his reasoning was wrong and Zaremba in 1911 and Lebesgue in 1924 gave counterexamples. However, if the domain is sufficiently nice there is a solution, as we will see below.

Definition 3.10.   Let $U \subset \mathbb{R}^d$ be a domain. We say that U satisfies the **Poincaré cone condition** at $x \in \partial U$ if there exists a cone V based at x with opening angle $\alpha > 0$, and $h > 0$ such that $V \cap \mathcal{B}(x, h) \subset U^c$.   ◇

The following lemma, which is illustrated by Figure 3.1, will prepare us to solve the Dirichlet problem for 'nice' domains. Recall that we denote, for any open or closed set $A \subset \mathbb{R}^d$, by $\tau(A)$ the first hitting time of the set A by Brownian motion,

$$\tau(A) = \inf\{t \geqslant 0: B(t) \in A\}.$$

Lemma 3.11 *Let $0 < \alpha < 2\pi$ and $C_0(\alpha) \subset \mathbb{R}^d$ be a cone based at the origin with opening angle α, and*

$$a = \sup_{x \in \mathrm{cl}\,\mathcal{B}(0,\frac{1}{2})} \mathbb{P}_x\{\tau(\partial\mathcal{B}(0,1)) < \tau(C_0(\alpha))\}.$$

Then $a < 1$ and, for any positive integer k and $h' > 0$, we have

$$\mathbb{P}_x\{\tau(\partial\mathcal{B}(z, h')) < \tau(C_z(\alpha))\} \leqslant a^k, \text{ for all } x, z \in \mathbb{R}^d \text{ with } |x - z| < 2^{-k}h',$$

where $C_z(\alpha)$ is a cone based at z with opening angle α.

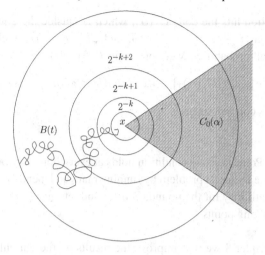

Fig. 3.1. Brownian motion avoiding a cone.

Proof. It is easy to verify $a < 1$ using, for example, Exercise 1.8. If $x \in \mathcal{B}(0, 2^{-k})$ then by the strong Markov property

$$\mathbb{P}_x\{\tau(\partial\mathcal{B}(0,1)) < \tau(C_0(\alpha))\}$$

$$\leqslant \prod_{i=0}^{k-1} \sup_{x \in \mathcal{B}(0, 2^{-k+i})} \mathbb{P}_x\{\tau(\partial\mathcal{B}(0, 2^{-k+i+1})) < \tau(C_0(\alpha))\} = a^k.$$

Therefore, for any positive integer k and $h' > 0$, we have by scaling $\mathbb{P}_x\{\tau(\partial\mathcal{B}(z, h')) < \tau(C_z(\alpha))\} \leqslant a^k$, for all x with $|x - z| < 2^{-k}h'$. ∎

Theorem 3.12 (Dirichlet Problem) *Suppose $U \subset \mathbb{R}^d$ is a bounded domain such that every boundary point satisfies the Poincaré cone condition, and suppose φ is a continuous function on ∂U. Let $\tau(\partial U) = \inf\{t > 0 \colon B(t) \in \partial U\}$, which is an almost surely finite stopping time. Then the function $u \colon \overline{U} \to \mathbb{R}$ given by*

$$u(x) = \mathbb{E}_x\big[\varphi(B(\tau(\partial U)))\big], \qquad \text{for } x \in \overline{U},$$

is the unique continuous function harmonic on U with $u(x) = \varphi(x)$ for all $x \in \partial U$.

Proof. The uniqueness claim follows from Corollary 3.7. The function u is bounded and hence harmonic on U by Theorem 3.8. It remains to show that the Poincaré cone condition implies that u is continuous on the boundary. Fix $z \in \partial U$, then there is a cone $C_z(\alpha)$ based at z with angle $\alpha > 0$ with $C_z(\alpha) \cap \mathcal{B}(z, h) \subset U^c$. By Lemma 3.11, for any positive integer k and $h' > 0$, we have

$$\mathbb{P}_x\{\tau(\partial\mathcal{B}(z, h')) < \tau(C_z(\alpha))\} \leqslant a^k$$

for all x with $|x - z| < 2^{-k}h'$. Given $\varepsilon > 0$, there is a $0 < \delta \leqslant h$ such that $|\varphi(y) - \varphi(z)| < \varepsilon$ for all $y \in \partial U$ with $|y - z| < \delta$. For all $x \in \overline{U}$ with $|z - x| < 2^{-k}\delta$,

$$|u(x) - u(z)| = \big|\mathbb{E}_x\varphi(B(\tau(\partial U))) - \varphi(z)\big| \leqslant \mathbb{E}_x\big|\varphi(B(\tau(\partial U))) - \varphi(z)\big|. \qquad (3.5)$$

If the Brownian motion hits the cone $C_z(\alpha)$, which is outside the domain U, before the sphere $\partial\mathcal{B}(z,\delta)$, then $|z - B(\tau(\partial U))| < \delta$, and $\varphi(B(\tau(\partial U)))$ is close to $\varphi(z)$. The complement has small probability. More precisely, (3.5) is bounded above by

$$2\|\varphi\|_\infty \mathbb{P}_x\{\tau(\partial\mathcal{B}(z,\delta)) < \tau(C_z(\alpha))\} + \varepsilon\mathbb{P}_x\{\tau(\partial U) < \tau(\partial\mathcal{B}(z,\delta))\} \leqslant 2\|\varphi\|_\infty a^k + \varepsilon.$$

This implies that u is continuous on $\overline{U}$.   ∎

Remark 3.13 If the Poincaré cone condition holds at every boundary point, one can simulate the solution of the Dirichlet problem by running many independent Brownian motions, starting in $x \in U$ until they hit the boundary of U and letting $u(x)$ be the average of the values of φ on the hitting points.   ◇

Remark 3.14 In Chapter 8 we will improve the results on the Dirichlet problem significantly and give sharp criteria for the existence of solutions.   ◇

To justify the introduction of conditions on the domain we now give an example where the function u of Theorem 3.12 fails to solve the Dirichlet problem.

Example 3.15 Take a solution $v\colon \mathcal{B}(0,1) \to \mathbb{R}$ of the Dirichlet problem on the planar disc $\mathcal{B}(0,1)$ with boundary condition $\varphi\colon \partial\mathcal{B}(0,1) \to \mathbb{R}$. Let $U = \{x \in \mathbb{R}^2 : 0 < |x| < 1\}$ be the punctured disc. We claim that $u(x) = \mathbb{E}_x\left[\varphi(B(\tau(\partial U)))\right]$ fails to solve the Dirichlet problem on U with boundary condition $\varphi\colon \partial\mathcal{B}(0,1) \cup \{0\} \to \mathbb{R}$ if $\varphi(0) \neq v(0)$. Indeed, as planar Brownian motion does not hit points, by Corollary 2.26, the first hitting time τ of $\partial U = \partial\mathcal{B}(0,1) \cup \{0\}$ agrees almost surely with the first hitting time of $\partial\mathcal{B}(0,1)$. Then, by Theorem 3.12, $u(0) = \mathbb{E}_0[\varphi(B(\tau))] = v(0) \neq \varphi(0)$.   ◇

We now show how the techniques we have developed so far can be used to prove a classical result from harmonic analysis, Liouville's theorem, by probabilistic means. The proof uses the reflection principle for higher-dimensional Brownian motion.

Theorem 3.16 (Liouville's theorem) *Any bounded harmonic function on $\mathbb{R}^d$ is constant.*

Proof.   Let $u\colon \mathbb{R}^d \to [-M, M]$ be a harmonic function, x, y two distinct points in $\mathbb{R}^d$, and H the hyperplane so that the reflection in H takes x to y. Let $\{B(t)\colon t \geqslant 0\}$ be Brownian motion started at x, and $\{\overline{B}(t)\colon t \geqslant 0\}$ its reflection in H. Let $\tau(H) = \min\{t\colon B(t) \in H\}$ and note that

$$\{B(t)\colon t \geqslant \tau(H)\} \stackrel{\mathrm{d}}{=} \{\overline{B}(t)\colon t \geqslant \tau(H)\}. \tag{3.6}$$

Harmonicity implies that $\mathbb{E}[u(B(t))] = u(x)$ and decomposing the above into $t < \tau(H)$ and $t \geqslant \tau(H)$ we get

$$u(x) = \mathbb{E}\left[u(B(t))\mathbf{1}_{\{t < \tau(H)\}}\right] + \mathbb{E}\left[u(B(t))\mathbf{1}_{\{t \geqslant \tau(H)\}}\right].$$

A similar equation holds for $u(y)$ when $B(t)$ is replaced by $\overline{B}(t)$. Now, using (3.6),

$$|u(x) - u(y)| = \left|\mathbb{E}\left[u(B(t))1_{\{t < \tau(H)\}}\right] - \mathbb{E}\left[u(\overline{B}(t))1_{\{t < \tau(H)\}}\right]\right|$$
$$\leqslant 2M\mathbb{P}\{t < \tau(H)\} \to 0 \quad \text{as } t \to \infty.$$

Thus $u(x) = u(y)$, and since x and y were chosen arbitrarily, u must be constant. ∎

Remark 3.17 Clearly, any linear function is harmonic. In Exercise 3.10, the reader will be asked to prove that any harmonic function in $\mathbb{R}^d$ with sublinear growth is constant. ◇

3.2 Recurrence and transience of Brownian motion

A Brownian motion $\{B(t) \colon t \geqslant 0\}$ in dimension d is called *transient* if

$$\lim_{t \uparrow \infty} |B(t)| = \infty \qquad \text{almost surely.}$$

Note that the event $\{\lim_{t \uparrow \infty} |B(t)| = \infty\}$ is a tail event and hence, by the zero-one law for tail events, it must have probability zero or one. In this section we decide in which dimensions d the Brownian motion is transient, and in which it is not. This question is intimately related to the exit probabilities of the Brownian motion from an annulus: Suppose the motion starts at a point x inside an annulus

$$A = \{x \in \mathbb{R}^d \colon r < |x| < R\}, \qquad \text{for } 0 < r < R < \infty.$$

What is the probability that the Brownian motion hits $\partial \mathcal{B}(0, r)$ before $\partial \mathcal{B}(0, R)$? The answer is given in terms of harmonic functions on the annulus and is therefore closely related to the Dirichlet problem.

To find explicit solutions $u \colon \mathrm{cl}\, A \to \mathbb{R}$ of the Dirichlet problem on an annulus it is first reasonable to assume that u is spherically symmetric, i.e. there is a function $\psi \colon [r, R] \to \mathbb{R}$ such that $u(x) = \psi(|x|^2)$. We can express derivatives of u in terms of ψ as

$$\partial_i \psi(|x|^2) = \psi'(|x|^2)2x_i \text{ and } \partial_{ii}\psi(|x|^2) = \psi''(|x|^2)4x_i^2 + 2\psi'(|x|^2).$$

Therefore, $\Delta u = 0$ means

$$0 = \sum_{i=1}^{d}\left(\psi''(|x|^2)4x_i^2 + 2\psi'(|x|^2)\right) = 4|x|^2\psi''(|x|^2) + 2d\psi'(|x|^2).$$

Letting $y = |x|^2 > 0$ we can write this as

$$\psi''(y) = \frac{-d}{2y}\psi'(y).$$

This is solved by every ψ satisfying $\psi'(y) = \text{const} \cdot y^{-d/2}$ and thus $\Delta u = 0$ holds on $\{|x| \neq 0\}$ for

$$u(x) = \begin{cases} |x| & \text{if } d = 1, \\ 2\log|x| & \text{if } d = 2, \\ |x|^{2-d} & \text{if } d \geqslant 3. \end{cases} \tag{3.7}$$

We write $u(r)$ for the value of $u(x)$ for all $x \in \partial \mathcal{B}(0, r)$. Now define stopping times

$$T_r = \tau(\partial \mathcal{B}(0, r)) = \inf\{t > 0 \colon |B(t)| = r\} \text{ for } r > 0,$$

and denote by $T = T_r \wedge T_R$ the first exit time from A. By Theorem 3.12 we have

$$u(x) = \mathbb{E}_x\left[u(B(T))\right] = u(r)\mathbb{P}_x\{T_r < T_R\} + u(R)(1 - \mathbb{P}_x\{T_r < T_R\}).$$

This formula can be solved

$$\mathbb{P}_x\{T_r < T_R\} = \frac{u(R) - u(x)}{u(R) - u(r)}$$

and we get an explicit solution for the exit problem.

Theorem 3.18 *Suppose* $\{B(t) \colon t \geq 0\}$ *is a Brownian motion in dimension* $d \geq 1$ *started in* $x \in A$, *which is an open annulus* A *with radii* $0 < r < R < \infty$. *Then,*

$$\mathbb{P}_x\{T_r < T_R\} = \begin{cases} \frac{R - |x|}{R - r} & \text{if } d = 1, \\[2mm] \frac{\log R - \log |x|}{\log R - \log r} & \text{if } d = 2, \\[2mm] \frac{R^{2-d} - |x|^{2-d}}{R^{2-d} - r^{2-d}} & \text{if } d \geq 3. \end{cases}$$

Letting $R \uparrow \infty$ in Theorem 3.18 leads to the following corollary.

Corollary 3.19 *For any* $x \notin \mathcal{B}(0, r)$, *we have*

$$\mathbb{P}_x\{T_r < \infty\} = \begin{cases} 1 & \text{if } d \leq 2, \\[2mm] \frac{r^{d-2}}{|x|^{d-2}} & \text{if } d \geq 3. \end{cases}$$

We now apply this to the problem of *recurrence* and *transience* of Brownian motion in various dimensions. Generally speaking, we call a Markov process $\{X(t) \colon t \geq 0\}$ with values in $\mathbb{R}^d$

- **point recurrent**, if, almost surely, for every $x \in \mathbb{R}^d$ there is a (random) sequence $t_n \uparrow \infty$ such that $X(t_n) = x$ for all $n \in \mathbb{N}$,
- **neighbourhood recurrent**, if, almost surely, for every $x \in \mathbb{R}^d$ and $\varepsilon > 0$, there exists a (random) sequence $t_n \uparrow \infty$ such that $X(t_n) \in \mathcal{B}(x, \varepsilon)$ for all $n \in \mathbb{N}$.
- **transient**, if it converges to infinity almost surely.

Theorem 3.20 *Brownian motion is*

- *point recurrent in dimension* $d = 1$,
- *neighbourhood recurrent, but not point recurrent, in* $d = 2$,
- *transient in dimension* $d \geq 3$.

Proof. We leave the case $d = 1$ as Exercise 3.4, and look at dimension $d = 2$. Fix $\varepsilon > 0$ and $x \in \mathbb{R}^d$. By Corollary 3.19 and shift-invariance the stopping time $t_1 = \inf\{t > 0 : B(t) \in \mathcal{B}(x, \varepsilon)\}$ is almost surely finite. Using the strong Markov property at time $t_1 + 1$ we see that this also applies to $t_2 = \inf\{t > t_1 + 1 : B(t) \in \mathcal{B}(x, \varepsilon)\}$, and continuing like this, we obtain a sequence of times $t_n \uparrow \infty$ such that, almost surely, $B(t_n) \in \mathcal{B}(x, \varepsilon)$ for all $n \in \mathbb{N}$. Taking an intersection over a countable family of balls $(\mathcal{B}(x_i, \varepsilon_i) : i = 1, 2, \ldots)$, forming a basis of the Euclidean topology, implies that in $d = 2$ Brownian motion is neighbourhood recurrent. Recall from Corollary 2.26 that planar Brownian motion does not hit points, hence it cannot be point recurrent.

It remains to show that Brownian motion is transient in dimensions $d \geqslant 3$. Look at the events $A_n := \{|B(t)| > n \text{ for all } t \geqslant T_{n^3}\}$. Recall from Proposition 1.23 that $T_{n^3} < \infty$ almost surely. By the strong Markov property, for every $n \geqslant |x|^{1/3}$,

$$\mathbb{P}_x(A_n^c) = \mathbb{E}_x\left[\mathbb{P}_{B(T_{n^3})}\{T_n < \infty\}\right] = \left(\frac{1}{n^2}\right)^{d-2}.$$

Note that the right hand side is summable, and hence the Borel–Cantelli lemma shows that only finitely many of the events A_n^c occur, which implies that $|B(t)|$ diverges to infinity, almost surely, and hence that Brownian motion in $d \geqslant 3$ is transient. ∎

Remark 3.21 Neighbourhood recurrence, in particular, implies that the path of a planar Brownian motion (running for an infinite amount of time) is dense in the plane. ◇

We now have a qualitative look at the transience of Brownian motion in $\mathbb{R}^d$, $d \geqslant 3$, and ask for the speed of escape to infinity. This material is slightly more advanced and can be skipped on first reading.

Consider a standard Brownian motion $\{B(t) : t \geqslant 0\}$ in $\mathbb{R}^d$, for $d \geqslant 3$, and fix a sequence $t_n \uparrow \infty$. For any $\varepsilon > 0$, by Fatou's lemma,

$$\mathbb{P}\{|B(t_n)| < \varepsilon\sqrt{t_n} \text{ infinitely often }\} \geqslant \limsup_{n \to \infty} \mathbb{P}\{|B(t_n)| < \varepsilon\sqrt{t_n}\} > 0.$$

By the zero-one law for tail events, see Theorem 2.9, the probability on the left hand side must therefore be one, whence

$$\liminf_{n \to \infty} \frac{|B(t_n)|}{\sqrt{t_n}} = 0, \qquad \text{almost surely.} \tag{3.8}$$

This statement is refined by the Dvoretzky–Erdős test.

Theorem* 3.22 (Dvoretzky–Erdős test) *Let $\{B(t) : t \geqslant 0\}$ be Brownian motion in $\mathbb{R}^d$ for $d \geqslant 3$ and $f : (0, \infty) \to (0, \infty)$ increasing. Then*

$$\int_1^\infty f(r)^{d-2} r^{-d/2}\, dr < \infty \qquad \textit{if and only if} \qquad \liminf_{t \uparrow \infty} \frac{|B(t)|}{f(t)} = \infty \textit{ almost surely.}$$

Conversely, if the integral diverges, then $\liminf_{t \uparrow \infty} |B(t)|/f(t) = 0$ almost surely.

For the proof we first recall two generally useful tools. The first is an easy case of the Paley–Zygmund inequality, see Exercise 3.5 for the full statement.

Lemma 3.23 (Paley–Zygmund inequality) *For any nonnegative random variable X with* $\mathbb{E}[X^2] < \infty$,

$$\mathbb{P}\{X > 0\} \geqslant \frac{\mathbb{E}[X]^2}{\mathbb{E}[X^2]} \, .$$

Proof.   The Cauchy–Schwarz inequality gives

$$\mathbb{E}[X] = \mathbb{E}[X \, 1\{X > 0\}] \leqslant \mathbb{E}[X^2]^{1/2} \left(\mathbb{P}\{X > 0\} \right)^{1/2},$$

and the required inequality follows immediately.   ∎

The second tool is a version of the Borel–Cantelli lemma, which allows some dependence of the events. This is known as the Kochen–Stone lemma, and is a consequence of the Paley–Zygmund inequality, see Exercise 3.6 or [KS64].

Lemma 3.24 *Suppose $E_1, E_2, \ldots$ are events with*

$$\sum_{n=1}^{\infty} \mathbb{P}(E_n) = \infty \qquad and \qquad \liminf_{k \to \infty} \frac{\sum_{m=1}^{k} \sum_{n=1}^{k} \mathbb{P}(E_n \cap E_m)}{\left(\sum_{n=1}^{k} \mathbb{P}(E_n) \right)^2} < \infty \, .$$

Then, with positive probability, infinitely many of the events take place.

A core estimate in the proof of the Dvoretzky–Erdős test is the following lemma, which is based on the hitting probabilities of the previous paragraphs.

Lemma 3.25 *There exists a constant $C_1 > 0$ depending only on the dimension d such that, for any $\rho > 0$, we have*

$$\sup_{x \in \mathbb{R}^d} \mathbb{P}_x \big\{ \text{ there exists } t > 1 \text{ with } |B(t)| \leqslant \rho \big\} \leqslant C_1 \, \rho^{d-2} \, .$$

Proof.   We use Corollary 3.19 for the probability that the motion started at time one hits $\mathcal{B}(0, \rho)$, to see that

$$\mathbb{P}_x \big\{ \text{ there exists } t > 1 \text{ with } |B(t)| \leqslant \rho \big\} \leqslant \mathbb{E}_0 \left[\left(\frac{\rho}{|B(1) + x|} \right)^{d-2} \right]$$

$$\leqslant \rho^{d-2} \frac{1}{(2\pi)^{d/2}} \int_{\mathbb{R}^d} |y + x|^{2-d} \, \exp \big\{ -\tfrac{|y|^2}{2} \big\} \, dy.$$

By considering the integration domains $|y + x| \geqslant |y|$ and $|y + x| \leqslant |y|$ separately, it is easy to see that the integral on the right is uniformly bounded in x.   ∎

Proof of Theorem 3.22.   Define events

$$A_n = \big\{ \text{ there exists } t \in (2^n, 2^{n+1}] \text{ with } |B(t)| \leqslant f(t) \big\} \, .$$

By Brownian scaling, monotonicity of f, and Lemma 3.25,

$$\mathbb{P}(A_n) \leqslant \mathbb{P}\{ \text{ there exists } t > 1 \text{ with } |B(t)| \leqslant f(2^{n+1})2^{-n/2}\}$$
$$\leqslant C_1 \left(f(2^{n+1}) \, 2^{-n/2}\right)^{d-2}.$$

Now assume that the integral converges, or equivalently, that

$$\sum_{n=1}^{\infty} \left(f(2^n) \, 2^{-n/2}\right)^{d-2} < \infty. \tag{3.9}$$

Then the Borel Cantelli lemma and (3.9) imply that, almost surely, the set $\{t > 0: |B(t)| \leqslant f(t)\}$ is bounded. Since (3.9) also applies to any constant multiple of f in place of f, it follows that $\liminf_{t\uparrow\infty} |B(t)|/f(t) = \infty$ almost surely.

For the converse, suppose that the integral diverges, whence

$$\sum_{n=1}^{\infty} \left(f(2^n) \, 2^{-n/2}\right)^{d-2} = \infty. \tag{3.10}$$

In view of (3.8), we may assume that $f(t) < \sqrt{t}$ for all large enough t. Changing f on a finite interval, we may assume that this inequality holds for all $t > 0$.

For $\rho \in (0,1)$, consider the random variable $I_\rho = \int_1^2 1\{|B(t)| \leqslant \rho\}\, dt$. Since the density of $|B(t)|$ on the unit ball is bounded from above and also away from zero for $t \in [1,2]$, we infer that

$$C_2 \rho^d \leqslant \mathbb{E}[I_\rho] \leqslant C_3 \rho^d$$

for suitable constants depending only on the dimension. To complement this by an estimate of the second moment, we use the Markov property to see that

$$\mathbb{E}[I_\rho^2] = 2\mathbb{E}\Big[\int_1^2 1\{|B(t)| \leqslant \rho\} \int_t^2 1\{|B(s)| \leqslant \rho\}\, ds\, dt\Big]$$
$$\leqslant 2\mathbb{E}\Big[\int_1^2 1\{|B(t)| \leqslant \rho\}\, \mathbb{E}_{B(t)} \int_0^\infty 1\{|\tilde{B}(s)| \leqslant \rho\}\, ds\, dt\Big],$$

where the inner expectation is with respect to a Brownian motion $\{\tilde{B}(t): t \geqslant 0\}$ started in the fixed point $B(t)$, whereas the outer expectation is with respect to $B(t)$. We analyse the dependence of the inner expectation on the starting point. Given $x \neq 0$, we let $T = \inf\{t > 0: |B(t)| = x\}$ and use the strong Markov property to see that

$$\mathbb{E}_0 \int_0^\infty 1\{|B(s)| \leqslant \rho\}\, ds \geqslant \mathbb{E}\int_T^\infty 1\{|B(s)| \leqslant \rho\}\, ds = \mathbb{E}_x \int_0^\infty 1\{|B(s)| \leqslant \rho\}\, ds,$$

so that the expectation is maximal if the process is started at the origin. Hence we obtain

$$\mathbb{E}[I_\rho^2] \leqslant 2C_3 \, \rho^d \, \mathbb{E}_0 \int_0^\infty 1\{|B(s)| \leqslant \rho\}\, ds.$$

Moreover, by Brownian scaling,

$$\mathbb{E}_0 \int_0^\infty 1\{|B(s)| \leqslant \rho\}\, ds = \rho^2 \int_0^\infty \mathbb{P}\{|B(s)| \leqslant 1\}\, ds$$
$$\leqslant \rho^2 \left(1 + \int_1^\infty \frac{\mathcal{L}(\mathcal{B}(0,1))}{(2\pi s)^{d/2}}\, ds\right) = C_4 \, \rho^2,$$

where C_4 is a finite constant. In summary, we have $\mathbb{E}[I_\rho^2] \leqslant 2C_3C_4\,\rho^{d+2}$. By the Paley–Zygmund inequality, for a suitable constant $C_5 > 0$,

$$\mathbb{P}\{I_\rho > 0\} \geqslant \frac{\mathbb{E}[I_\rho]^2}{\mathbb{E}[I_\rho^2]} \geqslant C_5\rho^{d-2}\,.$$

Now choose $\rho = f(2^n)2^{-n/2}$, which is smaller than one, as $f(t) < \sqrt{t}$. By Brownian scaling and monotonicity of f, we have

$$\mathbb{P}(A_n) \geqslant \mathbb{P}\{I_\rho > 0\} \geqslant C_5\left(f(2^n)\,2^{-n/2}\right)^{d-2},$$

so $\sum_n \mathbb{P}(A_n) = \infty$ by (3.10). For $m < n-1$, the Markov property at time 2^{n-1}, Brownian scaling and Lemma 3.25 yield that

$$\mathbb{P}[A_n \mid A_m] \leqslant \sup_{x \in \mathbb{R}^d} \mathbb{P}_x\Big\{ \text{ there exists } t > 1 \text{ with } |B(t)| \leqslant f(2^{n+1})2^{(1-n)/2}\Big\}$$
$$\leqslant C_1\left(f(2^{n+1})\,2^{(1-n)/2}\right)^{d-2}.$$

From this, and the assumption that $f(t) < \sqrt{t}$, we get that

$$\liminf_{k\to\infty} \frac{\sum_{m=1}^{k}\sum_{n=1}^{k}\mathbb{P}(A_n \cap A_m)}{\left(\sum_{n=1}^{k}\mathbb{P}(A_n)\right)^2} = 2\liminf_{k\to\infty} \frac{\sum_{m=1}^{k}\mathbb{P}(A_m)\sum_{n=m+2}^{k}\mathbb{P}[A_n \mid A_m]}{\left(\sum_{n=1}^{k}\mathbb{P}(A_n)\right)^2}$$
$$\leqslant 2\frac{C_1}{C_5}\liminf_{k\to\infty} \frac{\sum_{n=1}^{k}(f(2^{n+1})\,2^{(1-n)/2})^{d-2}}{\sum_{n=1}^{k}(f(2^n)\,2^{-n/2})^{d-2}} < \infty\,.$$

The Kochen–Stone lemma now yields that $\mathbb{P}\{A_n \text{ infinitely often}\} > 0$, whence by Theorem 2.9 this probability is 1. Thus the set $\{t > 0 \colon |B(t)| \leqslant f(t)\}$ is almost surely unbounded. Since (3.10) also applies to εf in place of f for any $\varepsilon > 0$, it follows that $\liminf_{t\uparrow\infty} |B(t)|/f(t) = 0$ almost surely. ∎

3.3 Occupation measures and Green's functions

We now address the following question: Given a bounded domain $U \subset \mathbb{R}^d$, how much time does Brownian motion spend in U? Our first result states that for a linear Brownian motion running for a finite amount of time, this time is comparable to the Lebesgue measure of U.

Theorem 3.26 *Let $\{B(s)\colon s \geqslant 0\}$ be a linear Brownian motion and $t > 0$. Define the occupation measure μ_t by*

$$\mu_t(A) = \int_0^t 1_A(B(s))\,ds \qquad \text{for } A \subset \mathbb{R} \text{ Borel.}$$

Then, almost surely, μ_t is absolutely continuous with respect to the Lebesgue measure.

Proof. Absolute continuity of μ_t with respect to the Lebesgue measure follows if

$$\liminf_{r \downarrow 0} \frac{\mu_t(\mathcal{B}(x,r))}{\mathcal{L}(\mathcal{B}(x,r))} < \infty \text{ for } \mu_t\text{-almost every } x \in \mathbb{R},$$

see for example Theorem 2.12 in [Ma95]. To see this we use first Fatou's lemma and then Fubini's theorem,

$$\mathbb{E} \int \liminf_{r \downarrow 0} \frac{\mu_t(\mathcal{B}(x,r))}{\mathcal{L}(\mathcal{B}(x,r))} \, d\mu_t(x) \leqslant \liminf_{r \downarrow 0} \frac{1}{2r} \mathbb{E} \int \mu_t(\mathcal{B}(x,r)) \, d\mu_t(x)$$

$$= \liminf_{r \downarrow 0} \frac{1}{2r} \int_0^t \int_0^t \mathbb{P}\{|B(s_1) - B(s_2)| \leqslant r\} \, ds_1 \, ds_2 \, .$$

Using that the density of a standard normal random variable X is bounded by one, we get

$$\mathbb{P}\{|B(s_1) - B(s_2)| \leqslant r\} = \mathbb{P}\{|X| \leqslant \tfrac{r}{\sqrt{|s_1-s_2|}}\} \leqslant \frac{2r}{\sqrt{|s_1-s_2|}},$$

and this implies that

$$\liminf_{r \downarrow 0} \frac{1}{2r} \int_0^t \int_0^t \mathbb{P}\{|B(s_1) - B(s_2)| \leqslant r\} \, ds_1 \, ds_2 \leqslant \int_0^t \int_0^t \frac{ds_1 \, ds_2}{\sqrt{|s_1 - s_2|}} < \infty.$$

This implies that, almost surely, μ_t is absolutely continuous with respect to $\mathcal{L}$. ∎

We now turn to higher dimensions $d \geqslant 2$. A first simple result shows that whether the overall time spent in a bounded set is finite or not depends just on transience or recurrence of the process.

Theorem 3.27 *Let $U \subset \mathbb{R}^d$ be a nonempty bounded open set and $x \in \mathbb{R}^d$ arbitrary.*

- *If $d = 2$, then $\mathbb{P}_x$-almost surely, $\displaystyle\int_0^\infty 1_U(B(t)) \, dt = \infty$.*

- *If $d \geqslant 3$, then $\displaystyle\mathbb{E}_x \int_0^\infty 1_U(B(t)) \, dt < \infty$.*

Proof. As U is contained in a ball and contains a ball, it suffices to show this for balls. By shifting, we can even restrict to balls $U = \mathcal{B}(0,r)$ centred in the origin. Let us start with the first claim. We let $d = 2$ and let $G = \mathcal{B}(0, 2r)$. Let $S_0 = 0$ and, for all $k \geqslant 0$, let

$$T_k = \inf\{t > S_k : B(t) \notin G\} \quad \text{and} \quad S_{k+1} = \inf\{t > T_k : B(t) \in U\}.$$

Recall that, almost surely, these stopping times are finite. From the strong Markov property we infer, for $k \geqslant 1$,

$$\mathbb{P}_x\left\{ \int_{S_k}^{T_k} 1_U(B(t)) \, dt \geqslant s \, \Big| \mathcal{F}^+(S_k) \right\} = \mathbb{P}_{B(S_k)}\left\{ \int_0^{T_1} 1_U(B(t)) \, dt \geqslant s \right\}$$

$$= \mathbb{E}_x\left[\mathbb{P}_{B(S_k)}\left\{ \int_0^{T_1} 1_U(B(t)) \, dt \geqslant s \right\} \right] = \mathbb{P}_x\left\{ \int_{S_k}^{T_k} 1_U(B(t)) \, dt \geqslant s \right\},$$

by rotation invariance. Hence the random variables $\{\int_{S_k}^{T_k} 1_U(B(t)) \, dt, : k = 1, 2, \ldots\}$ are independent and, as the second term does not depend on k, identically distributed. As they are not identically zero, but nonnegative, they have positive expectation and, by the strong law of large numbers we infer

$$\int_0^\infty 1_U(B(t))\,dt \geqslant \lim_{n\to\infty} \sum_{k=1}^n \int_{S_k}^{T_k} 1_U(B(t))\,dt = \infty,$$

which proves the first claim. For the second claim, we first look at Brownian motion started in the origin and obtain, making good use of Fubini's theorem and denoting by $\mathfrak{p}\colon [0,\infty) \times \mathbb{R}^d \times \mathbb{R}^d \to [0,1]$ the transition density of Brownian motion,

$$\mathbb{E}_0 \int_0^\infty 1_{\mathcal{B}(0,r)}(B(s))\,ds = \int_0^\infty \mathbb{P}_0\{B(s) \in \mathcal{B}(0,r)\}\,ds = \int_0^\infty \int_{\mathcal{B}(0,r)} \mathfrak{p}(s,0,y)\,dy\,ds$$

$$= \int_{\mathcal{B}(0,r)} \int_0^\infty \mathfrak{p}(s,0,y)\,ds\,dy$$

$$= \sigma(\partial\mathcal{B}(0,1)) \int_0^r \rho^{d-1} \int_0^\infty \left(\frac{1}{\sqrt{2\pi s}}\right)^d e^{\frac{-\rho^2}{2s}}\,ds\,d\rho.$$

Now we can use the substitution $t = \rho^2/s$ and obtain, using that $d \geqslant 3$ to ensure finiteness of the integral, for a suitable constant $C(d) < \infty$,

$$= C(d) \int_0^r \rho^{d-1} \rho^{2-d}\,d\rho = \frac{C(d)}{2} r^2 < \infty.$$

For start in an arbitrary $x \neq 0$, we look at a Brownian motion started in 0 and a stopping time T, which is the first hitting time of the sphere $\partial\mathcal{B}(0,|x|)$. Using spherical symmetry and the strong Markov property we obtain

$$\mathbb{E}_x \int_0^\infty 1_{\mathcal{B}(0,r)}(B(s))\,ds = \mathbb{E}_0 \int_T^\infty 1_{\mathcal{B}(0,r)}(B(s))\,ds$$

$$\leqslant \mathbb{E}_0 \int_0^\infty 1_{\mathcal{B}(0,r)}(B(s))\,ds < \infty. \qquad \blacksquare$$

In the case when Brownian motion is transient it is interesting to ask further for the expected time the process spends in a bounded open set. In order not to confine this discussion to the case $d \geqslant 3$ we introduce suitable stopping rules for Brownian motion in $d = 2$.

Definition 3.28. Suppose that $\{B(t)\colon 0 \leqslant t \leqslant T\}$ is a d-dimensional Brownian motion and one of the following three cases holds:

(1) $d \geqslant 3$ and $T = \infty$,
(2) $d \geqslant 2$ and T is an independent exponential time with parameter $\lambda > 0$,
(3) $d \geqslant 2$ and T is the first exit time from a bounded domain D.

We use the convention that $D = \mathbb{R}^d$ in cases $(1), (2)$. We refer to these three cases by saying that $\{B(t)\colon 0 \leqslant t \leqslant T\}$ is a **transient Brownian motion**. $\diamond$

Remark 3.29 For a transient Brownian motion $\{B(t)\colon 0 \leqslant t \leqslant T\}$, given $\mathcal{F}^+(t)$, on the event $\{B(t) = y,\, t < T\}$, the process $\{B(s + t)\colon 0 \leqslant s \leqslant T\}$ is again a transient Brownian motion of the same type, started in y. We do *not* consider Brownian motion stopped at a *fixed* time, because this model lacks this form of the Markov property. $\diamond$

Theorem 3.30 *For transient Brownian motion $\{B(t)\colon 0 \leqslant t \leqslant T\}$ there exists a transition (sub-)density $\mathfrak{p}^*\colon [0,\infty) \times \mathbb{R}^d \times \mathbb{R}^d \to [0,1]$ such that, for any $t > 0$,*

$$\mathbb{P}_x\{B(t) \in A \text{ and } t \leqslant T\} = \int_A \mathfrak{p}^*(t,x,y)\,dy \qquad \text{for every } A \subset \mathbb{R}^d \text{ Borel.}$$

Moreover, for all $t \geqslant 0$ and $\mathcal{L}$-almost every $x, y \in D$ we have $\mathfrak{p}^*(t,x,y) = \mathfrak{p}^*(t,y,x)$.

Proof. Fix t throughout the proof. For the existence of the density, by the Radon–Nikodým theorem, it suffices to check that $\mathbb{P}_x\{B(t) \in A \text{ and } t \leqslant T\} = 0$, if A is a Borel set of Lebesgue measure zero. This is obvious, by just dropping the requirement $t \leqslant T$, and recalling that $B(t)$ is normally distributed. If $d \geqslant 3$ and $T = \infty$, or if $d \geqslant 2$ and T is independent, exponentially distributed symmetry is obvious.

Hence we can now concentrate on the case $d \geqslant 2$ and a bounded domain D. We fix a compact set $K \subset D$ and define, for every $x \in K$ and $n \in \mathbb{N}$, a measure $\mu_x^{(n)}$ on the Borel sets $A \subset D$,

$$\mu_x^{(n)}(A) = \mathbb{P}_x\big\{B(\tfrac{kt}{2^n}) \in K \text{ for all } k = 0, \ldots, 2^n \text{ and } B(t) \in A\big\}.$$

Then $\mu_x^{(n)}$ has a density

$$\mathfrak{p}_n^*(t,x,y) = \int_K \cdots \int_K \prod_{i=1}^{2^n} \mathfrak{p}\big(\tfrac{t}{2^n}, z_{i-1}, z_i\big)\, dz_1 \ldots dz_{2^n - 1},$$

where $z_0 = x$, $z_{2^n} = y$ and $\mathfrak{p}$ is the transition density of d-dimensional Brownian motion. As $\mathfrak{p}$ is symmetric in the space variables, so is $\mathfrak{p}_n^*$ for every n. Note that $\mathfrak{p}_n^*$ is decreasing in n. From the monotone convergence theorem one can see that $\mathfrak{p}_K^*(t,x,y) := \lim \mathfrak{p}_n^*(t,x,y)$ is a transition subdensity of Brownian motion stopped upon leaving K. The symmetry of $\mathfrak{p}_n^*$ gives $\mathfrak{p}_K^*(t,x,y) = \mathfrak{p}_K^*(t,y,x)$. Choosing an increasing sequence of compact sets exhausting D and taking a monotone limit yields a symmetric version $\mathfrak{p}^*(t,x,y)$ of the transition density. ∎

In all of our three cases of transient Brownian motions we will from now on choose particular versions of the transition densities. Recall that $\mathfrak{p}$ denotes the transition kernel for the (unstopped) Brownian motion. Then,

(1) if $d \geqslant 3$ and $T = \infty$, we take $\mathfrak{p}^*(t,x,y) = \mathfrak{p}(t,x,y)$;

(2) if $d \geqslant 2$ and T is exponential with parameter $\lambda > 0$, we choose

$$\mathfrak{p}^*(t,x,y) = e^{-\lambda t}\,\mathfrak{p}(t,x,y);$$

(3) if $d \geqslant 2$ and T is the first exit time from D, we let

$$\mathfrak{p}^*(t,x,y) = \mathfrak{p}(t,x,y) - \mathbb{E}_x\big[\mathfrak{p}(t - T, B(T), y)\,\mathbf{1}\{T < t\}\big].$$

It is easy to verify that these $\mathfrak{p}^*$ are indeed transition densities as claimed.

Definition 3.31.   For transient Brownian motion $\{B(t) \colon 0 \leqslant t \leqslant T\}$ we define the **Green's function** $G \colon \mathbb{R}^d \times \mathbb{R}^d \to [0, \infty]$ by

$$G(x, y) = \int_0^\infty \mathfrak{p}^*(t, x, y) \, dt \,.$$

The Green's function is also called the **Green kernel**. Sometimes it is also called the *potential kernel*, but we shall reserve this terminology for a closely related concept, see Remark 8.21.   ◇

In probabilistic terms G is the density of the *expected* occupation measure for the transient Brownian motion started in x.

Theorem 3.32 *If* $f \colon \mathbb{R}^d \to [0, \infty]$ *is measurable, then*

$$\mathbb{E}_x \int_0^T f(B(t)) \, dt = \int f(y) \, G(x, y) \, dy.$$

Proof.   Fubini's theorem implies

$$\mathbb{E}_x \int_0^T f(B(t)) \, dt \quad = \quad \int_0^\infty \mathbb{E}_x \left[f(B(t)) \, 1_{\{t \leqslant T\}} \right] dt = \int_0^\infty \int \mathfrak{p}^*(t, x, y) \, f(y) \, dy \, dt$$

$$= \quad \int \int_0^\infty \mathfrak{p}^*(t, x, y) \, dt \, f(y) \, dy = \int G(x, y) f(y) \, dy,$$

by definition of the Green's function.   ∎

In case (1), i.e. if $T = \infty$, Green's function can be calculated explicitly.

Theorem 3.33 *If* $d \geqslant 3$ *and* $T = \infty$, *then*

$$G(x, y) = c(d) \, |x - y|^{2-d}, \qquad \textit{where } c(d) = \tfrac{\Gamma(d/2-1)}{2\pi^{d/2}}.$$

Proof.   Assume $d \geqslant 3$ and use the substitution $s = |x - y|^2 / 2t$ to obtain,

$$G(x, y) = \int_0^\infty \frac{1}{(2\pi t)^{d/2}} \, e^{-|x-y|^2/2t} \, dt = \int_\infty^0 \left(\frac{s}{\pi |x-y|^2} \right)^{d/2} e^{-s} \left(-\frac{|x-y|^2}{2s^2} \right) ds$$

$$= \frac{|x-y|^{2-d}}{2\pi^{d/2}} \int_0^\infty s^{(d/2)-2} \, e^{-s} \, ds = \frac{\Gamma(d/2-1)}{2\pi^{d/2}} \, |x-y|^{2-d},$$

where $\Gamma(x) = \int_0^\infty s^{x-1} e^{-s} \, ds$ is the Gamma function. This proves that G has the given form and the calculation above also shows that the integral is infinite if $d \leqslant 2$.   ∎

In case (2), if Brownian motion is stopped at an independent exponential time, one can find the asymptotics of $G(x, y)$ for $x \to y$.

Theorem 3.34 *If* $d = 2$ *and* T *is an independent exponential time with parameter* $\lambda > 0$, *then*

$$G(x, y) \sim -\frac{1}{\pi} \log |x - y| \qquad \textit{for } |x - y| \downarrow 0 \,.$$

Proof. From the explicit form of $\mathfrak{p}^*$ we get

$$G(x, y) = G_\lambda(x - y) := \int_0^\infty \frac{1}{2\pi t} \exp\left\{ -\frac{|x-y|^2}{2t} - \lambda t \right\} dt.$$

We thus get $G_\lambda(x - y) = G_1(\sqrt{\lambda}(x - y))$ and may assume without loss of generality that $\lambda = 1$. Then

$$G(x, y) = \frac{1}{2\pi} \int_0^\infty \frac{e^{-t}}{t} \int_{|x-y|^2/(2t)}^\infty e^{-s}\, ds\, dt = \frac{1}{2\pi} \int_0^\infty e^{-s} \int_{|x-y|^2/(2s)}^\infty \frac{e^{-t}}{t}\, dt\, ds.$$

For an upper bound we use that,

$$\int_{|x-y|^2/(2s)}^\infty \frac{e^{-t}}{t}\, dt \leqslant \begin{cases} \log \frac{2s}{|x-y|^2} + 1, & \text{if } |x-y|^2 \leqslant 2s, \\ 1, & \text{if } |x-y|^2 > 2s. \end{cases}$$

For $|x - y| \leqslant 1$ this gives, with $\tilde{\gamma} := \int_1^\infty e^{-s} \log s\, ds < \infty$, a bound of

$$G(x, y) \leqslant \frac{1}{2\pi} \left(1 + \log 2 + \tilde{\gamma} - 2\log|x - y| \right),$$

which is asymptotically equal to $-\frac{1}{\pi} \log|x - y|$. For a lower bound we use

$$\int_{|x-y|^2/(2s)}^\infty \frac{e^{-t}}{t}\, dt \geqslant \log \frac{2s}{|x - y|^2} - 1,$$

and thus with $0 < \gamma := -\int_0^\infty e^{-s} \log s\, ds$ denoting Euler's constant,

$$G(x, y) \geqslant \frac{1}{2\pi} \left(-1 + \log 2 - \gamma - 2\log|x - y| \right),$$

and again this is asymptotically equal to $-\frac{1}{\pi} \log|x - y|$. ∎

We now explore some of the major analytic properties of Green's function.

Theorem 3.35 *In all three cases of transient Brownian motion in $d \geqslant 2$, the Green's function $G \colon D \times D \to [0, \infty]$ has the following properties:*

   (i) *G is finite off and infinite on the diagonal $\Delta = \{(x, y) \colon x = y\}$.*

   (ii) *G is symmetric, i.e. $G(x, y) = G(y, x)$ for all $x, y \in D$.*

   (iii) *For any $y \in D$ the Green's function $G(\,\cdot\,, y)$ is subharmonic on $D \setminus \{y\}$. Moreover, in case (1) and (3) it is harmonic.*

This result is easy in the case $d \geqslant 3$, $T = \infty$, where the Green's function is explicitly known by Theorem 3.33. We prepare the proof in $d = 2$ by two lemmas of independent interest.

Lemma 3.36 *If $d = 2$, for $x, y, z \in \mathbb{R}^2$ with $|x - z| = 1$,*

$$-\frac{1}{\pi} \log |x - y| = \int_0^\infty \mathfrak{p}(s, x, y) - \mathfrak{p}(s, x, z) \, ds \, ,$$

where $\mathfrak{p}$ is the transition kernel for the (unstopped) Brownian motion.

Proof.   For $|x - z| = 1$, we obtain

$$\int_0^\infty \mathfrak{p}(t, x, y) - \mathfrak{p}(t, x, z) \, dt = \frac{1}{2\pi} \int_0^\infty \left(e^{-\frac{|x-y|^2}{2t}} - e^{-\frac{1}{2t}} \right) \frac{dt}{t}$$

$$= \frac{1}{2\pi} \int_0^\infty \left(\int_{|x-y|^2/(2t)}^{1/(2t)} e^{-s} \, ds \right) \frac{dt}{t} \, ,$$

and by changing the order of integration this equals

$$\frac{1}{2\pi} \int_0^\infty e^{-s} \left(\int_{|x-y|^2/(2s)}^{1/(2s)} \frac{dt}{t} \right) ds = -\frac{1}{\pi} \log |x - y| \, ,$$

which completes the proof.   ∎

Lemma 3.37 *Let $D \subset \mathbb{R}^2$ be a bounded domain and $x, y \in D$ and T the first exit time from D. Then, with $u(x) = 2 \log |x|$,*

$$G(x, y) = \frac{-1}{2\pi} u(x - y) - \mathbb{E}_x \left[\frac{-1}{2\pi} u\big(B(T) - y\big) \right].$$

Proof.   Recall that

$$\mathfrak{p}^*(t, x, y) = \mathfrak{p}(t, x, y) - \mathbb{E}_x \left[\mathfrak{p}(t - T, B(T), y) \mathbf{1}\{T < t\} \right].$$

As $\mathfrak{p}(t, x, x + v)$ does not depend on x, we can add

$$0 = -\mathfrak{p}(t, x, x + v) + \mathbb{E}_x \left[\mathfrak{p}(t, B(T), B(T) + v) \right]$$

on the right hand side. Integrating over t and using Lemma 3.36 yields the statement.   ∎

Proof of Theorem 3.35.   We first look at properties (i) and (ii). These are obvious in the case $d \geqslant 3$, $T = \infty$, by the explicit form of the Green's function uncovered in Theorem 3.33. In the case that T is an independent exponential time we can see from the explicit form of $\mathfrak{p}^*$ that the Green's function is symmetric and finite everywhere except on the diagonal. Moreover note for later reference that in this case twice differentiability is easy to check using dominated convergence.

We now focus on the case where the Brownian motion is stopped upon leaving a bounded domain D and look at the case $d = 2$ and $d \geqslant 3$ separately. First let $d = 2$. Lemma 3.37 gives, for $x \neq y$, that $G(x, y) < \infty$. However, we have

$$\mathbb{E}_x \left[-1/(2\pi) \, u(B(T) - x) \right] < \infty,$$

hence $G(x, x) = \infty$ by Lemma 3.37. If $x \in D$, then $G(x, \cdot)$ is continuous on $D \setminus \{x\}$, because the right hand side of the equation in Lemma 3.37 is continuous. Similarly, if $y \in D$ the right hand side is continuous in x on $D \setminus \{y\}$, as $\mathbb{E}_x [u(B(T) - y)]$ is harmonic in x. Hence $G(\,\cdot\,,x)$ is also continuous on $D \setminus \{x\}$. The symmetry follows from the almost-everywhere symmetry of $\mathfrak{p}^*(t,\,\cdot\,,\,\cdot\,)$ together with the continuity. If $d \geqslant 3$ the same proof works, replacing $-1/(2\pi)u(x-y)$ by $\ell(x,y) = c(d)|x-y|^{2-d}$. In fact the argument becomes easier because

$$\ell(x,y) = \int_0^\infty \mathfrak{p}(t,x,y)\, dt, \quad \text{for all } x, y \in \mathbb{R}^d,$$

and there is no need to subtract a 'renormalisation' term.

Next we investigate (sub-)harmonicity of the Green's function in all cases. Define

$$G_\varepsilon(x,y) := \int_{\mathcal{B}(y,\varepsilon)} G(x,z)\, dz, \quad \text{for } \mathcal{B}(y,\varepsilon) \subset D \text{ and } x \in D.$$

We first prove that $G_\varepsilon(\,\cdot\,,y)$ satisfies the mean value property of subharmonic functions on $D \setminus \mathcal{B}(y,\varepsilon)$, i.e.

$$G_\varepsilon(x,y) \leqslant \frac{1}{\mathcal{L}(\mathcal{B}(x,r))} \int_{\mathcal{B}(x,r)} G_\varepsilon(z,y)\, dz, \quad \text{for } 0 < r < |x-y| - \varepsilon. \quad (3.11)$$

Indeed, fix $x \neq y$ in D, let $0 < r < |x-y|$ and $\varepsilon < |x-y| - r$. Denote $\tau = \inf\{t\colon |B(t) - x| = r\}$. As a Brownian motion started in x spends no time in $\mathcal{B}(y,\varepsilon)$ before time τ, we can write

$$G_\varepsilon(x,y) = \mathbb{E}_x\left[\mathbb{1}\{\tau < T\} \int_\tau^T \mathbb{1}\{B(t) \in \mathcal{B}(y,\varepsilon)\}\, dt\right].$$

From the strong Markov property applied at time τ, we obtain

$$G_\varepsilon(x,y) = \mathbb{E}_x\left[\mathbb{1}\{\tau < T\}\mathbb{E}_{B(\tau)} \int_0^{\tilde{T}} \mathbb{1}\{\tilde{B}(t) \in \mathcal{B}(y,\varepsilon)\}\, dt\right],$$

where the inner expectation is with respect to a transient Brownian motion $\{\tilde{B}(t)\colon 0 \leqslant t \leqslant \tilde{T}\}$ with the same stopping rule, but started in the fixed point $B(\tau)$. By the strong Markov property and since, on the event $\tau < T$, the random variable $B(\tau)$ is uniformly distributed on $\partial\mathcal{B}(x,r)$, by rotational symmetry, we conclude,

$$G_\varepsilon(x,y) = \mathbb{P}_x\{\tau < T\} \int_{\partial\mathcal{B}(x,r)} G_\varepsilon(z,y)\, d\varpi_{x,r}(z) \leqslant \int_{\partial\mathcal{B}(x,r)} G_\varepsilon(z,y)\, d\varpi_{x,r}(z).$$

This implies (3.11) and it is also easy to see that in cases (1) and (3) we have equality in (3.11), as in these cases $\tau < T$ with probability one. Focusing on these two cases for the moment, we obtain using continuity of G, for $x, y \in D$ with $|x-y| > r$,

$$\begin{aligned}
G(x,y) &= \lim_{\varepsilon \downarrow 0} \frac{G_\varepsilon(x,y)}{\mathcal{L}(\mathcal{B}(y,\epsilon))} = \lim_{\varepsilon \downarrow 0} \frac{1}{\mathcal{L}(\mathcal{B}(x,r))} \int_{\mathcal{B}(x,r)} \frac{G_\varepsilon(z,y)}{\mathcal{L}(\mathcal{B}(y,\varepsilon))}\, dz \\
&= \frac{1}{\mathcal{L}(\mathcal{B}(x,r))} \int_{\mathcal{B}(x,r)} G(z,y)\, dz,
\end{aligned}$$

where the last equality follows from the bounded convergence theorem. This proves harmonicity in cases (1) and (3). In case (2) the same argument still gives (3.11), and we can infer that $G(\,\cdot\,,y)$ is subharmonic on $\mathbb{R}^d \setminus \{y\}$ as the function is twice differentiable. ∎

Remark 3.38 Let $K \subset \mathbb{R}^d$, for $d \geqslant 3$, be a compact set and μ be any measure on K. Then

$$u(x) = \int_K G(x, y) \, d\mu(y), \qquad \text{for } x \in K^c$$

is a harmonic function on K^c. This follows as, by Fubini's theorem, the mean value property of $G(\cdot, y)$ can be carried over to u. Physically, $u(x)$ is the electrostatic (or Newtonian) potential at x resulting from a charge represented by μ. The Green function $G(\cdot, y)$ can be interpreted as the electrostatic potential induced by a unit charge in the point y. $\diamond$

3.4 The harmonic measure

We have seen in the previous section that, for any compact set $K \subset \mathbb{R}^d$, $d \geqslant 3$, and any μ on K functions of the form $u(x) = \int G(x, y) \, d\mu(y)$ are positive harmonic functions on K^c. An interesting question is whether every positive harmonic function on K^c can be represented in such a way by a suitable measure μ on ∂K. The answer can be given in terms of the harmonic measure.

Definition 3.39. Let $\{B(t) : t \geqslant 0\}$ be a d-dimensional Brownian motion, $d \geqslant 2$, started in some point x and fix a closed set $A \subset \mathbb{R}^d$. Define a measure $\mu_A(x, \cdot)$ by

$$\mu_A(x, B) = \mathbb{P}\{B(\tau) \in B, \ \tau < \infty\} \quad \text{where } \tau = \inf\{t \geqslant 0 : B(t) \in A\},$$

for $B \subset A$ Borel. In other words, $\mu_A(x, \cdot)$ is the distribution of the first hitting point of A, and the total mass of the measure is the probability that a Brownian motion started in x ever hits the set A. If $x \notin A$ the harmonic measure is supported by ∂A. $\diamond$

The following corollary is an equivalent reformulation of Theorem 3.12.

Corollary 3.40 *If the Poincaré cone condition is satisfied at every point $x \in \partial U$ on the boundary of a bounded domain U, then the solution of the Dirichlet problem with boundary condition $\varphi \colon \partial U \to \mathbb{R}$, can be written as*

$$u(x) = \int \varphi(y) \, \mu_{\partial U}(x, dy) \quad \text{for all } x \in \overline{U}.$$

Remark 3.41 Of course, the harmonicity of u does not rely on the Poincaré cone condition. In fact, by Theorem 3.8, for any compact $A \subset \mathbb{R}^d$ and Borel set $B \subset \partial A$, the function $x \mapsto \mu_A(x, B)$ is harmonic on A^c. $\diamond$

Besides its value in the discussion of the Dirichlet problem, the harmonic measure is also interesting in its own right, as it intuitively weighs the points of A according to their accessibility from x. We now show that the measures $\mu_A(x, \cdot)$ for different values of $x \in A^c$ are mutually absolutely continuous. This is a form of the famous *Harnack principle*.

Theorem 3.42 (Harnack principle) *Suppose $A \subset \mathbb{R}^d$ is compact and x, y are in the unbounded component of A^c. Then $\mu_A(x, \cdot) \ll \mu_A(y, \cdot)$.*

Proof. Given $B \subset \partial A$ Borel, by Remark 3.41, the mapping $x \mapsto \mu_A(x, B)$ is a harmonic function on A^c. If it takes the value zero for some $y \in A^c$, then y is a minimum and the maximum principle, Theorem 3.5, together with the subsequent remark, imply that $\mu_A(x, B) = 0$ for all $x \in A^c$, as required. ∎

The Harnack principle allows to formulate the following definition.

Definition 3.43. A compact set A is called **nonpolar for Brownian motion**, or simply **nonpolar**, if $\mu_A(x, A) > 0$ for one (and hence for all) $x \in A^c$. Otherwise, the set A is called **polar for Brownian motion**. ◇

We now give an explicit formula for the harmonic measures on the unit sphere $\partial \mathcal{B}(0, 1)$. Note that if $x = 0$ then the distribution of $B(\tau)$ is (by symmetry) the uniform distribution, but if x is another point it is an interesting problem to determine this distribution in terms of a probability density.

Theorem 3.44 (Poisson's formula) *Suppose that $B \subset \partial \mathcal{B}(0, 1)$ is a Borel subset of the unit sphere for $d \geqslant 2$. Let ϖ denote the uniform distribution on the unit sphere. Then, for all $x \notin \partial \mathcal{B}(0, 1)$,*

$$\mu_{\partial \mathcal{B}(0,1)}(x, B) = \int_B \frac{|1 - |x|^2|}{|x - y|^d} \, d\varpi(y).$$

Remark 3.45 The density appearing in the theorem is usually called the *Poisson kernel* and appears frequently in potential theory. ◇

Proof. We start by looking at the case $|x| < 1$. Recall that τ denotes the first hitting time of the set $\partial \mathcal{B}(0, 1)$. To prove the theorem we indeed show that for every bounded measurable $f : \mathbb{R}^d \to \mathbb{R}$ we have

$$\mathbb{E}_x[f(B(\tau))] = \int_{\partial \mathcal{B}(0,1)} \frac{1 - |x|^2}{|x - y|^d} f(y) \, d\varpi(y), \tag{3.12}$$

which on the one hand implies the formula by choosing indicator functions, on the other hand, by the monotone class theorem, see e.g. Chapter 5, (1.5), in [Du95], it suffices to show this for smooth functions f. To prove (3.12) we recall Theorem 3.12, which tells us that we just have to show that the right hand side as a function in $x \in \mathcal{B}(0, 1)$ defines a solution of the Dirichlet problem on $\mathcal{B}(0, 1)$ with boundary value f.

Straightforward (double) differentiation shows that, for every $y \in \partial \mathcal{B}(0, 1)$, the mapping

$$x \mapsto \frac{1 - |x|^2}{|x - y|^d}$$

is harmonic on $\mathcal{B}(0, 1)$. Using the characterisation of harmonic functions via the mean value property, Theorem 3.2, we get for any ball $\mathcal{B}(z,r) \subset \mathcal{B}(0,1)$,

$$\frac{1}{\sigma_{z,r}(\partial\mathcal{B}(z,r))} \int_{\partial\mathcal{B}(z,r)} \left(\int_{\partial\mathcal{B}(0,1)} \frac{1-|x|^2}{|x-y|^d} f(y)\, d\varpi(y) \right) d\sigma_{z,r}(x)$$

$$= \int_{\partial\mathcal{B}(0,1)} \left(\frac{1}{\sigma_{z,r}(\partial\mathcal{B}(z,r))} \int_{\partial\mathcal{B}(z,r)} \frac{1-|x|^2}{|x-y|^d}\, d\sigma_{z,r}(x) \right) f(y)\, d\varpi(y)$$

$$= \int_{\partial\mathcal{B}(0,1)} \frac{1-|z|^2}{|z-y|^d} f(y)\, d\varpi(y),$$

by Fubini's theorem, which implies the required harmonicity. To check the boundary condition first look at the case $f \equiv 1$, in which case we have to show that, for all $x \in \mathcal{B}(0,1)$,

$$I(x) := \int_{\partial\mathcal{B}(0,1)} \frac{1-|x|^2}{|x-y|^d}\, \varpi(dy) \equiv 1.$$

Indeed, observe that $I(0) = 1$, I is invariant under rotation and $\Delta I = 0$ on $\mathcal{B}(0,1)$, by the first part. Now let $x \in \mathcal{B}(0,1)$ with $|x| = r < 1$ and let $\tau := \inf\{t : |B(t)| > r\}$. By Theorem 3.12,

$$I(0) = \mathbb{E}_0\big[I(B(\tau))\big] = I(x),$$

using rotation invariance in the second step. Hence $I \equiv 1$, as required.

Now we show that the right hand side in the theorem can be extended continuously to all points $y \in \partial\mathcal{B}(0,1)$ by $f(y)$. We write D_0 for $\partial\mathcal{B}(0,1)$ with a δ-neighbourhood $\mathcal{B}(y,\delta)$ removed and $D_1 = \partial\mathcal{B}(0,1) \setminus D_0 = \partial\mathcal{B}(0,1) \cap \mathcal{B}(y,\delta)$. We have, using that $I \equiv 1$, for all $x \in \mathcal{B}(y,\delta/2) \cap \mathcal{B}(0,1)$,

$$\left| f(y) - \int_{\partial\mathcal{B}(0,1)} \frac{1-|x|^2}{|x-z|^d} f(z)\, d\varpi(z) \right|$$

$$= \left| \int_{\partial\mathcal{B}(0,1)} \frac{1-|x|^2}{|x-z|^d} (f(y) - f(z))\, d\varpi(z) \right|$$

$$\leqslant 2\|f\|_\infty \int_{D_0} \frac{1-|x|^2}{|x-z|^d}\, d\varpi(z) + \sup_{z \in D_1} |f(y) - f(z)|.$$

For fixed $\delta > 0$ the first term goes to 0 as $x \to y$ by dominated convergence, whereas the second can be made arbitrarily small by choice of δ. This completes the proof if $x \in \mathcal{B}(0,1)$.

If $|x| > 1$ we use inversion at the unit circle to transfer the problem to the case studied before. Indeed, it is not hard to check that a function

$$u \colon \overline{\mathcal{B}(0,1)}^c \to \mathbb{R}$$

is harmonic if and only if its inversion

$$u^* \colon \mathcal{B}(0,1) \setminus \{0\} \to \mathbb{R}, \quad u^*(x) = u\big(\tfrac{x}{|x|^2}\big)|x|^{2-d},$$

is harmonic, see Exercise 3.2. Now suppose that $f \colon \partial\mathcal{B}(0,1) \to \mathbb{R}$ is a smooth function on the boundary. Then define a harmonic function

$$u \colon \overline{\mathcal{B}(0,1)}^c \to \mathbb{R}, \quad u(x) = \mathbb{E}_x\big[f\big(B(\tau(\partial\mathcal{B}(0,1)))\big)\,\mathbb{1}\{\tau(\partial\mathcal{B}(0,1)) < \infty\}\big].$$

Then $u^* : \mathcal{B}(0, 1) \setminus \{0\} \to \mathbb{R}$ is bounded and harmonic. By Exercise 3.11 we can extend it to the origin, so that the extension is harmonic on $\mathcal{B}(0, 1)$. In fact, this extension is obviously given by $u^*(0) = \int f \, d\varpi$. The harmonic extension is continuous on the closure, with boundary values given by f. Hence it agrees with the function of the first part, and $u = u^{**}$ must be its inversion, which gives the claimed formula. ∎

We now fix a compact nonpolar set $A \subset \mathbb{R}^d$, and look at the harmonic measure $\mu_A(x, \cdot)$ when $x \to \infty$. The first task is to make sure that the limit object is well-defined.

Theorem 3.46 *Let $A \subset \mathbb{R}^d$ be a compact, nonpolar set, then there exists a probability measure μ_A on A, given by*

$$\mu_A(B) = \lim_{x \to \infty} \mathbb{P}_x\big\{ B(\tau(A)) \in B \mid \tau(A) < \infty \big\} \quad \text{for } B \subset A \text{ Borel.}$$

This measure is called the **harmonic measure** *(from infinity).*

Remark 3.47 The harmonic measure weighs the points of A according to their accessibility from infinity. It is naturally supported by the *outer boundary* of A, which is the boundary of the infinite connected component of $\mathbb{R}^d \setminus A$. ◇

The proof is prepared by a lemma, which is yet another example how the strong Markov property can be exploited to great effect.

Lemma 3.48 *For $A \subset \mathbb{R}^d$ compact and nonpolar and every $\varepsilon > 0$, there exists a large $R > 0$ such that, for all $x \in \partial \mathcal{B}(0, R)$ and any hyperplane $H \subset \mathbb{R}^d$ containing the origin,*

$$\mathbb{P}_x\big\{ \tau(A) < \tau(H) \big\} < \varepsilon \, \mathbb{P}_x\big\{ \tau(A) < \infty \big\}.$$

Proof. Fix a radius $r > 0$ such that $A \subset \mathcal{B}(0, r)$. Suppose there exists a large radius $R > r$ such that, for all $x \in \partial \mathcal{B}(0, R)$,

$$\mathbb{P}_x\big\{ \tau(\mathcal{B}(0, r)) < \tau(H) \big\} < \varepsilon \, \mathbb{P}_x\big\{ \tau(\mathcal{B}(0, r)) < \infty \big\}. \tag{3.13}$$

Then, using the strong Markov property,

$$\mathbb{P}_x\big\{ \tau(A) < \tau(H) \big\} \leqslant \mathbb{E}_x\Big[\mathbf{1}\{\tau(\mathcal{B}(0, r)) < \tau(H)\} \, \mathbb{P}_{B(\tau(\mathcal{B}(0,r)))}\{\tau(A) < \infty\} \Big].$$

Now recall from Remark 3.41 that $x \mapsto \mathbb{P}_x\{\tau(A) < \infty\}$ is harmonic on A^c. Hence the ratio of any two values of this function on the compact set $\partial \mathcal{B}(0, r)$ is bounded by a fixed constant $C > 0$, independent of $\varepsilon > 0$. Therefore, using (3.13) in the second step,

$$\mathbb{P}_x\big\{ \tau(A) < \tau(H) \big\} \leqslant C \, \mathbb{E}_x\Big[\mathbf{1}\{\tau(\mathcal{B}(0, r)) < \tau(H)\} \min_{z \in \partial \mathcal{B}(0,r)} \mathbb{P}_z\{\tau(A) < \infty\} \Big]$$

$$< \varepsilon \, C \, \mathbb{P}_x\big\{ \tau(A) < \infty \big\},$$

from which the result follows.

It remains to show (3.13). Observe that there exists an absolute constant $q < 1$ such that, for any $x \in \partial \mathcal{B}(0,2)$ and hyperplane H,

$$\mathbb{P}_x\{\tau(\mathcal{B}(0,1)) < \tau(H)\} < q\,\mathbb{P}_x\{\tau(\mathcal{B}(0,1)) < \infty\}.$$

Let k be large enough to ensure that $q^k < \varepsilon$. Then, by the strong Markov property and Brownian scaling,

$$\sup_{x \in \partial \mathcal{B}(0,r2^k)} \mathbb{P}_x\{\tau(\mathcal{B}(0,r)) < \tau(H)\}$$
$$\leqslant \sup_{x \in \partial \mathcal{B}(0,r2^k)} \mathbb{E}_x\Big[1\{\tau(\mathcal{B}(0,r2^{k-1})) < \tau(H)\}$$
$$\times \mathbb{P}_{B(\tau(\mathcal{B}(0,r2^{k-1})))}\{\tau(\mathcal{B}(0,r)) < \tau(H)\}\Big]$$
$$\leqslant q \sup_{x \in \partial \mathcal{B}(0,r2^k)} \mathbb{P}_x\{\tau(\mathcal{B}(0,r2^{k-1})) < \infty\}$$
$$\times \sup_{x \in \partial \mathcal{B}(0,r2^{k-1})} \mathbb{P}_x\{\tau(\mathcal{B}(0,r)) < \tau(H)\}.$$

Iterating this and letting $R = r2^k$ gives

$$\sup_{x \in \partial \mathcal{B}(0,R)} \mathbb{P}_x\{\tau(\mathcal{B}(0,r)) < \tau(H)\} \leqslant q^k \prod_{j=1}^{k} \sup_{x \in \partial \mathcal{B}(0,r2^j)} \mathbb{P}_x\{\tau(\mathcal{B}(0,r2^{j-1})) < \infty\}$$
$$= q^k \sup_{x \in \partial \mathcal{B}(0,R)} \mathbb{P}_x\{\tau(\mathcal{B}(0,r)) < \infty\},$$

as required to complete the proof. ∎

Proof of Theorem 3.46.   Let $x, y \in \partial \mathcal{B}(0,r)$ and H be the hyperplane through the origin, which is orthogonal to $x - y$. If $\{B(t)\colon t \geqslant 0\}$ is a Brownian motion started in x, define $\{\overline{B}(t)\colon t \geqslant 0\}$ the Brownian motion started in y, obtained by defining $\overline{B}(t)$ as the reflection of $B(t)$ at H, for all times $t \leqslant \tau(H)$, and $\overline{B}(t) = B(t)$ for all $t \geqslant \tau(H)$. This coupling gives, for every $\varepsilon > 0$ and sufficiently large r,

$$\big|\mu_A(x,B) - \mu_A(y,B)\big| \leqslant \mathbb{P}_x\{\tau(A) < \tau(H)\} \leqslant \varepsilon \mu_A(x,A),$$

using Lemma 3.48 for the last inequality. In particular, we get $|\mu_A(x,A) - \mu_A(y,A)| \leqslant \varepsilon \mu_A(x,A)$. Next, let $|z| > r$ and apply the strong Markov property to obtain

$$\frac{\mu_A(x,B)}{\mu_A(x,A)} - \frac{\mu_A(z,B)}{\mu_A(z,A)} = \int \Big(\frac{\mu_A(x,B)}{\mu_{\mathcal{B}(0,r)}(z,\mathcal{B}(0,r))\mu_A(x,A)} - \frac{\mu_A(y,B)}{\mu_A(z,A)}\Big)\mu_{\mathcal{B}(0,r)}(z,dy)$$
$$= \frac{1}{\mu_A(z,A)} \int \Big(\mu_A(x,B)\frac{\mu_A(z,A)}{\mu_{\mathcal{B}(0,r)}(z,\mathcal{B}(0,r))\mu_A(x,A)} - \mu_A(y,B)\Big)\mu_{\mathcal{B}(0,r)}(z,dy)$$
$$\leqslant \frac{1}{\mu_A(z,A)} \int \big(\mu_A(x,B)(1+\varepsilon) - \mu_A(y,B)\big)\mu_{\mathcal{B}(0,r)}(z,dy),$$

where we used that

$$\mu_A(z,A) = \int \mu_{\mathcal{B}(0,r)}(z,dy)\mu_A(y,A) \leqslant (1+\varepsilon)\mu_{\mathcal{B}(0,r)}(z,\mathcal{B}(0,r))\mu_A(x,A).$$

This leads to the estimate

$$\frac{\mu_A(x,B)}{\mu_A(x,A)} - \frac{\mu_A(z,B)}{\mu_A(z,A)} \leqslant \varepsilon \frac{\mu_A(z,B)}{\mu_A(z,A)} + \varepsilon(1+\varepsilon)\frac{\mu_A(x,A)}{\mu_A(z,A)} \leqslant \varepsilon + \varepsilon(1+\varepsilon)^2.$$

Similarly, we obtain

$$\frac{\mu_A(x,B)}{\mu_A(x,A)} - \frac{\mu_A(z,B)}{\mu_A(z,A)} \geqslant \frac{1}{\mu_A(z,A)} \int \big(\mu_A(x,B)(1-\varepsilon) - \mu_A(y,B)\big)\,\mu_{\mathcal{B}(0,r)}(z,dy),$$

and from this

$$\frac{\mu_A(x,B)}{\mu_A(x,A)} - \frac{\mu_A(z,B)}{\mu_A(z,A)} \geqslant -\varepsilon\frac{\mu_A(z,B)}{\mu_A(z,A)} - \varepsilon(1+\varepsilon)\frac{\mu_A(x,A)}{\mu_A(z,A)} \geqslant -\varepsilon - \varepsilon(1+\varepsilon)^2.$$

As $\varepsilon > 0$ was arbitrary, this implies that $\mu_A(x,B)/\mu_A(x,A)$ converges as $x \to \infty$. ∎

Example 3.49 For any ball $\mathcal{B}(x,r)$ the harmonic measure $\mu_{\mathcal{B}(x,r)}$ is equal to the uniform distribution $\varpi_{x,r}$ on $\partial\mathcal{B}(x,r)$. Indeed, note that, for all $R > r$, we have

$$\varpi_{x,r}(\cdot) = C(R) \int_{\partial\mathcal{B}(x,R)} \mu_{\mathcal{B}(x,r)}(y,\cdot)\,d\varpi_{x,R}(y),$$

where $C(R)$ is a normalizing constant, because the two balls are concentric, and both sides of the equation are rotationally invariant finite measures on the sphere $\partial\mathcal{B}(x,r)$ and hence multiples of each other. Letting $R \uparrow \infty$, we obtain from Theorem 3.46, that $\varpi_{x,r} = \mu_{\mathcal{B}(x,r)}$. ◇

The following surprising theorem shows that the harmonic measure from infinity can also be obtained without this limiting procedure.

Theorem 3.50 *Let $A \subset \mathbb{R}^d$ be a nonpolar compact set, and suppose $\mathcal{B}(x,r) \supset A$, let $\varpi_{x,r}$ be the uniform distribution on $\partial\mathcal{B}(x,r)$. Then we have, for any Borel set $B \subset A$,*

$$\mu_A(B) = \frac{\int \mu_A(a,B)\,d\varpi_{x,r}(a)}{\int \mu_A(a,A)\,d\varpi_{x,r}(a)}.$$

Remark 3.51 The surprising fact here is that the right hand side does *not depend* on the choice of the ball $\mathcal{B}(x,r)$. ◇

The crucial observation behind this result is that, starting a Brownian motion in a uniformly chosen point on the boundary of a sphere, the first hitting point of any ball inside that sphere, if it exists, is again uniformly distributed, see Figure 3.2.

Lemma 3.52 *Let $\mathcal{B}(x,r) \subset \mathcal{B}(y,s)$ and $B \subset \partial\mathcal{B}(x,r)$ Borel. Then*

$$\frac{\int \mu_{\partial\mathcal{B}(x,r)}(a,B)\,d\varpi_{y,s}(a)}{\int \mu_{\partial\mathcal{B}(x,r)}(a,\partial\mathcal{B}(x,r))\,d\varpi_{y,s}(a)} = \varpi_{x,r}(B).$$

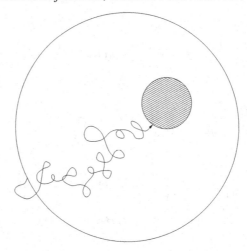

Fig. 3.2. Starting Brownian motion uniformly on the big circle, the distribution of the first hitting point on the small circle is also uniform.

Proof.   By Example 3.49 we have $\varpi_{y,s} = \mu_{\partial \mathcal{B}(y,s)}$ and hence, for the normalisation constant $c(R) := 1/\int \mu_{\partial \mathcal{B}(y,s)}(a, \partial \mathcal{B}(y,s))\, d\varpi_{x,R}(a)$, we have

$$\varpi_{y,s}(\,\cdot\,) = \lim_{R \uparrow \infty} c(R) \int \mu_{\partial \mathcal{B}(y,s)}(a, \,\cdot\,)\, d\varpi_{x,R}(a)\,.$$

Hence, for any $B \subset \partial \mathcal{B}(x,r)$ Borel, using the Markov property in the second step,

$$\int \mu_{\partial \mathcal{B}(x,r)}(a, B)\, d\varpi_{y,s}(a) = \lim_{R \uparrow \infty} c(R) \iint \mu_{\partial \mathcal{B}(x,r)}(a, B)\mu_{\partial \mathcal{B}(y,s)}(b, da)\, d\varpi_{x,R}(b)$$

$$= \lim_{R \uparrow \infty} c(R) \int \mu_{\partial \mathcal{B}(x,r)}(b, B)\, d\varpi_{x,R}(b)$$

$$= C\, \varpi_{x,r}(B)\,,$$

for a suitable constant C, because $\mathcal{B}(x, R)$ and $\mathcal{B}(x, r)$ are concentric. By substituting $B = \partial \mathcal{B}(x, r)$ into the equation, we see that the constant must be as claimed in the statement. ∎

Proof of Theorem 3.50.   Assume that $\mathcal{B}(x, r)$ and $\mathcal{B}(y, s)$ are two balls containing A. We may then find a ball $\mathcal{B}(z, t)$ containing both these balls. Using Lemma 3.52 and the strong Markov property applied to the first hitting of $\mathcal{B}(x, r)$ we obtain, for any $B \subset A$,

$$\int \mu_A(a, B)\, d\varpi_{x,r}(a) = c_1 \int\int \mu_A(a, B)\mu_{\mathcal{B}(x,r)}(b, da)\, d\varpi_{z,t}(b)$$

$$= c_1 \int \mu_A(b, B)\, d\varpi_{z,t}(b) = c_1 \int\int \mu_A(a, B)\mu_{\mathcal{B}(y,s)}(b, da)\, d\varpi_{z,t}(b)$$

$$= c_2 \int \mu_A(a, B)\, d\varpi_{y,s}(a),$$

for suitable constants c_1, c_2 depending only on the choice of the balls. Choosing $B = A$

gives the normalisation constant

$$c_2 = \frac{\int \mu_A(a, A)\, d\varpi_{x,r}(a)}{\int \mu_A(a, A)\, d\varpi_{y,s}(a)},$$

and this shows that the right hand side in Theorem 3.50 is independent of the choice of the enclosing ball. Hence it must stay constant as $r \to \infty$, which completes the proof. ∎

Exercises

Exercise 3.1. Show that, if $u \colon U \to \mathbb{R}$ is subharmonic, then

$$u(x) \leqslant \frac{1}{\mathcal{L}(\mathcal{B}(x, r))} \int_{\mathcal{B}(x,r)} u(y)\, dy \qquad \text{for any ball } \mathcal{B}(x, r) \subset U.$$

Conversely, show that any twice differentiable function $u \colon U \to \mathbb{R}$ satisfying (3.3) is subharmonic. Also give an example of a discontinuous function u satisfying (3.3).

Exercise 3.2. Let $d \geqslant 2$. Show that a function $u \colon \overline{\mathcal{B}(0,1)}^c \to \mathbb{R}$ is harmonic if and only if its inversion

$$u^* \colon \mathcal{B}(0,1) \setminus \{0\} \to \mathbb{R}, \quad u^*(x) = u\big(\tfrac{x}{|x|^2}\big)|x|^{2-d}$$

is harmonic.

Exercise 3.3. § Suppose $u \colon \mathcal{B}(x, r) \to \mathbb{R}$ is harmonic and bounded by M. Show that the kth order partial derivatives are bounded by a constant multiple of Mr^{-k}.

Exercise 3.4. Prove the case $d = 1$ in Theorem 3.20.

Exercise 3.5. § Prove the strong form of the *Paley–Zygmund inequality*:
For any nonnegative random variable X with $\mathbb{E}[X^2] < \infty$ and $\lambda \in [0, 1)$,

$$\mathbb{P}\{X > \lambda\, \mathbb{E}[X]\} \geqslant (1 - \lambda)^2\, \frac{\mathbb{E}[X]^2}{\mathbb{E}[X^2]}.$$

Exercise 3.6. Prove the *Kochen–Stone lemma*: Suppose $E_1, E_2, \ldots$ are events with

$$\sum_{n=1}^{\infty} \mathbb{P}(E_n) = \infty \qquad \text{and} \qquad \liminf_{k \to \infty} \frac{\sum_{m=1}^{k} \sum_{n=1}^{k} \mathbb{P}(E_n \cap E_m)}{\left(\sum_{n=1}^{k} \mathbb{P}(E_n)\right)^2} < \infty.$$

Then, with positive probability, infinitely many of the events take place.
Hint. Apply the Paley–Zygmund inequality to $X = \liminf_{n \to \infty} 1_{E_n}$.

Exercise 3.7. [§] Suppose that u is a radial harmonic function on the annulus

$$D = \{x \in \mathbb{R}^d : r < |x| < R\},$$

where radial means $u(x) = \tilde{u}(|x|)$ for some function $\tilde{u} \colon (r, R) \to \mathbb{R}$ and all x. Suppose further that u is continuous on $\bar{D}$. Show that,

- if $d \geqslant 3$, there exist constants a and b such that $u(x) = a + b|x|^{2-d}$;
- if $d = 2$, there exist constants a and b such that $u(x) = a + b \log |x|$.

Exercise 3.8. [§] Show that any positive harmonic function on $\mathbb{R}^d$ is constant.

Exercise 3.9. Let H be a hyperplane in $\mathbb{R}^d$ and let $\{B(t) \colon t \geqslant 0\}$ be a d-dimensional Brownian motion. For $z \in \mathbb{R}^d$, show that

$$\sup_{t>0} \mathbb{E}_z \big[|B(t)| \, 1\{t < \tau(H)\} \big] < \infty.$$

Hint. We may assume that H is the hyperplane $\{x_1 = 0\}$ and $z_1 > 0$. Bound the ℓ_2-norm by the ℓ_1-norm. If $B(t) = (B_1(t), ..., B_d(t))$, the estimate for $\mathbb{E}_z[|B_j(t)|1\{t < \tau(H)\}]$ when $j > 1$ follows from the tails of $\tau(H)$. The estimate for B_1 reduces to the one-dimensional setting, where the reflection principle yields the density of $B(t)1\{t < \tau(0)\}$.

Exercise 3.10. Let u be a harmonic function on $\mathbb{R}^d$ such that $\frac{|u(x)|}{|x|} \to 0$ as $x \to \infty$. Show that u is constant.

Hint. Follow the proof of Theorem 3.16, and use Exercise 3.9.

Exercise 3.11. [§] Let $D \subset \mathbb{R}^d$ be a domain and $x \in D$. Suppose $u \colon D \setminus \{x\} \to \mathbb{R}$ is bounded and harmonic. Show that there exists a unique harmonic continuation $u \colon D \to \mathbb{R}$.

Exercise 3.12. Let $f \colon (0,1) \to (0,\infty)$ with $t \mapsto f(t)/t$ decreasing. Then

$$\int_0^1 f(r)^{d-2} r^{-d/2} \, dr < \infty \qquad \text{if and only if} \qquad \liminf_{t \downarrow 0} \frac{|B(t)|}{f(t)} = \infty \text{ almost surely.}$$

Conversely, if the integral diverges, then $\liminf_{t \downarrow 0} |B(t)|/f(t) = 0$ almost surely.

Exercise 3.13. Show that, if $d \geqslant 3$ and T is an independent exponential time with parameter $\lambda > 0$, then

$$G(x,y) \sim c(d) \, |x-y|^{2-d} \qquad \text{for } |x-y| \downarrow 0,$$

where $c(d)$ is as in Theorem 3.33.

Exercise 3.14. [§] Show that if D is a bounded domain, then the Green's function

$$G \colon (D \times D) \setminus \Delta$$

is continuous, where $\Delta = \{(x,x) \colon x \in D\}$ is the diagonal.

Exercise 3.15. § Find the Green's function for the planar Brownian motion stopped when leaving the domain $\mathcal{B}(0, R)$.

Exercise 3.16. § Suppose $x, y \notin \overline{\mathcal{B}(0, r)}$ and $A \subset \mathcal{B}(0, r)$ is a compact, nonpolar set. Show that $\mu_A(x, \cdot)$ and $\mu_A(y, \cdot)$ are mutually absolutely continuous with a density bounded away from zero and infinity.

Exercise 3.17. § Suppose $K \subset \mathbb{R}^2$ is a compact set. The *Kallianpur–Robbins law* states that, for a standard planar Brownian motion $\{B_t : t \geqslant 0\}$,

$$\frac{\int_0^t 1_K(B_t)\, dt}{\log t} \xrightarrow{\text{d}} X, \qquad \text{as } t \uparrow \infty,$$

where X has an exponential distribution with mean $\frac{\mathcal{L}(K)}{2\pi}$.

(a) Fix radii $0 < r_1 < r_2$ and define stopping times $\tau_0 = 0$ and

$$\tau_{2k+i} = \inf\left\{t \geqslant \tau_{2k+i-1} : |B(t)| = r_i\right\} \quad \text{for integers } k \geqslant 0 \text{ and } i \in \{1, 2\}.$$

For any $R > r_2$ denote

$$N(R) = \sup\{k \in \mathbb{N} : \sup_{0 \leqslant t \leqslant \tau_{2k}} |B(t)| < R\}.$$

Show that

$$\frac{N(R)}{\log R} \xrightarrow{\text{d}} Y \qquad \text{as } R \uparrow \infty,$$

where Y has an exponential distribution with parameter $\log(r_2/r_1)$.

(b) Show that, for a Brownian motion $\{B(t) : t \geqslant 0\}$ started uniformly on $\partial\mathcal{B}(0, r_1)$ and stopped at the first time τ when they reach $\partial\mathcal{B}(0, r_2)$ we have

$$\mathbb{E} \int_0^\tau 1_K\big(B(s)\big)\, ds = \log\left(\tfrac{r_2}{r_1}\right) \tfrac{\mathcal{L}(K)}{\pi}.$$

(c) Use (a), (b) and the law of large numbers to show that, for $K = \mathcal{B}(0, 1)$,

$$\frac{\int_0^{T(R)} 1_K(B_t)\, dt}{\log R} \xrightarrow{\text{d}} X, \qquad \text{as } R \uparrow \infty,$$

where X has an exponential distribution with mean $\frac{\mathcal{L}(K)}{\pi}$.

(d) Use (c) to prove the Kallianpur–Robbins law in the case $K = \mathcal{B}(0, 1)$.

A modification of this technique can also be used to prove the Kallianpur–Robbins law for arbitrary compact sets K. If you want to try, see for example Section 3 in [Mö00] for a good hint.

Notes and comments

Gauss discusses the Dirichlet problem in [Ga40] in a paper on electrostatics. Examples which show that a solution may not exist for certain domains were given by Zaremba [Za11] and Lebesgue [Le24]. Zaremba's example is the punctured disc we discuss in Example 3.15, and Lebesgue's example is the thorn, which we will discuss in Example 8.40. For domains with smooth boundary the problem was solved by Poincaré [Po90]. The Dirichlet problem will be revisited in Chapter 8.

Bachelier [Ba00, Ba01] was the first to note a connection of Brownian motion and the Laplace operator. The first probabilistic approaches to the Dirichlet problem were made by Phillips and Wiener [PW23] and Courant, Friedrichs and Lewy [CFL28]. These proofs used probability in a discrete setting and approximation. The treatment of the Dirichlet problem using Brownian motion and the probabilistic definition of the harmonic measure are due to the pioneering work of Kakutani [Ka44a, Ka44b, Ka45]. Further relationships between Brownian motion and partial differential equations are the subject of the Feynman–Kac formulas explored later in this book, see Section 7.7.4, and can also be found in Durrett [Du84]. A current survey of probabilistic methods in analysis can be found in the book of Bass [Ba95], see also Rao [Ra77], Port and Stone [PS78] or Doob [Do84] for classical references.

Pólya [Po21] discovered that a simple symmetric random walk on $\mathbb{Z}^d$ is recurrent for $d \leqslant 2$ and transient otherwise. His result was later extended to Brownian motion by Lévy [Le40] and Kakutani [Ka44a]. Neighbourhood recurrence implies, in particular, that the path of a planar Brownian motion (running for an infinite amount of time) is dense in the plane. A more subtle question is whether in $d \geqslant 3$ all orthogonal projections of a d-dimensional Brownian motion are neighbourhood recurrent, or equivalently whether there is an infinite cylinder avoided by its range. In fact, an avoided cylinder does exist almost surely. This result is due to Adelman, Burdzy and Pemantle [ABP98]. The Dvoretzky–Erdős test is originally from [DE51] and more information and additional references can be found in Pruitt [Pr90]. There is also an analogous result for planar Brownian motion (with shrinking balls) which is due to Spitzer [Sp58].

Green introduced the function named after him in [Gr28]. Its probabilistic interpretation appears in Kac's paper [Ka51] and is investigated thoroughly by Hunt [Hu56]. Quite a lot can be said about the transition densities: $\mathfrak{p}^*(t, \cdot, \cdot)$ is jointly continuous on $\overline{D} \times \overline{D}$ and symmetric in the space variables. Moreover, $\mathfrak{p}^*(t, x, y)$ vanishes if either x or y is on the boundary of D, if this boundary is sufficiently regular. This is, of course, only difficult in case (3) and full proofs for this case can be found in Bass [Ba95] or in the classical book of Port and Stone [PS78].

Poisson's formula for the harmonic measure on a sphere is named after the French mathematician Siméon-Denis Poisson. The function u^* defined by inversion on a sphere, which we used in the proof, is also known as Kelvin transform of u, see also II.1 in Bass [Ba95].

The Kallianpur-Robbins law, first proved by Kallianpur and Robbins in [KR53], gives the limiting distribution of the scaled occupation times of recurrent Brownian motions. Exercise 3.17 gives the two-dimensional case, in which the limiting distribution is exponential, in the one-dimensional case the limiting distribution is a one-sided normal distribution. A substantial extension of this law was given by Darling and Kac in [DK57]. This leads to the study of additive functionals of Brownian motion, see Chapter X in [RY94]. A study of almost-sure Kallianpur-Robbins laws can be found in [Mö00].

4

Hausdorff dimension: Techniques and applications

Dimensions are a tool to measure the size of mathematical objects on a crude scale. For example, in classical geometry one can use dimension to see that a line segment (a one-dimensional object) is smaller than the surface of a ball (a two-dimensional object), but there is no difference between line-segments of different lengths. It may therefore come as a surprise that dimension is able to distinguish the size of so many objects in probability theory. In this chapter we first introduce a suitably general notion of dimension, the Hausdorff dimension. We then describe general techniques to calculate the Hausdorff dimension of arbitrary subsets of $\mathbb{R}^d$, and apply these techniques to the graph and zero set of Brownian motion in dimension one, and to the range of higher dimensional Brownian motion. Lots of further examples will follow in subsequent chapters.

4.1 Minkowski and Hausdorff dimension

4.1.1 The Minkowski dimension

How can we capture the dimension of a geometric object? One requirement for a useful definition of dimension is that it should be *intrinsic*. This means that it should be independent of an embedding of the object in an ambient space like $\mathbb{R}^d$. Intrinsic notions of dimension can be defined in arbitrary metric spaces.

Suppose E is a bounded metric space with metric ρ. Here bounded means that the diameter $|E| = \sup\{\rho(x,y) : x,y \in E\}$ of E is finite. The example we have in mind is a bounded subset of $\mathbb{R}^d$. The definition of Minkowski dimension is based on the notion of a *covering* of the metric space E. A **covering** of E is a finite or countable collection of sets

$$E_1, E_2, E_3, \ldots \text{ with } E \subset \bigcup_{i=1}^{\infty} E_i.$$

Define, for $\varepsilon > 0$,

$$M(E, \varepsilon) = \min \Big\{ k \geqslant 1: \text{ there exists a finite covering}$$

$$E_1, \ldots, E_k \text{ of } E \text{ with } |E_i| \leqslant \varepsilon \text{ for } i = 1, \ldots, k \Big\}, \tag{4.1}$$

where $|A|$ is the diameter of a set $A \subset E$. Intuitively, when E has dimension s the number $M(E, \varepsilon)$ should be of order ε^{-s}. This can be verified in simple cases like line segments, planar squares, etc. This intuition motivates the definition of *Minkowski dimension*.

Definition 4.1. For a bounded metric space E we define the **lower Minkowski dimension** as

$$\underline{\dim}_M E := \liminf_{\varepsilon \downarrow 0} \frac{\log M(E, \varepsilon)}{\log(1/\varepsilon)},$$

and the **upper Minkowski dimension** as

$$\overline{\dim}_M E := \limsup_{\varepsilon \downarrow 0} \frac{\log M(E, \varepsilon)}{\log(1/\varepsilon)}.$$

We always have $\underline{\dim}_M E \leqslant \overline{\dim}_M E$, but equality need not hold. If it holds we write

$$\dim_M E := \underline{\dim}_M E = \overline{\dim}_M E. \qquad \diamond$$

Remark 4.2 If E is a subset of the unit cube $[0, 1]^d \subset \mathbb{R}^d$ then let

$$\tilde{M}_n(E) = \#\{Q \in \mathfrak{D}_n : Q \cap E \neq \emptyset\}$$

be the number of dyadic cubes of side length 2^{-n} which hit E. Then there exists a constant $C(d) > 0$, not depending on E, such that

$$\tilde{M}_n(E) \geqslant M(E, \sqrt{d}\, 2^{-n}) \geqslant C(d)\, \tilde{M}_n(E).$$

Hence

$$\overline{\dim}_M E := \limsup_{n \uparrow \infty} \frac{\log \tilde{M}_n(E)}{n \log 2} \quad \text{and} \quad \underline{\dim}_M E := \liminf_{n \uparrow \infty} \frac{\log \tilde{M}_n(E)}{n \log 2}. \qquad \diamond$$

Example 4.3 In Exercise 4.1, we calculate the Minkowski dimension of a deterministic 'fractal', the (ternary) Cantor set,

$$C = \Big\{ \sum_{i=1}^{\infty} \frac{x_i}{3^i} : x_i \in \{0, 2\} \Big\} \subset [0, 1].$$

This set is obtained from the unit interval $[0, 1]$ by first removing the middle third, and then successively the middle third out of each remaining interval ad infinitum, see Figure 4.1 for the first three stages of the construction. $\qquad \diamond$

Remark 4.4 There is an unpleasant limitation of Minkowski dimension: Observe that singletons $S = \{x\}$ have Minkowski dimension 0, but we shall see in Exercise 4.2 that the set

$$E := \big\{ \tfrac{1}{n} : n \in \mathbb{N} \big\} \cup \{0\}$$

has positive dimension. Hence the Minkowski dimension does not have the **countable stability property**

$$\dim \bigcup_{k=1}^{\infty} E_k = \sup \big\{ \dim E_k : k \geqslant 1 \big\}.$$

This is one of the properties we expect from a reasonable concept of dimension. There are two ways out of this problem.

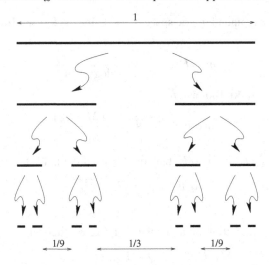

Fig. 4.1. The ternary Cantor set is obtained by removing the middle third from each interval. The figure shows the first three steps of the infinite procedure.

(i) One can use a notion of dimension taking variations of the size in the different sets in a covering into account. This captures finer details of the set and leads to the notion of *Hausdorff dimension*.

(ii) One can enforce the countable stability property by subdividing every set in countably many bounded pieces and taking the maximal dimension of them. The infimum over the numbers such obtained leads to the notion of *packing dimension*.

We follow the first route now, but come back to the second route later in the book. ◇

4.1.2 The Hausdorff dimension

The Hausdorff dimension and Hausdorff measure were introduced by Felix Hausdorff in 1919. Like the Minkowski dimension, Hausdorff dimension can be based on the notion of a covering of the metric space E. For the definition of the Minkowski dimension we have evaluated coverings crudely by counting the number of sets in the covering. Now we also allow infinite coverings and take the size of the covering sets, measured by their diameter, into account.

Looking back at the example of Exercise 4.2 one can see that the set $E = \{1/n : n \geqslant 1\} \cup \{0\}$ can be covered much more effectively, if we decrease the size of the balls as we move from right to left. In this example there is a big difference between evaluations of the covering which take into account that we use small sets in the covering, and the evaluation based on just counting the number of sets used to cover.

A very useful evaluation is the α-**value** of a covering. For every $\alpha \geqslant 0$ and covering $E_1, E_2, \ldots$ we say that the α-**value** of the covering is

$$\sum_{i=1}^{\infty} |E_i|^{\alpha}.$$

The terminology of the α-values of a covering allows to formulate a concept of dimension, which is sensitive to the effect that the fine features of this set occur in different scales at different places.

Definition 4.5. For every $\alpha \geqslant 0$ the α-**Hausdorff content** of a metric space E is defined as

$$\mathcal{H}_\infty^\alpha(E) = \inf \left\{ \sum_{i=1}^\infty |E_i|^\alpha : E_1, E_2, \ldots \text{ is a covering of } E \right\},$$

informally speaking the α-value of the most efficient covering. If $0 \leqslant \alpha \leqslant \beta$, and $\mathcal{H}_\infty^\alpha(E) = 0$, then also $\mathcal{H}_\infty^\beta(E) = 0$. Thus we can define

$$\dim E = \inf \left\{ \alpha \geqslant 0 : \mathcal{H}_\infty^\alpha(E) = 0 \right\} = \sup \left\{ \alpha \geqslant 0 : \mathcal{H}_\infty^\alpha(E) > 0 \right\},$$

the **Hausdorff dimension** of the set E. ◇

Remark 4.6 The Hausdorff dimension may, of course, be infinite. But it is easy to see that subsets of $\mathbb{R}^d$ have Hausdorff dimension no larger than d. Moreover, in Exercise 4.3 we show that for every bounded metric space, the Hausdorff dimension is bounded from above by the lower Minkowski dimension. Finally, in Exercise 4.4 we check that Hausdorff dimension has the countable stability property. ◇

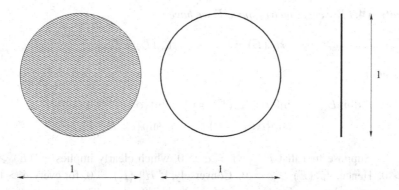

Fig. 4.2. The ball, sphere and line segment pictured here all have 1-Hausdorff content equal to one.

The concept of the α-Hausdorff content plays an important part in the definition of the Hausdorff dimension. However, it does not help distinguish the size of sets of the same dimension. For example, the three sets sketched in Figure 4.2 all have the same 1-Hausdorff content: the ball and the sphere on the left can be covered by a ball of diameter one, so that their 1-Hausdorff content is at most one, but the line segment on the right also does not permit a more effective covering and its 1-Hausdorff content is also 1. Therefore, one considers a refined concept, the *Hausdorff measure*. Here the idea is to consider only coverings by *small* sets.

Definition 4.7. Let X be a metric space and $E \subset X$. For every $\alpha \geqslant 0$ and $\delta > 0$ define

$$\mathcal{H}_\delta^\alpha(E) = \inf \left\{ \sum_{i=1}^{\infty} |E_i|^\alpha : E_1, E_2, E_3, \ldots \text{ cover } E, \text{ and } |E_i| \leqslant \delta \right\},$$

i.e. we are considering coverings of E by sets of diameter no more than δ. Then

$$\mathcal{H}^\alpha(E) = \sup_{\delta > 0} \mathcal{H}_\delta^\alpha(E) = \lim_{\delta \downarrow 0} \mathcal{H}_\delta^\alpha(E)$$

is the α-**Hausdorff measure** of the set E.   ◇

Remark 4.8 The α-Hausdorff measure has two obvious properties which, together with $\mathcal{H}^\alpha(\emptyset) = 0$, make it an *outer measure*. These are *countable subadditivity*,

$$\mathcal{H}^\alpha \left(\bigcup_{i=1}^{\infty} E_i \right) \leqslant \sum_{i=1}^{\infty} \mathcal{H}^\alpha(E_i), \qquad \text{for any sequence } E_1, E_2, E_3, \ldots \subset X,$$

and *monotonicity*,

$$\mathcal{H}^\alpha(E) \leqslant \mathcal{H}^\alpha(D), \qquad \text{if } E \subset D \subset X.$$ ◇

One can express the Hausdorff dimension in terms of the Hausdorff measure.

Proposition 4.9 *For every metric space E we have*

$$\mathcal{H}^\alpha(E) = 0 \quad \Leftrightarrow \quad \mathcal{H}_\infty^\alpha(E) = 0$$

and therefore

$$\begin{aligned}
\dim E &= \inf\{\alpha : \mathcal{H}^\alpha(E) = 0\} = \inf\{\alpha : \mathcal{H}^\alpha(E) < \infty\} \\
&= \sup\{\alpha : \mathcal{H}^\alpha(E) > 0\} = \sup\{\alpha : \mathcal{H}^\alpha(E) = \infty\}.
\end{aligned}$$

Proof.   Suppose first that $\mathcal{H}_\infty^\alpha(E) = c > 0$, which clearly implies $\mathcal{H}_\delta^\alpha(E) \geqslant c$ for all $\delta > 0$. Hence, $\mathcal{H}^\alpha(E) \geqslant c > 0$. Conversely, if $\mathcal{H}_\infty^\alpha(E) = 0$, for every $\delta > 0$ there exists a covering by sets $E_1, E_2, \ldots$ with $\sum_{k=1}^{\infty} |E_k|^\alpha < \delta$. These sets have diameter less than $\delta^{1/\alpha}$, hence $\mathcal{H}_{\delta^{1/\alpha}}^\alpha(E) < \delta$ and letting $\delta \downarrow 0$ yields $\mathcal{H}^\alpha(E) = 0$, proving the claimed equivalence. The equivalence readily implies that $\dim E = \inf\{\alpha : \mathcal{H}^\alpha(E) = 0\} = \sup\{\alpha : \mathcal{H}^\alpha(E) > 0\}$.

To verify the alternative representations it suffices to show that $\mathcal{H}^\alpha(E) < \infty$ implies $\mathcal{H}^\beta(E) = 0$ for all $\beta > \alpha$. So suppose $\mathcal{H}^\alpha(E) = C < \infty$. Given $\delta > 0$ there exists a covering by sets $E_1, E_2, \ldots$ with diameter less than δ and α-value not more than $C + 1$, whence $\mathcal{H}_\delta^\alpha(E) \leqslant C + 1$. Note that $\mathcal{H}_\delta^\beta(E) \leqslant \delta^{\beta-\alpha} \mathcal{H}_\delta^\alpha(E) \leqslant \delta^{\beta-\alpha}(C + 1)$. Letting $\delta \downarrow 0$ implies $\mathcal{H}^\beta(E) = 0$.   ∎

Remark 4.10 As Lipschitz maps increase the diameter of sets by at most a constant, the image of any set $A \subset E$ under a Lipschitz map has at most the Hausdorff dimension of A. This observation is particularly useful for projections. ◇

A natural generalisation of the last remark arises when we look at the effect of Hölder continuous maps on the Hausdorff dimension.

Definition 4.11. Let $0 < \alpha \leqslant 1$. A function $f \colon (E_1, \rho_1) \to (E_2, \rho_2)$ between metric spaces is called α-Hölder continuous if there exists a (global) constant $C > 0$ such that

$$\rho_2\big(f(x), f(y)\big) \leqslant C \, \rho_1 \big(x, y\big)^\alpha \qquad \text{for all } x, y \in E_1 \, .$$

A constant C as above is sometimes called **Hölder constant**. ◇

Remark 4.12 Hölder continuous maps allow some control on the Hausdorff measure of images: We show in Exercise 4.6 that, if $f \colon (E_1, \rho_1) \to (E_2, \rho_2)$ is surjective and α-Hölder continuous with constant C, then for any $\beta \geqslant 0$,

$$\mathcal{H}^\beta (E_2) \leqslant C^\beta \, \mathcal{H}^{\alpha\beta} (E_1),$$

and therefore $\dim(E_2) \leqslant \frac{1}{\alpha} \dim(E_1)$. ◇

4.1.3 Upper bounds on the Hausdorff dimension

We now give general upper bounds for the dimension of graph and range of functions, which are based on Hölder continuity.

Definition 4.13. For a function $f \colon A \to \mathbb{R}^d$, for $A \subset [0, \infty)$, we define the **graph** to be

$$\mathsf{Graph}_f (A) = \big\{ (t, f(t)) \colon t \in A \big\} \subset \mathbb{R}^{d+1},$$

and the **range** or **path** to be

$$\mathsf{Range}_f (A) = f(A) = \big\{ f(t) \colon t \in A \big\} \subset \mathbb{R}^d. \qquad ◇$$

Proposition 4.14 *Suppose $f \colon [0, 1] \to \mathbb{R}^d$ is an α-Hölder continuous function. Then*

(a) $\dim \big(\mathsf{Graph}_f [0, 1] \big) \leqslant 1 + (1 - \alpha) \big(d \wedge \frac{1}{\alpha} \big)$,

(b) *and, for any $A \subset [0, 1]$, we have* $\dim \mathsf{Range}_f (A) \leqslant \frac{\dim A}{\alpha}$.

Proof. Since f is α-Hölder continuous, there exists a constant C such that, if $s, t \in [0, 1]$ with $|t - s| \leqslant \varepsilon$, then $|f(t) - f(s)| \leqslant C\varepsilon^\alpha$. Cover $[0, 1]$ by no more than $\lceil 1/\varepsilon \rceil$ intervals of length ε. The image of each such interval is then contained in a ball of diameter $2C\varepsilon^\alpha$. One can now

- *either* cover each such ball by no more than a constant multiple of $\varepsilon^{d\alpha - d}$ balls of diameter ε,

- *or* use the fact that subintervals of length $(\varepsilon/C)^{1/\alpha}$ in the domain are mapped into balls of diameter ε to cover the image inside the ball by a constant multiple of $\varepsilon^{1-1/\alpha}$ balls of radius ε.

In both cases, look at the cover of the graph consisting of the product of intervals and corresponding balls of diameter ε. The first construction needs a constant multiple of $\varepsilon^{d\alpha-d-1}$ product sets, the second uses $\varepsilon^{-1/\alpha}$ product sets, all of which have diameter of order ε. This gives the upper bounds for (a), while (b) follows from Remark 4.12.   ∎

Remark 4.15 By countable stability of Hausdorff dimension, the statements of Proposition 4.14 remain true if $f\colon [0,\infty) \to \mathbb{R}^d$ is only *locally* α-Hölder continuous.   ◇

We now take a first look at dimensional properties of Brownian motion and harvest the results from our general discussion so far. We have shown in Corollary 1.20 that linear Brownian motion is everywhere locally α-Hölder continuous for any $\alpha < 1/2$, almost surely. This extends obviously to d-dimensional Brownian motion, and this allows us to get an upper bound on the Hausdorff dimension of its range and graph. For convenience, when referring to Brownian motion, we drop the reference to the function in the subindex of $\mathsf{Graph}_f(A)$ and $\mathsf{Range}_f(A)$.

Corollary 4.16 *For any fixed set $A \subset [0,\infty)$ the graph of a d-dimensional Brownian motion satisfies, almost surely,*

$$\dim\left(\mathsf{Graph}(A)\right) \leqslant \begin{cases} 3/2 & \text{if } d = 1, \\ 2 & \text{if } d \geqslant 2\,. \end{cases}$$

and its range satisfies, almost surely,

$$\dim \mathsf{Range}(A) \leqslant (2 \dim A) \wedge d.$$

Remark 4.17 The corresponding *lower bounds* for the Hausdorff dimension of $\mathsf{Graph}(A)$ and $\mathsf{Range}(A)$ are more subtle and will be discussed in Section 4.4.3, when we have more sophisticated tools at our disposal. Our upper bounds also hold for the Minkowski dimension, see Exercise 4.7, and corresponding lower bounds are easier than in the Hausdorff case and obtainable at this stage, see Exercise 4.10.   ◇

Corollary 4.16 does not make any statement about the 2-Hausdorff measure of the range, and any such statement requires more information than the Hölder exponent alone can provide, see for example Exercise 4.9. It is however not difficult to show that, for $d \geqslant 2$,

$$\mathcal{H}^2\big(B([0,1])\big) < \infty \qquad \text{almost surely.} \tag{4.2}$$

Indeed, for any $n \in \mathbb{N}$, we look at the covering of $B([0,1])$ by the closure of the balls

$$\mathcal{B}\Big(B(\tfrac{k}{n}), \max_{\frac{k}{n} \leqslant t \leqslant \frac{k+1}{n}} \big|B(t) - B(\tfrac{k}{n})\big|\Big), \quad k \in \{0,\dots,n-1\}.$$

By the uniform continuity of Brownian motion on the unit interval, the maximal diameter in these coverings goes to zero, as $n \to \infty$. Moreover, we have

$$\mathbb{E}\left[\left(\max_{\frac{k}{n} \leqslant t \leqslant \frac{k+1}{n}} |B(t) - B(\tfrac{k}{n})|\right)^2\right] = \mathbb{E}\left[\left(\max_{0 \leqslant t \leqslant \frac{1}{n}} |B(t)|\right)^2\right] = \frac{1}{n}\,\mathbb{E}\left[\left(\max_{0 \leqslant t \leqslant 1} |B(t)|\right)^2\right],$$

using Brownian scaling. The expectation on the right is finite by Proposition 2.43. Hence the expected 2-value of the nth covering is bounded from above by

$$4\mathbb{E}\left[\sum_{k=0}^{n-1} \left(\max_{\frac{k}{n} \leqslant t \leqslant \frac{k+1}{n}} |B(t) - B(\tfrac{k}{n})|\right)^2\right] = 4\,\mathbb{E}\left[\left(\max_{0 \leqslant t \leqslant 1} |B(t)|\right)^2\right],$$

which implies, by Fatou's lemma, that

$$\mathbb{E}\left[\liminf_{n \to \infty} 4 \sum_{k=0}^{n-1} \left(\max_{\frac{k}{n} \leqslant t \leqslant \frac{k+1}{n}} |B(t) - B(\tfrac{k}{n})|\right)^2\right] < \infty.$$

Hence the liminf is almost surely finite, which proves (4.2).

The next theorem improves upon (4.2) by showing that the 2-dimensional Hausdorff measure of the range of d-dimensional Brownian motion is zero for any $d \geqslant 2$. The proof is considerably more involved and may be skipped on first reading. It makes use of the fact that we have a 'natural' measure on the range at our disposal, which we can use as a tool to pick a good cover by cubes. The idea of using a natural measure supported by the 'fractal' for comparison purposes will also turn out to be crucial for the lower bounds for Hausdorff dimension, which we discuss in the next section.

Theorem* 4.18 *Let $\{B(t)\colon t \geqslant 0\}$ be a Brownian motion in dimension $d \geqslant 2$. Then, almost surely, for any set $A \subset [0, \infty)$ we have*

$$\mathcal{H}^2(\mathrm{Range}(A)) = 0.$$

Proof. It is sufficient to show that $\mathcal{H}^2(\mathrm{Range}[0, \infty)) = 0$ for $d \geqslant 3$, as 2-dimensional Brownian motion is the projection of 3-dimensional Brownian motion, and projections cannot increase the Hausdorff measure. Moreover it suffices to prove $\mathcal{H}^2(\mathrm{Range}[0, \infty) \cap \mathrm{Cube}) = 0$ almost surely, for any half-open cube $\mathrm{Cube} \subset \mathbb{R}^d$ of side length one at positive distance from the starting point of the Brownian motion. Without loss of generality we may assume that this cube is the unit cube $\mathrm{Cube} = [0, 1)^d$, and our Brownian motion is started at some $x \notin \mathrm{Cube}$.

Let $d \geqslant 3$, and recall the definition of the (locally finite) occupation measure μ, defined by

$$\mu(A) = \int_0^\infty \mathbf{1}_A(B(s))\,ds, \qquad \text{for } A \subset \mathbb{R}^d \text{ Borel}.$$

Let $\mathfrak{D}_k$ be the collection of all cubes $\prod_{i=1}^d [n_i 2^{-k}, (n_i + 1)2^{-k})$ where $n_1, \ldots, n_d \in \{0, \ldots, 2^k - 1\}$. We fix a threshold $m \in \mathbb{N}$ and let $M > m$. We call $D \in \mathfrak{D}_k$ with $k \geqslant m$ a *big* cube if

$$\mu(D) \geqslant \frac{1}{\varepsilon}\, 2^{-2k}.$$

The collection $\mathfrak{E}(M)$ consists of all maximal big cubes $D \in \mathfrak{D}_k$, $m \leqslant k < M$, i.e. all those which are not contained in another big cube, together with all cubes $D \in \mathfrak{D}_M$ which are not contained in a big cube, but intersect $\mathsf{Range}[0, \infty)$. Obviously $\mathfrak{E}(M)$ is a cover of $\mathsf{Range}[0, \infty) \cap \mathsf{Cube}$ by sets of diameter smaller than $\sqrt{d}2^{-m}$.

To find the expected 2-dimensional Hausdorff content of this cover, first look at a cube $D \in \mathfrak{D}_M$. We denote by $D = D_M \subset D_{M-1} \subset \cdots \subset D_m$ with $D_k \in \mathfrak{D}_k$ the ascending sequence of cubes containing D. Let D_k^* be the cube with the same centre as D_k and $3/2$ its side length, see Figure 4.3.

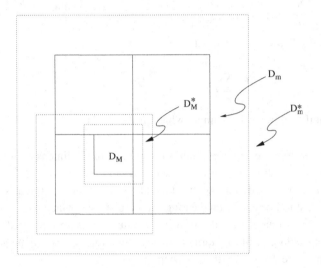

Fig. 4.3. Nested systems of cubes, cubes D_k^* indicated by dashed, D_k by solid boundaries.

Let $\tau(D)$ be the first hitting time of the cube D and $\tau_k = \inf\{t > \tau(D) : B(t) \notin D_k^*\}$ be the first exit time from D_k^* for $M > k \geqslant m$. For the cubes $\mathsf{Cube} = [0, 1)^d$ and $\mathsf{Child} = [0, \frac{1}{2})^d$ we also define the expanded cubes Cube^* and Child^* and the stopping time $\tau = \inf\{t > 0 : B(t) \notin \mathsf{Cube}^*\}$. Let

$$q := \sup_{y \in \mathsf{Child}^*} \mathbb{P}_y \left\{ \int_0^\tau 1_{\mathsf{Cube}}(B(s)) \, ds \leqslant \tfrac{1}{\varepsilon} \right\} < 1 \, .$$

By the strong Markov property applied to the stopping times $\tau_M < \ldots < \tau_{m+1}$ and Brownian scaling,

$$\mathbb{P}_x \left\{ \mu(D_k) \leqslant \tfrac{1}{\varepsilon} 2^{-2k} \text{ for all } M > k \geqslant m \, \middle| \, \tau(D) < \infty \right\}$$

$$\leqslant \mathbb{P}_x \left\{ \int_{\tau_{k+1}}^{\tau_k} 1_{D_k}(B(s)) \, ds \leqslant \tfrac{1}{\varepsilon} 2^{-2k} \text{ for all } M > k \geqslant m \, \middle| \, \tau(D) < \infty \right\}$$

$$\leqslant \prod_{k=m}^{M-1} \sup_{y \in D_{k+1}^*} \mathbb{P}_y \left\{ 2^{2k} \int_0^{\tilde{\tau}_k} 1_{D_k}(B(s)) \, ds \leqslant \tfrac{1}{\varepsilon} \right\} \leqslant q^{M-m} \, ,$$

where $\tilde{\tau}_k$ is the first exit time of the Brownian motion from D_k^* and the last inequality follows from Brownian scaling. Recall from Theorem 3.18 that $\mathbb{P}_x\{\tau(D) < \infty\} \leqslant c2^{-M(d-2)}$, for a constant $c > 0$ depending only on the dimension d and the fixed distance of x from the unit cube. Hence the probability that any given cube $D \in \mathfrak{D}_M$ is in our cover is

$$\mathbb{P}_x\left\{\mu(D_k) \leqslant \tfrac{1}{\varepsilon}2^{-2k} \text{ for all } M > k \geqslant m,\, \tau(D) < \infty\right\} \leqslant c2^{-M(d-2)}q^{M-m}\,.$$

Hence the expected 2-value from the cubes in $\mathfrak{C}(M) \cap \mathfrak{D}_M$ is

$$d2^{dM}2^{-2M}\mathbb{P}_x\left\{\mu(D_k) \leqslant \tfrac{1}{\varepsilon}2^{-2k} \text{ for all } M > k \geqslant m,\, \tau(D) < \infty\right\} \leqslant cd\,q^{M-m}\,. \quad (4.3)$$

The 2-value from the cubes in $\mathfrak{C}(M) \cap \bigcup_{k=M+1}^m \mathfrak{D}_k$ is bounded by

$$\sum_{k=m}^{M-1} d2^{-2k} \sum_{D \in \mathfrak{C}(M) \cap \mathfrak{D}_k} 1\{\mu(D) \geqslant 2^{-2k}\tfrac{1}{\varepsilon}\} \leqslant d\varepsilon \sum_{k=m}^{M-1} \sum_{D \in \mathfrak{C}(M) \cap \mathfrak{D}_k} \mu(D) \quad (4.4)$$

$$\leqslant d\varepsilon\,\mu(\mathsf{Cube})\,.$$

As $\mathbb{E}\mu(\mathsf{Cube}) < \infty$ by Theorem 3.27, we infer from (4.3) and (4.4) that the expected 2-value of our cover converges to zero for $\varepsilon \downarrow 0$ and a suitable choice $M = M(\varepsilon)$. Hence a subsequence converges to zero almost surely, and, as m was arbitrary, this ensures that $\mathcal{H}^2(\mathsf{Range}[0,\infty)) = 0$ almost surely. ∎

4.2 The mass distribution principle

From the definition of the Hausdorff dimension it is plausible that in many cases it is relatively easy to give an upper bound on the dimension: just find an efficient cover of the set and find an upper bound to its α-value. However it looks more difficult to give lower bounds, as we must obtain a lower bound on α-values of *all* covers of the set.

The mass distribution principle is a way around this problem, which is based on the existence of a nonzero measure on the set. The basic idea is that, if this measure distributes a positive amount of mass on a set E in such a manner that its local concentration is bounded from above, then the set must be large in a suitable sense. For the purpose of this method we call a measure μ on the Borel sets of a metric space E a **mass distribution** on E, if

$$0 < \mu(E) < \infty\,.$$

The intuition here is that a positive and finite mass is spread over the space E.

Theorem 4.19 (Mass distribution principle) *Suppose E is a metric space and $\alpha \geqslant 0$. If there is a mass distribution μ on E and constants $C > 0$ and $\delta > 0$ such that*

$$\mu(V) \leqslant C|V|^\alpha\,,$$

for all closed sets V with diameter $|V| \leqslant \delta$, then

$$\mathcal{H}^\alpha(E) \geqslant \frac{\mu(E)}{C} > 0,$$

and hence $\dim E \geqslant \alpha$.

Proof. Suppose that $U_1, U_2, \ldots$ is a cover of E by arbitrary sets with $|U_i| \leqslant \delta$. Let V_i be the closure of U_i and note that $|U_i| = |V_i|$. We have

$$0 < \mu(E) \leqslant \mu\Big(\bigcup_{i=1}^{\infty} U_i \Big) \leqslant \mu\Big(\bigcup_{i=1}^{\infty} V_i \Big) \leqslant \sum_{i=1}^{\infty} \mu(V_i) \leqslant C \sum_{i=1}^{\infty} |U_i|^{\alpha} \, .$$

Passing to the infimum over all such covers, and letting $\delta \downarrow 0$ gives the statement. ∎

We now apply this technique to find the Hausdorff dimension of the zero set of a linear Brownian motion. Recall that this is an uncountable set with no isolated points.

At first it is not clear what measure on Zeros would be suitable to apply the mass distribution principle. Here Lévy's theorem, see Theorem 2.34, comes to our rescue: Recall the definition of the maximum process $\{M(t) : t \geqslant 0\}$ associated with a Brownian motion from Chapter 2.2.3.

Definition 4.20. Let $\{B(t) : t \geqslant 0\}$ be a linear Brownian motion and $\{M(t) : t \geqslant 0\}$ the associated maximum process. A time $t \geqslant 0$ is a **record time** for the Brownian motion if $M(t) = B(t)$ and the set of all record times for the Brownian motion is denoted by Rec. ◇

Note that the record times are the zeros of the process $\{Y(t) : t \geqslant 0\}$ given by

$$Y(t) = M(t) - B(t).$$

By Theorem 2.34 this process is a reflected Brownian motion, and hence its zero set and the zero set of $\{B(t) : t \geqslant 0\}$ have the same distribution. A natural measure on Rec is given by the distribution function $\{M(t) : t \geqslant 0\}$, which allows us to get a lower bound for the Hausdorff dimension of Rec via the mass distribution principle.

Lemma 4.21 *Almost surely,* $\dim(\text{Rec} \cap [0,1]) \geqslant 1/2$ *and hence* $\dim(\text{Zeros} \cap [0,1]) \geqslant 1/2$.

Proof. The first equality follows from Theorem 2.34, so that we can focus in this proof on the record set. Since $t \mapsto M(t)$ is a non-decreasing and continuous function, we can regard it as a distribution function of a positive measure μ, with $\mu(a, b] = M(b) - M(a)$. This measure is supported on the (closed) set Rec of record times, see Exercise 4.12. We know that, with probability one, the Brownian motion is locally Hölder continuous with any exponent $\alpha < 1/2$. Thus there exists a (random) constant C_α, such that, almost surely,

$$M(b) - M(a) \leqslant \max_{0 \leqslant h \leqslant b-a} B(a+h) - B(a) \leqslant C_\alpha (b-a)^\alpha,$$

for all $a, b \in [0, 1]$. By the mass distribution principle, we get that, almost surely,

$$\dim(\text{Rec} \cap [0,1]) \geqslant \alpha.$$

Letting $\alpha \uparrow \frac{1}{2}$ finishes the proof. ∎

To get an upper bound on the Hausdorff dimension of Zeros we use a covering consisting of intervals. Define the collection $\mathfrak{D}_k$ of intervals $[j2^{-k}, (j+1)2^{-k})$ for $j = 0, \ldots, 2^k - 1$, and let $Z(I) = 1$ if there exists $t \in I$ with $B(t) = 0$, and $Z(I) = 0$ otherwise. To estimate the dimension of the zero set we need an estimate for the probability that $Z(I) = 1$, i.e. for the probability that a given interval contains a zero of Brownian motion.

Lemma 4.22 *There is an absolute constant C such that, for any $a, \varepsilon > 0$,*

$$\mathbb{P}\big\{ \text{ there exists } t \in (a, a + \varepsilon) \text{ with } B(t) = 0 \big\} \leqslant C\sqrt{\tfrac{\varepsilon}{a+\varepsilon}}.$$

Proof. Consider the event $A = \{|B(a+\varepsilon)| \leqslant \sqrt{\varepsilon}\}$. By the scaling property of Brownian motion, we can give the upper bound

$$\mathbb{P}(A) = \mathbb{P}\Big\{|B(1)| \leqslant \sqrt{\tfrac{\varepsilon}{a+\varepsilon}}\Big\} \leqslant 2\sqrt{\tfrac{\varepsilon}{a+\varepsilon}}.$$

Knowing that Brownian motion has a zero in $(a, a + \varepsilon)$ makes the event A very likely. Indeed, applying the strong Markov property at the stopping time $T = \inf\{t \geqslant a \colon B(t) = 0\}$, we have

$$\mathbb{P}(A) \geqslant \mathbb{P}\big(A \cap \{0 \in B[a, a + \varepsilon]\}\big)$$
$$\geqslant \mathbb{P}\{T \leqslant a + \varepsilon\} \min_{a \leqslant t \leqslant a+\varepsilon} \mathbb{P}\{|B(a + \varepsilon)| \leqslant \sqrt{\varepsilon} \,|\, B(t) = 0\}.$$

Clearly the minimum is achieved at $t = a$ and, using the scaling property of Brownian motion, we have $\mathbb{P}\{|B(a + \varepsilon)| \leqslant \sqrt{\varepsilon} \,|\, B(a) = 0\} = \mathbb{P}\{|B(1)| \leqslant 1\} =: c > 0$. Hence,

$$\mathbb{P}\{T \leqslant a + \varepsilon\} \leqslant \tfrac{2}{c}\sqrt{\tfrac{\varepsilon}{a+\varepsilon}},$$

and this completes the proof. ∎

Remark 4.23 This is only very crude information about the position of the zeros of a linear Brownian motion. Much more precise information is available, for example in the form of the arcsine law for the last sign-change, which we prove in the next chapter, and which (after a simple scaling) yields the precise value of the probability in Lemma 4.22. ◇

We have thus shown that, for any $\varepsilon > 0$ and sufficiently large integer k, we have

$$\mathbb{E}[Z(I)] \leqslant c_1 \, 2^{-k/2}, \qquad \text{for all } I \in \mathfrak{D}_k \text{ with } I \subset (\varepsilon, 1 - \varepsilon),$$

for some constant $c_1 > 0$. Hence the covering of the set $\{t \in (\varepsilon, 1 - \varepsilon) \colon B(t) = 0\}$ by all $I \in \mathfrak{D}_k$ with $I \cap (\varepsilon, 1 - \varepsilon) \neq \emptyset$ and $Z(I) = 1$ has an expected $\frac{1}{2}$-value of

$$\mathbb{E}\Big[\sum_{\substack{I \in \mathfrak{D}_k \\ I \cap (\varepsilon, 1-\varepsilon) \neq \emptyset}} Z(I) \, 2^{-k/2}\Big] = \sum_{\substack{I \in \mathfrak{D}_k \\ I \cap (\varepsilon, 1-\varepsilon) \neq \emptyset}} \mathbb{E}[Z(I)] \, 2^{-k/2} \leqslant c_1 \, 2^k \, 2^{-k/2} \, 2^{-k/2} = c_1.$$

We thus get, from Fatou's lemma,

$$\mathbb{E}\Big[\liminf_{k \to \infty} \sum_{\substack{I \in \mathfrak{D}_k \\ I \cap (\varepsilon, 1-\varepsilon) \neq \emptyset}} Z(I) \, 2^{-k/2}\Big] \leqslant \liminf_{k \to \infty} \mathbb{E}\Big[\sum_{\substack{I \in \mathfrak{D}_k \\ I \cap (\varepsilon, 1-\varepsilon) \neq \emptyset}} Z(I) \, 2^{-k/2}\Big] \leqslant c_1.$$

Hence the liminf is almost surely finite, which means that there exists a family of coverings with maximal diameter going to zero and bounded $\frac{1}{2}$-value. This implies that, almost surely,

$$\mathcal{H}^{\frac{1}{2}}\{t \in (\varepsilon, 1 - \varepsilon) \colon B(t) = 0\} < \infty,$$

and, in particular, that $\dim(\text{Zeros} \cap (\varepsilon, 1 - \varepsilon)) \leqslant \frac{1}{2}$. As $\varepsilon > 0$ was arbitrary, we obtain the same bound for the full zero set. Combining this estimate with Lemma 4.21 we have verified the following result.

Theorem 4.24 *Let $\{B(t) \colon 0 \leqslant t \leqslant 1\}$ be a linear Brownian motion. Then, with probability one, we have*

$$\dim\big(\text{Zeros} \cap [0, 1]\big) = \dim\big(\text{Rec} \cap [0, 1]\big) = \tfrac{1}{2}.$$

Remark 4.25 The Hausdorff measure $\mathcal{H}^{\frac{1}{2}}$ vanishes on the zero set of Brownian motion, see Exercise 4.14, just like that Hausdorff measure $\mathcal{H}^2$ vanishes on the range of Brownian motion, as seen in Theorem 4.18. Therefore another method is needed to construct a natural positive finite measure on the zero set. We encountered an indirect construction, via Lévy's identity, in the proof of Lemma 4.21. A powerful direct construction of the same measure, known as the *local time at zero*, will be the subject of Chapter 6.   ⋄

4.3 The energy method

The energy method is a technique to find a lower bound for the Hausdorff dimension, which is particularly interesting in applications to random fractals. It replaces the condition on the mass of all closed sets in the mass distribution principle by finiteness of an energy.

Definition 4.26. Suppose μ is a mass distribution on a metric space (E, ρ) and $\alpha \geqslant 0$. The α-**potential** of a point $x \in E$ with respect to μ is defined as

$$\phi_\alpha(x) = \int \frac{d\mu(y)}{\rho(x, y)^\alpha}.$$

In the case $E = \mathbb{R}^3$ and $\alpha = 1$, this is the Newton gravitational potential of the mass μ. The α-**energy** of μ is

$$I_\alpha(\mu) = \int \phi_\alpha(x)\, d\mu(x) = \iint \frac{d\mu(x)\, d\mu(y)}{\rho(x, y)^\alpha}.$$ ⋄

The simple idea of the energy method is the following: Mass distributions with $I_\alpha(\mu) < \infty$ spread the mass so that at each place the concentration is sufficiently small to overcome the singularity of the integrand. This is only possible on sets which are large in a suitable sense.

Theorem 4.27 (Energy method) *Let $\alpha \geqslant 0$ and μ be a mass distribution on a metric space E. Then, for every $\varepsilon > 0$, we have*

$$\mathcal{H}_\varepsilon^\alpha(E) \geqslant \frac{\mu(E)^2}{\iint_{\rho(x,y)<\varepsilon} \frac{d\mu(x)\,d\mu(y)}{\rho(x,y)^\alpha}}.$$

Hence, if $I_\alpha(\mu) < \infty$ then $\mathcal{H}^\alpha(E) = \infty$ and, in particular, $\dim E \geqslant \alpha$.

Remark 4.28 To get a lower bound on the dimension from this method it suffices to show finiteness of a single integral. In particular, in order to show for a random set E that $\dim E \geqslant \alpha$ almost surely, it suffices to show that $\mathbb{E}I_\alpha(\mu) < \infty$ for a (random) measure on E. ◇

Proof. If $\{A_n : n = 1, 2, \ldots\}$ is any pairwise disjoint covering of E consisting of sets of diameter $< \varepsilon$, then

$$\iint_{\rho(x,y)<\varepsilon} \frac{d\mu(x)\,d\mu(y)}{\rho(x,y)^\alpha} \geqslant \sum_{n=1}^\infty \iint_{A_n \times A_n} \frac{d\mu(x)\,d\mu(y)}{\rho(x,y)^\alpha} \geqslant \sum_{n=1}^\infty \frac{\mu(A_n)^2}{|A_n|^\alpha},$$

and moreover,

$$\mu(E) \leqslant \sum_{n=1}^\infty \mu(A_n) = \sum_{n=1}^\infty |A_n|^{\frac{\alpha}{2}} \frac{\mu(A_n)}{|A_n|^{\frac{\alpha}{2}}}.$$

Given $\delta > 0$ choose a covering as above such that additionally

$$\sum_{n=1}^\infty |A_n|^\alpha \leqslant \mathcal{H}_\varepsilon^\alpha(E) + \delta.$$

Using now the Cauchy–Schwarz inequality, we get

$$\mu(E)^2 \leqslant \sum_{n=1}^\infty |A_n|^\alpha \sum_{n=1}^\infty \frac{\mu(A_n)^2}{|A_n|^\alpha} \leqslant \left(\mathcal{H}_\varepsilon^\alpha(E) + \delta\right) \iint_{\rho(x,y)<\varepsilon} \frac{d\mu(x)\,d\mu(y)}{\rho(x,y)^\alpha}.$$

Letting $\delta \downarrow 0$ and dividing both sides by the integral gives the stated inequality. Further, letting $\varepsilon \downarrow 0$, if $\mathbb{E}I_\alpha(\mu) < \infty$ the integral converges to zero, so that $\mathcal{H}_\varepsilon^\alpha(E)$ diverges to infinity. ∎

We now apply the energy method to resolve questions left open in the first section of this chapter, namely the lower bounds for the Hausdorff dimension of the graph and range of Brownian motion.

The nowhere differentiability of linear Brownian motion established in the first chapter suggests that its graph may have dimension greater than one. For dimensions $d \geqslant 2$, it is interesting to look at the range of Brownian motion. We have seen that planar Brownian motion is neighbourhood recurrent, that is, it visits every neighbourhood in the plane infinitely often. In this sense, the range of planar Brownian motion is comparable to the plane itself and one can ask whether this is also true in the sense of dimension.

Theorem 4.29 (Taylor 1953) *Let $\{B(t): 0 \leqslant t \leqslant 1\}$ be d-dimensional Brownian motion.*

(a) *If $d = 1$, then $\dim \mathrm{Graph}[0,1] = 3/2$ almost surely.*

(b) *If $d \geqslant 2$, then $\dim \mathrm{Range}[0,1] = \dim \mathrm{Graph}[0,1] = 2$ almost surely.*

Recall that we already know the upper bounds from Corollary 4.16. We now look at lower bounds for the range of Brownian motion in $d \geqslant 2$.

Proof of Theorem 4.29(b).   A natural measure on $\mathrm{Range}[0,1]$ is the occupation measure μ defined by $\mu(A) = \mathcal{L}(B^{-1}(A) \cap [0,1])$, for all Borel sets $A \subset \mathbb{R}^d$, or, equivalently,

$$\int_{\mathbb{R}^d} f(x)\, d\mu(x) = \int_0^1 f\big(B(t)\big)\, dt,$$

for all bounded measurable functions f. We want to show that for any $0 < \alpha < 2$,

$$\mathbb{E} \iint \frac{d\mu(x)\, d\mu(y)}{|x - y|^\alpha} = \mathbb{E} \int_0^1 \int_0^1 \frac{ds\, dt}{|B(t) - B(s)|^\alpha} < \infty. \tag{4.5}$$

Let us evaluate the expectation

$$\mathbb{E}|B(t) - B(s)|^{-\alpha} = \mathbb{E}\big[(|t - s|^{1/2}|B(1)|)^{-\alpha}\big] = |t - s|^{-\alpha/2} \int_{\mathbb{R}^d} \frac{c_d}{|z|^\alpha} e^{-|z|^2/2}\, dz.$$

The integral can be evaluated using polar coordinates, but all we need is that it is a finite constant c depending on d and α only. Substituting this expression into (4.5) and using Fubini's theorem we get

$$\mathbb{E}I_\alpha(\mu) = c \int_0^1 \int_0^1 \frac{ds\, dt}{|t - s|^{\alpha/2}} \leqslant 2c \int_0^1 \frac{du}{u^{\alpha/2}} < \infty. \tag{4.6}$$

Therefore $I_\alpha(\mu) < \infty$ and hence $\dim \mathrm{Range}[0,1] > \alpha$, almost surely. The lower bound on the range follows by letting $\alpha \uparrow 2$. We also obtain a lower bound for the dimension of the graph: As the graph of a function can be projected onto the path, the dimension of the graph is at least the dimension of the path by Remark 4.10. Hence, if $d \geqslant 2$, almost surely $\dim \mathrm{Graph}[0,1] \geqslant 2$.   ∎

Now let us turn to linear Brownian motion and prove the first half of Taylor's theorem.

Proof of Theorem 4.29(a).   Again we use the energy method for a sharp lower bound. Recall that we have shown in Corollary 4.16 that $\dim \mathrm{Graph}[0,1] \leqslant 3/2$. Let $\alpha < 3/2$ and define a measure μ on the graph by

$$\mu(A) = \mathcal{L}_1(\{0 \leqslant t \leqslant 1 : (t, B(t)) \in A\}) \text{ for } A \subset [0,1] \times \mathbb{R} \text{ Borel.}$$

Changing variables, the α-energy of μ can be written as

$$\iint \frac{d\mu(x)\, d\mu(y)}{|x - y|^\alpha} = \int_0^1 \int_0^1 \frac{ds\, dt}{(|t - s|^2 + |B(t) - B(s)|^2)^{\alpha/2}}.$$

Bounding the integrand, taking expectations, and applying Fubini we get that

$$\mathbb{E}I_\alpha(\mu) \leqslant 2 \int_0^1 \mathbb{E}\left((t^2 + B(t)^2)^{-\alpha/2}\right) dt. \tag{4.7}$$

Let $\mathfrak{p}(z) = \sqrt{2\pi}^{-1} \exp(-z^2/2)$ denote the standard normal density. By scaling, the expectation above can be written as

$$2 \int_0^{+\infty} (t^2 + tz^2)^{-\alpha/2} \mathfrak{p}(z) \, dz. \tag{4.8}$$

Comparing the size of the summands in the integration suggests separating $z \leqslant \sqrt{t}$ from $z > \sqrt{t}$. Then we can bound (4.8) above by twice

$$\int_0^{\sqrt{t}} (t^2)^{-\alpha/2} dz + \int_{\sqrt{t}}^\infty (tz^2)^{-\alpha/2} \mathfrak{p}(z) \, dz = t^{\frac{1}{2}-\alpha} + t^{-\alpha/2} \int_{\sqrt{t}}^\infty z^{-\alpha} \mathfrak{p}(z) \, dz.$$

Furthermore, we separate the last integral at 1. We get

$$\int_{\sqrt{t}}^\infty z^{-\alpha} \mathfrak{p}(z) \, dz \leqslant 1 + \int_{\sqrt{t}}^1 z^{-\alpha} \, dz.$$

The latter integral is of order $t^{(1-\alpha)/2}$. Substituting these results into (4.7), we see that the expected energy is finite when $\alpha < 3/2$. The claim now follows from the energy method. ∎

4.4 Frostman's lemma and capacity

In this section we provide a converse to the mass distribution principle, i.e. starting from a lower bound on the Hausdorff measure we construct a mass distribution on a set. This is often useful, for example if one wants to relate the Hausdorff dimension of a set and its image under some transformation.

Theorem 4.30 (Frostman's lemma) *If $A \subset \mathbb{R}^d$ is a closed set such that $\mathcal{H}^\alpha(A) > 0$, then there exists a Borel probability measure μ supported on A and a constant $C > 0$ such that $\mu(D) \leqslant C|D|^\alpha$ for all Borel sets D.*

We now give a proof of Frostman's lemma, which is based on the representation of compact subsets of $\mathbb{R}^d$ by trees, an idea that we will encounter again in Chapter 9. The main ingredient in the proof is the max-flow min-cut theorem. See Section 12.4 in the appendix for definitions and notation associated with trees, flows on trees and statement and proof of the max-flow min-cut theorem.

Proof of Frostman's lemma. We may assume $A \subset [0,1]^d$. Any compact cube in $\mathbb{R}^d$ of side length s can be split into 2^d nonoverlapping compact cubes of side length $s/2$. We first create a tree with a root that we associate with the cube $[0,1]^d$. Every vertex in the tree has 2^d edges emanating from it, each leading to a vertex that is associated with one of the 2^d subcubes with half the side length of the original cube. We then erase the edges ending in vertices associated with subcubes that do not intersect A. In this way we construct a tree $T = (V, E)$ such that the rays in ∂T correspond to sequences of nested compact cubes, see Figure 4.4.

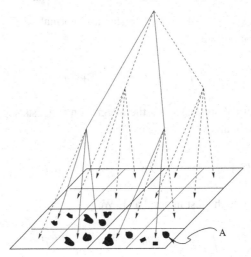

Fig. 4.4. The first two stages in the construction of the tree associated with the shaded set $A \subset [0,1]^2$. Dotted edges in the tree are erased.

There is a canonical mapping $\Phi \colon \partial T \to A$, which maps sequences of nested cubes to their intersection. Note that if $x \in A$, then there is an infinite path emanating from the root, all of whose vertices are associated with cubes that contain x and thus intersect A. Hence Φ is surjective.

For any edge e at level n define the capacity $C(e) = (d^{\frac{1}{2}} 2^{-n})^\alpha$. We now associate to every cutset Π a covering of A, consisting of those cubes associated with the initial vertices of the edges in the cutset. To see that the resulting collection of cubes is indeed a covering, let ξ be a ray. As Π is a cutset, it contains one of the edges in this ray, and the cube associated with the initial vertex of this edge contains the point $\Phi(\xi)$. Hence we indeed cover the entire set $\Phi(\partial T) = A$. This implies that

$$\inf\Big\{ \sum_{e \in \Pi} C(e) \colon \Pi \text{ a cutset} \Big\} \geqslant \inf\Big\{ \sum_j |A_j|^\alpha \colon A \subset \bigcup_j A_j \Big\},$$

and as $\mathcal{H}^\alpha_\infty(A) > 0$, by the equivalence in Proposition 4.9, this is bounded from zero. Thus, by the max-flow min-cut theorem, there exists a flow $\theta \colon E \to [0, \infty)$ of positive strength such that $\theta(e) \leqslant C(e)$ for all edges $e \in E$.

We now show how to define a suitable measure on the space of infinite paths. Given an edge $e \in E$ we associate a set $T(e) \subset \partial T$ consisting of all rays containing the edge e. Define

$$\widetilde{\nu}\big(T(e)\big) = \theta(e).$$

It is easily checked that the collection $\mathcal{C}(\partial T)$ of subsets $T(v) \subset \partial T$ for all $v \in T$ is a semi-algebra on ∂T. Recall that this means that if $A, B \in \mathcal{C}(\partial T)$, then $A \cap B \in \mathcal{C}(\partial T)$ and A^c is a finite disjoint union of sets in $\mathcal{C}(\partial T)$. Because the flow through any vertex is preserved, $\widetilde{\nu}$ is countably additive. Thus, using a measure extension theorem such as, for example A.1(1.3) in [Du95], we can extend $\widetilde{\nu}$ to a measure ν on the σ-algebra generated by $\mathcal{C}(\partial T)$.

We can now define a Borel measure $\mu = \nu \circ \Phi^{-1}$ on A, which satisfies $\mu(C) = \theta(e)$, where C is the cube associated with the initial vertex of the edge e. Suppose now that D is a Borel subset of $\mathbb{R}^d$ and n is the integer such that $2^{-n} < |D \cap [0,1]^d| \leqslant 2^{-(n-1)}$. Then $D \cap [0,1]^d$ can be covered with 3^d of the cubes in the above construction having side length 2^{-n}, or diameter $d^{\frac{1}{2}} 2^{-n}$. Using this bound, we have

$$\mu(D) \leqslant d^{\frac{\alpha}{2}} 3^d 2^{-n\alpha} \leqslant d^{\frac{\alpha}{2}} 3^d |D|^\alpha,$$

so we have a finite measure μ satisfying the requirement of the lemma. Normalising μ to get a probability measure completes the proof. ∎

Definition 4.31. We define the **Riesz α-capacity**, or simply the α-**capacity**, of a metric space (E, ρ) as

$$\text{Cap}_\alpha(E) := \sup \left\{ I_\alpha(\mu)^{-1} : \mu \text{ a mass distribution on } E \text{ with } \mu(E) = 1 \right\}.$$

In the case of the Euclidean space $E = \mathbb{R}^d$ with $d \geqslant 3$ and $\alpha = d - 2$ the Riesz α-capacity is also known as the **Newtonian capacity**. ◇

Theorem 4.27 states that a set of positive α-capacity has dimension at least α. We now show that, in this formulation the method is sharp. Our proof of this fact relies on Frostman's lemma and hence refers to closed subsets of Euclidean space.

Theorem 4.32 *For any closed set $A \subset \mathbb{R}^d$,*

$$\dim A = \sup \left\{ \alpha : \text{Cap}_\alpha(A) > 0 \right\}.$$

Proof. It only remains to show $\leqslant$, and for this purpose it suffices to show that if $\dim A > \alpha$, then there exists a Borel probability measure μ on A such that

$$I_\alpha(\mu) = \int_{\mathbb{R}^d} \int_{\mathbb{R}^d} \frac{d\mu(x)\, d\mu(y)}{|x - y|^\alpha} < \infty.$$

By our assumption for some sufficiently small $\beta > \alpha$ we have $\mathcal{H}^\beta(A) > 0$. By Frostman's lemma, there exists a nonzero Borel probability measure μ on A and a constant C such that $\mu(D) \leqslant C|D|^\beta$ for all Borel sets D. By restricting μ to a smaller set if necessary, we can make the support of μ have diameter less than one. Fix $x \in A$, and for $k \geqslant 1$ let $S_k(x) = \{y : 2^{-k} < |x - y| \leqslant 2^{1-k}\}$. Since μ has no atoms, we have

$$\int_{\mathbb{R}^d} \frac{d\mu(y)}{|x - y|^\alpha} = \sum_{k=1}^\infty \int_{S_k(x)} \frac{d\mu(y)}{|x - y|^\alpha} \leqslant \sum_{k=1}^\infty \mu(S_k(x)) 2^{k\alpha},$$

where the equality follows from the monotone convergence theorem and the inequality holds by the definition of the S_k. Also,

$$\sum_{k=1}^\infty \mu(S_k(x)) 2^{k\alpha} \leqslant C \sum_{k=1}^\infty |2^{2-k}|^\beta 2^{k\alpha} = C' \sum_{k=1}^\infty 2^{k(\alpha-\beta)},$$

where $C' = 2^{2\beta} C$.

Since $\beta > \alpha$, we have

$$I_\alpha(\mu) \leqslant C' \sum_{k=1}^{\infty} 2^{k(\alpha-\beta)} < \infty,$$

which proves the theorem.   ∎

In Corollary 4.16 we have seen that the image of a set $A \subset [0, \infty)$ under Brownian motion has at most twice the Hausdorff dimension of A. Naturally, the question arises whether this is a sharp estimate. The following result of McKean shows that, if $d \geqslant 2$, this is sharp for *any* set A, while in $d = 1$ it is sharp as long as $\dim A \leqslant \frac{1}{2}$.

Theorem 4.33 (McKean 1955) *Let $A \subset [0, \infty)$ be a closed subset and $\{B(t) : t \geqslant 0\}$ a d-dimensional Brownian motion. Then, almost surely,*

$$\dim B(A) = 2 \dim A \wedge d.$$

Proof.   The upper bound was verified in Corollary 4.16. For the lower bound let $\alpha < \dim(A) \wedge (d/2)$. By Theorem 4.32, there exists a Borel probability measure μ on A such that $I_\alpha(\mu) < \infty$. Denote by $\tilde{\mu}$ the measure on $\mathbb{R}^d$ defined by

$$\tilde{\mu}(D) = \mu(\{t \geqslant 0 : B(t) \in D\})$$

for all Borel sets $D \subset \mathbb{R}^d$. Then

$$\mathbb{E}[I_{2\alpha}(\tilde{\mu})] = \mathbb{E}\left[\iint \frac{d\tilde{\mu}(x)\, d\tilde{\mu}(y)}{|x - y|^{2\alpha}}\right] = \mathbb{E}\left[\int_0^\infty \int_0^\infty \frac{d\mu(t)\, d\mu(s)}{|B(t) - B(s)|^{2\alpha}}\right],$$

where the second equality can be verified by a change of variables. Note that the denominator on the right hand side has the same distribution as $|t - s|^\alpha |Z|^{2\alpha}$, where Z is a d-dimensional standard normal random variable. Since $2\alpha < d$, we have that

$$\mathbb{E}[|Z|^{-2\alpha}] = \frac{1}{(2\pi)^{d/2}} \int_{\mathbb{R}^d} |y|^{-2\alpha} e^{-|y|^2/2}\, dy < \infty.$$

Hence, using Fubini's theorem,

$$\mathbb{E}[I_{2\alpha}(\tilde{\mu})] = \int_0^\infty \int_0^\infty \mathbb{E}[|Z|^{-2\alpha}] \frac{d\mu(t)\, d\mu(s)}{|t - s|^\alpha} \leqslant \mathbb{E}[|Z|^{-2\alpha}] I_\alpha(\mu) < \infty.$$

Thus, $\mathbb{E}[I_{2\alpha}(\tilde{\mu})] < \infty$, and hence $I_{2\alpha}(\tilde{\mu}) < \infty$ almost surely. Moreover, $\tilde{\mu}$ is supported on $B(A)$ because μ is supported on A. It follows from Theorem 4.27 that $\dim B(A) \geqslant 2\alpha$ almost surely. By letting $\alpha \uparrow \dim(A) \wedge d/2$, we see that $\dim(B(A)) \geqslant 2 \dim(A) \wedge d$ almost surely. This completes the proof of Theorem 4.33.   ∎

Remark 4.34 We have indeed shown that, if $\text{Cap}_\alpha(A) > 0$, then $\text{Cap}_{2\alpha}(B(A)) > 0$ almost surely. The converse of this statement is also true and will be discussed later, see Theorem 9.36.   ◇

Remark 4.35 Later in the book, we shall be able to significantly improve McKean's theorem and show that for Brownian motion in dimension $d \geqslant 2$, almost surely, for any $A \subset [0, \infty)$, we have $\dim B(A) = 2 \dim(A)$. This result is Kaufman's theorem, see Theorem 9.28. Note the difference between the results of McKean and Kaufman: In Theorem 4.33, the null probability set depends on A, while Kaufman's theorem has a much stronger claim: it states dimension doubling simultaneously for *all* sets. This allows us to plug in random sets A, which may depend completely arbitrarily on the Brownian motion. For Kaufman's theorem, $d \geqslant 2$ is a necessary condition: we have seen that the zero set of one dimensional Brownian motion has dimension $1/2$, while its image is a single point. ◇

Exercises

Exercise 4.1. ⑤ Show that for the ternary Cantor set C, we have $\dim_M C = \frac{\log 2}{\log 3}$.

Exercise 4.2. ⑤ Let $E := \{1/n : n \in \mathbb{N}\} \cup \{0\}$. Show that $\dim_M E = \frac{1}{2}$.

Exercise 4.3. ⑤ Show that, for every bounded metric space, the Hausdorff dimension is bounded from above by the lower Minkowski dimension.

Exercise 4.4. ⑤ Show that Hausdorff dimension has the countable stability property.

Exercise 4.5. Show that, for the ternary Cantor set C we have $\dim C = \frac{\log 2}{\log 3}$.

Exercise 4.6. ⑤ Suppose $f: (E_1, \rho_1) \to (E_2, \rho_2)$ is surjective and α-Hölder continuous with constant C. Show that, for any $\beta \geqslant 0$,

$$\mathcal{H}^\beta(E_2) \leqslant C^\beta \, \mathcal{H}^{\alpha\beta}(E_1),$$

and therefore $\dim(E_2) \leqslant \frac{1}{\alpha} \dim(E_1)$.

Exercise 4.7. Suppose $f: [0, 1] \to \mathbb{R}^d$ is an α-Hölder continuous function. Show that

(a) $\overline{\dim}_M (\text{Graph}_f[0, 1]) \leqslant 1 + (1 - \alpha)\left(d \wedge \frac{1}{\alpha}\right)$,

(b) and, for any $A \subset [0, 1]$, we have $\overline{\dim}_M \text{Range}_f(A) \leqslant \frac{\overline{\dim}_M A}{\alpha}$.

Exercise 4.8. ⑤ For any integer $d \geqslant 1$ and $0 < \alpha < d$ construct a compact set $A \subset \mathbb{R}^d$ such that $\dim A = \alpha$.

Exercise 4.9. Construct a function $f: [0, 1] \to \mathbb{R}^d$ which is α-Hölder continuous for any $\alpha < \beta$, but has $\mathcal{H}^\beta(\text{Range}_f[0, 1]) = \infty$.

Exercise 4.10. A function $f: [0, 1] \to \mathbb{R}$ is called **reverse β-Hölder** for some $0 < \beta < 1$ if there exists a constant $C > 0$ such that for any interval $[t, s]$, there is a subinterval $[t_1, s_1] \subset [t, s]$, such that $|f(t_1) - f(s_1)| \geqslant C|t - s|^\beta$. Let $f: [0, 1] \to \mathbb{R}$ be reverse β-Hölder. Show that $\dim_M (\text{Graph}_f[0, 1]) \geqslant 2 - \beta$.

Exercise 4.11. Show that for $\{B(t): t \geqslant 0\}$ we have $\dim_M \text{Graph}[0,1] = \frac{3}{2}$ if $d = 1$, and $\dim_M \text{Graph}[0,1] = \dim_M B[0,1] = 2$ if $d \geqslant 2$, almost surely.

Exercise 4.12. Show that the set of record points of a linear Brownian motion satisfies, almost surely,

$$\text{Rec} = \{s \geqslant 0 : M(s+h) > M(s-h) \text{ for all } 0 < h < s\}.$$

Exercise 4.13. Show that $\dim_M \{0 \leqslant t \leqslant 1 : B(t) = 0\} = \frac{1}{2}$, almost surely.

Exercise 4.14. ⑤ Show that $\mathcal{H}^{1/2}(\text{Zeros}) = 0$, almost surely.

Exercise 4.15. For a Brownian path $\{B(t): t \geqslant 0\}$ in $\mathbb{R}^d$, $d \geqslant 2$, we denote by

$$W_\varepsilon(t) = \{x \in \mathbb{R}^d : |x - B(s)| < \varepsilon \text{ for some } 0 \leqslant s \leqslant t\}$$

the *Wiener sausage* of width $\varepsilon > 0$ up to time t.

(a) Show that, for a suitable constant $C > 0$, we have $\mathbb{E}\mathcal{L}(W_1(t)) \leqslant Ct$.

(b) Infer from the result of (a) that $\mathcal{H}^2(\text{Range}[0,1]) < \infty$, almost surely.

Notes and comments

Felix Hausdorff introduced the Hausdorff measure in his seminal paper [Ha19]. Credit should also be given to Carathéodory [Ca14] who introduced a general construction in which Hausdorff measure can be naturally embedded. The Hausdorff measure indeed defines a measure on the Borel sets, proofs can be found in [Ma95] and [Ro99]. If $X = \mathbb{R}^d$ and $\alpha = d$ the Hausdorff measure $\mathcal{H}^\alpha$ is a constant multiple of Lebesgue measure $\mathcal{L}_d$, moreover if α is an integer and X an embedded α-submanifold, then $\mathcal{H}^\alpha$ is (a constant multiple of) the surface measure. This idea can also be used to develop vector analysis on sets with much less smoothness than a differentiable manifold. For more about Hausdorff dimension and geometric questions related to it we strongly recommend Mattila [Ma95]. The classic text of Rogers [Ro99], which first appeared in 1970, is a thorough discussion of Hausdorff measures. Falconer [Fa97a, Fa97b] covers a range of applications and current developments, but with more focus on deterministic fractals.

The results on the Hausdorff dimension of graph and range of a Brownian motion are due to S.J. Taylor [Ta53, Ta55] and independently to Lévy [Le51] though the latter paper does not contain full proofs. Taylor also proved in [Ta55] that the dimension of the zero set of a Brownian motion in dimension one is $1/2$. Stronger results show that, almost surely, the Hausdorff dimension of *all* nondegenerate level sets is $1/2$. For this and much finer results see [Pe81]. A classical survey, which inspired a lot of activity in the area of Hausdorff dimension and stochastic processes is Taylor [Ta86] and a modern survey is Xiao [Xi04].

The energy method and Frostman's lemma all stem from Otto Frostman's famous 1935 thesis [Fr35], which lays the foundations of modern potential theory. The elegant quantitative proof of the energy method given here is due to Oded Schramm. Frostman's lemma was generalised to complete, separable metric spaces by Howroyd [Ho95] using a functional-analytic approach. The main difficulty arising in the proof is that, if $\mathcal{H}^\alpha(E) = \infty$, one has to find a subset $A \subset E$ with $0 < \mathcal{H}^\alpha(A) < \infty$, which is tricky to do in abstract metric spaces. Frostman's original proof uses, in a way, the same idea as the proof presented here, though the transfer to the tree setup is not done explicitly. Probability using trees became fashionable in the 1990s and indeed, this is the right way to look at many problems of Hausdorff dimension and fractal geometry. Recommended survey articles are by Pemantle [Pe95], Lyons [Ly96] or the chapter on random fractals in [KM09], more information can be found in [Pe99] and [LP05].

McKean's theorem is due to Henry McKean [McK55]. Its surprising extension by Kaufman is not as hard as one might think considering the wide applicability of the result. The original source is [Ka69], we discuss the result in depth in Chapter 9.

The concept of 'reverse Hölder' mappings only partially extends from Minkowski to Hausdorff dimension. If $f : [0, 1] \to \mathbb{R}$ is both β-Hölder and reverse β-Hölder for some $0 < \beta < 1$, it satisfies $\dim(\mathsf{Graph}_f[0, 1]) > 1$, see Przytycki and Urbański [PU89]. For example, the Weierstrass nowhere differentiable function $W(t) = \sum_{n=0}^\infty a^n \cos(b^n t)$, for $ab > 1, 0 < a < 1$, is β-Hölder and reverse β-Hölder for some $0 < \beta < 1$. The Hausdorff dimension of its graph is, however, not rigorously known in general.

There is a natural refinement of the notions of Hausdorff dimension and Hausdorff measure, which is based on evaluating sets by applying an arbitrary 'gauge' function φ to the diameter, rather than taking a power. Measuring sets using a gauge function not only allows much finer results, it also turns out that the natural measures on graph and range of Brownian paths, which we have encountered in this chapter, turn out to be Hausdorff measures for suitable gauge functions. Results in this direction are Ciesielski and Taylor [CT62], Ray [Ra63a], Taylor [Ta64] and we include elements of this discussion in Chapter 6, where the zero set of Brownian motion is considered.

The Wiener sausages, defined in Exercise 4.15, have been widely studied. In the early sixties, Kesten, Spitzer and Whitman, see e.g. p.252 in [IM74], showed that $\mathcal{L}(W_1(t))/t$, for $d \geqslant 3$, converges almost surely to the Newtonian capacity of the unit ball. This result indicates that covering of the Brownian path with balls of *fixed* size is not sufficient to show that its $\frac{1}{2}$-dimensional Hausdorff measure is zero. Spitzer [Sp64] showed that, for $d = 3$, the expected volume of the Wiener sausage satisfies

$$\mathbb{E}[\mathcal{L}(W_1(t))] = ct + \frac{4}{(2\pi)^{3/2}} c^2 \sqrt{t} + o(\sqrt{t}),$$

where $c = \mathrm{Cap}_1(\mathcal{B}(0, 1))$. A central limit theorem, which highlights the deep connection of the Wiener sausage to the self-intersections of the Brownian path is due to Le Gall [LG88b]. An integrated view of these results is given by Csáki and Hu [CHu07].

5

Brownian motion and random walk

In this chapter we discuss some aspects of the relation between random walk and Brownian motion. The first two sections aim to demonstrate the nature of this relation by examples, which are of interest in their own right. These are *first* the law of the iterated logarithm, which is easier to prove for Brownian motion and can be extended to random walks by an embedding argument, and *second* a proof that Brownian motion does not have points of increase, which is based on a combinatorial argument for a suitable class of random walks and then extended to Brownian motion. We then discuss the Skorokhod embedding problem systematically, and give a proof of the Donsker invariance principle based on the Skorokhod embedding. We give a variety of applications of Donsker's invariance principle, including the arcsine laws and Pitman's $2M - B$ theorem.

5.1 The law of the iterated logarithm

Suppose $\{B(t) : t \geqslant 0\}$ is a standard linear Brownian motion. Although at any given time t and for any open set $U \subset \mathbb{R}$ the probability of the event $\{B(t) \in U\}$ is positive, over a long time Brownian motion cannot grow arbitrarily fast. We have seen in Corollary 1.11 that, for any small $\varepsilon > 0$, almost surely, there exists $t_0 > 0$ such that $|B(t)| \leqslant \varepsilon t$ for all $t \geqslant t_0$, whereas Proposition 1.23 ensures that for every large k, almost surely, there exist arbitrarily large times t such that $|B(t)| \geqslant k\sqrt{t}$. It is therefore natural to ask for the asymptotic *smallest upper envelope* of the Brownian motion, i.e. for a function $\psi \colon (1, \infty) \to \mathbb{R}$ such that

$$\limsup_{t \to \infty} \frac{B(t)}{\psi(t)} = 1.$$

The law of the iterated logarithm (whose name comes from the answer to this question but is by now firmly established for this type of upper-envelope results) provides such a 'gauge' function, which determines the almost-sure *asymptotic growth* of a Brownian motion.

A similar problem arises for arbitrary random walks $\{S_n : n \geqslant 0\}$, where we ask for a sequence $(a_n : n \geqslant 0)$ such that

$$\limsup_{n \to \infty} \frac{S_n}{a_n} = 1.$$

These two questions are closely related, and we start with an answer to the first one.

118

Theorem 5.1 (Law of the Iterated Logarithm for Brownian motion) *Suppose* $\{B(t) : t \geqslant 0\}$ *is a standard linear Brownian motion. Then, almost surely,*

$$\limsup_{t \to \infty} \frac{B(t)}{\sqrt{2t \log \log(t)}} = 1.$$

Remark 5.2 By symmetry it follows that, almost surely,

$$\liminf_{t \to \infty} \frac{B(t)}{\sqrt{2t \log \log(t)}} = -1.$$

Hence, for any $\varepsilon > 0$, there exists t_0 such that $|B(t)| \leqslant (1+\varepsilon)\sqrt{2t \log \log(t)}$ for any $t \geqslant t_0$, while there exist arbitrarily large times t with $|B(t)| \geqslant (1-\varepsilon)\sqrt{2t \log \log(t)}$. ◇

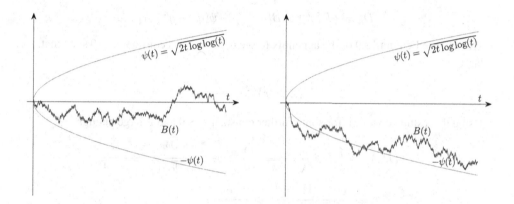

Fig. 5.1. Brownian motion and its asymptotic upper envelope $\psi(t) = \sqrt{2t \log \log(t)}$ at large times. In the picture on the left we see a typical Brownian path indicating that times where the path comes near to the envelope are very sparse. The picture on the right was chosen from a large number of samples so that the Brownian motion ends near the envelope. Due to the implicit conditioning on this event, the sample path of the motion has features untypical of Brownian paths. See the 'Notes and comments' section for more details.

Proof. The main idea is to scale by a geometric sequence. Let $\psi(t) = \sqrt{2t \log \log(t)}$. We first prove the upper bound. Fix $\varepsilon > 0$ and $q > 1$. Let

$$A_n = \left\{ \max_{0 \leqslant t \leqslant q^n} B(t) \geqslant (1+\varepsilon)\psi(q^n) \right\}.$$

By Theorem 2.21 the maximum of Brownian motion up to a fixed time t has the same distribution as $|B(t)|$. Therefore

$$\mathbb{P}(A_n) = \mathbb{P}\left\{ \frac{|B(q^n)|}{\sqrt{q^n}} \geqslant (1+\varepsilon)\frac{\psi(q^n)}{\sqrt{q^n}} \right\}.$$

We can use the tail estimate $\mathbb{P}\{Z > x\} \leqslant e^{-x^2/2}$ for a standard normally distributed Z and $x > 1$, see Lemma 12.9 in the appendix, to conclude that, for large n,

$$\mathbb{P}(A_n) \leqslant 2\exp\left(-(1+\varepsilon)^2 \log \log q^n\right) = \frac{2}{(n \log q)^{(1+\varepsilon)^2}}.$$

This is summable in n and hence, by the Borel–Cantelli lemma, we get that only finitely many of these events occur. For large t write $q^{n-1} \leqslant t < q^n$. We have

$$\frac{B(t)}{\psi(t)} = \frac{B(t)}{\psi(q^n)} \frac{\psi(q^n)}{q^n} \frac{t}{\psi(t)} \frac{q^n}{t} \leqslant (1+\varepsilon)q,$$

since $\psi(t)/t$ is decreasing in t, and thus

$$\limsup_{t\to\infty} \frac{B(t)}{\psi(t)} \leqslant (1+\varepsilon)q, \qquad \text{almost surely.}$$

Since this holds for any $\varepsilon > 0$ and $q > 1$ we have proved that $\limsup B(t)/\psi(t) \leqslant 1$ almost surely.

For the lower bound, fix $q > 1$. In order to use the Borel–Cantelli lemma in the other direction, we need to create a sequence of *independent* events. Let

$$D_n = \left\{ B(q^n) - B(q^{n-1}) \geqslant \psi(q^n - q^{n-1}) \right\}.$$

We now use Lemma 12.9 of the appendix to see that there is a constant $c > 0$ such that, for large x,

$$\mathbb{P}\{Z > x\} \geqslant \frac{ce^{-x^2/2}}{x}.$$

Using this estimate we get, for some further constant $\tilde{c} > 0$ and n large enough,

$$\mathbb{P}(D_n) = \mathbb{P}\left\{ Z \geqslant \frac{\psi(q^n - q^{n-1})}{\sqrt{q^n - q^{n-1}}} \right\} \geqslant c \frac{e^{-\log\log(q^n - q^{n-1})}}{\sqrt{2\log\log(q^n - q^{n-1})}}$$

$$\geqslant \frac{ce^{-\log(n\log q)}}{\sqrt{2\log(n\log q)}} > \frac{\tilde{c}}{n\log n},$$

and therefore $\sum_n \mathbb{P}(D_n) = \infty$. Thus for infinitely many n

$$B(q^n) \geqslant B(q^{n-1}) + \psi(q^n - q^{n-1}) \geqslant -2\psi(q^{n-1}) + \psi(q^n - q^{n-1}),$$

where the second inequality follows from applying the previously proved upper bound to $-B(q^{n-1})$. From the above we get that almost surely, for infinitely many n,

$$\frac{B(q^n)}{\psi(q^n)} \geqslant \frac{-2\psi(q^{n-1}) + \psi(q^n - q^{n-1})}{\psi(q^n)} \geqslant \frac{-2}{\sqrt{q}} + \frac{q^n - q^{n-1}}{q^n} = 1 - \frac{2}{\sqrt{q}} - \frac{1}{q}. \qquad (5.1)$$

Indeed, to obtain the second inequality first note that

$$\frac{\psi(q^{n-1})}{\psi(q^n)} = \frac{\psi(q^{n-1})}{\sqrt{q^{n-1}}} \frac{\sqrt{q^n}}{\psi(q^n)} \frac{1}{\sqrt{q}} \leqslant \frac{1}{\sqrt{q}},$$

since $\psi(t)/\sqrt{t}$ is increasing in t for large t. For the second term we just use the fact that $\psi(t)/t$ is decreasing in t. Now (5.1) implies that

$$\limsup_{t\to\infty} \frac{B(t)}{\psi(t)} \geqslant -\frac{2}{\sqrt{q}} + 1 - \frac{1}{q} \text{ almost surely,}$$

and letting $q \uparrow \infty$ concludes the proof of the lower bound. ∎

Corollary 5.3 *Suppose* $\{B(t): t \geqslant 0\}$ *is a standard Brownian motion. Then, almost surely,*

$$\limsup_{h \downarrow 0} \frac{|B(h)|}{\sqrt{2h \log \log(1/h)}} = 1.$$

Proof. By Theorem 1.9 the process $\{X(t): t \geqslant 0\}$ defined by $X(t) = tB(1/t)$ for $t > 0$ is a standard Brownian motion. Hence, using Theorem 5.1, we get

$$\limsup_{h \downarrow 0} \frac{|B(h)|}{\sqrt{2h \log \log(1/h)}} = \limsup_{t \uparrow \infty} \frac{|X(t)|}{\sqrt{2t \log \log t}} = 1. \qquad \blacksquare$$

The law of the iterated logarithm is a result which is easier to prove for Brownian motion than for random walks, as scaling arguments can be used to good effect in the proof. We now use an ad hoc argument to obtain a law of the iterated logarithm for simple random walks, i.e. the random walk with increments taking the values ± 1 with equal probability, from Theorem 5.1. A version for more general walks will follow with analogous arguments from the embedding techniques of Section 5.3, see Theorem 5.17.

Theorem 5.4 (Law of the Iterated Logarithm for simple random walk) *Let* $\{S_n : n \geqslant 0\}$ *be a simple random walk. Then, almost surely,*

$$\limsup_{n \to \infty} \frac{S_n}{\sqrt{2n \log \log n}} = 1.$$

We now start the technical work to transfer the result from Brownian motion to simple random walk. The next result shows that the limsup does not change if we only look along a sufficiently dense sequence of random times. We abbreviate $\psi(t) = \sqrt{2t \log \log(t)}$.

Lemma 5.5 *If* $\{T_n : n \geqslant 1\}$ *is a sequence of random times (not necessarily stopping times) satisfying* $T_n \to \infty$ *and* $T_{n+1}/T_n \to 1$ *almost surely, then*

$$\limsup_{n \to \infty} \frac{B(T_n)}{\psi(T_n)} = 1 \text{ almost surely.}$$

Furthermore, if $T_n/n \to a > 0$ *almost surely, then*

$$\limsup_{n \to \infty} \frac{B(T_n)}{\psi(an)} = 1 \text{ almost surely.}$$

Proof. The upper bound follows from the upper bound for continuous time without any conditions on $\{T_n : n \geqslant 1\}$. For the lower bound some restrictions are needed, which prevent us from choosing, for example, $T_0 = 0$ and $T_n = \inf\{t > T_{n-1} + 1: B(t) < \frac{1}{n}\}$. Our conditions $T_{n+1}/T_n \to 1$ and $T_n \to \infty$ make sure that the times are sufficiently dense to rule out this effect. Define, for fixed $q > 4$,

$$D_k = \left\{ B(q^k) - B(q^{k-1}) \geqslant \psi(q^k - q^{k-1}) \right\},$$

$$\Omega_k = \left\{ \min_{q^k \leqslant t \leqslant q^{k+1}} B(t) - B(q^k) \geqslant -\sqrt{q^k} \right\} \text{ and } D_k^* = D_k \cap \Omega_k.$$

Note that D_k and Ω_k are independent events. From Brownian scaling and Lemma 12.9 it is easy to see, as in the proof of Theorem 5.1, that, for a suitable constant $c > 0$,

$$\mathbb{P}(D_k) = \mathbb{P}\Big\{ B(1) \geqslant \frac{\psi(q^k - q^{k-1})}{\sqrt{q^k - q^{k-1}}} \Big\} \geqslant \frac{c}{k \log k}.$$

Moreover, by scaling, $\mathbb{P}(\Omega_k) =: c_q > 0$, and c_q that does not depend on k. As $\mathbb{P}(D_k^*) = c_q \mathbb{P}(D_k)$ the sum $\sum_k \mathbb{P}(D_{2k}^*)$ is infinite. As the events $\{D_{2k}^* : k \geqslant 1\}$ are independent, by the Borel–Cantelli lemma, for infinitely many (even) k,

$$\min_{q^k \leqslant t \leqslant q^{k+1}} B(t) \geqslant B(q^{k-1}) + \psi(q^k - q^{k-1}) - \sqrt{q^k}.$$

By Remark 5.2, for all sufficiently large k, we have $B(q^{k-1}) \geqslant -2\psi(q^{k-1})$ and, by easy asymptotics, $\psi(q^k - q^{k-1}) \geqslant \psi(q^k)(1 - \frac{1}{q})$. Hence, for infinitely many k,

$$\min_{q^k \leqslant t \leqslant q^{k+1}} B(t) \geqslant \psi(q^k - q^{k-1}) - 2\psi(q^{k-1}) - \sqrt{q^k} \geqslant \psi(q^k) \Big(1 - \frac{1}{q} - \frac{2}{\sqrt{q}} \Big) - \sqrt{q^k},$$

with the right hand side being positive by our choice of q. Now define $n(k) = \min\{n : T_n > q^k\}$. Since the ratios T_{n+1}/T_n tend to 1, it follows that for any fixed $\varepsilon > 0$, we have $q^k \leqslant T_{n(k)} < q^k (1 + \varepsilon)$ for all large k. Thus, for infinitely many k,

$$\frac{B(T_{n(k)})}{\psi(T_{n(k)})} \geqslant \frac{\psi(q^k)}{\psi(q^k(1 + \varepsilon))} \Big(1 - \frac{1}{q} - \frac{2}{\sqrt{q}} \Big) - \frac{\sqrt{q^k}}{\psi(q^k)}.$$

But since $\sqrt{q^k}/\psi(q^k) \to 0$ and $\psi(q^k)/\psi(q^k(1 + \varepsilon)) \to 1/\sqrt{1 + \varepsilon}$, we conclude that

$$\limsup_{n \to \infty} \frac{B(T_n)}{\psi(T_n)} \geqslant \frac{1}{\sqrt{1 + \varepsilon}} \Big(1 - \frac{1}{q} - \frac{2}{\sqrt{q}} \Big),$$

and since the left hand side does not depend on q and $\varepsilon > 0$ we can let $q \uparrow \infty$ and $\varepsilon \downarrow 0$ to arrive at the desired conclusion. For the last part, note that if $T_n/n \to a$ then $\psi(T_n)/\psi(an) \to 1$. ∎

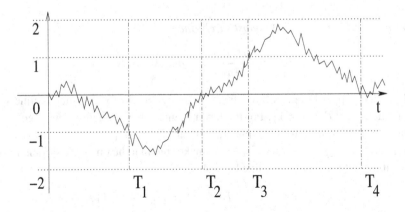

Fig. 5.2. Embedding simple random walk into Brownian motion

Proof of Theorem 5.4. To prove the law of the iterated logarithm for simple random walk, we let $T_0 = 0$ and, for $n \geqslant 1$,

$$T_n = \min\{t > T_{n-1} \colon |B(t) - B(T_{n-1})| = 1\}.$$

The times T_n are stopping times for Brownian motion and, hence, by the strong Markov property, the waiting times $T_n - T_{n-1}$ are independent and identically distributed random variables. Obviously, $\mathbb{P}\{B(T_n) - B(T_{n-1}) = 1\} = \mathbb{P}\{B(T_n) - B(T_{n-1}) = -1\} = \frac{1}{2}$, and therefore $\{B(T_n) \colon n \geqslant 0\}$ is a simple random walk. By Theorem 2.49, we have $\mathbb{E}[T_n - T_{n-1}] = 1$, and hence the law of large numbers ensures that T_n/n converges almost surely to 1, and the theorem follows from Lemma 5.5. ∎

Remark 5.6 The technique used to get Theorem 5.4 from Theorem 5.1 is based on finding an increasing sequence of stopping times $\{T_n \colon n \geqslant 0\}$ for the Brownian motion, such that $S_n = B(T_n)$ defines a simple random walk, while we keep some control over the size of T_n. This 'embedding technique' will be extended substantially in Section 5.3. ◇

5.2 Points of increase for random walk and Brownian motion

A point $t \in (0, \infty)$ is a local point of increase for the function $f \colon (0, \infty) \to \mathbb{R}$ if for some open interval (a, b) containing t we have $f(s) \leqslant f(t)$ for all $s \in (a, t)$ and $f(t) \leqslant f(s)$ for all $s \in (t, b)$. In this section we show that Brownian motion almost surely has no local points of increase. Our proof uses a combinatorial argument to derive a quantitative result for simple random walks, and then uses this result to study the case of Brownian motion. A crucial tool in the proof is an inequality of Harris [Ha60], which is of some independent interest.

Theorem 5.7 (Harris' inequality) *Suppose that $X = (X_1, \ldots, X_d)$ is a random variable with values in $\mathbb{R}^d$ and independent coordinates. Let $f, g \colon \mathbb{R}^d \to \mathbb{R}$ be measurable functions, which are non-decreasing in each coordinate. Then,*

$$\mathbb{E}[f(X)g(X)] \geqslant \mathbb{E}[f(X)]\,\mathbb{E}[g(X)]\,, \tag{5.2}$$

provided the above expectations are well-defined.

Proof. One can argue, using the monotone convergence theorem, that it suffices to prove the result when f and g are bounded. We assume f and g are bounded and proceed by induction on the dimension d. Suppose first that $d = 1$. Note that

$$(f(x) - f(y))(g(x) - g(y)) \geqslant 0\,, \qquad \text{for all } x, y \in \mathbb{R}.$$

Therefore, for Y an independent random variable with the same distribution as X,

$$
\begin{aligned}
0 \;&\leqslant\; \mathbb{E}\big[(f(X) - f(Y))(g(X) - g(Y))\big] \\
&=\; 2\mathbb{E}\big[f(X)g(X)\big] - 2\mathbb{E}\big[f(X)\big]\,\mathbb{E}\big[g(Y)\big]\,,
\end{aligned}
$$

and (5.2) follows easily. Now, suppose (5.2) holds for $d - 1$. Define

$$f_1(x_1) = \mathbb{E}\big[f(x_1, X_2, \ldots, X_d)\big]\,,$$

and define g_1 similarly. Note that $f_1(x_1)$ and $g_1(x_1)$ are non-decreasing functions of x_1. Since f and g are bounded, we may apply Fubini's theorem to write the left hand side of (5.2) as

$$\int_{\mathbb{R}} \mathbb{E}\big[f(x_1, X_2, \ldots, X_d)\, g(x_1, X_2, \ldots, X_d)\big]\, d\mu_1(x_1), \qquad (5.3)$$

where μ_1 denotes the law of X_1. The expectation in the integral is at least $f_1(x_1)g_1(x_1)$ by the induction hypothesis. Thus, using the result for the $d = 1$ case, we can bound (5.3) from below by $\mathbb{E}[f_1(X_1)]\,\mathbb{E}[g_1(X_1)]$, which equals the right hand side of (5.2), completing the proof. ∎

For the rest of this section, let $X_1, X_2, \ldots$ be independent random variables with

$$\mathbb{P}\{X_i = 1\} = \mathbb{P}\{X_i = -1\} = \tfrac{1}{2},$$

and let $S_k = \sum_{i=1}^{k} X_i$ be their partial sums. Denote

$$p_n = \mathbb{P}\{S_i \geqslant 0 \text{ for all } 1 \leqslant i \leqslant n\}. \qquad (5.4)$$

Then $\{S_n$ is a maximum among $S_0, S_1, \ldots S_n\}$ is precisely the event that the reversed random walk given by $S'_k = X_n + \ldots + X_{n-k+1}$ is nonnegative for all $k = 1, \ldots, n$. Hence this event also has probability p_n. The following lemma gives the order of magnitude of p_n, the proof will be given as Exercise 5.4.

Lemma 5.8 *There are positive constants C_1 and C_2 such that*

$$\frac{C_1}{\sqrt{n}} \leqslant \mathbb{P}\{S_i \geqslant 0 \text{ for all } 1 \leqslant i \leqslant n\} \leqslant \frac{C_2}{\sqrt{n}} \text{ for all } n \geqslant 1.$$

The next lemma expresses, in terms of the p_n defined in (5.4), the probability that S_j stays between 0 and S_n for j between 0 and n.

Lemma 5.9 *We have $p_n^2 \leqslant \mathbb{P}\{0 \leqslant S_j \leqslant S_n \text{ for all } 1 \leqslant j \leqslant n\} \leqslant p_{\lfloor n/2 \rfloor}^2$.*

Proof. The two events

$$\begin{aligned} A &= \{0 \leqslant S_j \text{ for all } j \leqslant \lfloor n/2 \rfloor\} \text{ and} \\ B &= \{S_j \leqslant S_n \text{ for } j \geqslant \lfloor n/2 \rfloor\} \end{aligned}$$

are independent, since A depends only on $X_1, \ldots, X_{\lfloor n/2 \rfloor}$ and B depends only on the remaining $X_{\lfloor n/2 \rfloor + 1}, \ldots, X_n$. Therefore,

$$\mathbb{P}\{0 \leqslant S_j \leqslant S_n \text{ for all } j \in \{0, \ldots, n\}\} \leqslant \mathbb{P}(A \cap B) = \mathbb{P}(A)\mathbb{P}(B) \leqslant p_{\lfloor n/2 \rfloor}^2,$$

which proves the upper bound.

For the lower bound, we let $f(x_1, \ldots, x_n) = 1$ if all the partial sums $x_1 + \ldots + x_k$ for $k = 1, \ldots, n$ are nonnegative, and $f(x_1, \ldots, x_n) = 0$ otherwise. Also, define $g(x_1, \ldots, x_n) =$

$f(x_n, \ldots, x_1)$. Then f and g are non-decreasing in each component. By Harris' inequality, for $X = (X_1, \ldots, X_n)$, we have $\mathbb{E}[f(X)g(X)] \geqslant \mathbb{E}[f(X)] \mathbb{E}[g(X)] = p_n^2$. Also,

$$
\begin{aligned}
\mathbb{E}[f(X)g(X)] &= \mathbb{P}\{X_1 + \ldots + X_j \geqslant 0 \text{ and } X_{j+1} + \ldots + X_n \geqslant 0 \text{ for all } j\} \\
&= \mathbb{P}\{0 \leqslant S_j \leqslant S_n \text{ for all } 1 \leqslant j \leqslant n\},
\end{aligned}
$$

which proves the lower bound. ∎

Definition 5.10.

(a) A sequence $s_0, s_1, \ldots, s_n$ of reals has a (global) **point of increase** at $k \in \{0, \ldots, n\}$, if $s_i \leqslant s_k$ for $i = 0, 1, \ldots, k-1$ and $s_k \leqslant s_j$ for $j = k+1, \ldots, n$.

(b) A real-valued function f has a **global point of increase in the interval** (a, b) if there is a point $t \in (a, b)$ such that $f(s) \leqslant f(t)$ for all $s \in (a, t)$ and $f(t) \leqslant f(s)$ for all $s \in (t, b)$. t is a **local point of increase** if it is a global point of increase in some interval. ◇

Theorem 5.11 *Let $S_0, S_1, \ldots, S_n$ be a simple random walk. Then*

$$
\mathbb{P}\{S_0, \ldots, S_n \text{ has a point of increase}\} \leqslant \frac{C}{\log n},
$$

for all $n > 1$, where C does not depend on n.

The key to Theorem 5.11 is the following upper bound, which holds for more general random walks. It will be proved as Exercise 5.5.

Lemma 5.12 *For any random walk $\{S_j : j \geqslant 0\}$ on the line,*

$$
\mathbb{P}\{S_0, \ldots, S_n \text{ has a point of increase}\} \leqslant 2 \frac{\sum_{k=0}^{n} p_k p_{n-k}}{\sum_{k=0}^{\lfloor n/2 \rfloor} p_k^2}. \tag{5.5}
$$

Remark 5.13 Equation (5.5) is easy to interpret: The expected number of points of increase by time n is the numerator in (5.5), and given that there is at least one point of increase in $[0, n/2]$, the expected number of these points in $[0, n]$ is bounded from below by the denominator. ◇

Proof of Theorem 5.11. To bound the numerator in (5.5), we can use symmetry to deduce from Lemma 5.8 that

$$
\sum_{k=0}^{n} p_k p_{n-k} \leqslant 2 + 2 \sum_{k=1}^{\lfloor n/2 \rfloor} p_k p_{n-k} \leqslant 2 + 2C_2^2 \sum_{k=1}^{\lfloor n/2 \rfloor} k^{-1/2}(n-k)^{-1/2}
$$

$$
\leqslant 2 + 4C_2^2 n^{-1/2} \sum_{k=1}^{\lfloor n/2 \rfloor} k^{-1/2},
$$

which is bounded above because the last sum is bounded by a constant multiple of $n^{1/2}$. Since Lemma 5.8 implies that the denominator in (5.5) is at least $C_1^2 \log \lfloor n/2 \rfloor$, this completes the proof.   ∎

We now see how we can use embedding ideas to pass from the result about *simple* random walks to the result about Brownian motion.

Theorem 5.14 *Brownian motion almost surely has no local points of increase.*

Proof.   To deduce this, it suffices to apply Theorem 5.11 to a *simple* random walk on the integers. Indeed, it clearly suffices to show that the Brownian motion $\{B(t): t \geqslant 0\}$ almost surely has no global points of increase in a fixed time interval (a, b) with rational endpoints. Sampling the Brownian motion when it visits a lattice yields a simple random walk; by refining the lattice, we may make this walk as long as we wish and capture all required detail.

More precisely, for any vertical spacing $h > 0$ define τ_0 to be the first $t > a$ such that $B(t)$ is an integral multiple of h, and for $i \geqslant 0$ let τ_{i+1} be the minimal $t \geqslant \tau_i$ such that $|B(t) - B(\tau_i)| = h$. Define $N_b = \max\{k \in \mathbb{Z} : \tau_k < b\}$. For integers i satisfying $0 \leqslant i \leqslant N_b$, define

$$S_i = \frac{B(\tau_i) - B(\tau_0)}{h}.$$

Then $\{S_i: i = 1, \ldots, N_b\}$ is a finite portion of a simple random walk. If the Brownian motion has a global point of increase $t_0 \in (a, b)$, and if k is an integer such that $\tau_{k-1} \leqslant t_0 \leqslant \tau_k$, then this random walk has points of increase at $k - 1$ and k. Similarly, if $t_0 < \tau_0$ or $t_0 > \tau_{N_b}$, then $k = 0$, resp. $k = N_b$, is a point of increase for the random walk. Therefore, for all n,

$$\mathbb{P}\{\{B(t): t \geqslant 0\} \text{ has a global point of increase in } (a, b)\}$$

$$\leqslant \mathbb{P}\{N_b \leqslant n\} + \sum_{m=n+1}^{\infty} \mathbb{P}\{S_0, \ldots, S_m \text{ has a point of increase and } N_b = m\}. \tag{5.6}$$

Note that $N_b \leqslant n$ implies $|B(b) - B(a)| \leqslant (n+1)h$, so

$$\mathbb{P}\{N_b \leqslant n\} \leqslant \mathbb{P}\{|B(b) - B(a)| \leqslant (n+1)h\} = \mathbb{P}\Big\{|Z| \leqslant \frac{(n+1)h}{\sqrt{b-a}}\Big\},$$

where Z has a standard normal distribution. Since $S_0, \ldots, S_m$, conditioned on $N_b = m$ is a finite portion of a simple random walk, it follows from Theorem 5.11 that for some constant C, we have

$$\sum_{m=n+1}^{\infty} \mathbb{P}\{S_0, \ldots, S_m \text{ has a point of increase, and } N_b = m\}$$

$$\leqslant \sum_{m=n+1}^{\infty} \mathbb{P}\{N_b = m\} \frac{C}{\log m} \leqslant \frac{C}{\log(n+1)}.$$

Thus, the probability in (5.6) can be made arbitrarily small by first taking n large and then picking $h > 0$ sufficiently small.   ∎

5.3 Skorokhod embedding and Donsker's invariance principle

In the proof of Theorem 5.4 we have made use of the fact that there exists a stopping time T for linear Brownian motion with the property that $\mathbb{E}[T] < \infty$ and the law of $B(T)$ is the uniform distribution on $\{-1, 1\}$. To use the same method for random walks $\{S_n : n \in \mathbb{N}\}$ with general increments, it would be necessary to find, for a given random variable X representing an increment, a stopping time T with $\mathbb{E}[T] < \infty$, such that $B(T)$ has the law of X. This problem is called the *Skorokhod embedding problem*. By Wald's lemmas, Theorem 2.44 and Theorem 2.48, for any integrable stopping time T, we have

$$\mathbb{E}\big[B(T)\big] = 0 \qquad \text{and} \qquad \mathbb{E}\big[B(T)^2\big] = \mathbb{E}[T] < \infty,$$

so that the Skorokhod embedding problem can only be solved for random variables X with mean zero and finite second moment. However, these are the only restrictions, as the following result shows.

Theorem 5.15 (Skorokhod embedding theorem) *Suppose that $\{B(t) : t \geqslant 0\}$ is a standard Brownian motion and that X is a real valued random variable with $E[X] = 0$ and $E[X^2] < \infty$. Then there exists a stopping time T, with respect to the natural filtration $(\mathcal{F}(t) : t \geqslant 0)$ of the Brownian motion, such that $B(T)$ has the law of X and $\mathbb{E}[T] = E[X^2]$.*

Example 5.16 Assume that X may take two values $a < b$. In order that $E[X] = 0$ we must have $a < 0 < b$ and $P\{X = a\} = b/(b-a)$ and $P\{X = b\} = -a/(b-a)$. We have seen in Theorem 2.49 that, for the stopping time $T = \inf\{t : B(t) \notin (a, b)\}$ the random variable $B(T)$ has the same distribution as X, and that $\mathbb{E}[T] = -ab$ is finite. ◇

Note that the Skorokhod embedding theorem allows us to use the arguments developed for the proof of the law of the iterated logarithm for simple random walks, Theorem 5.4, and obtain a much more general result.

Theorem 5.17 (Hartman–Wintner law of the iterated logarithm) *Let $\{S_n : n \in \mathbb{N}\}$ be a random walk with increments $S_n - S_{n-1}$ of zero mean and finite variance σ^2. Then*

$$\limsup_{n \to \infty} \frac{S_n}{\sqrt{2\sigma^2 \, n \log \log n}} = 1.$$

We now present two proofs of the Skorokhod embedding theorem, which actually represent different constructions of the required stopping times. Both approaches, Dubins' embedding, and the Azéma–Yor embedding are very elegant and have their own merits.

5.3.1 The Dubins' embedding theorem

The first one, due to Dubins [Du68], is particularly simple and based on the notion of binary splitting martingales. We say that a martingale $\{X_n : n \in \mathbb{N}\}$ is **binary splitting** if, whenever for some $x_0, \ldots, x_n \in \mathbb{R}$ the event

$$A(x_0, \ldots, x_n) := \{X_0 = x_0, X_1 = x_1, \ldots, X_n = x_n\}$$

has positive probability, the random variable X_{n+1} conditioned on $A(x_0, \ldots, x_n)$ is supported on at most two values.

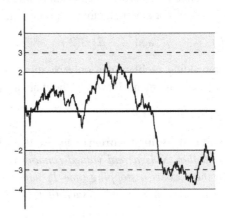

Fig. 5.3. Dubins' embedding for the uniform distribution on $\{-4, -2, 0, 2, 4\}$: First go until you hit $\{-3, 3\}$, in this picture you hit -3. Given that, continue until you hit either -2 or -4, in this picture you hit -2. Hence $B(T) = -2$ for this sample.

Lemma 5.18 *Let X be a random variable with $E[X^2] < \infty$. Then there is a binary splitting martingale $\{X_n : n \in \mathbb{N}\}$ such that $X_n \to X$ almost surely and in $\mathbf{L}^2$.*

Proof. We define the martingale $\{X_n : n \in \mathbb{N}\}$ and the associated filtration $(\mathcal{G}_n : n \in \mathbb{N})$ recursively. Let $\mathcal{G}_0$ be the trivial σ-algebra (consisting only of the empty set and the underlying probability space itself) and $X_0 = EX$. Define the random variable ξ_0 by

$$\xi_0 = \begin{cases} 1, & \text{if } X \geqslant X_0, \\ -1, & \text{if } X < X_0. \end{cases}$$

For any $n > 0$, let $\mathcal{G}_n = \sigma(\xi_0, \ldots, \xi_{n-1})$ and $X_n = E[X \mid \mathcal{G}_n]$. Also define the random variable ξ_n by

$$\xi_n = \begin{cases} 1, & \text{if } X \geqslant X_n, \\ -1, & \text{if } X < X_n. \end{cases}$$

Note that $\mathcal{G}_n$ is generated by a partition $\mathcal{P}_n$ of the underlying probability space into 2^n sets, each of which has the form $A(x_0, \ldots, x_n)$. As each element of $\mathcal{P}_n$ is a union of two elements of $\mathcal{P}_{n+1}$, the martingale $\{X_n : n \in \mathbb{N}\}$ is binary splitting. Also we have, for example as in (12.1) in the appendix, that

$$E[X^2] = E[(X - X_n)^2] + E[X_n^2] \geqslant E[X_n^2].$$

Hence $\{X_n : n \in \mathbb{N}\}$ is bounded in $\mathbf{L}^2$ and, from the convergence theorem for $\mathbf{L}^2$-bounded martingales and Lévy's upward theorem, see Theorems 12.28 and 12.25 in the appendix, we get

$$X_n \to X_\infty := E[X \,|\, \mathcal{G}_\infty] \qquad \text{almost surely and in } \mathbf{L}^2,$$

where $\mathcal{G}_\infty = \sigma\left(\bigcup_{i=0}^\infty \mathcal{G}_i\right)$. To conclude the proof we have to show that $X = X_\infty$ almost surely. We claim that, almost surely,

$$\lim_{n \uparrow \infty} \xi_n \left(X - X_{n+1}\right) = |X - X_\infty|. \tag{5.7}$$

Indeed, if $X(\omega) = X_\infty(\omega)$ this is easy. If $X(\omega) < X_\infty(\omega)$ then for some large enough N we have $X(\omega) < X_n(\omega)$ for any $n > N$, hence $\xi_n = -1$ and (5.7) holds. Similarly, if $X(\omega) > X_\infty(\omega)$ then $\xi_n = 1$ for $n > N$ and so (5.7) holds. Using that ξ_n is $\mathcal{G}_{n+1}$-measurable, we find that

$$E[\xi_n (X - X_{n+1})] = E[\xi_n E[X - X_{n+1} \,|\, \mathcal{G}_{n+1}]] = 0.$$

Recall that if $Y_n \to Y$ almost surely, and $\{Y_n : n = 0, 1, \cdots\}$ is $\mathbf{L}^2$-bounded, then $EY_n \to EY$ (see, for example, the discussion of uniform integrability in 12.3 of the appendix). Hence, as the left hand side of (5.7) is $\mathbf{L}^2$-bounded, we conclude that $E|X - X_\infty| = 0$. ∎

Proof of Theorem 5.15. From Lemma 5.18 we take a binary splitting martingale $\{X_n : n \in \mathbb{N}\}$ such that $X_n \to X$ almost surely and in $\mathbf{L}^2$. Recall from the example preceding this proof that if X is supported on a set of two elements $\{-a, b\}$ for some $a, b > 0$ then $T = \inf\{t : B(t) \in \{-a, b\}\}$ is the required stopping time. Hence, as X_n conditioned on $A(x_0, \ldots, x_{n-1})$ is supported on at most two values it is clear we can find a sequence of stopping times $T_0 \leqslant T_1 \leqslant \ldots$ such that $B(T_n)$ is distributed as X_n and $\mathbb{E}T_n = E[X_n^2]$. As T_n is an increasing sequence, we have $T_n \uparrow T$ almost surely for some stopping time T. Also, by the monotone convergence theorem

$$\mathbb{E}T = \lim_{n \uparrow \infty} \mathbb{E}T_n = \lim_{n \uparrow \infty} E[X_n^2] = E[X^2].$$

As $B(T_n)$ converges in distribution to X by construction, and converges almost surely to $B(T)$ by continuity of the Brownian sample paths, we get that $B(T)$ is distributed as X. ∎

5.3.2 The Azéma–Yor embedding theorem

In this section we discuss a second solution to the Skorokhod embedding problem with a more explicit construction of the stopping times.

Theorem* 5.19 (Azéma–Yor embedding theorem) *Suppose that X is a real valued random variable with $E[X] = 0$ and $E[X^2] < \infty$. Let*

$$\Psi(x) = E[X \,|\, X \geqslant x] \qquad \text{if } P\{X \geqslant x\} > 0,$$

and $\Psi(x) = 0$ otherwise. For a Brownian motion $\{B(t) : t \geqslant 0\}$ let $\{M(t) : t \geqslant 0\}$ be the maximum process and define a stopping time τ by

$$\tau = \inf\{t \geqslant 0 : M(t) \geqslant \Psi(B(t))\}.$$

Then $\mathbb{E}[\tau] = E[X^2]$ and $B(\tau)$ has the same law as X.

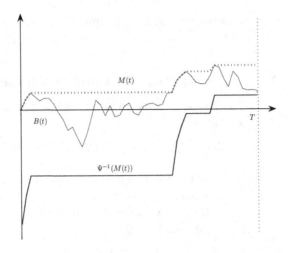

Fig. 5.4. The Azéma–Yor embedding: the path is stopped when the Brownian motion hits the level $\Psi^{-1}(M(t))$, where $\Psi^{-1}(x) = \sup\{b\colon \Psi(b) \leqslant x\}$.

We proceed in three steps. In the first step we formulate an embedding for random variables taking only finitely many values.

Lemma 5.20 *Suppose the random variable X with $\mathbb{E}X = 0$ takes only finitely many values*

$$x_1 < x_2 < \cdots < x_n.$$

Define $y_1 < y_2 < \cdots < y_{n-1}$ by $y_i = \Psi(x_{i+1})$, and define stopping times $T_0 = 0$ and

$$T_i = \inf\big\{t \geqslant T_{i-1}\colon B(t) \notin (x_i, y_i)\big\} \qquad for\ i \leqslant n-1.$$

Then $T = T_{n-1}$ satisfies $\mathbb{E}[T] = E[X^2]$ and $B(T)$ has the same law as X.

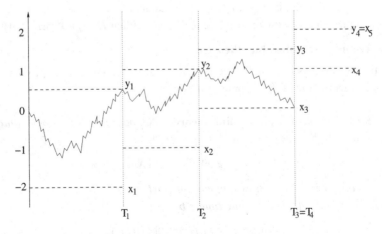

Fig. 5.5. The Azéma–Yor embedding for the uniform distribution on the set $\{-2, -1, 0, 1, 2\}$. The drawn path samples the value $B(T) = 0$ with $T = T_4$.

Proof. First observe that $y_i \geqslant x_{i+1}$ and equality holds if and only if $i = n - 1$. We have $\mathbb{E}[T_{n-1}] < \infty$, by Theorem 2.49, and $\mathbb{E}[T_{n-1}] = \mathbb{E}[B(T_{n-1})^2]$, from Theorem 2.48. For $i = 1, \ldots, n - 1$ define random variables

$$Y_i = \begin{cases} E[X \mid X \geqslant x_{i+1}] & \text{if } X \geqslant x_{i+1}, \\ X & \text{if } X \leqslant x_i. \end{cases}$$

Note that Y_1 has expectation zero and takes on the two values x_1, y_1. For $i \geqslant 2$, given $Y_{i-1} = y_{i-1}$, the random variable Y_i takes the values x_i, y_i and has expectation y_{i-1}. Given $Y_{i-1} = x_j, j \leqslant i - 1$ we have $Y_i = x_j$. Note that $Y_{n-1} = X$. We now argue that

$$(B(T_1), \ldots, B(T_{n-1})) \overset{d}{=} (Y_1, \ldots, Y_{n-1}).$$

Clearly, $B(T_1)$ can take only the values x_1, y_1 and has expectation zero, hence the law of $B(T_1)$ agrees with the law of Y_1. For $i \geqslant 2$, given $B(T_{i-1}) = y_{i-1}$, the random variable $B(T_i)$ takes the values x_i, y_i and has expectation y_{i-1}. Given $B(T_{i-1}) = x_j$ where $j \leqslant i - 1$, we have $B(T_i) = x_j$. Hence the two tuples have the same law and, in particular, $B(T_{n-1})$ has the same law as X. ∎

In the second step, we show that the stopping time we have constructed in Lemma 5.20 agrees with the stopping time τ in the Azéma–Yor embedding.

Lemma 5.21 *The stopping time T constructed in Lemma 5.20 and the stopping time τ in Theorem 5.19 are equal.*

Proof. Suppose that $B(T_{n-1}) = x_i$, and hence $\Psi(B(T_{n-1})) = y_{i-1}$. If $i \leqslant n - 1$, then i is minimal with the property that $B(T_i) = \cdots = B(T_{n-1})$, and thus $B(T_{i-1}) \neq B(T_i)$. Hence $M(T_{n-1}) \geqslant y_{i-1}$. If $i = n$ we also have $M(T_{n-1}) = x_n \geqslant y_{i-1}$, which implies in any case that $\tau \leqslant T$. Conversely, if $T_{i-1} \leqslant t < T_i$ then $B(t) \in (x_i, y_i)$ and this implies $M(t) < y_i \leqslant \Psi(B(t))$. Hence $\tau \geqslant T$, and altogether we have seen that $T = \tau$. ∎

This completes the proof of Theorem 5.19 for random variables taking finitely many values. The general case follows from a limiting procedure, which is left as Exercise 5.10.

5.3.3 The Donsker invariance principle

Let $\{X_n : n \geqslant 0\}$ be a sequence of independent and identically distributed random variables and assume that they are normalised, so that $\mathbb{E}[X_n] = 0$ and $\text{Var}(X_n) = 1$. This assumption is no loss of generality for X_n with finite variance, since we can always consider the normalisation

$$\frac{X_n - \mathbb{E}[X_n]}{\sqrt{\text{Var}(X_n)}}.$$

We look at the *random walk* generated by the sequence

$$S_n = \sum_{k=1}^{n} X_k,$$

and interpolate linearly between the integer points, i.e.

$$S(t) = S_{[t]} + (t - [t])(S_{[t]+1} - S_{[t]}).$$

This defines a random function $S \in \mathbf{C}[0, \infty)$. We now define a sequence $\{S_n^* : n \geqslant 1\}$ of random functions in $\mathbf{C}[0, 1]$ by

$$S_n^*(t) = \frac{S(nt)}{\sqrt{n}} \text{ for all } t \in [0, 1].$$

Theorem 5.22 (Donsker's invariance principle) *On the space* $\mathbf{C}[0, 1]$ *of continuous functions on the unit interval with the metric induced by the sup-norm, the sequence* $\{S_n^* : n \geqslant 1\}$ *converges in distribution to a standard Brownian motion* $\{B(t) : t \in [0, 1]\}$.

Remark 5.23 Donsker's invariance principle is also called the *functional central limit theorem*. The name *invariance principle* comes from the fact that the limit in Theorem 5.22 does not depend on the choice of the exact distribution of the normalised random variables X_n. ◇

The idea of the proof is to construct the random variables $X_1, X_2, X_3, \ldots$ on the same probability space as the Brownian motion in such a way that $\{S_n^* : n \geqslant 1\}$ is with high probability close to a scaling of this Brownian motion.

Lemma 5.24 *Suppose* $\{B(t) : t \geqslant 0\}$ *is a linear Brownian motion. Then, for any random variable* X *with mean zero and variance one, there exists a sequence of stopping times*

$$0 = T_0 \leqslant T_1 \leqslant T_2 \leqslant T_3 \leqslant \cdots$$

with respect to the Brownian motion, such that

(a) *the sequence* $\{B(T_n) : n \geqslant 0\}$ *has the distribution of the random walk with increments given by the law of* X,

(b) *the sequence of functions* $\{S_n^* : n \geqslant 0\}$ *constructed from this random walk satisfies*

$$\lim_{n \to \infty} \mathbb{P}\left\{ \sup_{0 \leqslant t \leqslant 1} \left| \frac{B(nt)}{\sqrt{n}} - S_n^*(t) \right| > \varepsilon \right\} = 0.$$

Proof. Using Skorokhod embedding, we define T_1 to be a stopping time with $\mathbb{E}[T_1] = 1$ such that $B(T_1) = X$ in distribution. By the strong Markov property,

$$\{B_2(t) : t \geqslant 0\} = \{B(T_1 + t) - B(T_1) : t \geqslant 0\}$$

is a Brownian motion and independent of $\mathcal{F}^+(T_1)$ and, in particular, of $(T_1, B(T_1))$. Hence we can define a stopping time T_2' for the Brownian motion $\{B_2(t) : t \geqslant 0\}$ such that $\mathbb{E}[T_2'] = 1$ and $B_2(T_2') = X$ in distribution. Then $T_2 = T_1 + T_2'$ is a stopping time for the original Brownian motion with $\mathbb{E}[T_2] = 2$, such that $B(T_2)$ is the second value in a random walk with increments given by the law of X. We can proceed inductively to get a sequence $0 = T_0 \leqslant T_1 \leqslant T_2 \leqslant T_3 < \ldots$ such that $S_n = B(T_n)$ is the embedded random walk, and $\mathbb{E}[T_n] = n$.

Abbreviate $W_n(t) = \frac{B(nt)}{\sqrt{n}}$ and let A_n be the event that there exists $t \in [0,1)$ such that $|S_n^*(t) - W_n(t)| > \varepsilon$. We have to show that $\mathbb{P}(A_n) \to 0$. Let $k = k(t)$ be the unique integer with $(k-1)/n \leqslant t < k/n$. Since S_n^* is linear on such an interval we have

$$A_n \subset \{ \text{ there exists } t \in [0,1) \text{ such that } |S_k/\sqrt{n} - W_n(t)| > \varepsilon \}$$
$$\cup \{ \text{ there exists } t \in [0,1) \text{ such that } |S_{k-1}/\sqrt{n} - W_n(t)| > \varepsilon \}.$$

As $S_k = B(T_k) = \sqrt{n} W_n(T_k/n)$, we obtain

$$A_n \subset A_n^* := \{ \text{ there exists } t \in [0,1) \text{ such that } |W_n(T_k/n)) - W_n(t)| > \varepsilon \}$$
$$\cup \{ \text{ there exists } t \in [0,1) \text{ such that } |W_n(T_{k-1}/n) - W_n(t)| > \varepsilon \}.$$

For given $0 < \delta < 1$ the event A_n^* is contained in

$$\{ \text{ there exist } s,t \in [0,2] \text{ such that } |s-t| < \delta, \ |W_n(s) - W_n(t)| > \varepsilon \} \tag{5.8}$$
$$\cup \{ \text{ there exists } t \in [0,1) \text{ such that } |T_k/n - t| \vee |T_{k-1}/n - t| \geqslant \delta \}. \tag{5.9}$$

Note that the probability of (5.8) does not depend on n. Choosing $\delta > 0$ small, we can make this probability as small as we wish, since Brownian motion is uniformly continuous on $[0,2]$. It remains to show that for arbitrary, fixed $\delta > 0$, the probability of (5.9) converges to zero as $n \to \infty$. To prove this we use that

$$\lim_{n \to \infty} \frac{T_n}{n} = \lim_{n \to \infty} \frac{1}{n} \sum_{k=1}^{n} (T_k - T_{k-1}) = 1 \text{ almost surely.}$$

This is Kolmogorov's law of large numbers for the sequence $\{T_k - T_{k-1}\}$ of independent identically distributed random variables with mean 1. Observe that for every sequence $\{a_n\}$ of reals one has

$$\lim_{n \to \infty} \frac{a_n}{n} = 1 \Rightarrow \lim_{n \to \infty} \sup_{0 \leqslant k \leqslant n} |a_k - k|/n = 0.$$

This is a matter of plain (deterministic) arithmetic and easily checked. Hence we have,

$$\lim_{n \to \infty} \mathbb{P}\left\{ \sup_{0 \leqslant k \leqslant n} \frac{|T_k - k|}{n} \geqslant \delta \right\} = 0. \tag{5.10}$$

Now recall that $t \in [(k-1)/n, k/n)$ and let $n > 2/\delta$. Then

$$\mathbb{P}\{ \text{ there exists } t \in [0,1] \text{ such that } |T_k/n - t| \vee |T_{k-1}/n - t| \geqslant \delta \}$$
$$\leqslant \ \mathbb{P}\left\{ \sup_{1 \leqslant k \leqslant n} \frac{(T_k - (k-1)) \vee (k - T_{k-1})}{n} \geqslant \delta \right\}$$
$$\leqslant \ \mathbb{P}\left\{ \sup_{1 \leqslant k \leqslant n} \frac{T_k - k}{n} \geqslant \delta/2 \right\} + \mathbb{P}\left\{ \sup_{1 \leqslant k \leqslant n} \frac{(k-1) - T_{k-1}}{n} \geqslant \delta/2 \right\},$$

and by (5.10) both summands converge to 0. ∎

Proof of the Donsker invariance principle. Choose the sequence of stopping times as in Lemma 5.24 and recall from the scaling property of Brownian motion that the random functions $\{W_n(t)\colon 0 \leqslant t \leqslant 1\}$ given by $W_n(t) = B(nt)/\sqrt{n}$ are standard Brownian motions. Suppose that $K \subset \mathbf{C}[0,1]$ is closed and define

$$K[\varepsilon] = \{f \in \mathbf{C}[0,1]\colon \|f - g\|_{\sup} \leqslant \varepsilon \text{ for some } g \in K\}.$$

Then $\mathbb{P}\{S_n^* \in K\} \leqslant \mathbb{P}\{W_n \in K[\varepsilon]\} + \mathbb{P}\{\|S_n^* - W_n\|_{\sup} > \varepsilon\}$. As $n \to \infty$, the second term goes to 0, whereas the first term does not depend on n and is equal to $\mathbb{P}\{B \in K[\varepsilon]\}$ for a Brownian motion B. As K is closed we have

$$\lim_{\varepsilon \downarrow 0} \mathbb{P}\{B \in K[\varepsilon]\} = \mathbb{P}\Big\{B \in \bigcap_{\varepsilon > 0} K[\varepsilon]\Big\} = \mathbb{P}\{B \in K\}.$$

Putting these facts together, we obtain $\limsup_{n \to \infty} \mathbb{P}\{S_n^* \in K\} \leqslant \mathbb{P}\{B \in K\}$, which is condition (ii) in the Portmanteau theorem, Theorem 12.6 in the appendix. Hence Donsker's invariance principle is proved. ∎

Below and in the following section we harvest a range of results for random walks, which we can transfer from Brownian motion by means of Donsker's invariance principle. Readers unfamiliar with the nature of convergence in distribution are recommended to look at the appendix, Chapter 12.1.

Theorem 5.25 *Suppose that $\{X_k\colon k \geqslant 1\}$ is a sequence of independent, identically distributed random variables with $\mathbb{E}[X_1] = 0$ and $0 < \mathbb{E}[X_1^2] = \sigma^2 < \infty$. Let $\{S_n\colon n \geqslant 0\}$ be the associated random walk and*

$$M_n = \max\{S_k\colon 0 \leqslant k \leqslant n\}$$

its maximal value up to time n. Then, for all $x \geqslant 0$,

$$\lim_{n \to \infty} \mathbb{P}\{M_n \geqslant x\sqrt{n}\} = \frac{2}{\sqrt{2\pi\sigma^2}} \int_x^\infty e^{-y^2/2\sigma^2} \, dy\,.$$

Proof. By scaling we can assume that $\sigma^2 = 1$. Suppose now that $g\colon \mathbb{R} \to \mathbb{R}$ is a continuous bounded function. Define a function $G\colon \mathbf{C}[0,1] \to \mathbb{R}$ by

$$G(f) = g\Big(\max_{x \in [0,1]} f(x)\Big),$$

and note that G is continuous and bounded. Then, by definition,

$$\mathbb{E}\Big[G(S_n^*)\Big] = \mathbb{E}\Big[g\Big(\max_{0 \leqslant t \leqslant 1} \frac{S(tn)}{\sqrt{n}}\Big)\Big] = \mathbb{E}\Big[g\Big(\frac{\max_{0 \leqslant k \leqslant n} S_k}{\sqrt{n}}\Big)\Big],$$

and

$$\mathbb{E}\Big[G(B)\Big] = \mathbb{E}\Big[g\Big(\max_{0 \leqslant t \leqslant 1} B(t)\Big)\Big].$$

Hence, by Donsker's invariance principle,

$$\lim_{n \to \infty} \mathbb{E}\Big[g\Big(\frac{M_n}{\sqrt{n}}\Big)\Big] = \mathbb{E}\Big[g\Big(\max_{0 \leqslant t \leqslant 1} B(t)\Big)\Big].$$

From the Portmanteau theorem, Theorem 12.6, and the reflection principle, Theorem 2.21, we infer

$$\lim_{n \to \infty} \mathbb{P}\{M_n \geqslant x\sqrt{n}\} = \mathbb{P}\{\max_{0 \leqslant t \leqslant 1} B(t) \geqslant x\} = 2\mathbb{P}\{B(1) \geqslant x\},$$

and the latter probability is the given integral. ∎

5.4 The arcsine laws for random walk and Brownian motion

We now discuss the two famous arcsine laws for Brownian motion and also for random walks. Their name comes from the **arcsine distribution**, which is the distribution on $(0,1)$ which has the density

$$\frac{1}{\pi\sqrt{x(1-x)}} \qquad \text{for } x \in (0,1).$$

The cumulative distribution function of an arcsine distributed random variable X is therefore given by

$$\mathbb{P}\{X \leqslant x\} = \frac{2}{\pi} \arcsin(\sqrt{x}) \qquad \text{for } x \in (0,1).$$

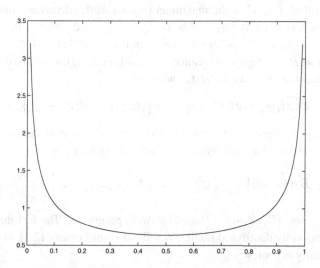

Fig. 5.6. The density of the arcsine distribution is concentrated near the boundary values 0 and 1.

The *first arcsine law* describes the law of the last passage over level zero by a Brownian motion or random walk running for finite time. In the case of a Brownian motion we shall find this law by a smart calculation, and then Donsker's invariance principle will allow us to transfer the result to random walks. Observe that the following result is surprising: the rightmost zero of Brownian motion in the interval $(0,1)$ is most likely to be near zero or one, see Figure 5.6.

Theorem 5.26 (First arcsine law for Brownian motion) *Let $\{B(t): t \geqslant 0\}$ be a standard linear Brownian motion. Then,*

 (a) *the random variable $L = \sup\{t \in [0,1]: B(t) = 0\}$, the last zero of Brownian motion in $[0,1]$, is arcsine distributed, and*

 (b) *the random variable $M^* \in [0,1]$, which is almost surely uniquely determined by*

$$B(M^*) = \max_{s \in [0,1]} B(s),$$

   is arcsine distributed.

Proof. By Theorem 2.11 Brownian motion has a unique maximum on the interval $[0,1]$, and hence the maximiser M^* is well-defined. Moreover, Theorem 2.34 shows that M^*, which is the last zero of the process $\{M(t) - B(t): t \geqslant 0\}$ has the same law as L. Hence it suffices to prove part (b).

Recall that $\{M(t): 0 \leqslant t \leqslant 1\}$ is defined by $M(t) = \max_{0 \leqslant s \leqslant t} B(s)$. For $s \in [0,1]$,

$$
\begin{aligned}
\mathbb{P}\{M^* < s\} &= \mathbb{P}\Big\{ \max_{0 \leqslant u \leqslant s} B(u) > \max_{s \leqslant v \leqslant 1} B(v) \Big\} \\
&= \mathbb{P}\Big\{ \max_{0 \leqslant u \leqslant s} B(u) - B(s) > \max_{s \leqslant v \leqslant 1} B(v) - B(s) \Big\} \\
&= \mathbb{P}\{M_1(s) > M_2(1-s)\},
\end{aligned}
$$

where $\{M_1(t): 0 \leqslant t \leqslant s\}$ is the maximum process of the Brownian motion $\{B_1(t): 0 \leqslant t \leqslant s\}$, which is given by $B_1(t) = B(s-t) - B(s)$, and $\{M_2(t): 0 \leqslant t \leqslant 1\}$ is the maximum process of the independent Brownian motion $\{B_2(t): 0 \leqslant t \leqslant 1-s\}$, which is given by $B_2(t) = B(s+t) - B(s)$. Since, by Theorem 2.21, for any fixed t, the random variable $M(t)$ has the same law as $|B(t)|$, we have

$$\mathbb{P}\{M_1(s) > M_2(1-s)\} = \mathbb{P}\{|B_1(s)| > |B_2(1-s)|\}.$$

Using the scaling invariance of Brownian motion we can express this in terms of a pair of two independent standard normal random variables Z_1 and Z_2, by

$$\mathbb{P}\{|B_1(s)| > |B_2(1-s)|\} = \mathbb{P}\{\sqrt{s}\,|Z_1| > \sqrt{1-s}\,|Z_2|\} = \mathbb{P}\Big\{ \frac{|Z_2|}{\sqrt{Z_1^2 + Z_2^2}} < \sqrt{s} \Big\}.$$

In polar coordinates, $(Z_1, Z_2) = (R\cos\theta, R\sin\theta)$ pointwise. The fact that the random variable θ is uniformly distributed on $[0, 2\pi]$ follows from Lemma 12.11 in the appendix. So the last quantity becomes

$$
\mathbb{P}\Big\{ \frac{|Z_2|}{\sqrt{Z_1^2 + Z_2^2}} < \sqrt{s} \Big\} = \mathbb{P}\{|\sin(\theta)| < \sqrt{s}\} = 4\mathbb{P}\{\theta < \arcsin(\sqrt{s})\}
$$

$$
= 4\left(\frac{\arcsin(\sqrt{s})}{2\pi} \right) = \frac{2}{\pi} \arcsin(\sqrt{s}).
$$

It follows by differentiating that M^* has density $(\pi\sqrt{s(1-s)})^{-1}$ for $s \in (0,1)$. ∎

For random walks the first arcsine law takes the form of a limit theorem, as the length of the walk tends to infinity.

Proposition 5.27 (Arcsine law for the last sign-change) *Suppose that $\{X_k : k \geqslant 1\}$ is a sequence of independent, identically distributed random variables with $\mathbb{E}[X_1] = 0$ and $0 < \mathbb{E}[X_1^2] = \sigma^2 < \infty$. Let $\{S_n : n \geqslant 0\}$ be the associated random walk and*

$$N_n = \max\{1 \leqslant k \leqslant n \,:\, S_k S_{k-1} \leqslant 0\}$$

the last time the random walk changes its sign before time n. Then, for all $x \in (0,1)$,

$$\lim_{n \to \infty} \mathbb{P}\{N_n \leqslant xn\} = \frac{2}{\pi} \arcsin(\sqrt{x}).$$

Proof. The strategy of proof is to use Theorem 5.26, and apply Donsker's invariance principle to extend the result to random walks. As N_n is unchanged under scaling of the random walk we may assume that $\sigma^2 = 1$. Define a bounded function g on $\mathbf{C}[0,1]$ by

$$g(f) = \max\{t \leqslant 1 \colon f(t) = 0\}.$$

It is clear that $g(S_n^*)$ differs from N_n/n by a term, which is bounded by $1/n$ and therefore vanishes asymptotically. Hence Donsker's invariance principle would imply convergence of N_n/n in distribution to $g(B) = \sup\{t \leqslant 1 \colon B(t) = 0\}$ — if g was continuous. g is *not* continuous, but we show that g is continuous on the set $\mathcal{C}$ of all $f \in \mathbf{C}[0,1]$ such that f takes positive and negative values in every neighbourhood of every zero and $f(1) \neq 0$. As, by Theorem 2.28, Brownian motion is almost surely in $\mathcal{C}$, we get from property (v) in the Portmanteau theorem, Theorem 12.6, and by Donsker's invariance principle, that, for every continuous bounded $h \colon \mathbb{R} \to \mathbb{R}$,

$$\lim_{n \to \infty} \mathbb{E}\left[h\left(\frac{N_n}{n}\right)\right] = \lim_{n \to \infty} \mathbb{E}\left[h \circ g(S_n^*)\right] = \mathbb{E}\left[h \circ g(B)\right] = \mathbb{E}\left[h(\sup\{t \leqslant 1 \colon B(t) = 0\})\right],$$

which completes the proof subject to the claim. To see that g is continuous on $\mathcal{C}$, let $\varepsilon > 0$ be given and $f \in \mathcal{C}$. Let

$$\delta_0 = \min_{t \in [g(f)+\varepsilon, 1]} |f(t)|,$$

and choose δ_1 such that $(-\delta_1, \delta_1) \subset f(g(f) - \varepsilon, g(f) + \varepsilon)$. Let $0 < \delta < \delta_0 \wedge \delta_1$. If now $\|h - f\|_\infty < \delta$, then h has no zero in $(g(f) + \varepsilon, 1]$, but has a zero in $(g(f) - \varepsilon, g(f) + \varepsilon)$, because there are $s, t \in (g(f) - \varepsilon, g(f) + \varepsilon)$ with $h(t) < 0$ and $h(s) > 0$. Thus $|g(h) - g(f)| < \varepsilon$. This shows that g is continuous on $\mathcal{C}$. ∎

There is a second arcsine law for Brownian motion, which describes the law of the random variable $\mathcal{L}\{t \in [0,1] \colon B(t) > 0\}$, the time spent by Brownian motion above the x-axis. This statement is much harder to derive directly for Brownian motion, though we will do this using more sophisticated tools in Chapter 8. At this stage we can use random walks to derive the result for Brownian motion.

Theorem 5.28 (Second arcsine law for Brownian motion) *Let $\{B(t) \colon t \geqslant 0\}$ be a standard linear Brownian motion. Then, $\mathcal{L}\{t \in [0,1] \colon B(t) > 0\}$, is arcsine distributed.*

The idea is to prove a direct relationship between the first maximum and the number of positive terms for a *simple* random walk by a combinatorial argument, and then transfer this to Brownian motion using Donsker's invariance principle.

Lemma 5.29 *Let $\{S_k\colon k = 1, \ldots, n\}$ be a simple, symmetric random walk on the integers. Then*

$$\#\{k \in \{1, \ldots, n\}\colon S_k > 0\} \stackrel{\mathrm{d}}{=} \min\{k \in \{0, \ldots, n\}\colon S_k = \max_{0 \leqslant j \leqslant n} S_j\}. \qquad (5.11)$$

Proof. Let $X_k = S_k - S_{k-1}$ for each $k \in \{1, \ldots, n\}$, with $S_0 := 0$. We rearrange the tuple $(X_1, \ldots, X_n)$ by

- placing first in *decreasing* order of k the terms X_k for which $S_k > 0$,
- and then in *increasing* order of k the X_k for which $S_k \leqslant 0$.

Denote the new tuple by $(Y_1, \ldots, Y_n) := T_n(X_1, \ldots, X_n)$. We first show that

$$(X_1, \ldots, X_n) \stackrel{\mathrm{d}}{=} (Y_1, \ldots, Y_n).$$

Note that, because the increments $(X_1, \ldots, X_n)$ are uniformly distributed on $\{-1, 1\}^n$, this is equivalent to showing that T_n is a bijection for every $n \in \mathbb{N}$. For $n = 1$ this is obviously true, and we continue by induction, assuming that T_k is a bijection for any $k \leqslant n - 1$. T_n is obviously a bijection on those tuples for which all partial sums are nonpositive. For all other tuples $(x_1, \ldots, x_n)$ let

$$\ell(x_1, \ldots, x_n) = \max\Big\{1 \leqslant k \leqslant n\colon \sum_{j=1}^{k} x_j > 0\Big\}.$$

Then, abbreviating $x = (x_1, \ldots, x_n)$,

$$T_n(x_1, \ldots, x_n) = \Big(x_{\ell(x)}, T_{\ell(x)-1}(x_1, \ldots, x_{\ell(x)-1}), x_{\ell(x)+1}, \ldots, x_n\Big).$$

Note that, if $y = T_n(x)$ then $\ell(x) = \ell(y)$, and therefore the inverse of T_n is given as

$$T_n^{-1}(y_1, \ldots, y_n) = \Big(T_{\ell(y)-1}^{-1}(y_2, \ldots, y_{\ell(y)}), y_1, y_{\ell(y)+1}, \ldots, y_n\Big),$$

proving that T_n is a bijection, as required.

Now $\{S_k(Y)\colon k = 1, \ldots, n\}$ given by $S_k(Y) = \sum_{j=1}^{k} Y_j$ is a random walk and we check by induction on n that

$$\#\{k \in \{1, \ldots, n\}\colon S_k(X) > 0\}$$
$$= \min\{k \in \{0, \ldots, n\}\colon S_k(Y) = \max_{0 \leqslant j \leqslant n} S_j(Y)\}. \qquad (5.12)$$

Indeed, this obviously holds for $n = 1$. Suppose it holds for fixed n. When X_{n+1} is appended there are two possibilities:

- Suppose $S_{n+1}(X) > 0$, so that

$$\#\{k \in \{1, \ldots, n+1\}\colon S_k(X) > 0\} = \#\{k \in \{1, \ldots, n\}\colon S_k(X) > 0\} + 1.$$

Denoting $(Y_1^*, \ldots, Y_{n+1}^*) = T_{n+1}(X_1, \ldots, X_{n+1})$ we have $Y_1^* = X_{n+1}$, and therefore

$$\min\{k \in \{0, \ldots, n+1\}\colon S_k(Y^*) = \max_{0 \leqslant j \leqslant n+1} S_j(Y^*)\}$$
$$= \min\{k \in \{0, \ldots, n\}\colon S_k(Y) = \max_{0 \leqslant j \leqslant n} S_j(Y)\} + 1.$$

In summary, appending the value X_{n+1} to $(X_1, \ldots, X_n)$ in this case, has led to the increase of both sides in Equation (5.12) by one.

- Suppose $S_{n+1}(X) \leqslant 0$, so that

$$\#\{k \in \{1, \ldots, n+1\} : S_k(X) > 0\} = \#\{k \in \{1, \ldots, n\} : S_k(X) > 0\}.$$

Then $Y_{n+1}^* = X_{n+1}$ and therefore

$$\min\{k \in \{0, \ldots, n+1\} : S_k(Y^*) = \max_{0 \leqslant j \leqslant n+1} S_j(Y^*)\}$$
$$= \min\{k \in \{0, \ldots, n\} : S_k(Y) = \max_{0 \leqslant j \leqslant n} S_j(Y)\}.$$

In summary, appending the value X_{n+1} to $(X_1, \ldots, X_n)$ in this case, has left both sides in Equation (5.12) unchanged.

This completes the induction step and proves the lemma. ∎

Proof of Theorem 5.28. Look at the right hand side of the equation (5.11), which divided by n can be written as $g(S_n^*)$ for the function $g \colon \mathbf{C}[0,1] \to [0,1]$ defined by

$$g(f) = \inf \left\{ t \in [0,1] : f(t) = \sup_{s \in [0,1]} f(s) \right\}.$$

The function g is continuous in every $f \in \mathbf{C}[0,1]$ which has a unique maximum, hence almost everywhere with respect to the distribution of Brownian motion. Hence, by Donsker's invariance principle and the Portmanteau theorem, Theorem 12.6, the right hand side in (5.11) divided by n converges to the distribution of $g(B)$, which by Theorem 5.26 is the arcsine distribution.

Similarly, by Exercise 5.11, the left hand side of (5.11) divided by n can be approximated in probability by $h(S_n^*)$ for the function $h \colon \mathbf{C}[0,1] \to [0,1]$ defined by

$$h(f) = \mathcal{L}\{t \in [0,1] : f(t) > 0\}.$$

It is not hard to see that the function h is continuous in every $f \in \mathbf{C}[0,1]$ with the property that

$$\lim_{\varepsilon \downarrow 0} \mathcal{L}\{t \in [0,1] : -\varepsilon \leqslant f(t) \leqslant \varepsilon\} = 0,$$

which again is equivalent to $\mathcal{L}\{t \in [0,1] : f(t) = 0\} = 0$, a property which Brownian motion has almost surely. Hence, again by Donsker's invariance principle and the Portmanteau theorem, the left hand side in (5.11) divided by n converges to the distribution of $h(B) = \mathcal{L}\{t \in [0,1] : B(t) > 0\}$, and this completes the argument. ∎

Remark 5.30 The proof of Theorem 5.28 can now be used literally to show that the second arcsine law holds for random walks $\{S_n : n \geqslant 0\}$ with mean zero and finite variance. Indeed, if $P_n = \#\{1 \leqslant k \leqslant n : S_k > 0\}$ is the number of positive values of the random walk before time n, then, for all $x \in (0,1)$,

$$\lim_{n \to \infty} \mathbb{P}\{P_n \leqslant xn\} = \frac{2}{\pi} \arcsin(\sqrt{x}).$$

◇

5.5 Pitman's $2M - B$ theorem

Pitman's $2M - B$ theorem describes an interesting relationship between the 3-dimensional Bessel process, which, loosely speaking, can be considered as a linear Brownian motion conditioned to avoid zero and a simple transformation of the Brownian path, namely the process

$$\{(2M(t) - B(t), M(t)): t \geqslant 0\} \quad \text{for } M(t) = \max_{0 \leqslant s \leqslant t} B(s).$$

Geometrically, the first component of this process is obtained by reflecting the Brownian path at each time in the level of the current maximum. We will obtain this result from a random walk analogue, using Donsker's invariance principle to pass to the Brownian motion case.

We start by discussing simple random walks conditioned to avoid zero, and its continuous-time analogue, the three-dimensional Bessel process. Consider a simple random walk on $\{0, 1, 2, \ldots, n\}$ conditioned to reach n before 0. By Bayes' rule, this conditioned process is a Markov chain with the following transition probabilities: $\widehat{p}(0, 1) = 1$ and for $1 \leqslant k < n$,

$$\widehat{p}(k, k+1) = (k+1)/2k \quad ; \quad \widehat{p}(k, k-1) = (k-1)/2k. \tag{5.13}$$

This is an instance of Doob's H-transform, see Exercise 5.13. Taking $n \to \infty$, this leads us to *define* the **simple random walk on** $\mathbb{N} = \{1, 2, \ldots\}$ **conditioned to avoid zero** (forever) as a Markov chain on $\mathbb{N}$ with transition probabilities as in (5.13) for all $k \geqslant 1$.

Lemma 5.31 *Let $\{S(j): j = 0, 1, \ldots\}$ be a simple random walk on $\mathbb{Z}$ and let $\{\tilde{\rho}(j): j = 0, 1, \ldots\}$ be a simple random walk on $\mathbb{N}$ conditioned to avoid zero. Then for $\ell \geqslant 1$ and any sequence $(x_0, \ldots, x_\ell)$ of positive integers, we have*

$$\mathbb{P}\{\tilde{\rho}(1) = x_1, \ldots, \tilde{\rho}(\ell) = x_\ell \mid \tilde{\rho}(0) = x_0\}$$
$$= \frac{x_\ell}{x_0} \mathbb{P}\{S(1) = x_1, \ldots, S(\ell) = x_\ell \mid S(0) = x_0\}.$$

Proof. We prove the result by induction on ℓ. The case $\ell = 1$ is just (5.13). Assume the lemma holds for $\ell - 1$ and let $(x_0, \ldots, x_\ell)$ be a sequence of positive integers such that $|x_j - x_{j-1}| = 1$ for $j = 1, \ldots, \ell$. Clearly, the probability on the right hand side of the equation is just $2^{-\ell}$. Moreover, using the induction hypothesis and the Markov property,

$$\mathbb{P}\{\tilde{\rho}(1) = x_1, \ldots, \tilde{\rho}(\ell) = x_\ell \mid \tilde{\rho}(0) = x_0\}$$
$$= \frac{x_{\ell-1}}{x_0} 2^{1-\ell} \, \mathbb{P}\{\tilde{\rho}(\ell) = x_\ell \mid \tilde{\rho}(\ell-1) = x_{\ell-1}\}$$
$$= \frac{x_{\ell-1}}{x_0} 2^{1-\ell} \frac{x_\ell}{2x_{\ell-1}} = \frac{x_\ell}{x_0} 2^{-\ell},$$

as required to complete the proof.   ∎

Define the **three-dimensional Bessel process** $\{\rho(t): t \geqslant 0\}$ by taking a 3-dimensional Brownian motion $\{W(t): t \geqslant 0\}$ and putting

$$\rho(t) = |W(t)|.$$

Fix $h > 0$ and assume $|W(0)| = h$. Define the stopping times $\{\tau_j^{(h)} : j = 0, 1, \ldots\}$ by $\tau_0^{(h)} = 0$ and, for $j \geqslant 0$,

$$\tau_{j+1}^{(h)} = \min \left\{ t > \tau_j^{(h)} : |\rho(t) - \rho(\tau_j^{(h)})| = h \right\}.$$

Given that $\rho(\tau_j^{(h)}) = kh$ for some $k > 0$, by Theorem 3.18, we have that

$$\rho(\tau_{j+1}^{(h)}) = \begin{cases} (k+1)h, & \text{with probability } \frac{k+1}{2k}, \\ (k-1)h, & \text{with probability } \frac{k-1}{2k}. \end{cases} \tag{5.14}$$

We abbreviate $\tau_j = \tau_j^{(1)}$. By (5.13) and (5.14), the sequence $\{\rho(\tau_j) : j = 0, 1, \ldots\}$ has the same distribution as the simple random walk on $\mathbb{N}$ conditioned to avoid zero, with the initial condition $\tilde{\rho}(0) = 1$.

Lemma 5.32 *The sequence $\{\tau_n - n : n \geqslant 0\}$ is a martingale and there exists $C > 0$ with*

$$\operatorname{Var}(\tau_n - n) \leqslant C n.$$

Proof.   If $\{B(t) : t \geqslant 0\}$ is standard linear Brownian motion, then we know from Lemma 2.47 that $\{B(t)^2 - t : t \geqslant 0\}$ is a martingale. As $\{\rho(t)^2 - 3t : t \geqslant 0\}$ is the sum of three independent copies of this martingale, it is also a martingale. Given that $\rho(\tau_{n-1}) = k$, optional sampling (recall Theorem 12.27 of the appendix) for this martingale at times τ_{n-1} and τ_n yields

$$k^2 - 3\tau_{n-1} = \frac{(k+1)^3}{2k} + \frac{(k-1)^3}{2k} - 3\mathbb{E}[\tau_n \mid \tau_{n-1}],$$

hence $\mathbb{E}[\tau_n - \tau_{n-1} \mid \tau_{n-1}] = 1$, so that $\{\tau_n - n : n \geqslant 0\}$ is a martingale. To bound its variance, consider the scalar product

$$Z := \left\langle W(t+1) - W(t), \frac{W(t)}{|W(t)|} \right\rangle.$$

Given $\mathcal{F}(t)$, the σ-algebra generated by $W(s)$, for $s \in [0, t]$, the distribution of Z is standard normal. This is clear if $W(t)$ is on a coordinate axis; and the general case follows by rotational symmetry of 3-dimensional Brownian motion. Moreover,

$$Z = \left\langle W(t+1), \frac{W(t)}{|W(t)|} \right\rangle - |W(t)| \leqslant |W(t+1)| - |W(t)|.$$

Therefore $\mathbb{P}\{|W(t+1)| - |W(t)| > 2 \mid \mathcal{F}(t)\} \geqslant \mathbb{P}\{Z > 2\}$. For any n,

$$\bigcup_{j=1}^{k} \left\{ |W(\tau_{n-1}+j)| - |W(\tau_{n-1}+j-1)| > 2 \right\} \subset \{\tau_n - \tau_{n-1} \leqslant k\},$$

so that, given τ_{n-1}, the difference $\tau_n - \tau_{n-1}$ is stochastically bounded from above by a geometric random variable with parameter $p := \mathbb{P}\{Z > 2\}$. Hence,

$$\operatorname{Var}(\tau_n - \tau_{n-1} - 1) \leqslant \mathbb{E}\left[(\tau_n - \tau_{n-1})^2\right] \leqslant \frac{2}{p}.$$

By orthogonality of martingale differences, see e.g. (12.1) in the appendix, we conclude that $\operatorname{Var}(\tau_n - n) \leqslant 2n/p$, which completes the proof. ∎

We use the following notation,

- $\{S(j)\colon j = 0, 1, \ldots\}$ is a simple random walk in $\mathbb{Z}$,

- $\{\tilde{M}(j)\colon j = 0, 1, \ldots\}$ defined by $\tilde{M}(j) = \max_{0 \leqslant a \leqslant j} S(a)$ is its maximum process;

- $\{\tilde{\rho}(j)\colon j = 0, 1, \ldots\}$ is a simple random walk on $\mathbb{N}$ conditioned to avoid zero,

- $\{\tilde{I}(j)\colon j = 0, 1, \ldots\}$ defined by $\tilde{I}(j) = \min_{k \geqslant j} \tilde{\rho}(k)$ is its future minimum process.

Let $\{I(t)\colon t \geqslant 0\}$ defined by $I(t) = \min_{s \geqslant t} \rho(s)$ be the future minimum process of the process $\{\rho(t)\colon t \geqslant 0\}$.

Proposition 5.33 *Let $\tilde{I}(0) = \tilde{\rho}(0) = 0$, and extend the processes $\{\tilde{\rho}(j)\colon j = 0, 1, \ldots\}$ and $\{\tilde{I}(j)\colon j = 0, 1, \ldots\}$ to $[0, \infty)$ by linear interpolation. Then*

$$\big\{ h\tilde{\rho}(t/h^2)\colon 0 \leqslant t \leqslant 1 \big\} \xrightarrow{\mathrm{d}} \big\{ \rho(t)\colon 0 \leqslant t \leqslant 1 \big\} \quad \text{as } h \downarrow 0, \tag{5.15}$$

and

$$\big\{ h\tilde{I}(t/h^2)\colon 0 \leqslant t \leqslant 1 \big\} \xrightarrow{\mathrm{d}} \big\{ I(t)\colon 0 \leqslant t \leqslant 1 \big\} \quad \text{as } h \downarrow 0, \tag{5.16}$$

where $\xrightarrow{\mathrm{d}}$ indicates convergence in law as random elements of $\mathbf{C}[0, 1]$.

Proof.   For any $h > 0$, Brownian scaling implies that the process $\{\tau_n^{(h)}\colon n = 0, 1, \ldots\}$ has the same law as the process $\{h^2 \tau_n\colon n = 0, 1, \ldots\}$. Doob's $\mathbf{L}^2$ maximal inequality, see Theorem 12.30, and Lemma 5.32 yield that

$$\mathbb{E}\Big[\max_{0 \leqslant j \leqslant n} (\tau_j - j)^2 \Big] \leqslant C\, n,$$

for a suitable constant $C > 0$. Therefore, taking $n = \lfloor h^{-2} t \rfloor$,

$$\mathbb{E}\Big[\max_{0 \leqslant t \leqslant 1} (\tau_{\lfloor h^{-2} t \rfloor}^{(h)} - h^2 \lfloor h^{-2} t \rfloor)^2 \Big] = h^4\, \mathbb{E}\Big[\max_{0 \leqslant t \leqslant 1} (\tau_{\lfloor h^{-2} t \rfloor} - \lfloor h^{-2} t \rfloor)^2 \Big] \leqslant C\, h^2,$$

whence also (for a slightly larger constant)

$$\mathbb{E}\Big[\max_{0 \leqslant t \leqslant 1} (\tau_{\lfloor h^{-2} t \rfloor}^{(h)} - t)^2 \Big] \leqslant C\, h^2. \tag{5.17}$$

Since $\{\rho(t)\colon 0 \leqslant t \leqslant 1\}$ is uniformly continuous almost surely, we infer that

$$\max_{0 \leqslant t \leqslant 1} |\rho(\tau_{\lfloor h^{-2} t \rfloor}^{(h)}) - \rho(t)| \to 0 \quad \text{in probability as } h \downarrow 0,$$

and similar reasoning gives the analogous result when $\lfloor \cdot \rfloor$ is replaced by $\lceil \cdot \rceil$. Since $\tilde{\rho}(t/h^2)$ is, by definition, a weighted average of $\tilde{\rho}(\lfloor h^{-2} t \rfloor)$ and $\tilde{\rho}(\lceil h^{-2} t \rceil)$, the proof of (5.15) is now concluded by recalling that $\{\rho(\tau_j^{(h)})\colon j = 0, 1, \ldots\}$ has the same distribution as $\{h\tilde{\rho}(j)\colon j = 0, 1, \ldots\}$. Similarly, $\{I(\tau_j^{(h)})\colon j = 0, 1, \ldots\}$ has the same distribution as $\{h\tilde{I}(j)\colon j = 0, 1, \ldots\}$, so (5.16) follows from (5.17) and the continuity of I. $\blacksquare$

Theorem 5.34 (Pitman's $2M - B$ theorem) *Let $\{B(t): t \geqslant 0\}$ be a linear Brownian motion and let $M(t) = \max_{0 \leqslant s \leqslant t} B(s)$ denote its maximum up to time t. Also let $\{\rho(t): t \geqslant 0\}$ be a three-dimensional Bessel process and let $\{I(t): t \geqslant 0\}$ be the corresponding future infimum process given by $I(t) = \inf_{s \geqslant t} \rho(s)$. Then*

$$\{(2M(t) - B(t), M(t)): t \geqslant 0\} \overset{\mathrm{d}}{=} \{(\rho(t), I(t)): t \geqslant 0\}.$$

In particular, $\{2M(t) - B(t): t \geqslant 0\}$ is a three-dimensional Bessel process.

Proof. Following the original paper [Pi75], we prove the theorem in the discrete setting, i.e. we show that, for $S(0) = \tilde{\rho}(0) = 0$,

$$\{(2\tilde{M}(j) - S(j), \tilde{M}(j)): j = 0, 1, \ldots\} \overset{\mathrm{d}}{=} \{(\tilde{\rho}(j), \tilde{I}(j)): j = 0, 1, \ldots\}. \qquad (5.18)$$

The theorem then follows directly by invoking Donsker's invariance principle and Proposition 5.33. First note that (5.18) is equivalent to

$$\{(S(j), \tilde{M}(j)): j = 0, 1, \ldots\} \overset{\mathrm{d}}{=} \{(2\tilde{I}(j) - \tilde{\rho}(j), \tilde{I}(j)): j = 0, 1, \ldots\},$$

which we establish by computing the transition probabilities. If $S(j) < \tilde{M}(j)$, then clearly

$$(S(j+1), \tilde{M}(j+1)) = \begin{cases} (S(j)+1, \tilde{M}(j)), & \text{with probability } \tfrac{1}{2}, \\ (S(j)-1, \tilde{M}(j)), & \text{with probability } \tfrac{1}{2}. \end{cases} \qquad (5.19)$$

If $S(j) = \tilde{M}(j)$, then

$$(S(j+1), \tilde{M}(j+1)) = \begin{cases} (S(j)+1, \tilde{M}(j)+1), & \text{with probability } \tfrac{1}{2}, \\ (S(j)-1, \tilde{M}(j)), & \text{with probability } \tfrac{1}{2}. \end{cases} \qquad (5.20)$$

We now compute the transition probabilities of $\{(2\tilde{I}(j) - \tilde{\rho}(j), \tilde{I}(j)): j = 0, 1, \ldots\}$. To this end, we first show that $\{\tilde{I}(j): j = 0, 1, \ldots\}$ is the maximum process of $\{2\tilde{I}(j) - \tilde{\rho}(j): j = 0, 1, \ldots\}$. Indeed, for all $j \leqslant k$, since $(\tilde{I} - \tilde{\rho})(j) \leqslant 0$, we have

$$2\tilde{I}(j) - \tilde{\rho}(j) = \tilde{I}(j) + (\tilde{I} - \tilde{\rho})(j) \leqslant \tilde{I}(k).$$

On the other hand, let j_* be the minimal $j_* \leqslant k$ such that $\tilde{I}(j_*) = \tilde{I}(k)$. Then $\tilde{\rho}(j_*) = \tilde{I}(j_*)$ and we infer that $(2\tilde{I} - \tilde{\rho})(j_*) = \tilde{I}(j_*) = I(k)$.

Assume now that $2\tilde{I}(j) - \tilde{\rho}(j) < \tilde{I}(j)$, i.e., $\tilde{\rho}(j) > \tilde{I}(j)$. Lemma 5.31 and the fact that $\{S(j): j = 0, 1, \ldots\}$ is recurrent imply that, for integers $k \geqslant i > 0$,

$$\mathbb{P}\{\exists j \text{ with } \tilde{\rho}(j) = i \,\big|\, \tilde{\rho}(0) = k\} = \frac{i}{k} \mathbb{P}\{\exists j \text{ with } S(j) = i \,\big|\, S(0) = k\} = \frac{i}{k}.$$

Thus, for $k \geqslant i > 0$,

$$\mathbb{P}\{\tilde{I}(j) = i \,\big|\, \tilde{\rho}(j) = k\}$$

$$= \mathbb{P}\{\exists j \text{ with } \tilde{\rho}(j) = i \,\big|\, \tilde{\rho}(0) = k\} - \mathbb{P}\{\exists j \text{ with } \tilde{\rho}(j) = i - 1 \,\big|\, \tilde{\rho}(0) = k\} = \frac{1}{k}.$$

Therefore,

$$\mathbb{P}\{\tilde{\rho}(j+1) = k - 1 \mid \tilde{\rho}(j) = k, \tilde{I}(j) = i\}$$

$$= \frac{\mathbb{P}\{\tilde{\rho}(j+1) = k - 1, \tilde{I}(j) = i \mid \tilde{\rho}(j) = k\}}{\mathbb{P}\{\tilde{I}(j) = i \mid \tilde{\rho}(j) = k\}} \qquad (5.21)$$

$$= \frac{\frac{k-1}{2k} \frac{1}{k-1}}{\frac{1}{k}} = \frac{1}{2}.$$

We conclude that if $2\tilde{I}(j) - \tilde{\rho}(j) < \tilde{I}(j)$, then

$$(2\tilde{I}(j+1) - \tilde{\rho}(j+1), \tilde{I}(j+1))$$

$$= \begin{cases} (2\tilde{I}(j) - \tilde{\rho}(j) + 1, \tilde{I}(j)), & \text{with probability } \frac{1}{2}, \\ (2\tilde{I}(j) - \tilde{\rho}(j) - 1, \tilde{I}(j)), & \text{with probability } \frac{1}{2}. \end{cases} \qquad (5.22)$$

Assume now that $\tilde{\rho}(j) = \tilde{I}(j) = k$. Then $\tilde{\rho}(j+1) = k + 1$, and a computation analogous to (5.21) shows that

$$\tilde{I}(j+1) = \begin{cases} \tilde{I}(j) + 1, & \text{with probability } \frac{1}{2}, \\ \tilde{I}(j), & \text{with probability } \frac{1}{2}. \end{cases} \qquad (5.23)$$

Indeed,

$$\mathbb{P}\{\tilde{I}(j+1) = k + 1 \mid \tilde{I}(j) = \tilde{\rho}(j) = k\}$$

$$= \frac{\mathbb{P}\{\tilde{\rho}(j+1) = k + 1 \mid \tilde{\rho}(j) = k\}\mathbb{P}\{\tilde{I}(j+1) = k + 1 \mid \tilde{\rho}(j+1) = k + 1\}}{\mathbb{P}\{\tilde{I}(j) = k \mid \tilde{\rho}(j) = k\}}$$

$$= \frac{\frac{k+1}{2k} \frac{1}{k+1}}{\frac{1}{k}} = \frac{1}{2}.$$

By (5.23), if $\tilde{\rho}(j) = \tilde{I}(j) = k$, then we have

$$(2\tilde{I}(j+1) - \tilde{\rho}(j+1), \tilde{I}(j+1))$$

$$= \begin{cases} (2\tilde{I}(j) - \tilde{\rho}(j) + 1, \tilde{I}(j) + 1), & \text{with probability } \frac{1}{2}, \\ (2\tilde{I}(j) - \tilde{\rho}(j) - 1, \tilde{I}(j)), & \text{with probability } \frac{1}{2}. \end{cases} \qquad (5.24)$$

Finally, comparing (5.19) and (5.20) to (5.22) and (5.24) completes the proof. ∎

We now use the combinatorial technique developed for Pitman's $2M - B$ theorem to prove a result of Ciesielski and Taylor [CT62], which relates the occupation times of a 3-dimensional Brownian motion to exit times of one-dimensional Brownian motion. An alternative proof based on a Feynman–Kac formula will be given in Section 7.4.

Theorem 5.35 (Ciesielski–Taylor identity) *Let $\{W(t) : t \geqslant 0\}$ be a 3-dimensional Brownian motion and let $T = \int_0^\infty 1\{|W(s)| \leqslant 1\} \, ds$ be the total amount of time it spends in the unit ball. Let $\{B(t) : t \geqslant 0\}$ be a 1-dimensional Brownian motion and let $\tau = \min\{t : |B(t)| = 1\}$. Then we have*

$$T \stackrel{\mathrm{d}}{=} \tau. \qquad (5.25)$$

Remark 5.36 The statement of Theorem 5.35 remains true if, for any $d \geqslant 3$, we let $\{W(t): t \geqslant 0\}$ be a d-dimensional Brownian motion and $\{B(t): t \geqslant 0\}$ be a $(d - 2)$-dimensional Brownian motion, but the proof we present here works only for $d = 3$. ◇

Lemma 5.37 *Let* $\{S(j): j = 0, \ldots, \gamma\}$ *be simple random walk started at* $S(0) = 0$ *and stopped at the random time* $\gamma = \min\{j: S(j) = n\}$. *Then* $\{n - S(\gamma - j): j = 0, \ldots, \gamma\}$ *has the same distribution as* $\{\tilde{\rho}(j): j = 0, \ldots, L\}$, *where* $\tilde{\rho}(0) = 0$ *and* $L = \max\{j : \tilde{\rho}(j) = n\}$. *In particular,* γ *and* L *have the same law.*

Proof. Fix $x_0 = 0$ and consider a possible path $(x_0, x_1, \ldots, x_\ell)$ for the simple random walk stopped at γ, where $|x_i - x_{i-1}| = 1$ for all $i \geqslant 1$ and $x_i < n$ for all $i < \ell$ with $x_\ell = n$. The probability that $\{S(j): j = 0, \ldots, \gamma\}$ takes this path is $2^{-\ell}$. The probability that $\{\tilde{\rho}(j): j = 1, \ldots, \ell\}$ takes the path $\{n - x_{\ell-j}: j = 1, \ldots, \ell\}$ is $2^{1-\ell}n$ by Lemma 5.31. Furthermore, conditioned on $\{\tilde{\rho}(j): j = 1, \ldots, \ell\}$ taking this path, the probability that $\tilde{I}(\ell + 1) = n + 1$ is

$$\mathbb{P}\{\tilde{\rho}(\ell+1) = n + 1 \mid \tilde{\rho}(\ell) = n\} \, \mathbb{P}\{\tilde{I}(\ell+1) = n + 1 \mid \tilde{\rho}(\ell+1) = n + 1\}$$
$$= \frac{n+1}{2n} \frac{1}{n+1} = \frac{1}{2n}.$$

Combining these facts yields the result. ∎

Proof of Theorem 5.35. We prove the theorem in the discrete setting, namely we denote $\tilde{\tau} = \min\{j \geqslant 0: |S(j)| = n\}$, and show that for $n \geqslant 1$,

$$\#\{i \geqslant 1 \,:\, \tilde{\rho}(i-1), \tilde{\rho}(i) \in \{0, \ldots, n\}\} \overset{\mathrm{d}}{=} \tilde{\tau}. \tag{5.26}$$

Dividing both sides of (5.26) by n^2 and letting $n \uparrow \infty$ yields (5.25), see Exercise 5.15.

As a warm up, observe that for $n = 1$ both sides of (5.26) are identically 1, and for $n = 2$ each side of (5.26) is a geometric random variable with parameter $\frac{1}{2}$, multiplied by 2. For the full argument let $\gamma = \min\{j: S(j) = n\}$ as in Lemma 5.37, which implies that

$$\#\{i \in \{1, \ldots, \gamma\}: S(i-1), S(i) \in \{0, \ldots, n\}\}$$
$$\overset{\mathrm{d}}{=} \#\{i \geqslant 1: \tilde{\rho}(i-1), \tilde{\rho}(i) \in \{0, \ldots, n\}\}.$$

But deleting the negative excursions between two points in which $\{S(i): i = 0, 1, \ldots\}$ is zero gives a reflected simple random walk with the law of $\{|S(i)|: i = 0, 1, \ldots\}$ and therefore

$$\#\{i \in \{1, \ldots, \gamma\}: S(i-1), S(i) \in \{0, \ldots, n\}\}$$
$$\overset{\mathrm{d}}{=} \#\{i \in \{1, \ldots, \tilde{\tau}\}: |S(i-1)|, |S(i)| \in \{0, \ldots, n\}\} = \tilde{\tau},$$

as required to prove (5.26). ∎

In a similar spirit the following theorem relates occupation times and exit times for a standard linear Brownian motion.

Theorem 5.38 *Let* $\{B(t) \colon t \geqslant 0\}$ *be a standard linear Brownian motion and, for* $a \geqslant 0$, *let* $\tau_a = \inf\{t \geqslant 0 \colon B(t) = a\}$ *and* $\sigma_a = \inf\{t \geqslant 0 \colon |B(t)| = a\}$. *Then*

$$\int_0^{\tau_a} 1\{0 \leqslant B(t) \leqslant a\}\,dt \overset{\mathrm{d}}{=} \sigma_a.$$

The key to the proof is the fact, due to David Williams, that removing the negative excursions from a standard linear Brownian motion $\{B(s) \colon s \geqslant 0\}$ leads to a reflected Brownian motion $\{|B(s)| \colon s \geqslant 0\}$.

Lemma 5.39 *Let* $s(t) = \int_0^t 1\{B(s) \geqslant 0\}\,ds$ *and let* $t(s) = \inf\{t \geqslant 0 \colon s(t) \geqslant s\}$ *its right-continuous inverse. Then*

$$\{B(t(s)) \colon s \geqslant 0\} \overset{\mathrm{d}}{=} \{|B(s)| \colon s \geqslant 0\}.$$

Proof.   Let $\{S(n) \colon n = 0, 1, \ldots\}$ be a simple random walk and consider $\{S_n^*(s) \colon s \geqslant 0\}$ defined as in Donsker's invariance principle. Removing the negative excursions from the simple random walk leads to a reflected simple random walk, therefore

$$\{S_n^*(t(s, S_n^*)) \colon s \geqslant 0\} \overset{\mathrm{d}}{=} \{|S_n^*(s)| \colon s \geqslant 0\}, \tag{5.27}$$

where $s(t, f) = \int_0^t 1\{f(s) \geqslant 0\}\,ds$ and $t(s, f) = \inf\{t \geqslant 0 \colon s(t, f) \geqslant s\}$. For every $t > 0$ the mapping $f \mapsto f(t(\cdot, f))$ is continuous in $f \in \mathbf{C}[0, t]$ with respect to the supremum norm provided that

$$\lim_{\varepsilon \downarrow 0} \mathcal{L}\{s \in [0, t] \colon -\varepsilon \leqslant f(s) \leqslant \varepsilon\} = 0,$$

a property which Brownian motion has almost surely. Hence Donsker's invariance principle gives the claim by letting $n \to \infty$ in (5.27). ∎

Proof of Theorem 5.38.   We obviously have that

$$\int_0^{\tau_a} 1\{0 \leqslant B(s) \leqslant a\}\,ds = \inf\{s \geqslant 0 \colon B(t(s)) = a\}.$$

By Lemma 5.39 we further have

$$\inf\{s \geqslant 0 \colon B(t(s)) = a\} \overset{\mathrm{d}}{=} \inf\{s \geqslant 0 \colon |B(s)| = a\} = \sigma_a,$$

which implies the result. ∎

Exercises

Exercise 5.1. Ⓢ Suppose $\{B(t) \colon t \geqslant 0\}$ is a standard linear Brownian motion. Show that

$$\limsup_{n \uparrow 0} \sup_{n \leqslant t < n+1} \frac{B(t) - B(n)}{\sqrt{2 \log n}} = 1 \qquad \text{almost surely.}$$

Exercise 5.2. Ⓢ Derive from Theorem 5.1 that, for a d-dimensional Brownian motion,

$$\limsup_{t \uparrow \infty} \frac{|B(t)|}{\sqrt{2t \log \log t}} = 1 \qquad \text{almost surely.}$$

Exercise 5.3. § Suppose $\{B(t): t \geqslant 0\}$ is a linear Brownian motion and τ the first hitting time of level 1. Show that, almost surely,

$$\limsup_{h \downarrow 0} \frac{B(\tau) - B(\tau - h)}{\sqrt{2h \log \log(1/h)}} \leqslant 1.$$

Exercise 5.4. § Let $\{S_k: k \geqslant 0\}$ be a simple, symmetric random walk on the integers. Show that there are positive constants C_1 and C_2 such that

$$\frac{C_1}{\sqrt{n}} \leqslant \mathbb{P}\{S_i \geqslant 0 \text{ for all } 1 \leqslant i \leqslant n\} \leqslant \frac{C_2}{\sqrt{n}} \quad \text{for all } n \geqslant 1.$$

Hint. For simple random walk a *reflection principle* holds in quite the same way as for Brownian motion. The key to the proof is to verify that

$$\mathbb{P}\{S_i \geqslant 0 \text{ for all } 1 \leqslant i \leqslant n\} = \mathbb{P}\{S_n \geqslant 0\} - \mathbb{P}\{S_n^* \leqslant -2\}$$

where S_n^* is the random walk reflected at the stopping time $\tau_{-1} = \min\{k: S_k = -1\}$.

Exercise 5.5. § Prove that, for any random walk $\{S_j: j \geqslant 0\}$ on the line,

$$\mathbb{P}\{S_0, \ldots, S_n \text{ has a point of increase}\} \leqslant 2 \frac{\sum_{k=0}^{n} p_k p_{n-k}}{\sum_{k=0}^{\lfloor n/2 \rfloor} p_k^2},$$

where $p_0, \ldots, p_n$ are as in (5.4).

Exercise 5.6. An event $A \subset \mathbb{R}^d$ is an **increasing event** if,

$$(x_1, \ldots, x_{i-1}, x_i, x_{i+1}, \ldots x_d) \in A \text{ and } \widetilde{x}_i \geqslant x_i$$
$$\implies \quad (x_1, \ldots, x_{i-1}, \widetilde{x}_i, x_{i+1}, \ldots x_d) \in A.$$

If A and B are increasing events, show that

$$\mathbb{P}(A \cap B) \geqslant \mathbb{P}(A)\mathbb{P}(B),$$

i.e. A and B are positively correlated.

Exercise 5.7. § Show that we can obtain a lower bound on the probability that a random walk has a point of increase that differs from the upper bound only by a constant factor. More precisely, for any random walk on the line,

$$\mathbb{P}\{S_0, \ldots, S_n \text{ has a point of increase}\} \geqslant \frac{\sum_{k=0}^{n} p_k p_{n-k}}{2 \sum_{k=0}^{\lfloor n/2 \rfloor} p_k^2},$$

where $p_0, \ldots, p_n$ are as in (5.4).

Exercise 5.8. Let $\{B(t): 0 \leqslant t \leqslant 1\}$ be a linear Brownian motion.

   (a) Use the Cameron–Martin theorem to show that, for any $F \in \mathbf{D}[0, 1]$, the process
$$\{B(t) + F(t): 0 \leqslant t \leqslant 1\}$$
   almost surely has no point of increase.

(b) Find a $F \in \mathbf{C}[0, 1]$ such that $\{B(t) + F(t) \colon 0 \leqslant t \leqslant 1\}$ has a point of increase.

Exercise 5.9. Suppose $X_1, \ldots, X_n$ are independent and identically distributed and consider their ordered relabelling given by $X_{(1)} \geqslant X_{(2)} \geqslant \ldots \geqslant X_{(n)}$. Show that

$$\mathbb{E}[X_{(i)} X_{(j)}] \geqslant \mathbb{E}[X_{(i)}] \mathbb{E}[X_{(j)}],$$

provided these expectations are well-defined.

Exercise 5.10. § Given a centred random variable X, show that there exist centred random variables X_n taking only finitely many values, such that X_n converges to X in law and, for $\Psi_n(x) = E[X_n \,|\, X_n \geqslant x]$, the embedding stopping times

$$\tau_n = \inf\{t \geqslant 0 \colon M(t) \geqslant \Psi_n(B(t))\}$$

converge almost surely to τ. Infer that $B(\tau)$ has the same law as X, and $\mathbb{E}[\tau] = \mathbb{E}[X^2]$.

Exercise 5.11. § Suppose that $\{S_n \colon n \geqslant 0\}$ is a random walk with mean zero and positive, finite variance. Define $\{S_n^*(t) \colon 0 \leqslant t \leqslant 1\}$ as in Donsker's invariance principle. Show that

$$\mathcal{L}\{t \in [0, 1] \colon S_n^*(t) > 0\} - \tfrac{1}{n} \#\big\{k \in \{1, \ldots, n\} \colon S_k > 0\big\}$$

converges to zero in probability.

Exercise 5.12. § Let $\{B(t) \colon t \geqslant 0\}$ be a standard linear Brownian motion and $a > 0$. Define stopping times $\tau_a = \inf\{t \geqslant 0 \colon B(t) = a\}$, $\tau_{a,0} = \inf\{t \geqslant \tau_a \colon B(t) = 0\}$ and a random time

$$\sigma_0 = \sup\{0 \leqslant t \leqslant \tau_a \colon B(t) = 0\}.$$

The process $\{e(t) \colon 0 \leqslant t \leqslant \tau^e\}$ given by

$$e(t) = B(\sigma_0 + t), \quad \tau^e = \tau_{a,0} - \sigma_0$$

is a *Brownian excursion* conditioned to hit level a, and τ^e is called its *lifetime*.

  (a) For any $0 < b \leqslant a$ denote by τ_b^e the first hitting time of level b by the excursion $\{e(t) \colon 0 \leqslant t \leqslant \tau^e\}$. Show that, for $0 < b < a$, the process $\{e(\tau_b^e + t) \colon 0 \leqslant t \leqslant \tau_a^e - \tau_b^e\}$ is a Brownian motion conditioned to hit level a before level zero.

  (b) Show that the time-reversed excursion $\{e(\tau^e - t) \colon 0 \leqslant t \leqslant \tau_e\}$ is also a Brownian excursion conditioned to hit level a.

Hint.   For (b) show an analogous statement for simple random walk and use Donsker's invariance principle to transfer the result to the Brownian motion case.

Exercise 5.13. Let $p(x, y)$ be the transition matrix of an irreducible Markov chain $\{X_j \colon j = 0, 1, \ldots\}$ on a finite state space V. For $a, b \in V$, consider the hitting time

$$T = T_{ab} = \min\{j \geqslant 0 \colon X_j \in \{a, b\}\}$$

and write $H(x) = \mathbb{P}_x\{X_T = b\}$. Show that the chain $\{X_j \colon j = 0, 1, \ldots\}$ conditioned to reach b before a and absorbed at b has the same law as the Markov chain $\{Y_j \colon j = 0, 1, \ldots\}$ on $V \setminus \{a\}$ with transition probabilities

$$\widehat{p}(x, y) = p(x, y) \frac{H(y)}{H(x)} \quad \text{for } x \neq b.$$

The chain $\{Y_j \colon j = 0, 1, \ldots\}$ is called the **Doob H-transform** of the original chain $\{X_j \colon j = 0, 1, \ldots\}$.

Exercise 5.14. Let $\{\rho(t) \colon t \geqslant 0\}$ be a three-dimensional Bessel process.

(a) Verify that the process $\{X(t) \colon t \geqslant 0\}$ given by $X(t) = \rho(t)^4 - 6t^2 \rho(t)^2 + 3\rho(t)^2$ is a martingale.

(b) Use (a) to derive a tighter bound for $\mathrm{Var}(\tau_n - \tau_{n-1})$ in Lemma 5.32.

Exercise 5.15. Let $\{S(j) \colon j = 0, 1, \ldots\}$ be a simple random walk on the integers started at $S(0) = 0$, and $\{\tilde{\rho}(j) \colon j = 0, 1, \ldots\}$ be a simple random walk on $\mathbb{N}$ conditioned to avoid zero, with $\tilde{\rho}(0) = 0$.

(a) Show that, as $n \uparrow \infty$,

$$\frac{1}{n^2} \min\{j \colon |S(j)| = n\} \xrightarrow{\mathrm{d}} \min\{t \geqslant 0 \colon |B(t)| = 1\},$$

where $\{B(t) \colon t \geqslant 0\}$ is a 1-dimensional Brownian motion.

(b) Show that, as $n \uparrow \infty$,

$$\frac{1}{n^2} \#\{i \geqslant 1 \colon \tilde{\rho}(i-1), \tilde{\rho}(i) \in \{0, \ldots, n\}\} \xrightarrow{\mathrm{d}} \int_0^\infty \mathbb{1}\{|W(s)| \leqslant 1\}\, ds,$$

where $\{W(t) \colon t \geqslant 0\}$ is a 3-dimensional Brownian motion.

Notes and comments

Historically, the law of the iterated logarithm was first proved for simple random walk by Khinchin [Kh23, Kh24] and later generalised to other random walks by Kolmogorov [Ko29] and Hartman and Wintner [HW41]. The original arguments of Kolmogorov, Hartman and Wintner were extremely difficult, and a lot of authors have since provided more accessible proofs, see, for example, de Acosta [dA83]. For Brownian motion the law of the iterated logarithm is also due to Khinchin [Kh33]. The idea of using embedding arguments to transfer the result from the Brownian motion to the random walk case is due to Strassen [St64]. For a survey of laws of the iterated logarithm, see Bingham [Bi86].

An extension of the law of the iterated logarithm is Strassen's law, which is first proved in [St64]. If a standard Brownian motion on the interval $[0, t]$ is rescaled by a factor $1/t$ in time and a factor $\sqrt{2t \log \log(1/t)}$ in space, the set of limit points in $C[0, 1]$ are exactly the functions f with $f(0) = 0$ and $\int_0^1 (f'(t))^2 \, dt \leqslant 1$. Strassen's law also explains the approximate form of the curve in the right half of Figure 5.1. Any function in this class with $f(1) = 1$ satisfies

$$1 \geqslant \int_0^1 (f'(t))^2 \, dt \geqslant \left(\int_0^1 f'(t) \, dt \right)^2 = 1,$$

which implies that $f'(t)$ is constant and thus $f(t) = t$ for all $t \in (0, 1)$. Therefore, for large t, the Brownian path conditioned on ending near to its upper envelope resembles a straight line in the sup-norm, as can be seen in Figure 5.1.

The nonincrease phenomenon, which is described in Theorem 5.11, holds for arbitrary symmetric random walks, and can thus be viewed as a combinatorial consequence of fluctuations in random sums. Indeed, our argument shows this — subject to a generalisation of Lemma 5.8. The latter result holds if the increments X_i have a symmetric distribution, or if the increments have mean zero and finite variance, see e.g. Section XII.8 in Feller [Fe66].

Dvoretzky, Erdős and Kakutani [DEK61] were the first to prove that Brownian motion almost surely has no local points of increase. Knight [Kn81] and Berman [Be83] noted that this follows from properties of the local time of Brownian motion; direct proofs were given by Adelman [Ad85] and Burdzy [Bu90]. The proof we give is taken from [Pe96c].

A higher-dimensional analogue of this question is whether, for Brownian motion in the plane, there exists a line such that the Brownian motion path, projected onto that line, has a global point of increase, or equivalently whether the Brownian motion path admits cut lines. We say a line ℓ is a *cut line* for the Brownian motion if, for some $t_0 \in (0, 1)$ with $B(t_0) \in \ell$, the points $B(t)$ lie on one side of ℓ for all $t \in [0, t_0)$ and on the other side of ℓ for all $t \in (t_0, 1]$. It was proved by Bass and Burdzy [BB97] that planar Brownian motion almost surely does *not* have cut lines. Burdzy [Bu89], with a correction to the proof in [Bu95], however showed that Brownian motion in the plane almost surely does have *cut points*, which are points $B(t)$ such that the Brownian motion path with the point $B(t)$ removed is disconnected. It was conjectured that the Hausdorff dimension of the set of cut points is $3/4$. This conjecture has recently been proved by Lawler, Schramm and Werner [LSW01c], see also the discussion in our appendix, Chapter 11.

For Brownian motion in three dimensions, there almost surely exist cut planes, where we say P is a *cut plane* if for some t, the set $\{B(s) \colon 0 < s < t\}$ lies on one side of the plane and the set $\{B(s) \colon 1 > s > t\}$ on the other side. This result, originally due to Pemantle, is also described in Bass and Burdzy [BB97]. An argument of Evans, which is closely related to material we discuss in the final section of Chapter 10, shows that the set of times corresponding to cut planes has Hausdorff dimension zero.

Pemantle [Pe97] has shown that the range of planar Brownian motion almost surely does not cover any straight line segment. Which curves can and which cannot be covered by a Brownian motion path is, in general, an open question. Also unknown is the minimal Hausdorff dimension of curves contained in the range of planar Brownian motion, though it is known that it contains a curve of Hausdorff dimension 4/3, namely its outer boundary, see Lawler, Schramm and Werner [LSW01c] and Chapter 11.

Harris' inequality was discovered by Harris [Ha60] and is also known as *FKG inequality* in recognition of the work of Fortuin, Kasteleyn and Ginibre [FKG71] who extended the original inequality beyond the case of product measures. 'Correlation inequalities' like these play an extremely important rôle in percolation theory and spatial statistical physics. Exercise 5.9 indicates the important rôle of this idea in the investigation of order statistics, see Lehmann [Le66] and Bickel [Bi67] for further discussion and applications.

The Skorokhod embedding problem is a classic, which still leads to some attractive research. The first embedding theorem is due to Skorokhod [Sk65]. The Russian original of this work appeared in 1961 and the Dubins embedding, which we have presented is not much younger, see [Du68]. Our presentation, based on the idea of binary splitting martingales, follows Ex. II.7, p 34 in Neveu [Ne75] and we thank Jim Pitman for directing us to this reference. Another classic embedding technique is Root's embedding, see [Ro69]. The Azéma–Yor embedding was first described in [AY79], but we follow Meilijson [Me83] in the proof. One of the attractive features of the Azéma–Yor embedding is that, among all stopping times T with $\mathbb{E}T < \infty$ which represent a given random variable X, it maximises the $\max_{0 \leqslant t \leqslant T} B(t)$. Generalisation of the embedding problem to more general classes of probability laws requires different forms of minimality for the embedding stopping time, or more general processes in which one embeds. A survey of current developments is [Ob04].

The idea of an invariance principle that allows to transfer limit theorems from special cases to general random walks can be traced to Erdős and Kac [EK46, EK47]. The first general result of this nature is due to Donsker [Do51] following an idea of Doob [Do49]. Our treatment of Donsker's invariance principle is close to that of Freedman [Fr83]. Besides the embedding technique there is also a popular alternative proof, which goes back to Prohorov [Pr56]. Suppose that a subsequence of $\{S_n^* : n \geqslant 1\}$ converges in distribution to a limit X. This limit is a continuous random function, which is easily seen to have stationary, independent increments, which have expectation zero and variance equal to their length. By a general result this implies that X is a Brownian motion. So Brownian motion is the only possible limit point of the sequence $\{S_n^* : n \geqslant 1\}$. The difficult part of this proof is now to show that every subsequence of $\{S_n^* : n \geqslant 1\}$ has a convergent subsubsequence, the *tightness property*.

Many interesting applications and extensions of Donsker's invariance principle can be found in [Bi99]. Central limit theorems also hold in the context of martingales, see Hall and Heyde [HH80] for an extensive treatment of this subject. An important class of extensions of Donsker's invariance principle are the strong approximation theorems which were provided by Skorokhod [Sk65] and Strassen [St64]. In these results the Brownian motion and the random walk are constructed on the same probability space in such a way that they are close almost surely. An optimal result in this direction is the famous paper of Komlós, Major and Tusnády [KMT75]. For an exposition of their work and applications, see [CR81], and an alternative, more transparent, treatment of the simple random walk case is given in [Ch07].

The arcsine laws for Brownian motion were first proved by Lévy in [Le39, Le48]. The proof of the first law, which we give here, follows Kallenberg [Ka02]. This law can also be proved by a direct calculation, which however is slightly longer, see for example Durrett [Du95]. Our proof of the second arcsine law goes back to an idea of Baxter [Ba62]. Arcsine laws also hold for symmetric random variables without any moment assumptions, see Feller [Fe66]. Some more recent developments related to arcsine laws can be found in Pitman and Yor [PY92] and [PY03].

Pitman's $2M - B$ theorem, often also called $2M - X$ theorem, is from [Pi75]. We follow Pitman's original approach with some small modifications. A closely related area is the subject of path decompositions due to Williams, see [Wi70, Wi74]. Lemma 5.39 offers a first glimpse: Removing the negative excursions from the path of a Brownian motion leads to a reflected Brownian motion. A nice treatment of Pitman's theorem and related path decomposition results is Le Gall [LG86c]. Hambly et al. [HMO01] discuss a generalisation of the discrete variant, whose proof is based in part on a reversibility argument that has a queueing interpretation. Further significant generalisations lead to interesting connections to families of non-colliding Brownian motions and eventually to random matrix theory, see e.g. Grabiner [Gr99]. The proof of the Ciesielski–Taylor identity is adapted from Pitman's paper [Pi75], but the idea goes back to Williams [Wi70].

6

Brownian local time

In this chapter we focus on linear Brownian motion and address the question how to measure the amount of time spent by a Brownian path at a given level. As we already know from Theorem 3.26 that the occupation times up to time t are absolutely continuous measures, their densities are a viable measure for the time spent at level a during the time interval $[0, t]$. We shall show that these densities make up a continuous random field $\{L^a(t): a \in \mathbb{R}, t \geqslant 0\}$, which is called the Brownian local time. Interesting information about the distribution of this process is contained in a theorem of Lévy (studying it as function of time) and the Ray–Knight theorem (studying it as function of the level). We finally show how to interpret local time as a family of Hausdorff measures.

6.1 The local time at zero

How can we measure the amount of time spent by a standard linear Brownian motion $\{B(t): t \geqslant 0\}$ at zero? We have already seen that, almost surely, the zero set has Hausdorff dimension $1/2$. Moreover, by Exercise 4.14, the $1/2$-dimensional Hausdorff measure of the zero set is zero, so Hausdorff measure as defined so far does not give an interesting answer.

We approach this problem by counting the number of downcrossings of a nested sequence of intervals decreasing to zero. More precisely, for a linear Brownian motion $\{B(t): t \geqslant 0\}$ with arbitrary starting point, given $a < b$, we define stopping times $\tau_0 = 0$ and, for $j \geqslant 1$,

$$\sigma_j = \inf\left\{t > \tau_{j-1}: B(t) = b\right\}, \qquad \tau_j = \inf\left\{t > \sigma_j: B(t) = a\right\}. \tag{6.1}$$

We call the random functions

$$B^{(j)}: [0, \tau_j - \sigma_j] \to \mathbb{R}, \qquad B^{(j)}(s) = B(\sigma_j + s)$$

the jth downcrossing of $[a, b]$. For every $t > 0$ we denote by

$$D(a, b, t) = \max\left\{j \in \mathbb{N}: \tau_j \leqslant t\right\}$$

the number of *downcrossings* of the interval $[a, b]$ before time t. Note that $D(a, b, t)$ is almost surely finite by the uniform continuity of Brownian motion on the compact interval $[0, t]$.

153

Theorem 6.1 (Downcrossing representation of the local time at zero) *There exists a stochastic process* $\{L(t)\colon t \geqslant 0\}$ *called the* **local time at zero** *such that for all sequences* $a_n \uparrow 0$ *and* $b_n \downarrow 0$ *with* $a_n < b_n$, *almost surely,*

$$\lim_{n \to \infty} 2(b_n - a_n) D(a_n, b_n, t) = L(t) \quad \text{for every } t > 0.$$

Moreover, this process is almost surely locally γ-*Hölder continuous for any* $\gamma < 1/2$.

Remark 6.2 To see why the normalisation in this formula is plausible recall from Theorem 5.38 that the time spent in the interval $[a_n, b_n]$ during a full downcrossing has the same law as the first exit time from $[a_n, 2b_n - a_n]$ by a Brownian motion started in b_n, which by Theorem 2.49 has a mean of $(b_n - a_n)^2$. By the law of large numbers the total time spent in $[a_n, b_n]$ is therefore approximately $2(b_n - a_n)^2 D(a_n, b_n, t)$ taking into account that about the same time is spent in up- and downcrossings. Therefore $L(t)$ plays the rôle of the density at zero of the occupation measure of Brownian motion.   ◇

In the following we will use two types of both geometric distributions: X is geometrically distributed on $\{1, 2, \ldots\}$ with success parameter p (or, equivalently, mean $\frac{1}{p}$) if

$$\mathbb{P}\{X = k\} = p\,(1 - p)^{k-1} \text{ for } k \in \{1, 2, \ldots\}.$$

Similarly, X is geometrically distributed on $\{0, 1, 2, \ldots\}$ with success parameter p if

$$\mathbb{P}\{X = k\} = p\,(1 - p)^{k} \text{ for } k \in \{0, 1, 2, \ldots\}.$$

If the type is not clear from the context we will always state the domain for clarification. The key ingredient of the proof of Theorem 6.1 is the following fact.

Lemma 6.3 *Suppose that* $a < m < b < c$ *and let* $\{B(t)\colon t \geqslant 0\}$ *be a linear Brownian motion, and* T *the first time when it hits level* c. *Let*

- *D be the number of downcrossings of the interval $[a, b]$ completed at time T,*

- *D_1 be the number of downcrossings of the interval $[a, m]$ completed at time T,*

- *D_u be the number of downcrossings of the interval $[m, b]$ completed at time T.*

There exist two independent sequences $X_0, X_1, \ldots$ *and* $Y_0, Y_1, \ldots$ *of independent nonnegative random variables, which are also independent of* D, *such that for* $j \geqslant 1$ *the random variables* X_j *are geometric on* $\{1, 2 \ldots\}$ *with mean* $(b - a)/(m - a)$ *and the random variables* Y_j *are geometric on* $\{1, 2 \ldots\}$ *with mean* $(b - a)/(b - m)$, *and*

$$D_1 = X_0 + \sum_{j=1}^{D} X_j \quad \text{and} \quad D_u = Y_0 + \sum_{j=1}^{D} Y_j.$$

Proof.   Recall the definition of the stopping times σ_j, τ_j from (6.1). For $j \geqslant 0$, define the jth downcrossings, resp. upcrossings, of $[a, b]$ by

$$B_\downarrow^{(j)} \colon [0, \tau_j - \sigma_j] \to \mathbb{R}, \qquad B_\downarrow^{(j)}(s) = B(\sigma_j + s), \text{ if } j \geqslant 1,$$
$$B_\uparrow^{(j)} \colon [0, \sigma_{j+1} - \tau_j] \to \mathbb{R}, \qquad B_\uparrow^{(j)}(s) = B(\tau_j + s).$$

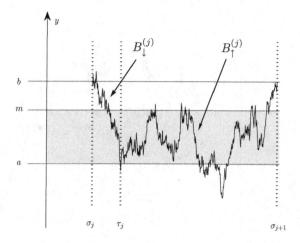

Fig. 6.1. The downcrossing of $[a, b]$ contains one downcrossing of $[a, m]$ and the following upcrossing of $[a, b]$ contains one further downcrossing of $[a, m]$.

By the strong Markov property all these pieces of the Brownian path are independent. Note that D depends only on the family $(B_\downarrow^{(j)} : j \geqslant 1)$ of downcrossings.

First look at D_1 and denote by X_0 the number of downcrossings of $[a, m]$ before the first downcrossing of $[a, b]$. The jth downcrossing of $[a, b]$ contains exactly one downcrossing of $[a, m]$ and the jth upcrossing of $[a, b]$ contains a random number $X_j - 1$ of downcrossings of $[a, m]$, which, by Theorem 2.49, satisfies

$$\mathbb{P}\{X_j = k\} = \left(\frac{m - a}{b - a}\right)\left(\frac{b - m}{b - a}\right)^{k-1} \qquad \text{for every } k \in \{1, \ldots\}.$$

In other words X_j is geometrically distributed on $\{1, 2, \ldots\}$ with success parameter given by $(m - a)/(b - a)$.

Second look at D_u and denote by Y_0 the number of downcrossings of $[m, b]$ after the last downcrossing of $[a, b]$. No downcrossings of $[m, b]$ can occur during an upcrossing of $[a, b]$. Fix a j and look at the downcrossing $B_\downarrow^{(j)} : [0, \infty) \to \mathbb{R}$ formally extended to have infinite lifetime by attaching an independent Brownian motion at the endpoint. Define stopping times $\tilde{\sigma}_0 = 0$ and, for $i \geqslant 1$,

$$\tilde{\tau}_i = \inf\left\{t > \tilde{\sigma}_{i-1} : B_\downarrow^{(j)}(t) = m\right\}, \qquad \tilde{\sigma}_i = \inf\left\{t > \tilde{\tau}_i : B_\downarrow^{(j)}(t) = b\right\}.$$

This subdivides the path of $B_\downarrow^{(j)}$ into downcrossing periods $[\tilde{\sigma}_{i-1}, \tilde{\tau}_i]$, and upcrossing periods $[\tilde{\tau}_i, \tilde{\sigma}_i]$ of $[m, b]$, such that the pieces

$$B_{\downarrow,i}^{(j)} : [0, \tilde{\tau}_i - \tilde{\sigma}_{i-1}] \to \mathbb{R}, \qquad B_{\downarrow,i}^{(j)}(s) = B(\tilde{\sigma}_{i-1} + s), \text{ for } i \geqslant 0,$$

$$B_{\uparrow,i}^{(j)} : [0, \tilde{\sigma}_i - \tilde{\tau}_i] \to \mathbb{R}, \qquad B_{\uparrow,i}^{(j)}(s) = B(\tilde{\tau}_i + s), \text{ for } i \geqslant 1,$$

are all independent. As $c > b$ the first hitting time of level c must lie in a downcrossing

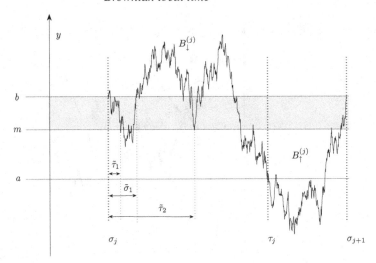

Fig. 6.2. The downcrossing of $[a, b]$ contains three downcrossings of $[m, b]$ and the following up-crossing of $[a, b]$ contains no further downcrossings of $[m, b]$.

period, while the lifetime $\tau_j - \sigma_j$ of $B_{\downarrow}^{(j)}$ expires when the lower boundary a is hit for the first time, which can only occur during an upcrossing period. By Theorem 2.49 the probability that a is hit during any upcrossing period equals $(b - m)/(b - a)$.

Hence the number of downcrossings of $[m, b]$ during the jth downcrossing of $[a, b]$ is a geometric random variable Y_j on $\{1, 2, \ldots\}$ with (success) parameter $(b - m)/(b - a)$, which completes the proof. ∎

For the proof of Theorem 6.1 we first prove the convergence for the case when the Brownian motion is stopped at the time $T = T_b$ when it first reaches some level $b > b_1$. This has the advantage that there cannot be any uncompleted upcrossings.

Lemma 6.4 *For any two sequences $a_n \uparrow 0$ and $b_n \downarrow 0$ with $a_n < b_n$, the discrete time stochastic process $\{2(b_n - a_n) D(a_n, b_n, T): n \in \mathbb{N}\}$ is a submartingale with respect to the natural filtration $(\mathcal{F}_n : n \in \mathbb{N})$.*

Proof. We may assume that, for each n, we have

$$either \quad (1)\; a_n = a_{n+1} \qquad or \quad (2)\; b_n = b_{n+1},$$

which is no loss of generality, as we may replace a step where both a_n and b_n are changed by two steps, where only one is changed at a time. The original sequence is then a subsequence of the modified one and inherits the submartingale property.

Now fix n and *first* assume that we are in case (1) $a_n = a_{n+1}$. By Lemma 6.3 for D_1, the total number $D(a_n, b_{n+1}, T)$ of downcrossings of $[a_n, b_{n+1}]$ given $\mathcal{F}_n$ is the sum of $D(a_n, b_n, T)$ independent geometric random variables with parameter $(b_{n+1} - a_n)/(b_n - $

a_n) plus a nonnegative contribution. Hence,

$$\mathbb{E}\big[(b_{n+1} - a_n)\, D(a_n, b_{n+1}, T) \,\big|\, \mathcal{F}_n\big] \geqslant (b_n - a_n)\, D(a_n, b_n, T),$$

which is the submartingale property (for the nth step).

Second assume that we are in case (2) $b_n = b_{n+1}$. Then Lemma 6.3 for D_u shows that the number of downcrossings of $[a_{n+1}, b_n]$ given $\mathcal{F}_n$ is the sum of $D(a_n, b_n, T)$ independent geometric random variables with parameter $(b_n - a_{n+1})/(b_n - a_n)$ plus a nonnegative contribution. Hence

$$\mathbb{E}\big[(b_n - a_{n+1})\, D(a_{n+1}, b_n, T) \,\big|\, \mathcal{F}_n\big] \geqslant (b_n - a_n)\, D(a_n, b_n, T),$$

and together with the first case this establishes that $\{2(b_n - a_n)\, D(a_n, b_n, T) : n \in \mathbb{N}\}$ is a submartingale with respect to its natural filtration. ∎

Lemma 6.5 *For any two sequences $a_n \uparrow 0$ and $b_n \downarrow 0$ with $a_n < b_n$ the limit*

$$L(T_b) := \lim_{n \to \infty} 2(b_n - a_n)\, D(a_n, b_n, T_b) \tag{6.2}$$

exists almost surely. It is not zero and does not depend on the choice of sequences.

Proof. Observe that $D(a_n, b_n, T_b)$ is a geometric random variable on $\{0, 1, \ldots\}$ with parameter $(b_n - a_n)/(b - a_n)$. Recall that the variance of a geometric random variable on $\{0, 1, \ldots\}$ with parameter p is $(1-p)/p^2$, and so its second moment is bounded by $2/p^2$. Hence

$$\mathbb{E}\big[4(b_n - a_n)^2\, D(a_n, b_n, T_b)^2\big] \leqslant 8\,(b - a_n)^2,$$

and thus the submartingale in Lemma 6.4 is $\mathbf{L}^2$-bounded. By the submartingale convergence theorem, see Theorem 12.21 in the appendix, the limit

$$\lim_{n \uparrow \infty} 2(b_n - a_n)\, D(a_n, b_n, T_b)$$

exists almost surely, and by Theorem 12.28 also in $\mathbf{L}^2$ ensuring that the limit is nonzero. Finally, note that the limit does not depend on the choice of the sequence $a_n \uparrow 0$ and $b_n \downarrow 0$ because if it did, then given two sequences with different limits in (6.2) we could construct a sequence of intervals alternating between the sequences, for which the limit in (6.2) would not exist. ∎

Lemma 6.6 *For any fixed time $t > 0$, almost surely, the limit*

$$L(t) := \lim_{n \to \infty} 2(b_n - a_n)\, D(a_n, b_n, t) \qquad \text{exists.}$$

Proof. We define an auxiliary Brownian motion $\{B_t(s) : s \geqslant 0\}$ by $B_t(s) = B(t + s)$. For any integer $b > b_1$ we denote by $D_t(a_n, b_n, T_b)$ the number of downcrossings of the interval $[a_n, b_n]$ by the auxiliary Brownian motion before it hits b. Then, almost surely,

$$L_t(T_b) := \lim_{n \uparrow \infty} 2(b_n - a_n)\, D_t(a_n, b_n, T_b),$$

exists by the previous lemma. Given $t > 0$ we fix a Brownian path such that this limit exists for all integers $b > b_1$. Pick b so large that $T_b > t$. Define

$$L(t) := L(T_b) - L_t(T_b).$$

To show that this is the required limit, observe that

$$D(a_n, b_n, T_b) - D_t(a_n, b_n, T_b) - 1 \leqslant D(a_n, b_n, t) \leqslant D(a_n, b_n, T_b) - D_t(a_n, b_n, T_b),$$

where the correction -1 on the left hand side arises from the possibility that t interrupts a downcrossing. Multiplying by $2(b_n - a_n)$ and taking a limit gives $L(T_b) - L_t(T_b)$ for both bounds, proving convergence. ∎

We now have to study the dependence of $L(t)$ on the time t in more detail. To simplify the notation we write

$$I_n(s, t) = 2(b_n - a_n)\left(D(a_n, b_n, t) - D(a_n, b_n, s)\right) \qquad \text{for all } 0 \leqslant s < t.$$

The following lemma contains a probability estimate, which is sufficient to get the convergence of the downcrossing numbers jointly for all times and to establish Hölder continuity.

Lemma 6.7 Let $\gamma < 1/2$ and $0 < \varepsilon < (1 - 2\gamma)/3$. Then, for all $t \geqslant 0$ and $0 < h < 1$, we have

$$\mathbb{P}\{L(t + h) - L(t) > h^\gamma\} \leqslant 2 \exp\{-\tfrac{1}{2} h^{-\varepsilon}\}.$$

Proof. As, by Fatou's lemma,

$$\mathbb{P}\{L(t+h)-L(t) > h^\gamma\} = \mathbb{P}\{\liminf_{n\to\infty} I_n(t, t+h) > h^\gamma\} \leqslant \liminf_{n\to\infty} \mathbb{P}\{I_n(t, t+h) > h^\gamma\}$$

we can focus on estimating $\mathbb{P}\{I_n(t, t + h) > h^\gamma\}$ for fixed large n. It suffices to estimate $\mathbb{P}_x\{I_n(0, h) > h^\gamma\}$ uniformly for all $x \in \mathbb{R}$. This probability is clearly maximal when $x = b_n$, so we may assume this. Let $T_h = \inf\{s > 0 \colon B(s) = b_n + h^{(1-\varepsilon)/2}\}$ and observe that

$$\{I_n(0, h) > h^\gamma\} \subset \{I_n(0, T_h) > h^\gamma\} \cup \{T_h < h\}.$$

The number of downcrossings of $[a_n, b_n]$ during the period before T_h is geometrically distributed on $\{0, 1, \dots\}$ with mean $(b_n - a_n)^{-1} h^{(1-\varepsilon)/2}$ and thus

$$\mathbb{P}_{b_n}\{I_n(0, T_h) > h^\gamma\} \leqslant \left(\frac{h^{(1-\varepsilon)/2}}{b_n - a_n + h^{(1-\varepsilon)/2}}\right)^{\lfloor \frac{1}{2(b_n - a_n)} h^\gamma \rfloor}$$

$$\xrightarrow{n\to\infty} \exp\left\{-\tfrac{1}{2} h^{\gamma - \frac{1}{2} + \frac{\varepsilon}{2}}\right\} \leqslant \exp\left\{-\tfrac{1}{2} h^{-\varepsilon}\right\}.$$

With $\{W(s) \colon s \geqslant 0\}$ denoting a standard linear Brownian motion,

$$\mathbb{P}_{b_n}\{T_h < h\} = \mathbb{P}\left\{\max_{0 \leqslant s \leqslant h} W(s) \geqslant h^{(1-\varepsilon)/2}\right\} \leqslant \sqrt{\tfrac{2}{\pi h^{-\varepsilon}}} \exp\left\{-\tfrac{1}{2} h^{-\varepsilon}\right\}$$

where we have used Remark 2.22 in the last step. The result follows by adding the last two displayed formulas. ∎

Lemma 6.8 *Almost surely,*

$$L(t) := \lim_{n \to \infty} 2(b_n - a_n) \, D(a_n, b_n, t)$$

exists for every $t \geqslant 0$.

Proof. It suffices to prove the simultaneous convergence for all $0 \leqslant t \leqslant 1$. We define a countable set of gridpoints

$$\mathcal{G} = \bigcup_{m \in \mathbb{N}} \mathcal{G}_m \cup \{1\}, \quad \text{for } \mathcal{G}_m = \left\{ \tfrac{k}{m} : k \in \{0, \ldots, m-1\} \right\}$$

and show that the stated convergence holds on the event

$$E_M = \bigcap_{t \in \mathcal{G}} \left\{ L(t) = \lim_{n \to \infty} 2(b_n - a_n) \, D(a_n, b_n, t) \text{ exists } \right\}$$

$$\cap \bigcap_{m > M} \bigcap_{t \in \mathcal{G}_m} \left\{ L(t + \tfrac{1}{m}) - L(t) \leqslant (1/m)^\gamma \right\}.$$

which, by choosing M suitably, has probability arbitrarily close to one by the previous two lemmas. Given any $t \in [0, 1)$ and a large m we find $t_1, t_2 \in \mathcal{G}_m$ with $t_2 - t_1 = \frac{1}{m}$ and $t \in [t_1, t_2]$. We obviously have

$$2(b_n - a_n) \, D(a_n, b_n, t_1) \leqslant 2(b_n - a_n) \, D(a_n, b_n, t) \leqslant 2(b_n - a_n) \, D(a_n, b_n, t_2).$$

Both bounds converge on E_M, and the difference of the limits is $L(t_2) - L(t_1)$, which is bounded by $m^{-\gamma}$ and thus can be made arbitrarily small by choosing a large m. ∎

Lemma 6.9 *For $\gamma < \frac{1}{2}$, almost surely, the process $\{L(t) : t \geqslant 0\}$ is locally γ-Hölder continuous.*

Proof. It suffices to look at $0 \leqslant t < 1$. We use the notation of the proof of the previous lemma and show that γ-Hölder continuity holds on the set constructed there. Indeed, whenever $0 \leqslant s < t < 1$ and $t - s < 1/M$ we pick $m \geqslant M$ such that

$$\frac{1}{m+1} \leqslant t - s < \frac{1}{m} .$$

We take $t_1 \leqslant s$ with $t_1 \in \mathcal{G}_m$ and $s - t_1 < 1/m$, and $t_2 \geqslant t$ with $t_2 \in \mathcal{G}_m$ and $t_2 - t < 1/m$. Note that $t_2 - t_1 \leqslant 2/m$ by construction and hence,

$$L(t) - L(s) \leqslant L(t_2) - L(t_1) \leqslant 2(1/m)^\gamma \leqslant 2 \left(\tfrac{m+1}{m} \right)^\gamma (t - s)^\gamma .$$

The result follows as the fraction on the right is bounded by 2. ∎

This completes the proof of the downcrossing representation, Theorem 6.1. It is easy to see from this representation that, almost surely, the local time at zero increases only on the zero set of the Brownian motion, see Exercise 6.1.

Observe that the increasing process $\{L(t): t \geqslant 0\}$ is *not* a Markov process. Heuristically, the size of the increment $L(t + h) - L(t)$ depends on the position of the first zero of the Brownian motion after time t, which is strongly dependent on the position of the last zero before time t. The last zero however is the position of the last point of increase of the local time process before time t, and therefore the path $\{L(s): 0 \leqslant s \leqslant t\}$ contains relevant information beyond its endpoint.

Nevertheless, we can describe the law of the local time process, thanks to the following famous theorem of Paul Lévy, which describes the law of the local time at zero in terms of the maximum process of Brownian motion. It opens the door to finer results on the local time at zero, like those presented in Section 6.4 of this chapter.

Theorem 6.10 (Lévy) *The local time at zero $\{L(t): t \geqslant 0\}$ and the maximum process $\{M(t): t \geqslant 0\}$ of a standard linear Brownian motion have the same distribution.*

Remark 6.11 In fact, a similar proof shows that the processes $\{(L(t), |B(t)|): t \geqslant 0\}$ and $\{(M(t), M(t) - B(t)): t \geqslant 0\}$ have the same distribution. Details are deferred to Exercise 6.2 as we present a different argument for this in Theorem 7.38. See also Exercise 6.5 for an alternative approach, which goes back to Lévy himself.   ◇

The proof of Theorem 6.10 uses the simple random walk embedded in the Brownian motion, a technique which we will exploit extensively. Define stopping times $\tau_0 := \tau_0^{(n)} := 0$ and

$$\tau_k := \tau_k^{(n)} := \inf \left\{ t > \tau_{k-1} : |B(t) - B(\tau_{k-1})| = 2^{-n} \right\}, \qquad \text{for } k \geqslant 1.$$

The nth embedded random walk $\{X_k^{(n)} : k = 1, 2, \ldots\}$ is defined by

$$X_k := X_k^{(n)} := 2^n B(\tau_k^{(n)}).$$

The length of the embedded random walk is

$$N := N^{(n)}(t) := \max\{k \in \mathbb{N} : \tau_k \leqslant t\},$$

which is easily seen to be independent of the actual walk.

Lemma 6.12 *For every $t > 0$, almost surely, $\displaystyle\lim_{n \to \infty} 2^{-2n} N^{(n)}(t) = t$.*

Proof.   First note that $\{\xi_k^{(n)} : k = 1, 2, \ldots\}$ defined by

$$\xi_k := \xi_k^{(n)} := \tau_k^{(n)} - \tau_{k-1}^{(n)}$$

is a sequence of independent random variables, for each n. By Theorem 2.49 the mean of ξ_k is 2^{-2n} and its variance is, by Brownian scaling, equal to $c2^{-4n}$ for some constant $c > 0$. (See, for example, Exercise 2.15 for instructions how to find the constant.) Define

$$S^{(n)}(t) = \sum_{k=1}^{\lceil 2^{2n} t \rceil} \xi_k^{(n)}.$$

Then $\mathbb{E}S^{(n)}(t) = \lceil 2^{2n}t\rceil 2^{-2n} \to t$ and $\mathrm{Var}\big(S^{(n)}(t)\big) = c2^{-4n}\lceil 2^{2n}t\rceil$, hence

$$\mathbb{E}\sum_{n=1}^{\infty} \big(S^{(n)}(t) - \mathbb{E}S^{(n)}(t)\big)^2 < \infty.$$

We infer that, almost surely, $\lim_{n\to\infty} S^{(n)}(t) = t$. For fixed $\varepsilon > 0$, we pick n_0 large so that

$$S^{(n)}(t-\varepsilon) \leqslant t \leqslant S^{(n)}(t+\varepsilon) \ \text{for all } n \geqslant n_0.$$

The sum over ξ_k up to $N^{(n)}(t)+1$ is at least t, by definition, and hence we get $N^{(n)}(t) + 1 \geqslant \lceil 2^{2n}(t-\varepsilon)\rceil$. Conversely, the sum over ξ_k up to $N^{(n)}(t)$ is at most t and hence $N^{(n)}(t) \leqslant \lceil 2^{2n}(t+\varepsilon)\rceil$. The result follows as $\varepsilon > 0$ was arbitrary. ∎

Lemma 6.13 *Almost surely, for every $t > 0$,*

$$\lim_{n\uparrow\infty} 2^{-n}\#\{k \in \{1,\ldots,N^{(n)}(t)\} : |X_{k-1}| = 0, |X_k| = 1\} = L(t).$$

Proof. By Theorem 6.1 applied to the sequences $a_n = -2^{-n}$ and $b_n = 0$ we have

$$\lim_{n\uparrow\infty} 2^{-n}\#\big\{k \in \{1,\ldots,N^{(n)}(t)\} : X_{k-1} = 0, X_k = -1\big\} = \tfrac{1}{2}L(t).$$

Applying Theorem 6.1 to the sequences $a_n = 0$ and $b_n = 2^{-n}$ we get

$$\lim_{n\uparrow\infty} 2^{-n}\#\big\{k \in \{1,\ldots,N^{(n)}(t)\} : X_{k-1} = 1, X_k = 0\big\} = \tfrac{1}{2}L(t).$$

As $\#\{k \leqslant N : X_{k-1} = 1, X_k = 0\}$ and $\#\{k \leqslant N : X_{k-1} = 0, X_k = 1\}$ differ by no more than one, the result follows by adding up the two displayed formulas. ∎

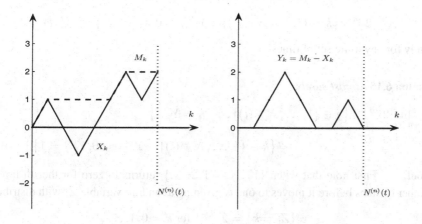

Fig. 6.3. On the left an embedded random walk $\{X_k : k \geqslant 0\}$ together with its maximum process $\{M_k : k \geqslant 0\}$. On the right the associated difference process $\{Y_k : k \geqslant 0\}$ defined by $Y_k = M_k - X_k$.

We define the maximum process $\{M_k^{(n)}\colon k = 1, 2, \ldots\}$ associated with the embedded random walk by

$$M_k = M_k^{(n)} = \max\big\{X_j^{(n)}\colon j \in \{0, \ldots, k\}\big\}\,.$$

Then the process $\{Y_k^{(n)}\colon k = 1, 2, \ldots\}$ defined by $Y_k := Y_k^{(n)} := M_k - X_k$ is a Markov chain with statespace $\{0, 1, 2, \ldots\}$ and the following transition mechanism

- if $j \neq 0$ then $\mathbb{P}\{Y_{k+1} = j + 1 \,|\, Y_k = j\} = \frac{1}{2} = \mathbb{P}\{Y_{k+1} = j - 1 \,|\, Y_k = j\}$,

- $\mathbb{P}\{Y_{k+1} = 0 \,|\, Y_k = 0\} = \frac{1}{2} = \mathbb{P}\{Y_{k+1} = 1 \,|\, Y_k = 0\}$.

One can recover the maximum process $\{M_k\colon k = 1, 2, \ldots\}$ from $\{Y_k\colon k = 1, 2, \ldots\}$ by counting the number of flat steps

$$M_k = \#\big\{j \in \{1, \ldots, k\}\colon Y_j = Y_{j-1}\big\}\,.$$

Hence we obtain, asymptotically, the maximum process of the Brownian motion as a limit of the number of flat steps in $\{Y_k^{(n)}\colon k = 1, 2, \ldots\}$.

Lemma 6.14 *For any time $t > 0$, almost surely,*

$$M(t) = \lim_{n \uparrow \infty} 2^{-n}\, \#\big\{j \in \{1, \ldots, N^{(n)}(t)\}\colon Y_j^{(n)} = Y_{j-1}^{(n)}\big\}\,.$$

Proof.   Note that $\#\big\{j \in \{1, \ldots, N^{(n)}(t)\}\colon Y_j = Y_{j-1}\big\}$ is the maximum of the random walk $\{X_k\colon k = 1, 2, \ldots, N^{(n)}(t)\}$ over its entire length. This maximum, multiplied by 2^{-n}, differs from $M(t)$ by no more than 2^{-n}, and this completes the argument.   ∎

Removing the flat steps in the process $\{Y_j^{(n)}\colon j = 1, 2, \ldots\}$ we obtain a process $\{\tilde{Y}_k^{(n)}\colon k = 1, 2, \ldots\}$, which has the same law as $\{|X_k|\colon k = 1, 2, \ldots\}$. By Lemma 6.13 we therefore have the convergence in distribution, as $n \uparrow \infty$,

$$2^{-n}\#\big\{k \in \{1, \ldots, N^{(n)}(t)\} \,:\, \tilde{Y}_{k-1}^{(n)} = 0, \tilde{Y}_k^{(n)} = 1\big\} \overset{\mathrm{d}}{\longrightarrow} L(t), \qquad (6.3)$$

jointly for any finite set of times.

Lemma 6.15 *Almost surely,*

$$\lim_{n \uparrow \infty} 2^{-n}\Big(\#\big\{j \in \{1, \ldots, N^{(n)}(t)\}\colon Y_{j-1}^{(n)} = Y_j^{(n)}\big\}$$

$$- \#\big\{k \in \{1, \ldots, N^{(n)}(t)\} \,:\, \tilde{Y}_{k-1}^{(n)} = 0, \tilde{Y}_k^{(n)} = 1\big\}\Big) = 0\,.$$

Proof.   First note that when $\{Y_j\colon j = 1, 2, \ldots\}$ returns to zero for the ith time, the number of steps before it moves to one is given by a random variable Z_i with distribution

$$\mathbb{P}\{Z_i = k\} = 2^{-k-1} \text{ for } k = 0, 1, \ldots.$$

Denoting by Z_0 the number of steps before it moves initially, the random variables $Z_0, Z_1, \ldots$ are independent and independent of the process $\{\tilde{Y}_k^{(n)}\colon k = 1, 2, \ldots\}$. Let

$$A^{(n)} = \#\big\{j \in \{1, \ldots, N^{(n)}(t)\}\colon Y_{j-1}^{(n)} = 1, Y_j^{(n)} = 0\big\}$$

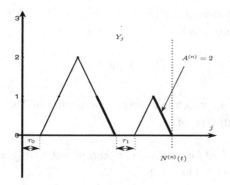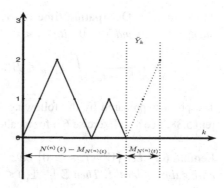

Fig. 6.4. On the left a sample of the processes $\{Y_j : 0 \leqslant j \leqslant N^{(n)}(t)\}$. On the right the associated $\{\tilde{Y}_k : 0 \leqslant k \leqslant N^{(n)}(t)\}$, which is obtained by removing the two flat steps and extending the path to its original length.

be the total number of returns to zero before time $N^{(n)}(t)$. With a possible small modification of the final value $Z_{A^{(n)}}$ we get, almost surely, as $n \uparrow \infty$,

$$2^{-n}\left(\# \left\{j \in \{1, \ldots, N^{(n)}(t)\} : Y_j^{(n)} = Y_{j-1}^{(n)}\right\}\right.$$
$$\left. - \#\{j \in \{1, \ldots, N^{(n)}(t)\} : Y_{j-1}^{(n)} = 0, Y_j^{(n)} = 1\}\right)$$
$$= 2^{-n} \sum_{i=0}^{A^{(n)}} (Z_i - 1) = \left(2^{-n} A^{(n)}\right) \frac{1}{A^{(n)}} \sum_{i=0}^{A^{(n)}} (Z_i - \mathbb{E}Z_i) \longrightarrow 0,$$

because the first factor converges by Lemma 6.13 and the second by the law of large numbers, irrespective of the actual value of $Z_{A^{(n)}}$. To study the effect of the removal of the flat pieces, recall that almost surely the length $N^{(n)}(t)$ of the walk is of order $2^{2n}t$, by Lemma 6.12, and the number of flat pieces is $M_{N^{(n)}(t)}$, which is of order 2^n, by Lemma 6.14. Hence, for all $\varepsilon > 0$, if n is large enough,

$$N^{(n)}(t - \varepsilon) + M_{N^{(n)}(t)} \leqslant N^{(n)}(t).$$

We infer from this that

$$2^{-n}\left(\#\{j \in \{1, \ldots, N^{(n)}(t)\} : \tilde{Y}_{j-1}^{(n)} = 0, \tilde{Y}_j^{(n)} = 1\}\right.$$
$$\left. - \#\{j \in \{1, \ldots, N^{(n)}(t)\} : Y_{j-1}^{(n)} = 0, Y_j^{(n)} = 1\}\right)$$
$$\leqslant 2^{-n} \#\{j \in \{N^{(n)}(t - \varepsilon) + 1, \ldots, N^{(n)}(t)\} : \tilde{Y}_{j-1}^{(n)} = 0, \tilde{Y}_j^{(n)} = 1\},$$

and the right hand side converges almost surely to a random variable, which has the law of $L(t) - L(t - \varepsilon)$ and hence can be made arbitrarily small by choice of $\varepsilon > 0$. ∎

Proof of Theorem 6.10. Note that both processes in Theorem 6.10 are continuous, so that it suffices to compare their finite dimensional distributions. Equality of these follows directly by combining Lemma 6.14, Equation (6.3) and Lemma 6.15. ∎

Theorem 6.16 (Occupation time representation of the local time at zero) *For all sequences* $a_n \uparrow 0$ *and* $b_n \downarrow 0$ *with* $a_n < b_n$, *almost surely,*

$$\lim_{n \to \infty} \frac{1}{b_n - a_n} \int_0^t 1\{a_n \leqslant B(s) \leqslant b_n\}\, ds = L(t) \quad \text{for every } t > 0.$$

The proof is prepared by the following lemma, which is a direct consequence of Theorem 5.38. See also Exercise 6.6 for an alternative proof.

Lemma 6.17 *Let* $\{W(s) \colon s \geqslant 0\}$ *be a standard linear Brownian motion and* τ_1 *its first hitting time of level* 1. *Then* $\mathbb{E} \int_0^{\tau_1} 1\{0 \leqslant W(s) \leqslant 1\}\, ds = 1.$

Proof. By Theorem 5.38 we have $\mathbb{E} \int_0^{\tau_1} 1\{0 \leqslant W(s) \leqslant 1\}\, ds = \mathbb{E}\sigma_1$, where σ_1 is the first exit time from $[-1, 1]$. By Theorem 2.49 we have $\mathbb{E}\sigma_1 = 1$. ∎

Proof of Theorem 6.16. Recall the stopping times τ_j defined for $a_n < b_n$ as in (6.1). For the proof of the lower bound note that

$$\int_0^t 1\{a_n \leqslant B(s) \leqslant b_n\}\, ds \geqslant \sum_{j=1}^{D(a_n, b_n, t)} \int_{\tau_{j-1}}^{\tau_j} 1\{a_n \leqslant B(s) \leqslant b_n\}\, ds.$$

By Brownian scaling

$$\int_{\tau_{j-1}}^{\tau_j} 1\{a_n \leqslant B(s) \leqslant b_n\}\, ds = (b_n - a_n)^2 \int_0^{\tau(j)} 1\{0 \leqslant W_j(s) \leqslant 1\}\, ds,$$

where $\{W_j(s) \colon s \geqslant 0\}$ are independent standard linear Brownian motions and $\tau(j) = \inf\{s > 0 \colon W_j(s) = 0$ and there exists $t < s$ with $W_j(t) = 1\}$. Hence

$$\frac{1}{b_n - a_n} \sum_{j=1}^{D(a_n, b_n, t)} \int_{\tau_{j-1}}^{\tau_j} 1\{a_n \leqslant B(s) \leqslant b_n\}\, ds$$

$$= (b_n - a_n) D(a_n, b_n, t) \left[\frac{1}{D(a_n, b_n, t)} \sum_{j=1}^{D(a_n, b_n, t)} \int_0^{\tau(j)} 1\{0 \leqslant W_j(s) \leqslant 1\}\, ds \right].$$

The first factor converges almost surely to $\frac{1}{2} L(t)$, by Theorem 6.1. From the law of large numbers we get for the second factor, almost surely,

$$\lim_{n \uparrow \infty} \frac{1}{D(a_n, b_n, t)} \sum_{j=1}^{D(a_n, b_n, t)} \int_0^{\tau(j)} 1\{0 \leqslant W(s) \leqslant 1\}\, ds = \mathbb{E} \int_0^{\tau} 1\{0 \leqslant W(s) \leqslant 1\}\, ds.$$

Applying Lemma 6.17 first to $\{W(s) \colon s \geqslant 0\}$, and then to $\{1 - W(s + \tau_1) \colon s \geqslant 0\}$ yields

$$\mathbb{E} \int_0^{\tau} 1\{0 \leqslant W(s) \leqslant 1\}\, ds = 2.$$

This verifies the lower bound. The upper bound can be obtained by including the period $[\tau_j, \tau_{j+1}]$ for $j = D(a_n, b_n, t)$ in the summation and using the same arguments as for the lower bound. This completes the proof of Theorem 6.16. ∎

6.2 A random walk approach to the local time process

Given a level $a \in \mathbb{R}$ the construction of the previous chapter allows us to define the *local time at level a* for a linear Brownian motion $\{B(t): t \geqslant 0\}$. Indeed, simply let $\{L^a(t): t \geqslant 0\}$ be the local time at zero of the auxiliary Brownian motion $\{B^a(t): t \geqslant 0\}$ defined by $B^a(t) = B(t) - a$. Using Theorem 6.16 it is not hard to show that $\{L^a(t): a \in \mathbb{R}\}$ is the density of the occupation measure μ_t introduced in Theorem 3.26.

Theorem 6.18 *For linear Brownian motion $\{B(t): t \geqslant 0\}$, almost surely, for any bounded measurable $g: \mathbb{R} \to \mathbb{R}$ and $t > 0$,*

$$\int g(a)\, d\mu_t(a) = \int_0^t g(B(s))\, ds = \int_{-\infty}^\infty g(a)\, L^a(t)\, da.$$

Proof. First, observe that for the statement it suffices to have $\{L^a(t): t \geqslant 0\}$ defined for $\mathcal{L}$-almost every a. Second, we may assume that t is fixed. Indeed, it suffices to verify the second equality for a countable family of bounded measurable $g: \mathbb{R} \to \mathbb{R}$, for example the indicator functions of rational intervals. Having fixed such a g both sides are continuous in t. For fixed t, we know from Theorem 3.26 that $\mu_t \ll \mathcal{L}$ almost surely, hence a density f exists by the Radon–Nikodým theorem and may be obtained as

$$f(a) = \lim_{\varepsilon \downarrow 0} \frac{1}{2\varepsilon} \int_0^t 1\{a - \varepsilon \leqslant B(s) \leqslant a + \varepsilon\}\, ds,$$

which equals $L^a(t)$ by Theorem 6.16, almost surely for $\mathcal{L}$-almost every a. ∎

A major result about linear Brownian motion is that the density $\{L^a(t): a \in \mathbb{R}\}$ of the occupation measures can be chosen to be continuous, a fact which we now prove. To explore $L^a(t)$ as a function of the levels a we extend the downcrossing representation to hold *simultaneously* at all levels a.

Given $a \in \mathbb{R}$ and a large integer n we let $I(a, n)$ be the unique dyadic interval such that $a \in I(a, n) = [j(a)2^{-n}, (j(a) + 1)2^{-n})$. For a standard Brownian motion $\{B(t): t \geqslant 0\}$ we denote by $D^{(n)}(a, t)$ the number of downcrossings of the interval $I(a, n)$ before time t. In the notation of the previous section we can write

$$D^{(n)}(a, t) := \#\{k \in \{0, \ldots, N^{(n)}(t) - 1\}: X_k^{(n)} = j(a) + 1,\ X_{k+1}^{(n)} = j(a)\}.$$

Theorem 6.19 (Trotter's theorem) *Let $\{B(t): t \geqslant 0\}$ be a standard linear Brownian motion and let $D^{(n)}(a, t)$ be the number of downcrossings before time t of the nth stage dyadic interval containing a. Then, almost surely,*

$$L^a(t) = \lim_{n \to \infty} 2^{-n+1} D^{(n)}(a, t) \qquad \textit{exists for all } a \in \mathbb{R} \textit{ and } t \geqslant 0.$$

Moreover, for every $\gamma < \frac{1}{2}$, the random field

$$\{L^a(t): a \in \mathbb{R}, t \geqslant 0\}$$

is almost surely locally γ-Hölder continuous.

Remark 6.20 Note that $\{L^a(t)\colon a \in \mathbb{R},\ t \geqslant 0\}$ is a stochastic process depending on more than one parameter, and to emphasise this fact we use the notion **random field**.   ◇

The proof uses the following estimate for the sum of independent geometric random variables with mean two, which we prove as Exercise 6.7.

Lemma 6.21 *Let $X_1, X_2, \ldots$ be independent geometrically distributed random variables on $\{1, 2, \ldots\}$ with mean 2. Then, for sufficiently small $\varepsilon > 0$, for all nonnegative integers $k \leqslant m$,*

$$\mathbb{P}\Big\{\Big|\sum_{j=1}^{k}(X_j - 2)\Big| \geqslant \varepsilon m\Big\} \leqslant 4 \exp\big\{-\tfrac{1}{5}\varepsilon^2 m\big\}.$$

The following lemma is the heart of the proof of Theorem 6.19.

Lemma 6.22 *Suppose that $a < b$ and let $\{B(t)\colon 0 \leqslant t \leqslant T\}$ be a linear Brownian motion stopped at the time T when it first hits a given level above b. Let*

- *D be the number of downcrossings of the interval $[a, b]$,*

- *D_{l} be the number of downcrossings of the interval $\big[a, \frac{a+b}{2}\big]$,*

- *D_{u} be the number of downcrossings of the interval $\big[\frac{a+b}{2}, b\big]$.*

Then, for sufficiently small $\varepsilon > 0$, for all nonnegative integers $k \leqslant m$,

$$\mathbb{P}\big\{\big|D - \tfrac{1}{2}D_{\mathrm{l}}\big| > \varepsilon m \text{ or } \big|D - \tfrac{1}{2}D_{\mathrm{u}}\big| > \varepsilon m \,\big|\, D = k\big\} \leqslant 12 \exp\big\{-\tfrac{1}{5}\varepsilon^2 m\big\}.$$

Proof.   By Lemma 6.3 we have that, given $\{D = k\}$, there exist independent random variables $X_0, X_1, X_2 \ldots$, such that

$$D_{\mathrm{l}} = X_0 + \sum_{j=1}^{k} X_j\,,$$

and $X_1, X_2, \ldots$ are geometrically distributed on $\{1, 2, \ldots\}$ with mean 2. An inspection of the proof of Theorem 6.3 reveals that X_0 is either zero or also geometrically distributed with mean 2, depending on the starting point of the Brownian motion.

Using Lemma 6.21 and Chebyshev's inequality, we get, if $\varepsilon > 0$ is small enough,

$$\mathbb{P}\big\{\big|\tfrac{1}{2}D_{\mathrm{l}} - D\big| > \varepsilon m \,\big|\, D = k\big\} \leqslant \mathbb{P}\Big\{\Big|\sum_{j=1}^{k}(X_j - 2)\Big| > \varepsilon m \,\Big|\, D = k\Big\} + \mathbb{P}\big\{X_0 > \varepsilon m\big\}$$

$$\leqslant 4 \exp\big\{-\tfrac{\varepsilon^2}{5}m\big\} + 2\exp\{-\varepsilon m \log 2\} \leqslant 6 \exp\big\{-\tfrac{\varepsilon^2}{5}m\big\}.$$

The argument is analogous for D_{u}, and this completes the proof.   ∎

We now fix $\gamma < \tfrac{1}{2}$ and a large integer N. We stop the Brownian motion at time T_N when it first hits level N, and abbreviate $D^{(n)}(a) := D^{(n)}(a, T_N)$. We denote the nth dyadic grid by $\mathcal{D}_n := \mathcal{D}_n(N) := \{k2^{-n}\colon k \in \{-N2^n, -N2^n + 1, \ldots, N2^n - 1\}\}$.

Lemma 6.23 *Denote by $\Omega(m)$ the event that, for all $n \geqslant m$,*

    (a) $\left|D^{(n)}(a) - \frac{1}{2}D^{(n+1)}(a)\right| \leqslant 2^{n(1-\gamma)}$ *for all $a \in [-N, N)$,*

    (b) $\left|D^{(n)}(a) - D^{(n)}(b)\right| \leqslant 2\,2^{n(1-\gamma)}$ *for all $a, b \in [-N, N)$ with $|a - b| \leqslant 2^{-n}$.*

Then

$$\lim_{m \uparrow \infty} \mathbb{P}\big(\Omega(m)\big) = 1\,.$$

Proof. The event in item (a) follows by combining the following three events,

    (i) $\left|D^{(n)}(a) - \frac{1}{2}D^{(n+1)}(a)\right| \leqslant \frac{1}{n^2}2^{-n\gamma}D^{(n)}(a)$ for $a \in [-N, N)$ with $D^{(n)}(a) \geqslant 2^n$,

    (ii) $\left|D^{(n)}(a) - \frac{1}{2}D^{(n+1)}(a)\right| \leqslant 2^{n(1-\gamma)}$ for all $a \in [-N, N)$ with $D^{(n)}(a) < 2^n$,

    (iii) $D^{(n)}(a) \leqslant n^2 2^n$ for all $a \in [-N, N)$.

We observe that it is equivalent to show (i),(ii) for all $a \in \mathcal{D}_{n+1}$ and (iii) for all $a \in \mathcal{D}_n$. To estimate the probability of (i) we use Lemma 6.22 with $\varepsilon = \frac{1}{n^2}\,2^{-n\gamma}$ and $m = k$. We get that

$$\sum_{n=m}^{\infty} \sum_{a \in \mathcal{D}_{n+1}} \mathbb{P}\left\{\left|D^{(n)}(a) - \tfrac{1}{2}D^{(n+1)}(a)\right| > \tfrac{1}{n^2}\,2^{-n\gamma}D^{(n)}(a) \text{ and } D^{(n)}(a) \geqslant 2^n\right\}$$

$$\leqslant \sum_{n=m}^{\infty} \sum_{a \in \mathcal{D}_{n+1}} 12 \exp\left\{-\tfrac{1}{5n^4}\,2^{n(1-2\gamma)}\right\}$$

$$\leqslant (48N) \sum_{n=m}^{\infty} 2^n \exp\left\{-\tfrac{1}{5n^4}\,2^{n(1-2\gamma)}\right\} \overset{m\to\infty}{\longrightarrow} 0\,.$$

For event (ii) we get from Lemma 6.22 with $\varepsilon = 2^{-\gamma n}$ and $m = 2^n > k$. This gives that

$$\sum_{n=m}^{\infty} \sum_{a \in \mathcal{D}_{n+1}} \mathbb{P}\left\{\left|D^{(n)}(a) - \tfrac{1}{2}D^{(n+1)}(a)\right| > 2^{n(1-\gamma)} \text{ and } D^{(n)}(a) < 2^n\right\}$$

$$\leqslant \sum_{n=m}^{\infty} \sum_{a \in \mathcal{D}_{n+1}} 12 \exp\left\{-\tfrac{1}{5}\,2^{n(1-2\gamma)}\right\}$$

$$\leqslant (48N) \sum_{n=m}^{\infty} 2^n \exp\left\{-\tfrac{1}{5}\,2^{n(1-2\gamma)}\right\} \overset{m\to\infty}{\longrightarrow} 0\,.$$

For event (iii) we use that, given that the walk hits $j(a)$, the random variable $D^{(n)}(a)$ is geometrically distributed with parameter $\frac{2^{-n}}{N-a} \geqslant \frac{2^{-n}}{2N}$. We therefore obtain, for some sequence $\delta_n \to 0$,

$$\mathbb{P}\left\{D^{(n)}(a) > n^2 2^n\right\} \leqslant \left(1 - \tfrac{2^{-n}}{2N}\right)^{n^2 2^n - 1} \leqslant \exp\{-n^2 \tfrac{1-\delta_n}{2N}\},$$

hence, for sufficiently large m,

$$\sum_{n=m}^{\infty} \sum_{a \in \mathcal{D}_n} \mathbb{P}\left\{D^{(n)}(a) > n^2 2^n\right\} \leqslant \sum_{n=m}^{\infty} (2N)2^n \exp\left\{-n^2 \tfrac{1-\delta_n}{2N}\right\} \overset{m\to\infty}{\longrightarrow} 0\,.$$

This completes the estimates needed for item (a).

The event in item (b) need only be checked for all $a, b \in \mathcal{D}_n$ with $|a - b| = 2^{-n}$. Note that $D^{(n)}(a)$, resp. $D^{(n)}(b)$, are the number of downcrossings of the lower, resp. upper, half of an interval of length 2^{-n+1}, which may or may not be dyadic. Denote by $\tilde{D}^{(n-1)}(a) = \tilde{D}^{(n-1)}(b)$ the number of downcrossings of this interval. Then

$$\mathbb{P}\{|D^{(n)}(a) - D^{(n)}(b)| > 2\,2^{n(1-\gamma)}\}$$
$$\leqslant \mathbb{P}\{|D^{(n)}(a) - \tfrac{1}{2}\tilde{D}^{(n-1)}(a)| > 2^{n(1-\gamma)}\} + \mathbb{P}\{|D^{(n)}(b) - \tfrac{1}{2}\tilde{D}^{(n-1)}(b)| > 2^{n(1-\gamma)}\},$$

and summability of these probabilities over all $a, b \in \mathcal{D}_n$ with $|a - b| = 2^{-n}$ and $n \geqslant m$ has been established in the proof of item (a). This completes the proof. ∎

Lemma 6.24 *On the set $\Omega(m)$ we have that*

$$L^a(T_N) := \lim_{n \to \infty} 2^{-n+1} D^{(n)}(a)$$

exists for every $a \in [-N, N)$.

Proof. We show that the sequence defined by $2^{-n+1} D^{(n)}(a)$, for $n \in \mathbb{N}$, is a Cauchy sequence. Indeed, by item (a) in the definition of the set $\Omega(m)$, for any $a \in [-N, N]$ and $n \geqslant m$, we get that

$$\left|2^{-n+1} D^{(n)}(a) - 2^{-n} D^{(n+1)}(a)\right| \leqslant 2^{-n\gamma+1}.$$

Thus, for any $n \geqslant m$,

$$\sup_{k \geqslant n} \left|2^{-n+1} D^{(n)}(a) - 2^{-k+1} D^{(k)}(a)\right|$$
$$\leqslant \sum_{k=n}^{\infty} \left|2^{-k+1} D^{(k)}(a) - 2^{-k} D^{(k+1)}(a)\right| \leqslant \sum_{k=n}^{\infty} 2^{-k\gamma+1} \overset{n \to \infty}{\longrightarrow} 0,$$

and thus the sequence is a Cauchy sequence and therefore convergent. ∎

Lemma 6.25 *On $\Omega(m)$ the process $\{L^a(T_N) : a \in [-N, N)\}$ is γ-Hölder continuous.*

Proof. Fix $a, b \in [-N, N)$ with $2^{-n-1} \leqslant a - b \leqslant 2^{-n}$ for some $n \geqslant m$. Then, using item (a) and item (b) in the definition of $\Omega(m)$, for all $k \geqslant n$,

$$\left|2^{-k+1} D^{(k)}(a) - 2^{-k+1} D^{(k)}(b)\right| \leqslant \left|2^{-n+1} D^{(n)}(a) - 2^{-n+1} D^{(n)}(b)\right|$$
$$+ \sum_{j=n}^{k-1} \left|2^{-j} D^{(j+1)}(a) - 2^{-j+1} D^{(j)}(a)\right| + \sum_{j=n}^{k-1} \left|2^{-j} D^{(j+1)}(b) - 2^{-j+1} D^{(j)}(b)\right|$$
$$\leqslant 4\,2^{-n\gamma} + 4 \sum_{j=n}^{\infty} 2^{-j\gamma},$$

Letting $k \uparrow \infty$, we get

$$|L^a(T_N) - L^b(T_N)| \leqslant \left(4 + \tfrac{4}{1-2^{-\gamma}}\right) 2^{-n\gamma} \leqslant \left(2^{2+\gamma} + \tfrac{2^{2+\gamma}}{1-2^{-\gamma}}\right) |a - b|^\gamma,$$

which completes the proof. ∎

Lemma 6.26 *For any fixed time $t > 0$, almost surely, the limit*

$$L^a(t) := \lim_{n \to \infty} 2^{-n+1} D^{(n)}(a) \qquad \text{exists for all } a \in \mathbb{R}$$

and moreover $\{L^a(t) \colon a \in \mathbb{R}\}$ is γ-Hölder continuous.

Proof. Given $t > 0$ define the auxiliary Brownian motion $\{B_t(s) \colon s \geqslant 0\}$ by $B_t(s) = B(t + s)$ and denote by $D_t^{(n)}(a)$ the number of downcrossings associated to the auxiliary Brownian motion. Then, almost surely, $L_t^a(T_N) := \lim_{n \uparrow \infty} 2^{-n+1} D_t^{(n)}(a)$ exists for all $a \in \mathbb{R}$ and integers N. On this event we pick N so large that $T_N > t$. Define $L^a(t) := L^a(T_N) - L_t^a(T_N)$, and observe that $\{L^a(t) \colon a \in \mathbb{R}\}$ defined like this is γ-Hölder continuous by Lemma 6.25. It remains to show that this definition agrees with the one stated in the lemma. To this end, observe that

$$D^{(n)}(a, T_N) - D_t^{(n)}(a, T_N) - 1 \leqslant D^{(n)}(a, t) \leqslant D^{(n)}(a, T_N) - D_t^{(n)}(a, T_N).$$

Multiplying by 2^{-n+1} and taking a limit proves the claimed convergence. ∎

Lemma 6.27 *Almost surely,*

$$L^a(t) := \lim_{n \to \infty} 2^{-n+1} D^{(n)}(a, t)$$

exists for every $t \geqslant 0$ and $a \in \mathbb{R}$ and $\{L^a(t) \colon a \in \mathbb{R}, t \geqslant 0\}$ is γ-Hölder continuous.

Proof. It suffices to look at $t \in [0, N)$ and $a \in [-N, N)$. Recall the definition of the dyadic points $\mathcal{D}_n$ in $[-N, N)$ and additionally define dyadic points in $[0, N)$ by

$$\mathcal{H}_m = \{k2^{-m} \colon k \in \{0, \ldots, N2^m - 1\}\}, \quad \mathcal{H} = \bigcup_{m=1}^{\infty} \mathcal{H}_m.$$

We show that the claimed statements hold on the set

$$\bigcap_{t \in \mathcal{H}} \{L^a(t) \text{ exists for all } a \in [-N, N) \text{ and } a \mapsto L^a(t) \text{ is } \gamma\text{-Hölder continuous}\}$$

$$\cap \bigcap_{m > M} \bigcap_{t \in \mathcal{H}_m} \bigcap_{a \in \mathcal{D}_m} \{L^a(t + 2^{-m}) - L^a(t) \leqslant 2^{-m\gamma}\},$$

which, by choosing M suitably, has probability arbitrarily close to one by Lemma 6.26 and Lemma 6.7.

Given any $t \in [0, N)$ and $a \in [-N, N]$, for any large m, we find $t_1, t_2 \in \mathcal{H}_m$ with $t_2 - t_1 = 2^{-m}$ and $t \in [t_1, t_2]$. We have

$$2^{-n+1} D^{(n)}(a, t_1) \leqslant 2^{-n+1} D(a, t) \leqslant 2^{-n+1} D(a, t_2).$$

Both bounds converge on our set, and the difference of the limits is $L^a(t_2) - L^a(t_1)$. We can then find $b \in \mathcal{H}_k$ for $k \geqslant M$ with $|L^a(t_1) - L^b(t_1)| < 2^{-m\gamma}$ and $|L^a(t_2) - L^b(t_2)| < 2^{-m\gamma}$ and get

$$0 \leqslant L^a(t_2) - L^a(t_1) \leqslant |L^a(t_2) - L^b(t_2)| + |L^b(t_2) - L^b(t_1)| + |L^a(t_1) - L^b(t_1)|$$
$$\leqslant 3 \times 2^{-m\gamma},$$

which can be made arbitrarily small by choice of m, proving simultaneous convergence.

For the proof of continuity, suppose $a, b \in [-N, N]$ and $s, t \in [0, N)$ with $2^{-m} \leqslant |a - b| \leqslant 2^{-m}$ and $2^{-m} \leqslant t - s \leqslant 2^{-m}$ for some $m \geqslant M$. We pick $s_1, s_2 \in \mathcal{H}_m$ and $t_1, t_2 \in \mathcal{H}_m$ such that $s - 2^{-m} < s_1 \leqslant s \leqslant s_2 < s + 2^{-m}$ and $t - 2^{-m} < t_1 \leqslant t \leqslant t_2 < t + 2^{-m}$, and $a_1, b_1 \in \mathcal{D}_m$ with $|a - a_1| \leqslant 2^{-m}$ and $|b - b_1| \leqslant 2^{-m}$. Then

$$L^a(t) - L^b(s) \leqslant L^a(t_2) - L^b(s_1)$$
$$\leqslant |L^a(t_2) - L^{a_1}(t_2)| + |L^{a_1}(t_2) - L^{a_1}(s_1)| + |L^{a_1}(s_1) - L^b(s_1)|,$$

$$L^a(s) - L^b(t) \leqslant L^a(s_2) - L^b(t_1)$$
$$\leqslant |L^a(s_2) - L^{a_1}(s_2)| + |L^{a_1}(s_2) - L^{a_1}(t_1)| + |L^{a_1}(t_1) - L^b(t_1)|,$$

and all contributions on the right are bounded by constant multiples of $2^{-m\gamma}$, by the construction of our set. This completes the proof of γ-Hölder continuity. ∎

This completes the proof of Trotter's theorem, Theorem 6.19.

6.3 The Ray–Knight theorem

We now have a closer look at the distributions of local times $L^x(T)$ as a function of the level x in the case that Brownian motion is started at an arbitrary point and stopped at the time T when it first hits level zero. The following remarkable distributional identity goes back to the work of Ray and Knight.

Theorem 6.28 (Ray–Knight theorem) *Suppose $a > 0$ and $\{B(t) \colon 0 \leqslant t \leqslant T\}$ is a linear Brownian motion started at a and stopped at time $T = \inf\{t \geqslant 0 \colon B(t) = 0\}$, when it reaches level zero for the first time. Then*

$$\{L^x(T) \colon 0 \leqslant x \leqslant a\} \stackrel{\mathrm{d}}{=} \{|W(x)|^2 \colon 0 \leqslant x \leqslant a\},$$

where $\{W(x) \colon x \geqslant 0\}$ is a standard planar Brownian motion.

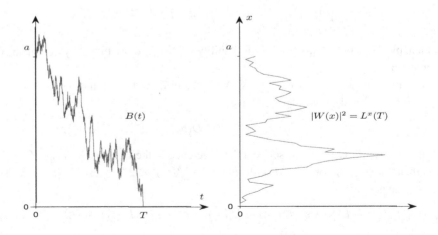

Fig. 6.5. The Brownian path on the left, and its local time as a function of the level, on the right.

Remark 6.29 The process $\{|W(x)|^2 : x \geqslant 0\}$ of squared norms of a planar Brownian motion is called the squared **two-dimensional Bessel process**. For any fixed x, the random variable $|W(x)|^2$ is exponentially distributed with mean $2x$, see Lemma 12.16 in the appendix. ◇

We carry out the proof of the Ray–Knight theorem in three steps. As a warm-up, we look at one point $0 < x \leqslant a$. Recall from the downcrossing representation, Theorem 6.1, that

$$\lim_{n \to \infty} \tfrac{2}{n} D_n(x) = L^x(T) \qquad \text{almost surely},$$

where $D_n(x)$ denotes the number of downcrossings of the interval $[x - 1/n, x]$ before time T. Recall that basic facts about convergence in distribution, indicated with $\xrightarrow{\text{d}}$, are collected in Section 12.1 of the appendix.

Lemma 6.30 *For any $0 < x \leqslant a$, we have $\tfrac{2}{n} D_n(x) \xrightarrow{\text{d}} |W(x)|^2$ as $n \uparrow \infty$.*

Proof. By the strong Markov property and the exit probabilities from an interval described in Theorem 2.49, it is clear that, provided $n > 1/x$, the random variable $D_n(x)$ is geometrically distributed with (success) parameter $1/(nx)$, i.e. $\mathbb{P}\{D_n(x) = k\} = \tfrac{1}{nx}(1 - \tfrac{1}{nx})^{k-1}$ for all $k \in \{1, 2, \ldots\}$. Hence, as $n \to \infty$, we obtain that

$$\mathbb{P}\{D_n(x) > ny/2\} = \left(1 - \tfrac{1}{nx}\right)^{\lfloor ny/2 \rfloor} \longrightarrow e^{-y/(2x)},$$

and the result follows, as $|W(x)|^2$ is exponentially distributed with mean $2x$. ∎

Lemma 6.30 is the 'one-point version' of Theorem 6.28. The essence of the Ray–Knight theorem is captured in the 'two-point version', which we prove next. We fix two points x and $x + h$ with $0 < x < x + h < a$. The next three lemmas are the crucial ingredients for the proof of Theorem 6.28.

Lemma 6.31 *Let $0 < x < x + h < a$. Then, for all $n > h$, we have*

$$D_n(x + h) = D + \sum_{j=1}^{D_n(x)} I_j N_j,$$

where

- *$D = D^{(n)}$ is the number of downcrossings of the interval $[x + h - \tfrac{1}{n}, x + h]$ before the Brownian motion hits level x,*
- *for any $j \in \mathbb{N}$ the random variable $I_j = I_j^{(n)}$ is Bernoulli distributed with mean $\tfrac{1}{nh+1}$,*
- *for any $j \in \mathbb{N}$ the random variable $N_j = N_j^{(n)}$ is geometrically distributed with mean $nh + 1$,*

and all these random variables are independent of each other and of $D_n(x)$.

Proof. The decomposition of $D_n(x + h)$ is based on counting the number of downcrossings of the interval $[x + h - 1/n, x + h]$ that have taken place between the stopping times in the sequence

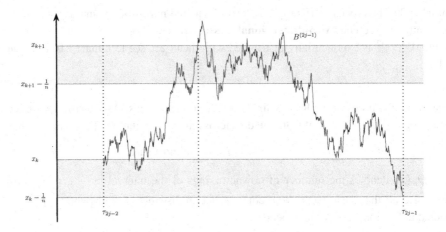

Fig. 6.6. The random variables I_j and N_j depend only on the pieces $B^{(2j-1)}$ for $j \geqslant 1$. For this sample $I_j = 1$ as the path hits $x + h$ before $x - \frac{1}{n}$ and $N_j = 2$, because the path downcrosses $[x + h - \frac{1}{n}, x + h]$ twice before hitting $x - \frac{1}{n}$.

$$\tau_0 = \inf\left\{t > 0\colon B(t) = x\right\}, \qquad \tau_1 = \inf\left\{t > \tau_0\colon B(t) = x - \tfrac{1}{n}\right\},$$
$$\tau_{2j} = \inf\left\{t > \tau_{2j-1}\colon B(t) = x\right\}, \qquad \tau_{2j+1} = \inf\left\{t > \tau_{2j}\colon B(t) = x - \tfrac{1}{n}\right\},$$

for $j \geqslant 1$. By the strong Markov property the pieces

$$B^{(0)}\colon [0, \tau_0] \to \mathbb{R}, \qquad B^{(0)}(s) = B(s)$$
$$B^{(j)}\colon [0, \tau_j - \tau_{j-1}] \to \mathbb{R}, \qquad B^{(j)}(s) = B(\tau_{j-1} + s),\, j \geqslant 1,$$

are all independent. The crucial observation of the proof is that the vector $D_n(x)$ is a function of the pieces $B^{(2j)}$ for $j \geqslant 1$, whereas we shall define the random variables D, $I_1, I_2, \ldots$ and $N_1, N_2 \ldots$ depending only on the other pieces $B^{(0)}$ and $B^{(2j-1)}$ for $j \geqslant 1$. First, let D be the number of downcrossings of $[x + h - 1/n, x + h]$ during the time interval $[0, \tau_0]$. Then fix $j \geqslant 1$ and hence a piece $B^{(2j-1)}$. Define I_j to be the indicator of the event that $B^{(2j-1)}$ reaches level $x + h$ during its lifetime. By Theorem 2.49 this event has probability $1/(nh + 1)$. Observe that the number of downcrossings by $B^{(2j-1)}$ is zero if the event fails. If the event holds, we define N_j as the number of downcrossings of $[x + h - 1/n, x + h]$ by $B^{(2j-1)}$, which is a geometric random variable with mean $nh + 1$ by the strong Markov property and Theorem 2.49.

The claimed decomposition follows now from the fact that the pieces $B^{(2j)}$ for $j \geqslant 1$ do not upcross the interval $[x+h-1/n, x+h]$ by definition and that $B^{(2j-1)}$ for $j = 1, \ldots, D_n(x)$ are exactly the pieces that take place before the Brownian motion reaches level zero. ∎

Lemma 6.32 *Suppose* nu_n *are nonnegative, even integers and* $u_n \to u$. *Then*

$$\frac{2}{n} D^{(n)} + \frac{2}{n} \sum_{j=1}^{\frac{nu_n}{2}} I_j^{(n)} N_j^{(n)} \xrightarrow{\mathrm{d}} \tilde{X}^2 + \tilde{Y}^2 + 2 \sum_{j=1}^{M} \tilde{Z}_j \qquad as \; n \uparrow \infty,$$

where $\tilde{X}, \tilde{Y}$ *are normally distributed with mean zero and variance* h, *the random variable* M *is Poisson distributed with parameter* $u/(2h)$ *and* $\tilde{Z}_1, \tilde{Z}_2, \ldots$ *are exponentially distributed with mean* h, *and all these random variables are independent.*

Proof. By Lemma 6.30, we have, for $\tilde{X}, \tilde{Y}$ as defined in the lemma,

$$\frac{2}{n} D^{(n)} \xrightarrow{\mathrm{d}} |W(h)|^2 \overset{\mathrm{d}}{=} \tilde{X}^2 + \tilde{Y}^2 \qquad as \; n \uparrow \infty.$$

Moreover, we observe that

$$\frac{2}{n} \sum_{j=1}^{\frac{nu_n}{2}} I_j^{(n)} N_j^{(n)} \overset{\mathrm{d}}{=} \frac{2}{n} \sum_{j=1}^{B_n} N_j^{(n)},$$

where B_n is binomial with parameters $nu_n/2 \in \{0, 1, \ldots\}$ and $1/(nh+1) \in (0,1)$ and independent of $N_1^{(n)}, N_2^{(n)}, \ldots$. We now show that, when $n \uparrow \infty$, the random variables B_n converge in distribution to M and the random variables $\frac{1}{n} N_j^{(n)}$ converge to $\tilde{Z}_j$, as defined in the lemma. For this purpose it suffices to show convergence of the Laplace transforms, see Proposition 12.8 in the appendix.

First note that, for $\lambda, \theta > 0$, we have

$$\mathbb{E} \exp\left\{ -\lambda \tilde{Z}_j \right\} = \tfrac{1}{\lambda h + 1}, \quad \mathbb{E}\left[\theta^M \right] = \exp\left\{ -\tfrac{u(1-\theta)}{2h} \right\},$$

and hence

$$\mathbb{E} \exp\left\{ -\lambda \sum_{j=1}^{M} \tilde{Z}_j \right\} = \mathbb{E}\left(\tfrac{1}{\lambda h + 1} \right)^M = \exp\left\{ -\tfrac{u}{2h} \tfrac{\lambda h}{\lambda h + 1} \right\} = \exp\left\{ -\tfrac{u\lambda}{2\lambda h + 2} \right\}.$$

Convergence of $\frac{1}{n} N_j^{(n)}$ is best seen using tail probabilities

$$\mathbb{P}\left\{ \tfrac{1}{n} N_j^{(n)} > a \right\} = \left(1 - \tfrac{1}{nh+1} \right)^{\lfloor na \rfloor} \longrightarrow \exp\left\{ -\tfrac{a}{h} \right\} = \mathbb{P}\{ \tilde{Z}_j > a \}.$$

Hence, for a suitable sequence $\delta_n \to 0$,

$$\mathbb{E} \exp\left\{ -\lambda \tfrac{1}{n} N_j^{(n)} \right\} = \frac{1 + \delta_n}{\lambda h + 1}.$$

For the binomial distributions we have

$$\mathbb{E}\left[\theta^{B_n} \right] = \left(\tfrac{\theta}{nh+1} + \left(1 - \tfrac{1}{nh+1}\right) \right)^{nu_n/2} \longrightarrow \exp\left\{ -\tfrac{u(1-\theta)}{2h} \right\},$$

and thus

$$\lim_{n \uparrow \infty} \mathbb{E} \exp\left\{ -\lambda \tfrac{1}{n} \sum_{j=1}^{B_n} N_j^{(n)} \right\} = \lim_{n \uparrow \infty} \mathbb{E}\left[\left(\tfrac{1+\delta_n}{\lambda h + 1} \right)^{B_n} \right] = \lim_{n \uparrow \infty} \exp\left\{ -\tfrac{u}{2h} \tfrac{\lambda h - \delta_n}{\lambda h + 1} \right\}$$

$$= \exp\left\{ -\tfrac{u\lambda}{2\lambda h + 2} \right\} = \mathbb{E} \exp\left\{ -\lambda \sum_{j=1}^{M} \tilde{Z}_j \right\}. \qquad \blacksquare$$

Lemma 6.33 *Suppose X is standard normally distributed, $Z_1, Z_2, \ldots$ standard exponentially distributed and N Poisson distributed with parameter $\ell^2/2$ for some $\ell > 0$. If all these random variables are independent, then*

$$(X + \ell)^2 \overset{\mathrm{d}}{=} X^2 + 2\sum_{j=1}^{N} Z_j \,.$$

Proof.   It suffices to show that the Laplace transforms of the random variables on the two sides of the equation agree. Let $\lambda > 0$. Completing the square, we find

$$
\begin{aligned}
\mathbb{E}\exp\{-\lambda\,(X+\ell)^2\} &= \frac{1}{\sqrt{2\pi}}\int \exp\{-\lambda\,(x+\ell)^2 - x^2/2\}\,dx \\
&= \frac{1}{\sqrt{2\pi}}\int \exp\Big\{ -\tfrac{1}{2}\Big(\sqrt{2\lambda+1}\,x + \tfrac{2\lambda\ell}{\sqrt{2\lambda+1}}\Big)^2 - \lambda\ell^2 + \tfrac{2\lambda^2\ell^2}{2\lambda+1} \Big\}\,dx \\
&= \frac{1}{\sqrt{2\lambda+1}}\,\exp\Big\{ -\tfrac{\lambda\ell^2}{2\lambda+1} \Big\}\,.
\end{aligned}
$$

From the special case $\ell = 0$ we get $\mathbb{E}\exp\{-\lambda X^2\} = \frac{1}{\sqrt{2\lambda+1}}$. For any $\theta > 0$,

$$\mathbb{E}[\theta^N] = \exp\{-\ell^2/2\}\sum_{k=0}^{\infty} \frac{(\ell^2\theta/2)^k}{k!} = \exp\{(\theta - 1)\ell^2/2\}\,.$$

Using this and that $\mathbb{E}\exp\{-2\lambda Z_j\} = \frac{1}{2\lambda+1}$ we get

$$\mathbb{E}\exp\Big\{ -\lambda\,\Big(X^2 + 2\sum_{j=1}^{N} Z_j\Big)\Big\} = \frac{1}{\sqrt{2\lambda+1}}\,\mathbb{E}\Big(\frac{1}{2\lambda+1}\Big)^N = \frac{1}{\sqrt{2\lambda+1}}\,\exp\Big\{ -\tfrac{\lambda\ell^2}{2\lambda+1} \Big\},$$

which completes the proof.   ∎

Remark 6.34 An alternative proof of Lemma 6.33 will be given in Exercise 6.8.   ◇

By combining the previous three lemmas we obtain the following convergence result for the conditional distribution of $D_n(x + h)$ given $D_n(x)$, which is the 'two-point version' of the Ray–Knight theorem.

Lemma 6.35 *Suppose nu_n are nonnegative, even integers and $u_n \to u$. For any $\lambda \geqslant 0$,*

$$\lim_{n\to\infty} \mathbb{E}\big[\exp\big\{ -\lambda\tfrac{2}{n}D_n(x+h)\big\} \,\big|\, \tfrac{2}{n}D_n(x) = u_n\big] = \mathbb{E}_{(0,\sqrt{u})}\big[\exp\big\{ -\lambda|W(h)|^2\big\}\big],$$

where $\{W(x)\colon x \geqslant 0\}$ denotes a planar Brownian motion started in $(0, \sqrt{u}) \in \mathbb{R}^2$.

Proof. Combining Lemmas 6.31 and 6.32 we get

$$
\lim_{n \to \infty} \mathbb{E}\big[\exp\big\{ -\lambda \tfrac{2}{n} D_n(x+h) \big\} \,\big|\, \tfrac{2}{n} D_n(x) = u_n \big]
$$

$$
= \mathbb{E}\Big[\exp\Big\{ -\lambda\Big(\tilde{X}^2 + \tilde{Y}^2 + 2\sum_{j=1}^{M} \tilde{Z}_j \Big)\Big\}\Big]
$$

$$
= \mathbb{E}\Big[\exp\Big\{ -\lambda h\Big(X^2 + Y^2 + 2\sum_{j=1}^{M} Z_j \Big)\Big\}\Big],
$$

where X, Y are standard normally distributed, $Z_1, Z_2, \ldots$ are standard exponentially distributed and M is Poisson distributed with parameter $\ell^2/2$, for $\ell = \sqrt{u/h}$. By Lemma 6.33 the right hand side can thus be rewritten as

$$
\mathbb{E}\big[\exp\big\{ -\lambda h\big((X + \sqrt{u/h})^2 + Y^2 \big)\big\}\big] = \mathbb{E}_{(0, \sqrt{u})}\big[\exp\big\{ -\lambda |W(h)|^2 \big\}\big],
$$

which proves the lemma. ∎

Now we complete the proof of Theorem 6.28. Note that, as both $\{L^x(T) : x \geq 0\}$ and $\{|W(x)|^2 : x \geq 0\}$ are continuous processes, it suffices to show that, for any

$$
0 < x_1 < \cdots < x_m < a
$$

the vectors

$$
\big(L^{x_1}(T), \ldots, L^{x_m}(T)\big) \qquad \text{and} \qquad \big(|W(x_1)|^2, \ldots, |W(x_m)|^2 \big)
$$

have the same distribution. The Markov property of the downcrossing numbers, which approximate the local times, allows us to reduce this problem to the study of the 'two-point version'.

Lemma 6.36 *For all sufficiently large integers n, the process*

$$
\big\{ D_n(x_k) : k = 1, \ldots, m \big\}
$$

is a (possibly inhomogeneous) Markov chain.

Proof. Fix $k \in \{2, \ldots, m\}$. By Lemma 6.31 applied to $x = x_{k-1}$ and $h = x_k - x_{k-1}$ we can write $D_n(x_k)$ as a function of $D_n(x_{k-1})$ and various random variables, which by construction, are independent of $D_n(x_1), \ldots, D_n(x_{k-1})$. This establishes the Markov property. ∎

Note that, by rotational invariance of planar Brownian motion, $\{|W(x_k)|^2 : k = 1, \ldots, m\}$ is a Markov chain with transition probabilities given by

$$
\mathbb{E}\big[\exp\{-\lambda |W(x_{k+1})|^2\} \,\big|\, |W(x_k)|^2 = u \big] = \mathbb{E}_{(0, \sqrt{u})}\big[\exp\{-\lambda |W(x_{k+1} - x_k)|^2\}\big],
$$

for all $\lambda > 0$. The following general fact about the convergence of families of Markov chains ensures that we have done enough to complete the proof of Theorem 6.28.

Lemma 6.37 *Suppose, for $n = 1, 2, \ldots$, that $\{X_k^{(n)} : k = 1, \ldots, m\}$ is a Markov chain with discrete state space $\Omega_n \subset [0, \infty)$ and that $\{X_k : k = 1, \ldots, m\}$ is a Markov chain with state space $[0, \infty)$. Suppose further that*

(1) $(X_1^{(n)}, \ldots, X_m^{(n)})$ converges almost surely to some random vector $(Y_1, \ldots, Y_m)$,

(2) $X_1^{(n)} \xrightarrow{\text{d}} X_1$ as $n \uparrow \infty$,

(3) for all $k = 1, \ldots, m - 1$, $\lambda > 0$ and $y_n \in \Omega_n$ with $y_n \to y$, we have

$$\lim_{n \to \infty} \mathbb{E}\big[\exp\{-\lambda X_{k+1}^{(n)}\} \,\big|\, X_k^{(n)} = y_n\big] = \mathbb{E}\big[\exp\{-\lambda X_{k+1}\} \,\big|\, X_k = y\big].$$

Then

$$(X_1^{(n)}, \ldots, X_m^{(n)}) \xrightarrow{\text{d}} (X_1, \ldots, X_m)$$

and, in particular, the vectors $(X_1, \ldots, X_m)$ and $(Y_1, \ldots, Y_m)$ have the same distribution.

Proof.   Recall from Proposition 12.8 in the appendix that it suffices to show that the Laplace transforms converge. Let $\lambda_1, \ldots, \lambda_m \geq 0$. By assumption (2) we have $X_1^{(n)} \xrightarrow{\text{d}} X_1$ and hence we may assume, by way of induction, that for some fixed $k = 1, \ldots, m - 1$, we have

$$(X_1^{(n)}, \ldots, X_k^{(n)}) \xrightarrow{\text{d}} (X_1, \ldots, X_k).$$

This implies, in particular, that $(X_1, \ldots, X_k)$ and $(Y_1, \ldots, Y_k)$ have the same distribution. Define

$$\Phi_n : \Omega_n \to [0, 1], \qquad \Phi_n(y) = \mathbb{E}\big[\exp\{-\lambda_{k+1} X_{k+1}^{(n)}\} \,\big|\, X_k^{(n)} = y\big]$$

and

$$\Phi : [0, \infty) \to [0, 1], \qquad \Phi(y) = \mathbb{E}\big[\exp\{-\lambda_{k+1} X_{k+1}\} \,\big|\, X_k = y\big].$$

Then, combining assumption (1) and (3), $\Phi_n(X_k^{(n)}) \to \Phi(Y_k)$ almost surely. Hence, using this and once more assumption (1),

$$\mathbb{E}\Big[\exp\Big\{-\sum_{j=1}^{k+1} \lambda_j X_j^{(n)}\Big\}\Big] = \mathbb{E}\Big[\exp\Big\{-\sum_{j=1}^{k} \lambda_j X_j^{(n)}\Big\} \Phi_n(X_k^{(n)})\Big]$$

$$\to \mathbb{E}\Big[\exp\Big\{-\sum_{j=1}^{k} \lambda_j Y_j\Big\} \Phi(Y_k)\Big].$$

As the vectors $(X_1, \ldots, X_k)$ and $(Y_1, \ldots, Y_k)$ have the same distribution the limit can be rewritten as

$$\mathbb{E}\Big[\exp\Big\{-\sum_{j=1}^{k} \lambda_j X_j\Big\} \Phi(X_k)\Big] = \mathbb{E}\Big[\exp\Big\{-\sum_{j=1}^{k+1} \lambda_j X_j\Big\}\Big],$$

and this completes the induction step.

Finally, as $(X_1^{(n)}, \ldots, X_m^{(n)})$ converges almost surely, and hence also in distribution to $(Y_1, \ldots, Y_m)$, this vector must have the same distribution as $(X_1, \ldots, X_m)$. This completes the proof. ∎

Proof of Theorem 6.28. We use Lemma 6.37 with $X_k^{(n)} = \frac{2}{n} D_n(x_k)$, $X_k = |W(x_k)|^2$ and $Y_k = L^{x_k}(T)$. Then assumption (1) is satisfied by the downcrossing representation, assumption (2) follows from Lemma 6.30 and assumption (3) from Lemma 6.35. Lemma 6.37 thus gives that the random vector $(L^{x_1}(T), \ldots, L^{x_m}(T))$ and the random vector $(|W(x_1)|^2, \ldots, |W(x_m)|^2)$ have the same distribution, concluding the proof. ∎

As an easy application of the Ray–Knight theorem, we answer the question whether, almost surely, *simultaneously* for all levels $x \in [0, a)$ the local times at level x are positive.

Theorem 6.38 (Ray's theorem) *Suppose $a > 0$ and $\{B(t): 0 \leqslant t \leqslant T_a\}$ is a linear Brownian motion started at zero and stopped at time $T_a = \inf\{t \geqslant 0: B(t) = a\}$, when it reaches level a for the first time. Then, almost surely, $L^x(T_a) > 0$ for all $0 \leqslant x < a$.*

Proof. The statement can be reworded as saying that the process $\{L^{a-x}(T_a): 0 < x \leqslant a\}$ almost surely does not hit zero. By the Ray–Knight theorem (applied to the Brownian motion $\{a - B(t): t \geqslant 0\}$) this process agrees with $\{|W(x)|^2: 0 < x \leqslant a\}$ for a standard planar Brownian motion $\{W(x): x \geqslant 0\}$ which, by Theorem 3.20, never returns to the origin. ∎

Ray's theorem can be exploited to give a result on the Hausdorff dimension of the level sets of the Brownian motion, which holds *simultaneously* for all levels $a \in \mathbb{R}$. We prepare the proof by a lemma.

Lemma 6.39 *Almost surely, for all $a \in \mathbb{R}$, we have*

$$\{t > 0: B(t) = a \text{ and } t \text{ is not locally extremal}\}$$
$$= \{t > 0: L^a(t+h) - L^a(t-h) > 0 \text{ for all } h > 0\}.$$

Proof. The inclusion '⊃' follows directly from Trotter's theorem and the uniqueness of local extrema, see Theorem 2.11. For the inclusion '⊂' we note that, by the strong Markov property and Ray's theorem, almost surely for any rational $q \geqslant 0$ and $\varepsilon > 0$ and stopping time $\tau_q(\varepsilon) := \inf\{t > q: B(t) = B(q) + \varepsilon\}$ we have

$$L^x(\tau_q(\varepsilon)) - L^x(q) > 0 \text{ for all } B(q) \leqslant x < B(q) + \varepsilon.$$

Suppose $B(t) = x$ and $h > 0$. If t is neither a local minimiser from the left nor a local maximiser, there exist a rational $q \in (t-h, t)$ with $B(q) \leqslant x < B(q)+\varepsilon$ and $\tau_q(\varepsilon) < t+h$. From the monotonicity of local time we infer that $L^x(t + h) - L^x(t - h) > 0$. A similar argument for the time-reversed Brownian motion can be given to deal with those t which are neither a local minimiser from the right nor a local maximiser. ∎

Theorem 6.40 *Almost surely,* $\dim\{t \geqslant 0 : B(t) = a\} \geqslant \frac{1}{2}$, *for all* $a \in \mathbb{R}$.

Proof.   Obviously, it suffices to show that, for every fixed $a > 0$, almost surely,

$$\dim\left\{0 \leqslant t < T_a : B(t) = x\right\} \geqslant \tfrac{1}{2} \qquad \text{for all } 0 \leqslant x < a.$$

This can be achieved using the mass distribution principle. Considering the increasing function $L^x : [0, T^a) \to [0, \infty)$ as distribution function of a measure ℓ^x, we infer from Lemma 6.39 that, almost surely, for every $x \in [0, a)$, the measure ℓ^x is a mass distribution on the set $\{0 \leqslant t < T_a : B(t) = x\}$. By Theorem 6.19, for any $\gamma < 1/2$, almost surely, there exists a (random) $C > 0$ such that, for all $x \in [0, a)$, $t \in [0, T_a)$ and $\varepsilon \in (0, 1)$,

$$\ell^x(t - \varepsilon, t + \varepsilon) \leqslant |L^x(t + \varepsilon) - L^x(t - \varepsilon)| \leqslant C\,(2\varepsilon)^\gamma .$$

The claim therefore follows from the mass distribution principle, Theorem 4.19.   ∎

Remark 6.41 Equality holds in Theorem 6.40. We will obtain the full result later as an easy corollary of Kaufman's dimension doubling theorem, see Theorem 9.28.   ◇

6.4 Brownian local time as a Hausdorff measure

In this section we show that the local time $L^0(t)$ can be obtained as an intrinsically defined measure of the random set Zeros $\cap\,[0, t]$. The only family of intrinsically defined measures on metric spaces we have encountered so far in this book is the family of α-dimensional Hausdorff measures. As the α-dimensional Hausdorff measure of the zero set is always either zero (if $\alpha \geqslant \frac{1}{2}$) or infinity (if $\alpha < \frac{1}{2}$) we need to look out for an alternative construction.

We need not look very far. The definition of Hausdorff dimension still makes sense if we evaluate coverings by applying, instead of a simple power, an arbitrary non-decreasing function to the diameters of the sets in a covering.

Definition 6.42.   A non-decreasing function $\phi : [0, \varepsilon) \to [0, \infty)$ with $\phi(0) = 0$ defined on a nonempty interval $[0, \varepsilon)$ is called a **(Hausdorff) gauge function**.
Let X be a metric space and $E \subset X$. For every gauge function ϕ and $\delta > 0$ define

$$\mathcal{H}^\phi_\delta(E) = \inf\left\{ \sum_{i=1}^\infty \phi(|E_i|) : E_1, E_2, E_3, \ldots \text{ cover } E, \text{ and } |E_i| \leqslant \delta \right\}.$$

Then

$$\mathcal{H}^\phi(E) = \sup_{\delta > 0} \mathcal{H}^\phi_\delta(E) = \lim_{\delta \downarrow 0} \mathcal{H}^\phi_\delta(E)$$

is the **generalised ϕ-Hausdorff measure** of the set E.   ◇

Theorem* 6.43 *There exists a constant $c > 0$ such that, almost surely, for all $t > 0$,*

$$L^0(t) = \mathcal{H}^\varphi\left(\mathsf{Zeros} \cap [0, t]\right),$$

for the gauge function $\varphi(r) = c\sqrt{r \log \log(1/r)}$.

The remainder of this section is devoted to the proof of this theorem. The material developed here will not be used in the remainder of the book. An important tool in the proof is the following classical theorem of Rogers and Taylor.

Proposition 6.44 (Rogers–Taylor Theorem) *Let μ be a Borel measure on $\mathbb{R}^d$ and let ϕ be a Hausdorff gauge function.*

   (i) *If $\Lambda \subset \mathbb{R}^d$ is a Borel set and*

$$\limsup_{r \downarrow 0} \frac{\mu \mathcal{B}(x, r)}{\phi(r)} < \alpha$$

   for all $x \in \Lambda$, then $\mathcal{H}^\phi(\Lambda) \geqslant \alpha^{-1}\mu(\Lambda)$.

   (ii) *If $\Lambda \subset \mathbb{R}^d$ is a Borel set and*

$$\limsup_{r \downarrow 0} \frac{\mu \mathcal{B}(x, r)}{\phi(r)} > \theta$$

   for all $x \in \Lambda$, then $\mathcal{H}^\phi(\Lambda) \leqslant \kappa_d \theta^{-1}\mu(V)$ for any open set $V \subset \mathbb{R}^d$ that contains Λ, where κ_d depends only on d.

Moreover, in $d = 1$ one can also obtain an analogue of (i) for one-sided intervals.

   (iii) *If $\Lambda \subset \mathbb{R}$ is a closed set and*

$$A := \left\{ t \in \Lambda : \limsup_{r \downarrow 0} \frac{\mu[t, t + r]}{\phi(r)} < \alpha \right\},$$

   then $\mathcal{H}^\phi(A) \geqslant \alpha^{-1}\mu(A)$.

Remark 6.45 If μ is finite on compact sets, then $\mu(\Lambda)$ is the infimum of $\mu(V)$ over all open sets $V \supset \Lambda$, see for example Section 2.18 in [Ru87]. Hence $\mu(V)$ can be replaced by $\mu(\Lambda)$ on the right hand side of the inequality in (ii). ◇

Proof. **(i)** We write

$$\Lambda_\varepsilon = \left\{ x \in \Lambda : \sup_{r \in (0, \varepsilon)} \frac{\mu \mathcal{B}(x, r)}{\phi(r)} < \alpha \right\}$$

and note that $\mu(\Lambda_\varepsilon) \to \mu(\Lambda)$ as $\varepsilon \downarrow 0$.

Fix $\varepsilon > 0$ and consider a cover $\{A_j\}$ of Λ_ε. Suppose that A_j intersects Λ_ε and $r_j =$

$|A_j| < \varepsilon$ for all j. Choose $x_j \in A_j \cap \Lambda_\varepsilon$ for each j. Then $\mu \mathcal{B}(x_j, r_j) < \alpha \phi(r_j)$ for every j, whence

$$\sum_{j \geqslant 1} \phi(r_j) \geqslant \alpha^{-1} \sum_{j \geqslant 1} \mu \mathcal{B}(x_j, r_j) \geqslant \alpha^{-1} \mu(\Lambda_\varepsilon).$$

Thus $\mathcal{H}_\varepsilon^\phi(\Lambda) \geqslant \mathcal{H}_\varepsilon^\phi(\Lambda_\varepsilon) \geqslant \alpha^{-1} \mu(\Lambda_\varepsilon)$. Letting $\varepsilon \downarrow 0$ proves (i).

(ii) Let $\varepsilon > 0$. For each $x \in \Lambda$, choose a positive $r_x < \varepsilon$ such that $\mathcal{B}(x, 2r_x) \subset V$ and $\mu \mathcal{B}(x, r_x) > \theta \phi(r_x)$; then among the dyadic cubes of diameter at most r_x that intersect $\mathcal{B}(x, r_x)$, let Q_x be a cube with $\mu(Q_x)$ maximal. (We consider here dyadic cubes of the form $\prod_{i=1}^d [a_i / 2^m, (a_i + 1)/2^m)$ where a_i are integers). In particular, $Q_x \subset V$ and $|Q_x| > r_x / 2$ so the side-length of Q_x is at least $r_x / (2\sqrt{d})$. Let $N_d = 1 + 8\lceil \sqrt{d} \rceil$ and let Q_x^* be the cube with the same center z_x as Q_x, scaled by N_d (i.e., $Q_x^* = z_x + N_d(Q_x - z_x)$). Observe that Q_x^* contains $\mathcal{B}(x, r_x)$, so $\mathcal{B}(x, r_x)$ is covered by at most N_d^d dyadic cubes that are translates of Q_x. Therefore, for every $x \in \Lambda$, we have

$$\mu(Q_x) \geqslant N_d^{-d} \mu \mathcal{B}(x, r_x) > N_d^{-d} \theta \phi(r_x).$$

Let $\{Q_{x(j)} : j \geqslant 1\}$ be any enumeration of the maximal dyadic cubes among $\{Q_x : x \in \Lambda\}$. Then

$$\mu(V) \geqslant \sum_{j \geqslant 1} \mu(Q_{x(j)}) \geqslant N_d^{-d} \theta \sum_{j \geqslant 1} \phi(r_{x(j)}).$$

The collection of cubes $\{Q_{x(j)}^* : j \geqslant 1\}$ forms a cover of Λ. Since each of these cubes is covered by N_d^d cubes of diameter at most $r_{x(j)}$, we infer that

$$\mathcal{H}_\varepsilon^\phi(\Lambda) \leqslant N_d^d \sum_{j \geqslant 1} \phi(r_{x(j)}) \leqslant N_d^{2d} \theta^{-1} \mu(V).$$

Letting $\varepsilon \downarrow 0$ proves (ii).

(iii) Without loss of generality we may assume that μ has no atoms. Given $\varepsilon > 0$ we find $\delta > 0$ such that

$$A_\delta(\alpha) = \left\{ t \in \Lambda : \sup_{h < \delta} \frac{\mu[t, t+h]}{\varphi(h)} \leqslant \alpha - \delta \right\}$$

satisfies $\mu(A_\delta(\alpha)) > (1 - \varepsilon)\mu(A)$. Observe that $A_\delta(\alpha)$ is closed. Given a cover $\{\tilde{I}_j\}$ of A with $|\tilde{I}_j| < \delta$ we look at $I_j = [a_j, b_j]$ where a_j is the maximum and b_j the minimum of the compact set $\mathrm{cl}\, \tilde{I}_j \cap A_\delta(\alpha)$. Then $\{I_j\}$ covers $A_\delta(\alpha)$ and hence

$$\sum_{j \geqslant 1} \varphi(|\tilde{I}_j|) \geqslant \sum_{j \geqslant 1} \varphi(|I_j|) \geqslant (\alpha - \delta)^{-1} \sum_{j \geqslant 1} \mu(I_j)$$

$$\geqslant (\alpha - \delta)^{-1} \mu(A_\delta(\alpha)) \geqslant (\alpha - \delta)^{-1} (1 - \varepsilon)\mu(A),$$

and (iii) follows for $\delta \downarrow 0$, as $\varepsilon > 0$ was arbitrary.   ∎

For the proof of Theorem 6.43 we first note that, by Theorem 6.10, it is equivalent to show that, for the maximum process $\{M(t) : t \geqslant 0\}$ of a Brownian motion $\{B(t) : t \geqslant 0\}$, we have, almost surely,

$$M(t) = \mathcal{H}^\varphi \big(\mathrm{Rec} \cap [0, t] \big) \qquad \text{for all } t \geqslant 0, \tag{6.4}$$

where Rec denotes the set of record points of the Brownian motion. To show this, recall from Exercise 4.12 that $\text{Rec} = \{s \geqslant 0 \colon M(s+h) > M(s-h) \text{ for all } 0 < h < s\}$. We define the measure μ on Rec as given by the distribution function M, i.e.

$$\mu(a,b] = M(b) - M(a) \qquad \text{for all intervals } (a,b] \subset \mathbb{R}.$$

Then μ is also the image measure of the Lebesgue measure on $[0,\infty)$ under the mapping

$$a \mapsto T_a := \inf\{s \geqslant 0 \colon B(s) = a\}.$$

The main part is to show that, for closed sets $\Lambda \subset [0,\infty)$,

$$c\,\mu(\Lambda) \leqslant \mathcal{H}^\phi\big(\Lambda \cap \text{Rec}\big) \leqslant C\,\mu(\Lambda), \tag{6.5}$$

where $\phi(r) = \sqrt{r \log\log(1/r)}$ and c, C are positive constants.

The easier direction, the lower bound for the Hausdorff measure, follows from part **(iii)** of the Rogers–Taylor theorem and the upper bound in the law of the iterated logarithm. Indeed, for any level $a > 0$ let $T_a = \inf\{s \geqslant 0 \colon B(t) = a\}$. Observe that

$$\limsup_{r \downarrow 0} \frac{M(T_a + r) - M(T_a)}{\sqrt{2r \log\log(1/r)}} = \limsup_{r \downarrow 0} \frac{B(T_a + r) - B(T_a)}{\sqrt{2r \log\log(1/r)}},$$

where we use that $M(T_a) = B(T_a)$ and that for any $r > 0$ there exists $0 < \tilde{r} < r$ with $M(T_a + r) = B(T_a + \tilde{r})$. Combining this with Corollary 5.3 applied to the standard Brownian motion $\{B(T_a + t) - B(T_a) \colon t \geqslant 0\}$ we get, almost surely,

$$\limsup_{r \downarrow 0} \frac{M(T_a + r) - M(T_a)}{\sqrt{2r \log\log(1/r)}} = 1.$$

Defining the set

$$A = \big\{s \in \text{Rec} \colon \limsup_{r \downarrow 0} \mu[s, s+r]/\phi(r) \leqslant \sqrt{2}\big\},$$

this means that, for every $a > 0$, we have $T_a \in A$ almost surely. By Fubini's theorem,

$$\mathbb{E}\mu(A^c) = \mathbb{E}\int_0^\infty 1\{T_a \notin A\}\,da = \int_0^\infty \mathbb{P}\{T_a \notin A\}\,da = 0,$$

and hence, almost surely, $\mu(A^c) = 0$. By part **(iii)** of the Rogers–Taylor theorem, for every closed set $\Lambda \subset [0,\infty)$,

$$\mathcal{H}^\phi(\Lambda \cap \text{Rec}) \geqslant \mathcal{H}^\phi(\Lambda \cap A) \geqslant \tfrac{1}{\sqrt{2}}\,\mu(\Lambda \cap A) = \tfrac{1}{\sqrt{2}}\,\mu(\Lambda),$$

showing the left inequality in (6.5).

For the harder direction, the upper bound for the Hausdorff measure, it is important to note that the lower bound in Corollary 5.3 does not suffice. Instead, we need a law of the iterated logarithm which holds simultaneously for $\mathcal{H}^\phi$-almost all record times. Recall that $\phi(r) = \sqrt{r \log\log(1/r)}$.

Lemma 6.46 *For every $\vartheta > 0$ small enough, almost surely,*

$$\mathcal{H}^\phi\Big\{s \in \text{Rec} \colon \limsup_{h \downarrow 0} \frac{M(s+h) - M(s-h)}{\phi(h)} < \vartheta\Big\} = 0.$$

Proof.   We only need to prove that, for some $\tilde{\theta} > 0$, the set

$$\Lambda(\tilde{\theta}) = \Big\{ s \in \mathsf{Rec} \cap (0,1) \colon \limsup_{h \downarrow 0} \frac{M(s+h)-M(s-h)}{\phi(h)} < \tilde{\theta} \Big\}$$

satisfies $\mathcal{H}^\phi(\Lambda(\tilde{\theta})) = 0$. Moreover, denoting

$$\Lambda_\delta(\tilde{\theta}) = \Big\{ s \in \mathsf{Rec} \cap [\delta, 1-\delta] \colon \sup_{h < \delta} \frac{M(s+h)-M(s-h)}{\phi(h)} < \tilde{\theta} \Big\},$$

we have

$$\Lambda(\tilde{\theta}) = \bigcup_{\delta > 0} \Lambda_\delta(\tilde{\theta}) .$$

It thus suffices to show, for fixed $\delta > 0$, that, almost surely,

$$\liminf_{n \uparrow \infty} \mathcal{H}^\phi_{1/n}(\Lambda_\delta(\tilde{\theta})) = 0 .$$

Fix $\delta > 0$ and a positive integer n such that $1/\sqrt{n} < \delta$. For parameters

$$A > 1, \theta > \tilde{\theta} \text{ and } q > 2,$$

which we choose later, we say that an interval of the form $I = [(k-1)/n, k/n]$ with $k \in \{1, \ldots, n\}$ is *good* if

   (i) I contains a record point, in other words,

$$\tau := \inf\Big\{ t \geq \tfrac{k-1}{n} \colon B(t) = M(t) \Big\} \leq \tfrac{k}{n} ,$$

and either of the following two conditions hold,

   (ii) there exists $j \geq 0$ with $1 \leq q^{j+1} \leq \sqrt{n}$ such that

$$B\big(\tau + \tfrac{q^j}{n}\big) - B(\tau) < -A\,\phi\big(\tfrac{q^j}{n}\big) ;$$

   (iii) for all $j \geq 0$ with $1 \leq q^{j+1} \leq \sqrt{n}$ we have that

$$B\big(\tau + \tfrac{q^{j+1}}{n}\big) - B\big(\tau + \tfrac{q^j}{n}\big) < \theta\,\phi\big(\tfrac{q^{j+1}-q^j}{n}\big) .$$

We now argue pathwise, and show that, given $A > 1$, $\theta > \tilde{\theta}$ we can find $q > 2$ such that the good intervals cover the set $\Lambda_\delta(\tilde{\theta})$. Indeed, suppose that I is not good but contains a minimal record point $\tau \in [(k-1)/n, k/n]$. Then there exists $j \geq 0$ with $1 \leq q^{j+1} \leq \sqrt{n}$ such that

$$B\big(\tau + \tfrac{q^j}{n}\big) - B(\tau) \geq -A\,\phi\big(\tfrac{q^j}{n}\big) \quad \text{and} \quad B\big(\tau + \tfrac{q^{j+1}}{n}\big) - B\big(\tau + \tfrac{q^j}{n}\big) \geq \theta\,\phi\big(\tfrac{q^{j+1}-q^j}{n}\big) .$$

This implies that, for any $t \in [(k-1)/n, k/n] \cap \mathsf{Rec}$,

$$M\big(t + \tfrac{q^{j+1}}{n}\big) - M\big(t - \tfrac{q^{j+1}}{n}\big) \geq M\big(\tau + \tfrac{q^{j+1}}{n}\big) - M(\tau) \geq B\big(\tau + \tfrac{q^{j+1}}{n}\big) - B(\tau)$$
$$\geq \theta\,\phi\big(\tfrac{q^{j+1}-q^j}{n}\big) - A\,\phi\big(\tfrac{q^j}{n}\big) \geq \tilde{\theta}\,\phi\big(\tfrac{q^{j+1}}{n}\big),$$

if q is chosen large enough. Hence the interval I does not intersect $\Lambda_\delta(\tilde{\theta})$ and therefore the good intervals cover this set.

Next we show that, for any $A > \sqrt{2} > \theta$ and suitably chosen $C > 0$, for every $I = [(k-1)/n, k/n]$ with $I \cap [\delta, 1-\delta] \neq \emptyset$,

$$\mathbb{P}\left\{ \left[\tfrac{k-1}{n}, \tfrac{k}{n}\right] \text{ is good } \right\} \leqslant C \frac{1}{\sqrt{n}} \left(\frac{1}{\log n} \right)^{\frac{A^2}{2}-1}. \tag{6.6}$$

By Lemma 4.22 in conjunction with Theorem 2.34 we get, for some constant $C_0 > 0$ depending only on $\delta > 0$,

$$\mathbb{P}\left\{ \tau < \tfrac{k}{n} \right\} \leqslant C_0 \frac{1}{\sqrt{n}}.$$

We further get, for some constant $C_1 > 0$, for all j with $q^{j+1} \leqslant \sqrt{n}$,

$$\mathbb{P}\left\{ B\left(\tau + \tfrac{q^j}{n}\right) - B(\tau) < -A\,\phi\left(\tfrac{q^j}{n}\right) \right\} \leqslant \mathbb{P}\left\{ B(1) < -A\sqrt{\log\log(n/q^j)} \right\}$$

$$\leqslant \exp\left\{ -\tfrac{A^2}{2} \log\log(\sqrt{n}) \right\} \leqslant C_1 \left(\frac{1}{\log n} \right)^{\frac{A^2}{2}}.$$

Using the independence of these events and summing over all $j \geqslant 0$ with $1 \leqslant q^{j+1} \leqslant \sqrt{n}$, of which there are no more than $C_2 \log n$, we get that

$$\mathbb{P}\left\{ \left[\tfrac{k-1}{n}, \tfrac{k}{n}\right] \text{ satisfies (i) and (ii) } \right\} \leqslant C_0 C_1 C_2 \frac{1}{\sqrt{n}} \left(\frac{1}{\log n} \right)^{\frac{A^2}{2}-1}. \tag{6.7}$$

To estimate the probability that $[(k-1)/n, k/n]$ satisfies (i) and (iii) we first note, for sufficiently large n, that

$$\mathbb{P}\left\{ B\left(\tfrac{q^{j+1}-q^j}{n}\right) < \theta\,\phi\left(\tfrac{q^{j+1}-q^j}{n}\right) \right\} \leqslant \mathbb{P}\left\{ B(1) < \theta\sqrt{\log\log\left(\tfrac{n}{q-1}\right)} \right\}$$

$$\leqslant 1 - \frac{\exp\left\{ -\tfrac{\theta^2}{2} \log\log\left(\tfrac{n}{q-1}\right) \right\}}{\theta\sqrt{\log\log\left(\tfrac{n}{q-1}\right)}},$$

using Lemma 12.9 of the appendix. From this we infer that, for suitable $c_3 > 0$,

$$\mathbb{P}\left\{ B\left(\tau + \tfrac{q^{j+1}}{n}\right) - B\left(\tau + \tfrac{q^j}{n}\right) < \theta\,\phi\left(\tfrac{q^{j+1}-q^j}{n}\right) \text{ for all } 1 \leqslant q^{j+1} \leqslant \sqrt{n} \right\}$$

$$\leqslant \prod_{j \leqslant \frac{\log n}{2\log q}} \left(1 - \frac{\exp\{-\tfrac{\theta^2}{2}\log\log n\}}{\theta\sqrt{\log\log n}} \right) \leqslant \left(1 - \frac{1}{\theta\,(\log n)^{\frac{\theta^2}{2}}(\log\log n)^{\frac{1}{2}}} \right)^{\frac{\log n}{2\log q}}$$

$$\leqslant \exp\left\{ -c_3 \frac{(\log n)^{1-\frac{\theta^2}{2}}}{(\log\log n)^{\frac{1}{2}}} \right\}.$$

Combining this with the estimate for $\tau < k/n$ we get that

$$\mathbb{P}\left\{ \left[\tfrac{k-1}{n}, \tfrac{k}{n}\right] \text{ satisfies (i) and (iii) } \right\} \leqslant C_0 \frac{1}{\sqrt{n}} \exp\left\{ -c_3 \frac{(\log n)^{1-\frac{\theta^2}{2}}}{(\log\log n)^{\frac{1}{2}}} \right\}. \tag{6.8}$$

As $\theta < \sqrt{2}$, the right hand side in (6.8) is of smaller order than the right hand side in (6.7) and hence we have shown (6.6).

Finally, we look at the expected ϕ-values of our covering. We obtain that

$$\mathbb{E}\mathcal{H}^\phi_{1/n}(\Lambda_\delta(\tilde{\theta})) \leqslant \sum_{k=\lceil \delta n \rceil}^{\lceil n(1-\delta) \rceil} \phi(1/n)\mathbb{P}\left\{ \left[\tfrac{k-1}{n}, \tfrac{k}{n}\right] \text{ is good } \right\} \leqslant C\,\frac{\sqrt{\log\log n}}{(\log n)^{A^2/2-1}} \longrightarrow 0,$$

and, by Fatou's lemma we get, almost surely,

$$\liminf_{n\uparrow\infty} \mathcal{H}^\phi_{1/n}(\Lambda_\delta(\tilde{\theta})) = 0,$$

as required to complete the proof.   ∎

The right inequality in (6.5) now follows easily from Lemma 6.46 and part (ii) of the Rogers–Taylor theorem. We define the set

$$A = \left\{ s \in \mathsf{Rec}\colon \limsup_{r\downarrow 0} \mu\mathcal{B}(s,r)/\phi(r) \geqslant \vartheta \right\},$$

and note that $\mathcal{H}^\phi(\mathsf{Rec}\cap A^c) = 0$, for ϑ sufficiently small. By part (ii) of the Rogers–Taylor theorem we get, for every Borel set $\Lambda \subset [0,\infty)$,

$$\mathcal{H}^\phi(\Lambda \cap \mathsf{Rec}) = \mathcal{H}^\phi(\Lambda \cap A) \leqslant \kappa_1\vartheta^{-1}\,\mu(\Lambda\cap A) \leqslant \kappa_1\vartheta^{-1}\,\mu(\Lambda).$$

This implies the right inequality and hence completes the proof of (6.5).

To complete the proof of Theorem 6.43 we look at the process $\{X(a)\colon a \geqslant 0\}$ defined by

$$X(a) = \mathcal{H}^\phi\left(\mathsf{Rec} \cap [0,T_a]\right).$$

The next lemma will help us to show that this process is, in a suitable sense, degenerate.

Lemma 6.47 *Suppose $\{Y(t)\colon t \geqslant 0\}$ is a stochastic process starting in zero with the following properties,*

- *the paths are almost surely continuous,*
- *the increments are independent, nonnegative and stationary,*
- *there exists a $C > 0$ such that, almost surely, $Y(t) \leqslant Ct$ for all $t > 0$.*

Then there exists $\tilde{c} \geqslant 0$ such that, almost surely, $Y(t) = \tilde{c}t$ for every $t \geqslant 0$.

Proof.   We first look at the function $m\colon [0,\infty) \to [0,\infty)$ defined by $m(t) = \mathbb{E}Y(t)$. This function is continuous, as the paths of $\{Y(t)\colon t \geqslant 0\}$ are continuous and bounded on compact sets. Further, because the process $\{Y(t)\colon t \geqslant 0\}$ has independent and stationary increments, the function m is linear and hence there exists $\tilde{c} \geqslant 0$ with $m(t) = \tilde{c}t$. It thus suffices to show that the variance of $Y(t)$ is zero. Indeed, for every $n > 0$, we have

$$\mathrm{Var}\,Y(t) = \sum_{k=1}^{n} \mathrm{Var}\left(Y\big(\tfrac{kt}{n}\big) - Y\big(\tfrac{(k-1)t}{n}\big)\right) = n\,\mathrm{Var}\,Y\big(\tfrac{t}{n}\big) \leqslant n\,\mathbb{E}\big[Y\big(\tfrac{t}{n}\big)^2\big]$$
$$\leqslant n\,C^2\big(\tfrac{t}{n}\big)^2 \xrightarrow{n\to\infty} 0,$$

and hence $Y(t) = \mathbb{E}Y(t) = \tilde{c}t$ as claimed.   ∎

Let us check that $\{X(a) \colon a \geqslant 0\}$ satisfies the conditions of Lemma 6.47. We first note that

$$X(a + h) - X(a) = \mathcal{H}^\phi\left(\mathsf{Rec} \cap [0, T_{a+h}]\right) - \mathcal{H}^\phi\left(\mathsf{Rec} \cap [0, T_a]\right)$$
$$= \mathcal{H}^\phi\left(\mathsf{Rec} \cap [T_a, T_{a+h}]\right),$$

as can be seen easily from the definition of the Hausdorff measure $\mathcal{H}^\phi$.

Using this, continuity of the paths follows from the fact that, by (6.5),

$$\mathcal{H}^\phi\left(\mathsf{Rec} \cap [T_a, T_{a+h}]\right) \leqslant C\left(M(T_{a+h}) - M(T_a)\right) = C\,h.$$

The strong Markov property implies that the increments are independent and stationary, and they are obviously nonnegative. And finally, by (6.5), almost surely, for any $a \geqslant 0$,

$$X(a) = \mathcal{H}^\phi\left(\mathsf{Rec} \cap [0, T_a]\right) \leqslant C\,M(T_a) = C\,a.$$

Lemma 6.47 thus implies that there exists $\tilde{c} \geqslant 0$ with

$$\mathcal{H}^\phi\left(\mathsf{Rec} \cap [0, T_a]\right) = \tilde{c}\,a = \tilde{c}\,M(T_a)$$

for all $a \geqslant 0$. It remains to show that this holds not only for the stopping times T_a, but in fact for all elements of Rec.

Lemma 6.48 *Almost surely, the set* $\{T_a \colon a \in \mathbb{R}\}$ *is dense in* Rec.

Proof.   Obviously, $\{T_a \colon a \in \mathbb{R}\} \subset \mathsf{Rec}$. Conversely, if $t \in \mathsf{Rec}$, then either $B(s) < B(t)$ for all $0 \leqslant s < t$, in which case $t = T_a$ for $a = B(t)$, or there exists a minimal $s < t$ with $B(s) = B(t)$. In the latter case $s = T_a$ for $a = B(t)$ by definition.

Because, by Theorem 2.11, every local maximum is a strict local maximum and no two local maxima are the same, we have

$$t = \lim_{\substack{b \to a \\ b > a}} T_b,$$

in particular t is in the closure of the set $\{T_a \colon a \in \mathbb{R}\}$.   ∎

Using this lemma and continuity of both sides, we infer that, almost surely, $\mathcal{H}^\phi(\mathsf{Rec} \cap [0, t]) = \tilde{c}\,M(t)$ for all $t \in \mathsf{Rec}$. For general $t \geqslant 0$ we let $\tau = \max(\mathsf{Rec} \cap [0, t])$ and note that

$$\mathcal{H}^\phi\left(\mathsf{Rec} \cap [0, t]\right) = \mathcal{H}^\phi\left(\mathsf{Rec} \cap [0, \tau]\right) = \tilde{c}\,M(\tau) = \tilde{c}\,M(t).$$

By the lower bound in (6.5) we must have $\tilde{c} > 0$ and hence we can put $c = 1/\tilde{c}$ and get

$$M(t) = c\,\mathcal{H}^\phi\left(\mathsf{Rec} \cap [0, t]\right) = \mathcal{H}^{c\phi}\left(\mathsf{Rec} \cap [0, t]\right),$$

as required to complete the proof of Theorem 6.43.

Exercises

Exercise 6.1. Using the downcrossing representation of the local time process $\{L(t) \colon t \geqslant 0\}$ given in Theorem 6.1, show that, almost surely, $L(s) = L(t)$ for every interval (s, t) not containing a zero of the Brownian motion. In other words, the local time at zero increases only on the zero set of the Brownian motion.

Exercise 6.2. Show, by reviewing the argument in the proof of Theorem 6.10, that for a standard linear Brownian motion the processes $\{(|B(t)|, L(t)) \colon t \geqslant 0\}$ and $\{(M(t) - B(t), M(t)) \colon t \geqslant 0\}$ have the same distribution.
Hint. In Theorem 7.38 we give a proof of this result using stochastic integration.

Exercise 6.3. Show that, for a standard Brownian motion, $\mathbb{E} L(t) = \sqrt{\dfrac{2t}{\pi}}$.

Exercise 6.4. Show that $\mathbb{P}_0\{L(t) > 0 \text{ for every } t > 0\} = 1$.
Hint. This follows easily from Theorem 6.10.

Exercise 6.5. Derive Theorem 6.10 from Theorem 2.34.
Hint. Show that the maximum process $\{M(t) \colon t \geqslant 0\}$ can be computed from $\{M(t) - B(t) \colon t \geqslant 0\}$ by counting downcrossings, so that $\{L(t) \colon t \geqslant 0\}$ is the same measurable function of $\{|B(t)| \colon t \geqslant 0\}$ as $\{M(t) \colon t \geqslant 0\}$ is of $\{M(t) - B(t) \colon t \geqslant 0\}$.

Exercise 6.6. ⑤ Let $\{W(s) \colon s \geqslant 0\}$ be a standard linear Brownian motion and τ_1 its first hitting time of level 1. Use Exercise 2.17 to show that

$$\mathbb{E} \int_0^{\tau_1} 1\{0 \leqslant W(s) \leqslant 1\} \, ds = 1.$$

Exercise 6.7. ⑤ Suppose $X_1, X_2, \ldots$ are independent geometrically distributed random variables on $\{1, 2, \ldots\}$ with mean 2. Then, for sufficiently small $\varepsilon > 0$, for all nonnegative integers $k \leqslant m$,

$$\mathbb{P}\left\{ \left| \sum_{j=1}^{k} (X_j - 2) \right| \geqslant \varepsilon m \right\} \leqslant 4 \exp\left\{ -\tfrac{1}{5} \varepsilon^2 m \right\}.$$

Exercise 6.8. ⑤ Give an alternative proof of Lemma 6.33 by computing the densities of the random variables $(X + \ell)^2$ and $X^2 + 2 \sum_{j=1}^{N} Z_j$.

Exercise 6.9. Use the Ray–Knight theorem and Lévy's theorem, Theorem 6.10, to show that, for a suitable constant $c > 0$, the function

$$\varphi(h) = c\sqrt{h \log(1/h)} \qquad \text{for } 0 < h < 1,$$

is a modulus of continuity for the random field $\{L^a(t) \colon a \in \mathbb{R}, t \geqslant 0\}$.

Exercise 6.10. Let X be a metric space and φ_1, φ_2 two gauge functions such that

$$0 < \mathcal{H}^{\varphi_1}(X), \mathcal{H}^{\varphi_2}(X) < \infty.$$

Show that

$$\limsup_{\varepsilon \downarrow 0} \frac{\varphi_1(\varepsilon)}{\varphi_2(\varepsilon)} > 0 \quad \text{and} \quad \liminf_{\varepsilon \downarrow 0} \frac{\varphi_1(\varepsilon)}{\varphi_2(\varepsilon)} < \infty.$$

Exercise 6.11. Show that, for $\phi(r) = \sqrt{r \log \log(1/r)}$, almost surely,

$$\int_A dL(t) = 0$$

simultaneously for all sets $A \subset [0, \infty)$ with $\mathcal{H}^\phi(A) = 0$.

Notes and comments

The study of local times is crucial for the Brownian motion in dimension one and good references are Revuz and Yor [RY94] and the survey article Borodin [Bo89]. Brownian local times were first introduced by Paul Lévy in [Le48] and a thorough investigation is initiated in a paper by Trotter [Tr58] who showed that there is a version of local time continuous in time and space. An alternative construction of local times can be given in terms of stochastic integrals, using Tanaka's formula as a definition. We shall explore this direction in Section 7.3.

A crucial aspect which is not covered by our treatment is the relation of local times to excursion theory and point processes, which allows a discussion of more general Markov processes. An excellent reference for this is Williams [Wi77], his treatment appears also in Rogers and Williams [RW00a]. Greenwood and Pitman [GP80] show how to use the same kind of argument to construct local time for a recurrent point of a strong Markov process. The basic insight comes from Lévy and Itô's theory of Poisson point processes of excursions, see Pitman and Yor [PY07] for a recent review. Walsh [Wa78] also discusses downcrossings and local time, leading to the Ray Knight theorem.

The equality for the upcrossing numbers in Lemma 6.3 agrees with the functional equation for a branching process with immigration. The relationship between local times and branching processes, which is underlying our entire treatment, can be exploited and extended in various ways. One example of this can be found in Neveu and Pitman [NP89], for more recent progress in this direction, see Le Gall and Le Jan, [LL98]. A good source for further reading is the discussion of Lévy processes and trees by Duquesne and Le Gall in [DL02]. For an introduction into branching processes with and without immigration, see the classical book of Athreya and Ney [AN04].

In a similar spirit, a result which is often called the second Ray–Knight theorem describes the process $\{L_T^a : a > 0\}$ when $T = \inf\{t > 0 : L_t^0 = x\}$, see [RY94] or the original papers by Ray and Knight cited above. The resulting process is a Feller diffusion, which is the canonical process describing critical branching with initial mass x. The local times of Brownian motion can therefore be used to encode the branching information for a variety of processes describing the evolution of particles which undergo critical branching and spatial migration. For more information on this powerful link between Brownian motion and the world of spatial branching processes, see for example Le Gall [LG99].

The concept of local times can be extended to a variety of processes like continuous semimartingales, see e.g. [RY94], or Markov processes [BG68]. The idea of introducing local times as densities of occupation measure has been fruitful in a variety of contexts, in particular in the introduction of local times on the intersection of Brownian paths. Important papers in this direction are Geman and Horowitz [GH80] and Geman, Horowitz and Rosen [GHR84].

The Ray–Knight theorem was discovered by D. Ray and F. Knight independently by different methods in 1963. The proof of Knight uses discretisation, see [Kn63] for the original paper and [Kn81] for more information. Ray's approach to Theorem 6.28 is less intuitive but more versatile, and is based on the Feynman–Kac formula, see [Ra63b] for the original paper. Our presentation is simpler than Knight's method. The distributional identity at its core, see Lemma 6.33, is yet to be explained probabilistically. The analytic proof given in Exercise 6.8 is due to H. Robbins and E.J.G. Pitman [RP49].

Extensions of the Ray–Knight theorem includes a characterisation of $\{L^x(T) : x \geqslant 0\}$ for parameters exceeding a. This is best discussed in the framework of Brownian excursion theory, see for example [RY94]. The Ray–Knight theorem can be extended into a deep relationship between the local times of symmetric Markov processes and an associated Gaussian process, which is the subject of the famous Dynkin isomorphism theorem. See Eisenbaum [Ei94] or the comprehensive monograph by Marcus and Rosen [MR06] for more on this subject.

According to Taylor [Ta86], Hausdorff measures with arbitrary gauge functions were introduced by A.S. Besicovitch. General theory of outer measures, as presented in Rogers [Ro99] shows that $\mathcal{H}^\phi$ indeed defines a measures on the Borel sets of a metric space. The fact that, for $\phi(r) = \sqrt{2r \log\log(1/r)}$, the local time at zero agrees with a constant multiple of the ϕ-Hausdorff measure of the zero set is due to Taylor and Wendel [TW66]. Perkins [Pe81] showed that the constant is one and further that the local times $L^a(t)$ agree with the ϕ-Hausdorff measure of the set $\{s \in [0, t] : B(s) = a\}$ simultaneously for all levels a and times t. His proof uses nonstandard analysis.

The Rogers–Taylor theorem is due to C.A. Rogers and S.J. Taylor in [RT61]. The original statement is slightly more general as it allows to replace $\mu(V)$ by $\mu(\Lambda)$ on the right hand side without any regularity condition on μ. Most proofs in the literature of the harder half, statement (ii) in our formulation, use the Besicovitch covering theorem. We give a self-contained proof using dyadic cubes instead.

Other natural measures related to Brownian motion can also be shown to agree with Hausdorff measures with suitable gauge functions. The most notable example is the occupation measure, whose gauge function is

$$\varphi(r) = \begin{cases} c_d \, r^2 \, \log\log(1/r) & \text{if } d \geqslant 3, \\ c_2 \, r^2 \, \log(1/r) \, \log\log\log(1/r) & \text{if } d = 2. \end{cases}$$

This result is due to Ciesielski and Taylor [CT62] in the first case, and to Ray [Ra63a] and Taylor [Ta64] in the second case. A stimulating survey of this subject is Le Gall [LG85].

7

Stochastic integrals and applications

In this chapter we first construct an integral with respect to Brownian motion. Amongst the applications are the conformal invariance of Brownian motion, a short look at windings of Brownian motion, the Tanaka formula for Brownian local times, and the Feynman–Kac formula.

7.1 Stochastic integrals with respect to Brownian motion

7.1.1 Construction of the stochastic integral

We look at a Brownian motion in dimension one $\{B(t) \colon t \geqslant 0\}$ considered as a random continuous function. As we have found in Theorem 1.35, this function is almost surely of unbounded variation, which is why we cannot use *Lebesgue–Stieltjes integration* to define integrals of the form $\int_0^t f(s)\, dB(s)$. There is however an escape from this dilemma, if one is willing to take advantage of the fact that Brownian motions are *random* functions and therefore one can make use of weaker forms of limits. This is the idea of *stochastic integration*.

Before explaining the procedure, we have a look at a reasonable class of integrands, as we would like to go beyond the Paley–Wiener integral constructed in Lemma 1.41 and admit random functions as integrands. A suitable class of random integrands is the class of *progressively measurable processes*. We denote by $(\Omega, \mathcal{A}, \mathbb{P})$ the probability space on which our Brownian motion $\{B(t) \colon t \geqslant 0\}$ is defined and suppose that $(\mathcal{F}(t) \colon t \geqslant 0)$ is a filtration to which the Brownian motion is adapted such that the strong Markov property holds.

Because we also want the integral up to time t to be adapted to our filtration, we assume that the filtration $(\mathcal{F}(t) \colon t \geqslant 0)$ is **complete**, i.e. contains all sets of probability zero in $\mathcal{A}$. Note that every filtration can be completed simply by adding all these sets and their complements, and that the completion preserves the strong Markov property.

Definition 7.1. A process $\{X(t, \omega) \colon t \geqslant 0, \omega \in \Omega\}$ is called **progressively measurable** if for each $t \geqslant 0$ the mapping $X \colon [0, t] \times \Omega \to \mathbb{R}$ is measurable with respect to the σ-algebra $\mathfrak{B}([0, t]) \otimes \mathcal{F}(t)$. ◇

Lemma 7.2 *Any processes* $\{X(t)\colon t \geqslant 0\}$, *which is adapted and either right- or left-continuous is also progressively measurable.*

Proof.　Assume that $\{X(t)\colon t \geqslant 0\}$ is right-continuous. Fix $t > 0$. For a positive integer n and $0 \leqslant s \leqslant t$ define $X_n(0,\omega) = X(0,\omega)$ and

$$X_n(s,\omega) = X\left(\tfrac{(k+1)t}{2^n},\omega\right), \qquad \text{for } kt2^{-n} < s \leqslant (k+1)t2^{-n}.$$

The mapping $(s,\omega) \mapsto X_n(s,\omega)$ is $\mathfrak{B}([0,t]) \otimes \mathcal{F}(t)$ measurable. By right-continuity we have $\lim_{n\uparrow\infty} X_n(s,\omega) = X(s,\omega)$ for all $s \in [0,t]$ and $\omega \in \Omega$, hence the limit map $(s,\omega) \mapsto X(s,\omega)$ is also $\mathfrak{B}([0,t]) \otimes \mathcal{F}(t)$ measurable, proving progressive measurability. The left-continuous case is analogous. ∎

The construction of the integrals is quite straightforward. We start by integrating progressively measurable step processes $\{H(t,\omega)\colon t \geqslant 0,\ \omega \in \Omega\}$ of the form

$$H(t,\omega) = \sum_{i=1}^{k} A_i(\omega)\mathbf{1}_{(t_i,t_{i+1}]}(t), \text{ for } 0 \leqslant t_1 \leqslant \ldots \leqslant t_{k+1}, \text{ and } \mathcal{F}(t_i)\text{-measurable } A_i.$$

In complete analogy to the classical case we define the integral as

$$\int_0^\infty H(s)\,dB(s) := \sum_{i=1}^{k} A_i\big(B(t_{i+1}) - B(t_i)\big).$$

Now let H be a progressively measurable process satisfying $\mathbb{E}\int_0^\infty H(s)^2\,ds < \infty$. Suppose H can be approximated by a family of progressively measurable step processes H_n, $n \geqslant 1$, then we define

$$\int_0^\infty H(s)\,dB(s) := \lim_{n\to\infty} \int_0^\infty H_n(s)\,dB(s). \tag{7.1}$$

At this stage we focus on $\mathbf{L}^2$-convergence, though we shall see later that the stochastic integral can also be constructed as an almost sure limit, see Remark 7.7. For the approximation of H by progressively measurable step processes we look at the norm

$$\|H\|_2^2 := \mathbb{E}\int_0^\infty H(s)^2\,ds.$$

What we have to show now to complete the definition is that,

(1) every progressively measurable process satisfying $\mathbb{E}\int_0^\infty H(s)^2\,ds < \infty$ can be approximated in the $\|\cdot\|_2$ norm by progressively measurable step processes,

(2) for each approximating sequence the limit in (7.1) exists in the $\mathbf{L}^2$-sense,

(3) and this limit does not depend on the choice of the approximating step processes.

This is what we check now, beginning with item (1).

Lemma 7.3 *For every progressively measurable process* $\{H(s,\omega)\colon s \geqslant 0,\ \omega \in \Omega\}$ *satisfying* $\mathbb{E}\int_0^\infty H(s)^2\,ds < \infty$ *there exists a sequence* $\{H_n\colon n \in \mathbb{N}\}$ *of progressively measurable step processes such that* $\lim_{n\to\infty} \|H_n - H\|_2 = 0$.

Proof.   We approximate the progressively measurable process successively by

- a bounded progressively measurable process,
- a bounded, almost surely continuous progressively measurable process,
- and finally, by a progressively measurable step process.

Let $\{H(s,\omega)\colon s \geqslant 0,\ \omega \in \Omega\}$ be a progressively measurable process with $\|H\|_2 < \infty$. We *first* define the cut-off at a fixed time $n > 0$ by letting $H_n(s,\omega) = H(s,\omega)$ for $s \leqslant n$ and $H_n(s,\omega) = 0$ otherwise. Clearly $\lim_{n\uparrow\infty} \|H_n - H\|_2 = 0$.

Second, we approximate any progressively measurable H on a finite interval by truncating its values, i.e. for large n we define H_n by letting $H_n(s,\omega) = H(s,\omega) \wedge n$. Clearly H_n is progressively measurable and $\lim_{n\uparrow\infty} \|H_n - H\|_2 = 0$.

Third, we approximate any uniformly bounded progressively measurable H by a bounded, almost-surely continuous, progressively measurable process. Let $h = 1/n$ and, using the convention $H(s,\omega) = H(0,\omega)$ for $s < 0$ we define

$$H_n(s,\omega) = \frac{1}{h} \int_{s-h}^{s} H(t,\omega)\, dt.$$

Because we only take an average over the past, H_n is again progressively measurable. It is almost surely continuous and it is a well-known fact that, for every $\omega \in \Omega$ and almost every $s \in [0,t]$,

$$\lim_{h\downarrow 0} \frac{1}{h} \int_{s-h}^{s} H(t,\omega)\, dt = H(s,\omega).$$

Since H is uniformly bounded (and using progressive measurability) we can take expectations and an average over time, and obtain from the bounded convergence theorem that

$$\lim_{n\uparrow\infty} \|H_n - H\|_2 = 0.$$

Finally, a bounded, almost-surely continuous, progressively measurable process can be approximated by a step process H_n by taking $H_n(s,\omega) = H(j/n,\omega)$ for $j/n \leqslant s < (j+1)/n$. These functions are again progressively measurable and one easily sees $\lim_{n\uparrow\infty} \|H_n - H\|_2 = 0$. This completes the approximation. ∎

The following lemma describes the crucial property of the integral of step processes.

Lemma 7.4 *Let H be a progressively measurable step process and $\mathbb{E} \int_0^\infty H(s)^2\, ds < \infty$, then*

$$\mathbb{E}\left[\left(\int_0^\infty H(s)\, dB(s) \right)^2 \right] = \mathbb{E} \int_0^\infty H(s)^2\, ds.$$

Proof.   We use the Markov property to see that, for every progressively measurable step process $H = \sum_{i=1}^{k} A_i \mathbf{1}_{(a_i,a_{i+1}]}$,

$$\mathbb{E}\left[\left(\int_0^\infty H(s)\, dB(s) \right)^2 \right] = \mathbb{E}\left[\sum_{i,j=1}^{k} A_i A_j \left(B(a_{i+1}) - B(a_i) \right) \left(B(a_{j+1}) - B(a_j) \right) \right]$$

$$= 2 \sum_{i=1}^{k} \sum_{j=i+1}^{k} \mathbb{E}\Big[A_i A_j \big(B(a_{i+1}) - B(a_i)\big) \mathbb{E}\big[B(a_{j+1}) - B(a_j) \,\big|\, \mathcal{F}(a_j)\big]\Big]$$

$$+ \sum_{i=1}^{k} \mathbb{E}\Big[A_i^2 \big(B(a_{i+1}) - B(a_i)\big)^2\Big]$$

$$= \sum_{i=1}^{k} \mathbb{E}\big[A_i^2\big]\,(a_{i+1} - a_i) = \mathbb{E} \int_0^\infty H(s)^2 \, ds. \qquad\blacksquare$$

Corollary 7.5 *Suppose* $\{H_n : n \in \mathbb{N}\}$ *is a sequence of progressively measurable step processes such that*

$$\mathbb{E} \int_0^\infty \big(H_n(s) - H_m(s)\big)^2 \, ds \longrightarrow 0, \ \textit{as } n, m \to \infty,$$

then

$$\mathbb{E}\Bigg[\bigg(\int_0^\infty H_n(s) - H_m(s)\,dB(s)\bigg)^2\Bigg] \longrightarrow 0, \ \textit{as } n, m \to \infty.$$

Proof. Because the difference of two step processes is again a step process, Lemma 7.4 can be applied to $H_n - H_m$ and this gives the statement. $\qquad\blacksquare$

The following theorem addresses issues (2) and (3), thus completing our construction of the stochastic integral.

Theorem 7.6 *Suppose* $\{H_n : n \in \mathbb{N}\}$ *is a sequence of progressively measurable step processes and* H *a progressively measurable process such that*

$$\lim_{n \to \infty} \mathbb{E} \int_0^\infty \big(H_n(s) - H(s)\big)^2 \, ds = 0,$$

then

$$\lim_{n \to \infty} \int_0^\infty H_n(s)\,dB(s) =: \int_0^\infty H(s)\,dB(s)$$

exists as a limit in the $\mathbf{L}^2$*-sense and is independent of the choice of* $\{H_n : n \in \mathbb{N}\}$*. Moreover, we have*

$$\mathbb{E}\Bigg[\bigg(\int_0^\infty H(s)\,dB(s)\bigg)^2\Bigg] = \mathbb{E} \int_0^\infty H(s)^2 \, ds. \qquad (7.2)$$

Remark 7.7 If the sequence of step processes is chosen such that

$$\sum_{n=1}^\infty \mathbb{E} \int_0^\infty \big(H_n(s) - H(s)\big)^2 \, ds < \infty,$$

then, by (7.2), we get $\sum_{n=1}^\infty \mathbb{E}[(\int_0^\infty H_n(s) - H(s)\,dB(s))^2] < \infty$, and therefore, almost surely,

$$\sum_{n=1}^\infty \bigg[\int_0^\infty H_n(s)\,dB(s) - \int_0^\infty H(s)\,dB(s)\bigg]^2 < \infty.$$

This implies that, almost surely,

$$\lim_{n \to \infty} \int_0^\infty H_n(s) \, dB(s) = \int_0^\infty H(s) \, dB(s). \qquad \diamond$$

Proof of Theorem 7.6. By the triangle inequality $\{H_n : n \in \mathbb{N}\}$ satisfies the assumption of Corollary 7.5, and hence $\{\int_0^\infty H_n(s) \, dB(s) : n \in \mathbb{N}\}$ is a Cauchy sequence in $\mathbf{L}^2$. By completeness of this space, the limit exists, and Corollary 7.5 also shows that the limit is independent of the choice of the approximating sequence. The last statement follows from Lemma 7.4, applied to H_n, by taking the limit $n \to \infty$. ∎

Finally, we describe the stochastic integral as a stochastic process in time. The crucial properties of this process are continuity and the martingale property.

Definition 7.8. Suppose $\{H(s,\omega) : s \geqslant 0, \omega \in \Omega\}$ is progressively measurable with $\mathbb{E} \int_0^t H(s,\omega)^2 \, ds < \infty$. Define the progressively measurable process $\{H^t(s,\omega) : s \geqslant 0, \omega \in \Omega\}$ by

$$H^t(s,\omega) = H(s,\omega) \, \mathbf{1}\{s \leqslant t\}.$$

Then the **stochastic integral up to** t is defined as,

$$\int_0^t H(s) \, dB(s) := \int_0^\infty H^t(s) \, dB(s). \qquad \diamond$$

Remark 7.9 Provided they both exist, the Paley–Wiener integral agrees with the stochastic integral just defined, see Exercise 7.1 for more details. $\diamond$

Definition 7.10. We say that a stochastic process $\{X(t) : t \geqslant 0\}$ is a **modification** of a process $\{Y(t) : t \geqslant 0\}$ if, for every $t \geqslant 0$, we have $\mathbb{P}\{X(t) = Y(t)\} = 1$. $\diamond$

The next result shows that we can modify stochastic integrals in such a way that they become almost surely continuous in time. From this point on when referring to the process $\{\int_0^t H(s) \, dB(s) : t \geqslant 0\}$ we will always refer to this modification.

Theorem 7.11 *Suppose the process $\{H(s,\omega) : s \geqslant 0, \omega \in \Omega\}$ is progressively measurable with*

$$\mathbb{E} \int_0^t H(s,\omega)^2 \, ds < \infty \quad \text{for any } t \geqslant 0.$$

Then there exists an almost surely continuous modification of $\{\int_0^t H(s) \, dB(s) : t \geqslant 0\}$. Moreover, this process is a martingale and hence

$$\mathbb{E} \int_0^t H(s) \, dB(s) = 0 \quad \text{for every } t \geqslant 0.$$

Proof. Fix a large integer t_0 and let H_n be a sequence of step processes such that $\|H_n - H^{t_0}\|_2 \to 0$, and therefore

$$\mathbb{E}\left[\left(\int_0^\infty \left(H_n(s) - H^{t_0}(s)\right) dB(s)\right)^2\right] \to 0.$$

Obviously, for any $s \leqslant t$ the random variable $\int_0^s H_n(u) \, dB(u)$ is $\mathcal{F}(s)$-measurable and $\int_s^t H_n(u) \, dB(u)$ is independent of $\mathcal{F}(s)$, meaning that the process

$$\left\{\int_0^t H_n(u) \, dB(u) \colon 0 \leqslant t \leqslant t_0\right\}$$

is a martingale, for every n. For any $0 \leqslant t \leqslant t_0$ define

$$X(t) = \mathbb{E}\left[\int_0^{t_0} H(s) \, dB(s) \,\Big|\, \mathcal{F}(t)\right],$$

so that $\{X(t) \colon 0 \leqslant t \leqslant t_0\}$ is also a martingale and

$$X(t_0) = \int_0^{t_0} H(s) \, dB(s).$$

By Doob's maximal inequality, Proposition 2.43, for $p = 2$,

$$\mathbb{E}\left[\sup_{0 \leqslant t \leqslant t_0} \left(\int_0^t H_n(s) \, dB(s) - X(t)\right)^2\right] \leqslant 4\,\mathbb{E}\left[\left(\int_0^{t_0} \left(H_n(s) - H(s)\right) dB(s)\right)^2\right],$$

which converges to zero, as $n \to \infty$. This implies, in particular, that almost surely, the process $\{X(t) \colon 0 \leqslant t \leqslant t_0\}$ is a uniform limit of continuous processes, and hence continuous. For fixed $0 \leqslant t \leqslant t_0$, by taking $\mathbf{L}^2$-limits from the step process approximation, the random variable $\int_0^t H(s) \, dB(s)$ is $\mathcal{F}(t)$-measurable and $\int_t^{t_0} H(s) \, dB(s)$ is independent of $\mathcal{F}(t)$ with zero expectation. Therefore $\int_0^t H(s) \, dB(s)$ is a conditional expectation of $X(t_0)$ given $\mathcal{F}(t)$, hence coinciding with $X(t)$ almost surely. ∎

We now have a basic stochastic integral at our disposal. Obviously, a lot of bells and whistles can be added to this construction, but we refrain from doing so and keep focused on the essential properties and eventually on the applications to Brownian motion.

7.1.2 Itô's formula

For stochastic integration Itô's formula plays the same rôle as the fundamental theorem of calculus for classical integration. Let f be continuously differentiable and $x \colon [0, \infty) \to \mathbb{R}$, then the fundamental theorem can be written as

$$f(x(t)) - f(x(0)) = \int_0^t f'(x(s)) \, dx(s),$$

and this formula holds when x is continuous and of bounded variation. Itô's formula offers an analogue of this for the case that x is a Brownian motion. The crucial difference is that a third term enters, which makes the existence of a second derivative of f necessary. The next result, a key step in the derivation of this formula, is an extension of Exercise 1.16.

Theorem 7.12 *Suppose* $f \colon \mathbb{R} \to \mathbb{R}$ *is continuous,* $t > 0$, *and* $0 = t_1^{(n)} < \ldots < t_n^{(n)} = t$ *are partitions of the interval* $[0, t]$, *such that the mesh converges to zero. Then, in probability,*

$$\sum_{j=1}^{n-1} f\big(B(t_j^{(n)})\big) \big(B(t_{j+1}^{(n)}) - B(t_j^{(n)})\big)^2 \longrightarrow \int_0^t f(B(s))\, ds \,.$$

Proof.   Let T be the first exit time from a compact interval. It suffices to prove the statement for Brownian motion stopped at T, as the interval may be chosen to make $\mathbb{P}\{T < t\}$ arbitrarily small. By continuity of f and the definition of the Riemann integral, almost surely,

$$\lim_{n \to \infty} \sum_{j=1}^{n-1} f\big(B(t_j^{(n)} \wedge T)\big) \big(t_{j+1}^{(n)} \wedge T - t_j^{(n)} \wedge T\big) = \int_0^{t \wedge T} f(B(s))\, ds \,.$$

It thus suffices to show that

$$\lim_{n \to \infty} \mathbb{E}\Big(\sum_{j=1}^{n-1} f\big(B(t_j^{(n)} \wedge T)\big) \big(\big(B(t_{j+1}^{(n)} \wedge T) - B(t_j^{(n)} \wedge T)\big)^2 - \big(t_{j+1}^{(n)} \wedge T - t_j^{(n)} \wedge T\big) \big) \Big)^2 = 0.$$

Recall that $\{B(t)^2 - t \colon t \geqslant 0\}$ is a martingale, by Lemma 2.47, and hence, for all $r \leqslant s$,

$$\mathbb{E}\big[(B(s) - B(r))^2 - (s - r) \,\big|\, \mathcal{F}(r) \big] = 0 \,.$$

This allows us to simplify the previous expression as follows,

$$\mathbb{E}\Big[\Big(\sum_{j=1}^{n-1} f\big(B(t_j^{(n)} \wedge T)\big) \big(\big(B(t_{j+1}^{(n)} \wedge T) - B(t_j^{(n)} \wedge T)\big)^2 - \big(t_{j+1}^{(n)} \wedge T - t_j^{(n)} \wedge T\big) \big) \Big)^2 \Big]$$

$$= \sum_{j=1}^{n-1} \mathbb{E}\Big[f\big(B(t_j^{(n)} \wedge T)\big)^2 \big(\big(B(t_{j+1}^{(n)} \wedge T) - B(t_j^{(n)} \wedge T)\big)^2 - \big(t_{j+1}^{(n)} \wedge T - t_j^{(n)} \wedge T\big) \big)^2 \Big].$$

We can now bound f by its maximum on the compact interval, and multiplying out the square and dropping a negative cross term we get an upper bound, which is a constant multiple of

$$\sum_{j=1}^{n-1} \mathbb{E}\Big[\big(B(t_{j+1}^{(n)} \wedge T) - B(t_j^{(n)} \wedge T)\big)^4 \Big] + \sum_{j=1}^{n-1} \mathbb{E}\Big[\big(t_{j+1}^{(n)} \wedge T - t_j^{(n)} \wedge T\big)^2 \Big]. \qquad (7.3)$$

Using Brownian scaling on the first term, we see that this expression is bounded by a constant multiple of

$$\sum_{j=1}^{n-1} \big(t_{j+1}^{(n)} - t_j^{(n)}\big)^2 \leqslant t\, \Delta(n),$$

where $\Delta(n)$ denotes the mesh, which goes to zero. This completes the proof. ∎

We are now able to formulate and prove a first version of Itô's formula.

Theorem 7.13 (Itô's formula I) *Let* $f: \mathbb{R} \to \mathbb{R}$ *be twice continuously differentiable such that* $\mathbb{E} \int_0^t f'(B(s))^2 \, ds < \infty$ *for some* $t > 0$. *Then, almost surely, for all* $0 \leqslant s \leqslant t$,

$$f(B(s)) - f(B(0)) = \int_0^s f'(B(u)) \, dB(u) + \frac{1}{2} \int_0^s f''(B(u)) \, du.$$

Proof. We denote the modulus of continuity of f'' on $[-M, M]$ by

$$\omega(\delta, M) := \sup_{\substack{x, y \in [-M, M] \\ |x-y| < \delta}} f''(x) - f''(y).$$

Then, by Taylor's formula, for any $x, y \in [-M, M]$ with $|x - y| < \delta$,

$$\left| f(y) - f(x) - f'(x)(y-x) - \tfrac{1}{2} f''(x)(y-x)^2 \right| \leqslant \omega(\delta, M)(y-x)^2.$$

Now, for any sequence $0 = t_1 < \ldots < t_n = t$ with $\delta_B := \max_{1 \leqslant i \leqslant n-1} |B(t_{i+1}) - B(t_i)|$ and $M_B = \max_{0 \leqslant s \leqslant t} |B(s)|$, we get

$$\left| \sum_{i=1}^{n-1} \big(f(B(t_{i+1})) - f(B(t_i)) \big) - \sum_{i=1}^{n-1} f'(B(t_i)) \big(B(t_{i+1}) - B(t_i) \big) \right.$$

$$\left. - \sum_{i=1}^{n-1} \tfrac{1}{2} f''(B(t_i)) \big(B(t_{i+1}) - B(t_i) \big)^2 \right| \leqslant \omega(\delta_B, M_B) \sum_{i=1}^{n-1} \big(B(t_{i+1}) - B(t_i) \big)^2.$$

Note that the first sum is simply $f(B(t)) - f(B(0))$. By the definition of the stochastic integral and Theorem 7.12 we can choose a sequence of partitions with mesh going to zero, such that, almost surely, the first subtracted term on the left converges to $\int_0^t f'(B(s)) \, dB(s)$, the second subtracted term converges to $\frac{1}{2} \int_0^t f''(B(s)) \, ds$, and the sum on the right hand side converges to t. By continuity of the Brownian path $\omega(\delta_B, M_B)$ converges almost surely to zero. This proves Itô's formula for fixed t, or indeed almost surely for all rational times $0 \leqslant s \leqslant t$. As all the terms in Itô's formula are continuous almost surely, we get the result simultaneously for all $0 \leqslant s \leqslant t$. ∎

Next, we provide an enhanced version of Itô's formula, which allows the function f to depend not only on the position of Brownian motion, but also on a second argument, which is assumed to be increasing in time.

Theorem 7.14 (Itô's formula II) *Suppose* $\{\zeta(s): s \geqslant 0\}$ *is an increasing, continuous adapted stochastic process. Let* $f: \mathbb{R} \times \mathbb{R} \to \mathbb{R}$ *be twice continuously differentiable in the* x-*coordinate, and once continuously differentiable in the* y-*coordinate. Assume that*

$$\mathbb{E} \int_0^t \big[\partial_x f(B(s), \zeta(s)) \big]^2 \, ds < \infty,$$

for some $t > 0$. *Then, almost surely, for all* $0 \leqslant s \leqslant t$,

$$f(B(s), \zeta(s)) - f(B(0), \zeta(0)) = \int_0^s \partial_x f(B(u), \zeta(u)) \, dB(u)$$

$$+ \int_0^s \partial_y f(B(u), \zeta(u)) \, d\zeta(u) + \frac{1}{2} \int_0^s \partial_{xx} f(B(u), \zeta(u)) \, du.$$

Proof. To begin with, we inspect the proof of Theorem 7.12 and see that it carries over without difficulty to the situation, when f is allowed to depend additionally on an adapted process $\{\zeta(s): s \geqslant 0\}$, i.e. we have for any partitions $0 = t_1^{(n)} < \ldots < t_n^{(n)} = t$ with mesh going to zero, in probability,

$$\sum_{j=1}^{n-1} f\big(B\big(t_j^{(n)}\big), \zeta\big(t_j^{(n)}\big)\big) \big(B\big(t_{j+1}^{(n)}\big) - B\big(t_j^{(n)}\big)\big)^2 \longrightarrow \int_0^t f(B(s), \zeta(s))\, ds. \qquad (7.4)$$

We denote the modulus of continuity of $\partial_y f$ by

$$\omega_1(\delta, M) = \sup_{\substack{-M \leqslant x_1, x_2, y_1, y_2 \leqslant M \\ |x_1 - x_2| \vee |y_1 - y_2| < \delta}} \big|\partial_y f(x_1, y_1) - \partial_y f(x_2, y_2)\big|,$$

and the modulus of continuity of $\partial_{xx} f$ by

$$\omega_2(\delta, M) = \sup_{\substack{-M \leqslant x_1, x_2, y_1, y_2 \leqslant M \\ |x_1 - x_2| \vee |y_1 - y_2| < \delta}} \big|\partial_{xx} f(x_1, y_1) - \partial_{xx} f(x_2, y_2)\big|.$$

Now take $x, x_0, y, y_0 \in [-M, M]$ with $|x - x_0| \vee |y - y_0| < \delta$. By the mean value theorem, there exists a value $\tilde{y} \in [-M, M]$ with the property that $|\tilde{y} - y| \vee |\tilde{y} - y_0| < \delta$ such that

$$f(x, y) - f(x, y_0) = \partial_y f(x, \tilde{y})\, (y - y_0),$$

and hence

$$\big|f(x, y) - f(x, y_0) - \partial_y f(x_0, y_0)\, (y - y_0)\big| \leqslant \omega_1(M, \delta)\, (y - y_0).$$

Taylor's formula implies that

$$\big|f(x, y_0) - f(x_0, y_0) - \partial_x f(x_0, y_0)(x - x_0) - \tfrac{1}{2}\partial_{xx} f(x_0, y_0)(x - x_0)^2\big| \leqslant \omega_2(\delta, M)(x - x_0)^2.$$

Combining the last two formulas using the triangle inequality, we get that

$$\begin{aligned}
\big|f(x, y) - f(x_0, y_0) &- \partial_y f(x_0, y_0)\, (y - y_0) \\
&- \partial_x f(x_0, y_0)\, (x - x_0) - \tfrac{1}{2}\partial_{xx} f(x_0, y_0)(x - x_0)^2\big| \\
&\leqslant \omega_1(\delta, M)\, (y - y_0) + \omega_2(\delta, M)(x - x_0)^2.
\end{aligned} \qquad (7.5)$$

Now, for any sequence $0 = t_1 < \ldots < t_n = t$ define

$$\delta := \max_{1 \leqslant i \leqslant n-1} \big|B(t_{i+1}) - B(t_i)\big| \wedge \max_{1 \leqslant i \leqslant n-1} \big|\zeta(t_{i+1}) - \zeta(t_i)\big|,$$

and

$$M := \max_{0 \leqslant s \leqslant t} |B(s)| \wedge \max_{0 \leqslant s \leqslant t} |\zeta(s)|.$$

We get from (7.5),

$$\left| f\big(B(t), \zeta(t)\big) - f\big(B(0), \zeta(0)\big) - \sum_{i=1}^{n-1} \partial_x f\big(B(t_i), \zeta(t_i)\big) \big(B(t_{i+1}) - B(t_i)\big) \right.$$

$$- \sum_{i=1}^{n-1} \partial_y f\big(B(t_i), \zeta(t_i)\big) \big(\zeta(t_{i+1}) - \zeta(t_i)\big)$$

$$\left. - \tfrac{1}{2} \sum_{i=1}^{n-1} \partial_{xx} f\big(B(t_i), \zeta(t_i)\big) \big(B(t_{i+1}) - B(t_i)\big)^2 \right|$$

$$\leqslant \omega_1(\delta, M)\big(\zeta(t) - \zeta(0)\big) + \omega_2(\delta, M) \sum_{i=1}^{n-1} \big(B(t_{i+1}) - B(t_i)\big)^2.$$

We can choose a sequence of partitions with mesh going to zero, such that, almost surely, the following convergence statements hold,

- the first sum on the left converges to $\int_0^t \partial_x f\big(B(s), \zeta(s)\big) \, dB(s)$ by the definition of the stochastic integral,
- the second sum on the left converges to $\int_0^t \partial_y f\big(B(s), \zeta(s)\big) \, d\zeta(s)$ by definition of the Stieltjes integral,
- the third sum on the left converges to $\frac{1}{2} \int_0^t \partial_{xx} f\big(B(s), \zeta(s)\big) \, ds$ by (7.4),
- the sum on the right hand side converges to t by Theorem 7.12.

By continuity of the Brownian path $\omega_1(\delta, M)$ and $\omega_2(\delta, M)$ converge almost surely to zero. This proves the enhanced Itô's formula for fixed t, and looking at rationals and exploiting continuity as before, we get the result simultaneously for all $0 \leqslant s \leqslant t$. . ∎

With exactly the same technique, we obtain a version of Itô's formula for higher dimensional Brownian motion. The detailed proof will be an exercise, see Exercise 7.4. To give a pleasant formulation, we introduce some notation for functions $f \colon \mathbb{R}^{d+m} \to \mathbb{R}$, where we interpret the argument as two vectors, $x \in \mathbb{R}^d$ and $y \in \mathbb{R}^m$. We write ∂_j for the partial derivative in direction of the jth coordinate, and

$$\nabla_x f = (\partial_1 f, \dots, \partial_d f) \quad \text{and} \quad \nabla_y f = (\partial_{d+1} f, \dots, \partial_{d+m} f)$$

for the vector of derivatives in the directions of x, respectively y. For integrals we use the scalar product notation

$$\int_0^t \nabla_x f\big(B(u), \zeta(u)\big) \cdot dB(u) = \sum_{i=1}^d \int_0^t \partial_i f\big(B(u), \zeta(u)\big) \, dB_i(u),$$

and

$$\int_0^t \nabla_y f\big(B(u), \zeta(u)\big) \cdot d\zeta(u) = \sum_{i=1}^m \int_0^t \partial_{d+i} f\big(B(u), \zeta(u)\big) \, d\zeta_i(u).$$

Finally, for the Laplacian in the x-variable we write

$$\Delta_x f = \sum_{j=1}^d \partial_{jj} f.$$

Theorem 7.15 (Multidimensional Itô's formula) *Let $\{B(t): t \geqslant 0\}$ be a d-dimensional Brownian motion and suppose $\{\zeta(s): s \geqslant 0\}$ is a continuous, adapted stochastic process with values in $\mathbb{R}^m$ and increasing components. Let $f: \mathbb{R}^{d+m} \to \mathbb{R}$ be such that the partial derivatives $\partial_i f$ and $\partial_{jk} f$ exist for all $1 \leqslant j, k \leqslant d, d+1 \leqslant i \leqslant d+m$ and are continuous. If, for some $t > 0$,*

$$\mathbb{E} \int_0^t \left| \nabla_x f\big(B(s), \zeta(s)\big) \right|^2 ds < \infty,$$

then, almost surely, for all $0 \leqslant s \leqslant t$,

$$\begin{aligned}
f\big(B(s), \zeta(s)\big) - f\big(B(0), \zeta(0)\big) &= \int_0^s \nabla_x f\big(B(u), \zeta(u)\big) \cdot dB(u) \\
&+ \int_0^s \nabla_y f\big(B(u), \zeta(u)\big) \cdot d\zeta(u) + \tfrac{1}{2} \int_0^s \Delta_x f\big(B(u), \zeta(u)\big) \, du.
\end{aligned} \tag{7.6}$$

Remark 7.16 As the Itô formula holds almost surely simultaneously for all times $s \in [0, t]$, it also holds for stopping times bounded by t. Suppose now that $f: U \to \mathbb{R}$ satisfies the differentiability conditions on an open set U, and $K \subset U$ is compact. Take a smooth function $g: \mathbb{R}^m \to [0, 1]$ with compact support inside U, such that $g \equiv 1$ on K. Then $f^* = fg: \mathbb{R}^m \to \mathbb{R}$ satisfies $f^* = f$ on K and all relevant derivatives are bounded, so that the conditions of Theorem 7.15 are satisfied. Let T be the first exit time from K. Applying Theorem 7.15 to f^* yields (7.6) for f, almost surely, for all times $s \wedge T$, for $s \leqslant t$. ◇

To appreciate the following discussion, we introduce a localisation of the notion of a martingale.

Definition 7.17. An adapted stochastic process $\{X(t): 0 \leqslant t \leqslant T\}$ is called a **local martingale** if there exist stopping times T_n, which are almost surely increasing to T, such that $\{X(t \wedge T_n): t \geqslant 0\}$ is a martingale, for every n. ◇

The following theorem is a substantial extension of Corollary 2.53.

Theorem 7.18 *Let $D \subset \mathbb{R}^d$ be a domain and $f: D \to \mathbb{R}$ be harmonic on D. Suppose that $\{B(t): 0 \leqslant t \leqslant T\}$ is a Brownian motion started inside D and stopped at the time T when it first exits the domain D.*

(a) The process $\{f(B(t)): 0 \leqslant t \leqslant T\}$ is a local martingale.

(b) If we have

$$\mathbb{E} \int_0^{t \wedge T} \left| \nabla f(B(s)) \right|^2 ds < \infty \qquad \text{for all } t > 0,$$

then $\{f(B(t \wedge T)): t \geqslant 0\}$ is a martingale.

Proof. Suppose that K_n, $n \in \mathbb{N}$, is an increasing sequence of compact sets whose union is D, and let T_n be the associated exit times. By Theorem 7.15 in conjunction with Remark 7.16,

$$f\big(B(t \wedge T_n)\big) = f\big(B(0)\big) + \int_0^{t \wedge T_n} \nabla f\big(B(s)\big) \cdot dB(s),$$

whence $\{f(B(t \wedge T_n)): t \geqslant 0\}$ is a martingale, which proves (a).
Obviously, almost surely,

$$f(B(t \wedge T)) = \lim_{n \uparrow \infty} f(B(t \wedge T_n)). \tag{7.7}$$

For any $t \geqslant 0$, the process $\{f(B(t \wedge T_n)): n \in \mathbb{N}\}$ is a discrete-time martingale by the optional stopping theorem. By our integrability assumption,

$$\mathbb{E}\Big[f\big(B(t \wedge T_n)\big)^2\Big] = \mathbb{E} \int_0^{T_n \wedge t} |\nabla f(B(s))|^2 \, ds \leqslant \mathbb{E} \int_0^{T \wedge t} |\nabla f(B(s))|^2 \, ds < \infty,$$

so that the martingale is $\mathbf{L}^2$-bounded and convergence in (7.7) holds in the $\mathbf{L}^1$-sense. Taking limits in the equation

$$\mathbb{E}\big[f\big(B(t \wedge T_m)\big) \mid \mathcal{F}(s \wedge T_n)\big] = f\big(B(s \wedge T_n)\big), \qquad \text{for } m \geqslant n \text{ and } t \geqslant s,$$

first for $m \uparrow \infty$, then $n \uparrow \infty$, gives

$$\mathbb{E}\big[f\big(B(t \wedge T)\big) \mid \mathcal{F}(s \wedge T)\big] = f\big(B(s \wedge T)\big), \qquad \text{for } t \geqslant s.$$

This shows that $\{f(B(t \wedge T)): t \geqslant 0\}$ is a martingale and completes the proof. ∎

Example 7.19 The radially symmetric functions (related to the radial potential),

$$f(x) = \begin{cases} \log|x| & \text{if } d = 2, \\ |x|^{2-d} & \text{if } d \geqslant 3. \end{cases}$$

are harmonic on the domain $\mathbb{R}^d \setminus \{0\}$. For a d-dimensional Brownian motion $\{B(t): t \geqslant 0\}$ with $B(0) \neq 0$, the process $\{f(B(t)): t \geqslant 0\}$ is however not a martingale. Indeed, it is a straightforward calculation to verify that

$$\lim_{t \uparrow \infty} \mathbb{E} \log |B(t)| = \infty, \qquad \text{if } d = 2,$$

and

$$\lim_{t \uparrow \infty} \mathbb{E}[|B(t)|^{2-d}] = 0, \qquad \text{if } d \geqslant 3,$$

contradicting the martingale property. Hence the integrability condition in Theorem 7.18(b) cannot be dropped without replacement, in other words a local martingale is not necessarily a martingale. ◇

7.2 Conformal invariance and winding numbers

We now focus on planar Brownian motion $\{B(t): t \geqslant 0\}$ and formulate an invariance property which is at the heart of the rôle of Brownian motion in the context of planar random curves. Throughout this section we use the identification of $\mathbb{R}^2$ and $\mathbb{C}$ and use complex notation when it is convenient.

To motivate the main result suppose that $f \colon \mathbb{C} \to \mathbb{C}$ is **analytic**, i.e. everywhere complex differentiable, and write $f = f_1 + \mathrm{i} f_2$ for the decomposition of f into a real and an imaginary part. Then, by the Cauchy–Riemann equations $\partial_1 f_1 = \partial_2 f_2$ and $\partial_2 f_1 = -\partial_1 f_2$, we have $\Delta f_1 = \Delta f_2 = 0$. Then Itô's formula (if applicable) states that almost surely, for every $t \geqslant 0$,

$$f\bigl(B(t)\bigr) = \int_0^t f'\bigl(B(s)\bigr)\, dB(s),$$

where $dB(s)$ is short for $dB_1(s) + \mathrm{i}\, dB_2(s)$ with $B(s) = B_1(s) + \mathrm{i} B_2(s)$. The right hand side defines a continuous process with independent increments, and it is at least plausible that they are Gaussian. Moreover, its expectation vanishes and

$$\mathbb{E}\left[\left(\int_0^t f'\bigl(B(s)\bigr)\, dB(s)\right)^2\right] = \mathbb{E}\int_0^t \bigl|f'\bigl(B(s)\bigr)\bigr|^2\, ds,$$

suggesting that $\{f(B(t)) \colon t \geqslant 0\}$ is a Brownian motion 'travelling' with the modified speed

$$t \mapsto \int_0^t |f'(B(s))|^2\, ds.$$

To turn this heuristic into a powerful theorem we allow the function to be an analytic map $f \colon U \to V$ between domains in the plane. Recall that such a map is called **conformal** if it is a bijection.

Theorem 7.20 *Let U be a domain in the complex plane, $x \in U$, and let $f \colon U \to V$ be analytic. Let $\{B(t) \colon t \geqslant 0\}$ be a planar Brownian motion started in x and*

$$\tau_U = \inf\bigl\{t \geqslant 0 \colon B(t) \notin U\bigr\}$$

its first exit time from the domain U. Then the process $\{f(B(t)) \colon 0 \leqslant t \leqslant \tau_U\}$ is a time-changed Brownian motion, i.e. there exists a planar Brownian motion $\{\widetilde{B}(t) \colon t \geqslant 0\}$ such that, for any $t \in [0, \tau_U)$,

$$f\bigl(B(t)\bigr) = \widetilde{B}(\zeta(t)), \qquad \text{where} \qquad \zeta(t) = \int_0^t \bigl|f'\bigl(B(s)\bigr)\bigr|^2\, ds.$$

If, additionally, f is conformal, then $\zeta(\tau_U)$ is the first exit time from V by $\{\widetilde{B}(t) \colon t \geqslant 0\}$.

Remark 7.21 Note that, as f is complex differentiable, the derivative $Df(x)$ is just multiplication by a complex number $f'(x)$, and f can be approximated locally around x by its tangent $z \mapsto f(x) + f'(x)(z - x)$. The derivative of the time change is

$$\partial_t \zeta(t) = |f'(B(t))|^2 = \bigl(\partial_1 f_1(B(t))\bigr)^2 + \bigl(\partial_2 f_1(B(t))\bigr)^2. \qquad \diamond$$

Remark 7.22 The famous *Riemann mapping theorem* states that for any pair of simply connected open sets $U, V \subsetneq \mathbb{C}$ there exists a conformal mapping $f \colon U \to V$, see, e.g., [Ru87] or [Ah78]. This ensures that there are plenty of examples for Theorem 7.20. $\qquad \diamond$

Proof. Note first that the derivative of f is nonzero except for an at most countable set of points, which does not have a limit point in U. As this set is not hit by Brownian motion, we may remove it from U and note that the resulting set is still open. We may therefore assume that f has nonvanishing derivative everywhere on U.

We may also assume, without loss of generality, that f is a mapping between *bounded* domains. Otherwise choose $U_n \subset K_n \subset U$ such that U_n is open with $\bigcup U_n = U$ and K_n is compact, which implies that $V_n = f(U_n)$ is bounded. Then the process $\{f(B(t)): t \leqslant \tau_{U_n}\}$ is a time-changed Brownian motion for all n, and this extends immediately to the process $\{f(B(t)): t \leqslant \tau_U\}$.

The main argument of the proof is based on stochastic integration. Recall that the Cauchy–Riemann equations imply that the vectors ∇f_1 and ∇f_2 are orthogonal and $|\nabla f_1| = |\nabla f_2| = |f'|$. We start by defining for each $t \geqslant 0$, a stopping time

$$\sigma(t) = \inf \{s \geqslant 0 \,:\, \zeta(s) \geqslant t\},$$

which represents the inverse of the time change. Let $\{\widetilde{B}(t): t \geqslant 0\}$ be a Brownian motion independent of $\{B(t): t \geqslant 0\}$, and define a process $\{W(t): t \geqslant 0\}$ by

$$W(t) = f\big(B(\sigma(t) \wedge \tau_U)\big) + \widetilde{B}(t) - \widetilde{B}(t \wedge \zeta(\tau_U)), \quad \text{for } t \geqslant 0.$$

In rough words, at the random time $\zeta(\tau_U)$ an independent Brownian motion is attached at the endpoint of the process $\{f(B(\sigma(t))): 0 \leqslant t \leqslant \zeta(\tau_U)\}$. Denote by $\mathcal{G}(t)$ the σ-algebra generated by $\{W(s): s \leqslant t\}$. It suffices to prove that the process $\{W(t): t \geqslant 0\}$ is a Brownian motion.

It is obvious that the process is continuous almost surely and hence it suffices to show that its finite dimensional distributions coincide with those of a Brownian motion. Recalling the Laplace transform of the bivariate normal distribution, this is equivalent to showing that, for any $0 \leqslant s \leqslant t$ and $\lambda \in \mathbb{C}$,

$$\mathbb{E}\big[e^{\langle \lambda, W(t)\rangle} \,\big|\, \mathcal{G}(s)\big] = \exp\big(\tfrac{1}{2}|\lambda|^2(t-s) + \langle \lambda, W(s)\rangle\big).$$

where we have used $\langle \cdot, \cdot \rangle$ to denote the scalar product. This follows directly once we show that, for $x \in U$,

$$\mathbb{E}\big[e^{\langle \lambda, W(t)\rangle} \,\big|\, W(s) = f(x)\big] = \exp\big(\tfrac{1}{2}|\lambda|^2(t-s) + \langle \lambda, f(x)\rangle\big). \qquad (7.8)$$

For simplicity of notation we may assume $s = 0$. For the proof we first evaluate the expectation with respect to the independent Brownian motion $\{\widetilde{B}(t): t \geqslant 0\}$ inside, which gives

$$\mathbb{E}\big[e^{\langle \lambda, W(t)\rangle} \,\big|\, W(0) = f(x)\big]$$
$$= \mathbb{E}_x \exp\big(\langle \lambda, f(B(\sigma(t) \wedge \tau_U))\rangle + \tfrac{1}{2}|\lambda|^2\,(t - \zeta(\sigma(t) \wedge \tau_U))\big).$$

We use the multidimensional Itô's formula for the bounded mapping

$$F(x, u) = \exp\big(\langle \lambda, f(x)\rangle + \tfrac{1}{2}|\lambda|^2(t - u)\big),$$

which is defined on $U \times (-1, \infty)$, see Remark 7.16. To prepare this, note that $\partial_{ii} e^g = [\partial_{ii} g + (\partial_i g)^2 e^g]$ and hence

$$\Delta e^g = \big[\Delta g + |\nabla g|^2\big] e^g. \qquad (7.9)$$

For $g = \langle \lambda, f \rangle$ we have $\nabla g = \sum_{i=1}^{2} \lambda_i \nabla f_i$, which implies $|\nabla g|^2 = |\lambda|^2 \, |f'|^2$ as the vectors ∇f_i are orthogonal with norm $|f'|$. Moreover, $\Delta g = 0$ by the analyticity of f. Applying (7.9) gives

$$\Delta \, e^{\langle \lambda, f(x) \rangle} = |\lambda|^2 \, |f'(x)|^2 \, e^{\langle \lambda, f(x) \rangle} .$$

Moreover, we have

$$\partial_u \exp \left(\tfrac{1}{2} |\lambda|^2 \, (t - u) \right) = -\tfrac{1}{2} |\lambda|^2 \, \exp \left(\tfrac{1}{2} |\lambda|^2 \, (t - u) \right) .$$

We now let $U_n = \{ x \in U : |x - y| \geqslant \frac{1}{n} \text{ for all } y \in \partial U \}$. Then $|f'(x)|$ is bounded away from zero on U_n and therefore the stopping time $T = \sigma(t) \wedge \tau_{U_n}$ is bounded. The multidimensional version of Itô's formula gives, almost surely,

$$F\big(B(T), \zeta(T)\big) = F\big(B(0), \zeta(0)\big) + \int_0^T \nabla_x F\big(B(s), \zeta(s)\big) \cdot dB(s)$$

$$+ \int_0^T \partial_u F\big(B(s), \zeta(s)\big) \, d\zeta(s) + \tfrac{1}{2} \int_0^T \Delta_x F\big(B(s), \zeta(s)\big) \, ds .$$

Looking back at the two preparatory displays and recalling that $d\zeta(u) = |f'(B(u))|^2 \, du$ we see that the two terms in the second line cancel each other. Making use of bounded convergence and the fact that the stochastic integral has zero expectation, see Exercise 7.2, we obtain that

$$\mathbb{E} \big[e^{\langle \lambda, W(t) \rangle} \mid W(0) = f(x) \big] = \mathbb{E}_x \big[F\big(B(\sigma(t) \wedge \tau_U), \zeta(\sigma(t) \wedge \tau_U)\big) \big]$$

$$= \lim_{n \to \infty} \mathbb{E}_x \big[F\big(B(T), \zeta(T)\big) \big] = F(x, 0) = \exp \left(\tfrac{1}{2} |\lambda|^2 t + \langle \lambda, f(x) \rangle \right) .$$

This shows (7.8) and thus completes the proof of the main statement. It remains to note that, if f is conformal then as $t \uparrow \tau_u$ the point $f(B(t))$ converges to a boundary point of V. Hence $\zeta(\tau_U)$ is the first exit time from V by the process $\{ \widetilde{B}(t) : t \geqslant 0 \}$. ∎

As a first application we look at harmonic measure and exploit its conformal invariance in order to give an explicit formula in an interesting special case.

Theorem 7.23 *Suppose $U, V \subset \mathbb{R}^2$ are domains and $f : \bar{U} \to \bar{V}$ is continuous and maps U conformally into V.*

(a) *If $x \in U$, then $\mu_{\partial U}(x, \cdot) \circ f^{-1} = \mu_{\partial V}(f(x), \cdot)$.*

(b) *Suppose additionally that $U = K^c$ and $V = L^c$ are the complements of nonpolar compact sets and $\lim_{x \to \infty} f(x) = \infty$. Then*

$$\mu_K \circ f^{-1} = \mu_L .$$

Proof.   (a) follows from Theorem 7.20 together with the continuity of f on $\bar{U}$, which ensures that the first hitting point of ∂U by a Brownian motion is mapped onto the first hitting point of ∂V by its conformal image. For (b) take the limit $x \to \infty$ and recall Theorem 3.46. ∎

Example 7.24 We find the harmonic measure from infinity on the unit interval

$$[0,1] = \{x + iy \colon y = 0, 0 \leqslant x \leqslant 1\}.$$

The starting point is the harmonic measure on the circle $\partial \mathcal{B}(0,1)$, which we know is the uniform distribution ϖ. Let U be the complement of the unit ball $\mathcal{B}(0,1)$ and V the complement of the interval $[-1,1]$, and take the conformal mapping

$$f \colon U \to V, \quad f(z) = \frac{1}{2}\left(z + \frac{1}{z}\right),$$

which satisfies our conditions. Hence $\varpi \circ f^{-1}$ is the harmonic measure on $[-1,1]$. If $z = x + iy = \cos\theta + i\sin\theta \in \partial \mathcal{B}(0,1)$, then $|f'(z)|^2 = \sin^2\theta$, and hence $|f'(z)| = |y| = \sqrt{1 - x^2}$. Recalling that every $x \in [-1,1]$ has two preimages, we get that the density of $\varpi \circ f^{-1}$ at $x = \cos\theta$ is

$$\frac{2}{2\pi|f'(e^{i\theta})|} = \frac{1}{\pi}\frac{1}{\sqrt{1 - x^2}}.$$

Mapping V via $z \mapsto z^2$ onto the complement of $[0,1]$, noting that $|f'(z)| = 2|z|$ and that again we have two preimages, we obtain that the harmonic measure on $[0,1]$ is

$$d\mu_{[0,1]}(x) = \frac{1}{\pi}\frac{1}{\sqrt{x(1-x)}}\,dx,$$

which is the Beta$(\frac{1}{2}, \frac{1}{2})$ distribution. ◇

As a further important application of conformal invariance we calculate the probability that a planar Brownian motion exits a cone before leaving a disc, see Figure 7.1.

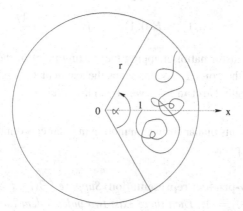

Fig. 7.1. The Brownian path does not exit the cone before leaving the disc.

Theorem 7.25 *Let $\alpha \in (0, 2\pi]$ and denote by $W[\alpha]$ an open cone with vertex in the origin, symmetric about the x-axis, with opening angle α. Let $\{B(t) \colon t \geqslant 0\}$ be planar Brownian motion started in $x = (1, 0)$, and denote $T(r) = \inf\{t \geqslant 0 \colon |B(t)| = r\}$. Then, for $r > 1$,*

$$\mathbb{P}\{B[0, T(r)] \subset W[\alpha]\} = \frac{2}{\pi}\arctan\left(\frac{2r^{\frac{\pi}{\alpha}}}{r^{\frac{2\pi}{\alpha}} - 1}\right).$$

Proof. For ease of notation we identify $\mathbb{R}^2$ with the complex plane. In the first step we use the conformal map $f: W[\alpha] \rightarrow W[\pi]$ defined by $f(x) = x^{\pi/\alpha}$ to map the cone onto a halfspace. Let $B^* = f \circ B$, which by conformal invariance is a time-changed Brownian motion started in the point $B^*(0) = 1$. We thus have that

$$\{B[0, T(r)] \subset W[\alpha]\} = \{B^*[0, T(r^{\pi/\alpha})] \subset W[\pi]\}.$$

It therefore suffices to show the result in the case $\alpha = \pi$. So let $\{B(t): t \geqslant 0\}$ be a Brownian motion started in $B(0) = 1$ and look at the stopping time $S = \min\{t \geqslant 0: \mathfrak{Re}(B(t)) \leqslant 0\}$. We use reflection on the imaginary axis, i.e. for $f(x, y) = (-x, y)$ we let

$$\widetilde{B}(t) = \begin{cases} B(t) & \text{if } t \leqslant S, \\ f(B(t)) & \text{if } t \geqslant S. \end{cases}$$

Then $\widetilde{B}$ is a Brownian motion started in $\widetilde{B}(0) = 1$ and, denoting $\widetilde{T}(r) = \inf\{t \geqslant 0: |\widetilde{B}(t)| = r\}$, we have

$$\mathbb{P}\{\mathfrak{Re}(B(T(r))) > 0\}$$
$$= \mathbb{P}\{\mathfrak{Re}(B(T(r))) > 0, T(r) < S\} + \mathbb{P}\{\mathfrak{Re}(B(T(r))) > 0, T(r) > S\}$$
$$= \mathbb{P}\{T(r) < S\} + \mathbb{P}\{\mathfrak{Re}(\widetilde{B}(\widetilde{T}(r))) < 0\}.$$

As $\{T(r) < S\}$ is the event whose probability we need to bound, it just remains to find

$$\mathbb{P}\{\mathfrak{Re}(B(T(r))) > 0\} - \mathbb{P}\{\mathfrak{Re}(B(T(r))) < 0\}.$$

By Brownian scaling we may assume that the Brownian motion is started at $B(0) = 1/r$ and $T = \min\{t \geqslant 0: |B(t)| = 1\}$. We apply the conformal map

$$f: \mathcal{B}(0, 1) \rightarrow \mathcal{B}(0, 1), \qquad f(z) = \frac{z - 1/r}{1 - z/r},$$

which is a Möbius transformation mapping the starting point of the Brownian motion to the origin and fixing the point 1. As this maps the segment $\{z \in \partial \mathcal{B}(0, 1): \mathfrak{Re}(z) < 0\}$ onto a segment of length $2 \arctan \frac{r^2 - 1}{2r}$ we obtain the result. ∎

The next result represents planar Brownian motion in polar coordinates. Again we identify $\mathbb{R}^2$ with the complex plane.

Theorem 7.26 (Skew-product representation) *Suppose $\{B(t): t \geqslant 0\}$ is a planar Brownian motion with $B(0) = 1$. Then there exist two independent linear Brownian motions $\{W_i(t): t \geqslant 0\}$, for $i = 1, 2$, such that*

$$B(t) = \exp\left(W_1(H(t)) + \mathrm{i}\, W_2(H(t))\right), \text{ for all } t \geqslant 0,$$

where

$$H(t) = \int_0^t \frac{ds}{|B(s)|^2} = \inf\left\{u \geqslant 0: \int_0^u \exp(2W_1(s))\, ds > t\right\}.$$

Remark 7.27 By the result, both the logarithm of the radius, and the continuous determination of the angle of a planar Brownian motion are time-changed Brownian motions.

The time-change itself depends only on the radius of the motion and ensures that the angle changes slowly away from the origin, but rapidly near the origin. ◇

Proof. Note first that $H(t)$ itself is well-defined by Corollary 2.26. Moreover, the claimed equality for $H(t)$ follows easily from the fact that both sides have the same value at $t = 0$ and the same derivative.

As the continuous processes $\{W_1(t): t \geqslant 0\}$ and $\{W_2(t): t \geqslant 0\}$ can be constructed uniquely from $\{B(t): t \geqslant 0\}$ and vice versa we may start with a planar Brownian motion $\{W(t): t \geqslant 0\}$ and let $W(t) = W_1(t) + i W_2(t)$ be its decomposition into real and imaginary part. It suffices to show that the process $\{B(t): t \geqslant 0\}$ constructed from this pair of linear Brownian motions is a planar Brownian motion. By Theorem 7.20,

$$\exp\big(W(t)\big) = B(\zeta(t)), \tag{7.10}$$

where $\{B(t): t \geqslant 0\}$ is a planar Brownian motion and

$$\zeta(t) = \int_0^t \exp(2W_1(s))\, ds.$$

By definition H is the inverse function of ζ. Hence, using (7.10) for $t = H(s)$, we get

$$B(s) = \exp\big(W(H(s))\big) = \exp\big(W_1(H(s)) + i W_2(H(s))\big),$$

which is the desired result. ∎

Example 7.28 By the skew-product representation, for a planar Brownian motion $\{B(t): t \geqslant 0\}$, we have $\log|B(t)| = W_1(H(t))$ and hence the process $\{\log|B(t)| : t \geqslant 0\}$ is a time-changed Brownian motion in dimension one. However, recall from Example 7.19 that it is *not* a martingale. ◇

For further applications, we need to study the asymptotics of the random clock $H(t)$ more carefully. To state the next result let $\{W_1(t): t \geqslant 0\}$ be a linear Brownian motion as in Theorem 7.26 and, for $a > 0$, let $\{W_1^a(t): t \geqslant 0\}$ be the Brownian motion given by $W_1^a(t) = a^{-1}W_1(a^2 t)$. For each such Brownian motion we look at the first hitting time of level b, defined as $T_b^a = \inf\{t \geqslant 0 : W_1^a(t) = b\}$.

Theorem 7.29 *For every $\varepsilon > 0$ we have*

$$\lim_{t \to \infty} \mathbb{P}\left\{\left|\frac{4H(t)}{(\log t)^2} - T_1^{\frac{1}{2}\log t}\right| > \varepsilon\right\} = 0.$$

The proof uses the following simple fact, sometimes known as *Laplace's method*.

Lemma 7.30 *For any continuous $f : [0, t] \to \mathbb{R}$ and $t > 0$,*

$$\lim_{a \uparrow \infty} \frac{1}{a} \log \int_0^t \exp(af(v))\, dv = \max_{0 \leqslant s \leqslant t} f(s).$$

Proof.   The upper bound is obvious, by replacing f by its maximum. For the lower bound, let $s \in [0, t]$ be a point where the maximum is taken. We use continuity to find, for any $\varepsilon > 0$, some $0 < \delta < 1$ such that $f(r) \geqslant f(s) - \varepsilon$ for all $r \in (s - \delta, s + \delta)$. Restricting the limit to this interval gives a lower bound of $\max_{0 \leqslant s \leqslant t} f(s) - \varepsilon$, and the result follows as $\varepsilon > 0$ was arbitrary.   ∎

Proof of Theorem 7.29.   Recall that $W_1(0) = 0$. We abbreviate $a = a(t) = \frac{1}{2} \log t$. As we have, for any $\delta > 0$,

$$\lim_{\varepsilon \downarrow 0} \mathbb{P}\left\{ T_{1+\varepsilon}^{\frac{1}{2} \log t} - T_{1-\varepsilon}^{\frac{1}{2} \log t} > \delta \right\} = \lim_{\varepsilon \downarrow 0} \mathbb{P}\left\{ T_{1+\varepsilon}^1 - T_{1-\varepsilon}^1 > \delta \right\} = 0$$

it suffices to show that

$$\lim_{t \uparrow \infty} \mathbb{P}\left\{ \frac{4H(t)}{(\log t)^2} > T_{1+\varepsilon}^{\frac{1}{2} \log t} \right\} = 0, \quad \text{and} \quad \lim_{t \uparrow \infty} \mathbb{P}\left\{ \frac{4H(t)}{(\log t)^2} < T_{1-\varepsilon}^{\frac{1}{2} \log t} \right\} = 0.$$

We first show that

$$\lim_{t \uparrow \infty} \mathbb{P}\left\{ \frac{4H(t)}{(\log t)^2} > T_{1+\varepsilon}^{\frac{1}{2} \log t} \right\} = 0. \tag{7.11}$$

We have

$$\left\{ \frac{4H(t)}{(\log t)^2} > T_{1+\varepsilon}^{\frac{1}{2} \log t} \right\} = \left\{ \int_0^{a^2 T_{1+\varepsilon}^a} \exp(2W_1(u)) \, du < t \right\}$$

$$= \left\{ \frac{1}{2a} \log \int_0^{a^2 T_{1+\varepsilon}^a} \exp(2W_1(u)) \, du < 1 \right\},$$

recalling that $a = \frac{1}{2} \log t$. Note now that

$$\frac{1}{2a} \log \int_0^{a^2 T_{1+\varepsilon}^a} \exp(2W_1(u)) \, du = \frac{\log a}{a} + \frac{1}{2a} \log \int_0^{T_{1+\varepsilon}^a} \exp(2aW_1^a(u)) \, du,$$

and the right hand side has the same distribution as

$$\frac{\log a}{a} + \frac{1}{2a} \log \int_0^{T_{1+\varepsilon}^1} \exp(2aW_1(u)) \, du.$$

Laplace's method gives that, almost surely,

$$\lim_{a \uparrow \infty} \frac{1}{2a} \log \int_0^{T_{1+\varepsilon}^1} \exp(2aW_1(u)) \, du = \sup_{0 \leqslant s \leqslant T_{1+\varepsilon}^1} W_1(s) = 1 + \varepsilon.$$

Hence,

$$\lim_{a \uparrow \infty} \mathbb{P}\left\{ \left| \frac{\log a}{a} + \frac{1}{2a} \log \int_0^{T_{1+\varepsilon}^1} \exp(2aW_1(u)) \, du - (1 + \varepsilon) \right| > \varepsilon \right\} = 0.$$

This proves (7.11). In the same way one can show that

$$\lim_{t \uparrow \infty} \mathbb{P}\left\{ \frac{4H(t)}{(\log t)^2} < T_{1-\varepsilon}^{\frac{1}{2} \log t} \right\} = 0,$$

and this completes the proof.   ∎

Remark 7.31 As $\{W_1^a(t)\colon t \geqslant 0\}$ is a Brownian motion for every $a > 0$, the law of T_1^a does not depend on $a > 0$. Therefore, Theorem 7.29 implies that

$$\frac{4H(t)}{(\log t)^2} \xrightarrow{\mathrm{d}} T_1,$$

where $T_1 = \inf\{s \geqslant 0\colon W(s) = 1\}$. The distribution of T_1 is, by Theorem 2.35 given by the density $(2\pi s^3)^{-1/2} \exp(-1/(2s))$.   ◇

We are now able to determine the asymptotic law of the winding numbers $\theta(t) = W_2(H(t))$ of a planar Brownian motion, as $t \to \infty$.

Theorem 7.32 (Spitzer's law) *For any $x \in \mathbb{R}$,*

$$\lim_{t \to \infty} \mathbb{P}\Big\{\frac{2}{\log t} \, \theta(t) \leqslant x\Big\} = \int_{-\infty}^{x} \frac{dy}{\pi(1 + y^2)}.$$

In other words, the law of $\frac{2\theta(t)}{\log t}$ converges to a standard symmetric Cauchy distribution.

Proof.   We define $\{W_2^a(t)\colon t \geqslant 0\}$ by $W_2^a(t) = (1/a)W_2(a^2 t)$. Then,

$$a^{-1}\theta(t) = a^{-1} W_2(H(t)) = W_2^a(a^{-2} H(t)).$$

By Theorem 7.29 and the uniform continuity of $\{W_2^a(t)\colon t \geqslant 0\}$ on compact sets we get, for $a = a(t) = \frac{1}{2}\log t$,

$$\lim_{t \to \infty} \mathbb{P}\Big\{\Big|\frac{2\theta(t)}{\log t} - W_2^a\big(T_1^a\big)\Big| > \varepsilon\Big\} = \lim_{t \to \infty} \mathbb{P}\Big\{\Big|W_2^a\big(\tfrac{4H(t)}{(\log t)^2}\big) - W_2^a\big(T_1^a\big)\Big| > \varepsilon\Big\} = 0.$$

The law of the random variable $W_2^a(T_1^a)$ does not depend on the choice of a. By Theorem 2.37, see also Exercise 7.5, it is Cauchy distributed.   ■

7.3 Tanaka's formula and Brownian local time

In this section we establish a deep connection between Itô's formula and Brownian local times for linear Brownian motion $\{B(t)\colon t \geqslant 0\}$. The basic idea is to give an analogue of Itô's formula for the function $f\colon \mathbb{R} \to \mathbb{R}$, $f(t) = |t - a|$. Note that this function is not twice continuously differentiable, so Itô's formula cannot be applied directly.

To see what we are aiming at, let's apply Itô's formula informally. We have in the distributional sense that $f'(x) = \mathrm{sign}(x - a)$ and $f''(x) = 2\delta_a$. Hence Itô's formula would give

$$|B(t) - a| - |B(0) - a| = \int_0^t \mathrm{sign}(B(s) - a)\, dB(s) + \int_0^t \delta_a(B(s))\, ds,$$

The last integral can be interpreted as the time spent by Brownian motion at level a and hence it is natural to expect that it is the local time $L^a(t)$. Tanaka's formula confirms this intuition.

Theorem 7.33 (Tanaka's formula) *Let $\{B(t) : t \geqslant 0\}$ be linear Brownian motion. Then, for every $a \in \mathbb{R}$, almost surely, for all $t > 0$,*

$$|B(t) - a| - |B(0) - a| = \int_0^t \operatorname{sign}(B(s) - a)\, dB(s) + L^a(t),$$

where $\operatorname{sign} x = 1_{\{x > 0\}} - 1_{\{x < 0\}}$.

Remark 7.34 There is an easy analogue of Tanaka's formula for simple random walk on the integers, see Exercise 7.8.   ⋄

Tanaka's formula can be used to generalise Itô's formula to functions which are not twice continuously differentiable.

Corollary 7.35 *Suppose that $f \colon \mathbb{R} \to \mathbb{R}$ is twice differentiable such that f' has compact support, but do not assume that f'' is continuous. Then*

$$f(B(t)) - f(B(0)) = \int_0^t f'(B(s))\, dB(s) + \tfrac{1}{2} \int_0^t f''(B(s))\, ds \,.$$

Proof.   Under our assumptions on f there exist constants b, c such that

$$f'(x) = \tfrac{1}{2} \int \operatorname{sign}(x - a)\, f''(a)\, da + c \text{ and } f(x) = \tfrac{1}{2} \int |x - a|\, f''(a)\, da + cx + b.$$

Integrating Tanaka's formula with respect to $\tfrac{1}{2} f''(a)\, da$ and exchanging this integral with the stochastic integral, which is justified by Exercise 7.9, gives

$$f(B(t)) - f(B(0)) = \int_0^t f'(B(s))\, dB(s) + \tfrac{1}{2} \int L^a(t)\, f''(a)\, da \,.$$

By Theorem 6.18 the last term equals $\tfrac{1}{2} \int_0^t f''(B(s))\, ds$.   ∎

For the proof of Tanaka's formula we define, for fixed $a \in \mathbb{R}$,

$$\tilde{L}^a(t) := |B(t) - a| - |B(0) - a| - \int_0^t \operatorname{sign}(B(s) - a)\, dB(s) \quad \text{for } t \geqslant 0,$$

and show that this represents the density at point a of the occupation measure.

Lemma 7.36 *For every $t \geqslant 0$ and $a \in \mathbb{R}$,*

$$\tilde{L}^a(t) = \lim_{\varepsilon \downarrow 0} \frac{1}{\varepsilon} \int_0^t 1_{(a, a + \varepsilon)}(B(s))\, ds, \quad \text{in probability.}$$

Proof.   Using the strong Markov property the statement can be reduced to the case $a = 0$. The main idea of the proof is now to use convolution to make $|x|$ smooth, and then use Itô's formula for the smooth function. For this purpose, recall that, for any $\delta > 0$ we can find smooth functions $g, h \colon \mathbb{R} \to [0, 1]$ with compact support such that $g \leqslant 1_{(0,1)} \leqslant h$ and $\int g = 1 - \delta$, $\int h = 1 + \delta$. This reduces the problem to showing that

$$\tilde{L}^0(t) = \lim_{\varepsilon \downarrow 0} \frac{1}{\varepsilon} \int_0^t f(\varepsilon^{-1} B(s))\, ds, \quad \text{in probability,}$$

for $f \colon \mathbb{R} \to [0,1]$ smooth, with compact support in $[-1, 2]$ and $\int f = 1$. Let

$$f_\varepsilon(x) = \varepsilon^{-1} \int |x - a| \, f(\varepsilon^{-1} a) \, da = \int |x - \varepsilon a| \, f(a) \, da.$$

The function f_ε is smooth. Moreover, $f_\varepsilon'(x) = \int \operatorname{sign}(x - \varepsilon a) f(a) \, da$ and $f_\varepsilon''(x) = 2\varepsilon^{-1} f(\varepsilon^{-1} x)$.

Itô's formula gives

$$f_\varepsilon(B(t)) - f_\varepsilon(B(0)) - \int_0^t f_\varepsilon'(B(s)) \, dB(s) = \varepsilon^{-1} \int_0^t f(\varepsilon^{-1} B(s)) \, ds. \qquad (7.12)$$

Now we let $\varepsilon \downarrow 0$ for each term. From the definition of f_ε we infer that $|f_\varepsilon(x) - |x|| \leqslant 3\varepsilon$. In other words, $f_\varepsilon(x) \to |x|$ uniformly and this ensures convergence in probability of the first two terms on the left hand side of (7.12). To deal with the third term, we observe that, for $x \neq 0$,

$$f_\varepsilon'(x) = \int \operatorname{sign}(x - \varepsilon a) \, f(a) \, da \longrightarrow \operatorname{sign}(x) \text{ as } \varepsilon \downarrow 0.$$

Now we use the isometry property (7.2) to infer that

$$\mathbb{E}\!\left[\left(\int_0^t \operatorname{sign}(B(s)) \, dB(s) - \int_0^t f_\varepsilon'(B(s)) \, dB(s)\right)^2\right]$$

$$= \mathbb{E} \int_0^t (\operatorname{sign}(B(s)) - f_\varepsilon'(B(s)))^2 \, ds.$$

The right hand side converges to zero by the bounded convergence theorem. Hence we have shown that, in probability,

$$\lim_{\varepsilon \downarrow 0} \varepsilon^{-1} \int_0^t g(\varepsilon^{-1} B(s)) \, ds = \lim_{\varepsilon \downarrow 0} f_\varepsilon(B(t)) - f_\varepsilon(B(0)) - \int_0^t f_\varepsilon'(B(s)) \, dB(s)$$

$$= |B(t)| - |B(0)| - \int_0^t \operatorname{sign}(B(s)) \, dB(s) = \tilde{L}^0(t). \qquad \blacksquare$$

Proof of Theorem 7.33.   First fix $t \geqslant 0$ and recall from Theorem 6.19 that, almost surely, the occupation measure μ_t given by $\mu_t(A) = \int_0^t 1_A(B(s)) \, ds$ has a continuous density given by $\{L^a(t) \colon a \in \mathbb{R}\}$. Therefore, for every $a \in \mathbb{R}$, we have

$$L^a(t) = \lim_{\varepsilon \downarrow 0} \frac{\mu_t(a, a + \varepsilon)}{\varepsilon} = \lim_{\varepsilon \downarrow 0} \frac{1}{\varepsilon} \int_0^t 1_{(a, a+\varepsilon)}(B(s)) \, ds.$$

On the other hand, given $a \in \mathbb{R}$, by Lemma 7.36 there exists a sequence $\varepsilon_n \downarrow 0$ such that, almost surely,

$$\tilde{L}^a(t) = \lim_{n \uparrow \infty} \frac{1}{\varepsilon_n} \int_0^t 1_{(a, a+\varepsilon_n)}(B(s)) \, ds.$$

Hence, for every $a \in \mathbb{R}$ and $t \geqslant 0$, we have $L^a(t) = \tilde{L}^a(t)$ almost surely. Finally, for any $a \in \mathbb{R}$, both the local time $\{L^a(t) \colon t \geqslant 0\}$ and $\{\tilde{L}^a(t) \colon t \geqslant 0\}$ are almost surely continuous and therefore they agree. $\blacksquare$

Corollary 7.37 *For every $a \in \mathbb{R}$, almost surely, for all $t \geqslant 0$,*

$$\tfrac{1}{2} L^a(t) = (B(t) - a)^+ - (B(0) - a)^+ - \int_0^t 1_{\{B(s) > a\}} \, dB(s),$$

and

$$\tfrac{1}{2} L^a(t) = (B(t) - a)^- - (B(0) - a)^- + \int_0^t 1_{\{B(s) \leqslant a\}} \, dB(s).$$

Proof. The right sides in these formulas add up to $L^a(t)$, while their difference is zero. ∎

We now use Tanaka's formula to prove Lévy's theorem describing the joint law of the modulus and local time of a Brownian motion.

Theorem 7.38 (Lévy) *The processes*

$$\{(|B(t)|, L^0(t)) : t \geqslant 0\} \qquad and \qquad \{(M(t) - B(t), M(t)) : t \geqslant 0\}$$

have the same distribution.

Remark 7.39 This result extends both Theorem 2.34 where it was shown that the processes $\{|B(t)| : t \geqslant 0\}$ and $\{M(t) - B(t) : t \geqslant 0\}$ have the same distribution, and Theorem 6.10 where it was shown that $\{L^0(t) : t \geqslant 0\}$ and $\{M(t) : t \geqslant 0\}$ have the same distribution. Exercise 6.2 suggests an alternative proof using random walk methods. ◇

As a preparation for the proof we find the law of the process given by integrating the sign of a Brownian motion with respect to that Brownian motion.

Lemma 7.40 *For every $a \in \mathbb{R}$, the process $\{W(t) : t \geqslant 0\}$ given by*

$$W(t) = \int_0^t \operatorname{sign}(B(s) - a) \, dB(s)$$

is a standard Brownian motion.

Proof. Assume, without loss of generality, that $a < 0$. Suppose that $T = \inf\{t > 0 : B(t) = a\}$. Then $W(t) = B(t)$ for all $t \leqslant T$ and hence $\{W(t) : 0 \leqslant t \leqslant T\}$ is a (stopped) Brownian motion. By the strong Markov property the process $\{\widetilde{B}(t) : t \geqslant 0\}$ given by $\widetilde{B}(t) = B(t + T) - a$ is a Brownian motion started in the origin, which is independent of $\{W(t) : 0 \leqslant t \leqslant T\}$. As

$$W(t + T) = W(T) + \int_T^{t+T} \operatorname{sign}(B(s) - a) \, dB(s) = B(T) + \int_0^t \operatorname{sign}(\widetilde{B}(s)) \, d\widetilde{B}(s),$$

it suffices to show that the second term is a Brownian motion to complete the proof. Hence we may henceforth assume that $a = 0$. Now fix $0 \leqslant s < t$ and recall that $W(t) - W(s)$ is independent of $\mathcal{F}(s)$. For the proof it hence suffices to show that $W(t) - W(s)$ has a centred normal distribution with variance $t - s$. Choose $s = t_1^{(n)} < \ldots < t_n^{(n)} = t$ with mesh $\Delta(n) \downarrow 0$, and approximate the progressively measurable process $H(u) = \operatorname{sign}(B(u))$ by the step processes

$$H_n(u) = \operatorname{sign}\left(B(t_j^{(n)})\right) \qquad \text{if } t_j^{(n)} < u \leqslant t_{j+1}^{(n)}.$$

It follows from the fact that the zero set of Brownian motion is a closed set of measure zero, that $\lim \mathbb{E} \int_s^t (H_n(u) - H(u))^2 \, du = 0$, and hence

$$W(t) - W(s) = \int_s^t H(u) \, dB(u) = \mathbf{L}^2 - \lim_{n \to \infty} \int_s^t H_n(u) \, dB(u)$$

$$= \mathbf{L}^2 - \lim_{n \to \infty} \sum_{j=1}^{n-1} \operatorname{sign}\big(B(t_j^{(n)})\big) \big(B(t_{j+1}^{(n)}) - B(t_j^{(n)})\big).$$

From the independence of the Brownian increments and elementary properties of the normal distribution, one can see that the random variables on the right all have a centred normal distribution with variance $t - s$. Hence this also applies to $W(t) - W(s)$. ∎

Proof of Theorem 7.38. By Tanaka's formula we have

$$|B(t)| = \int_0^t \operatorname{sign}(B(s)) \, dB(s) + L^0(t) = W(t) + L^0(t).$$

Define a standard Brownian motion $\{\tilde{W}(t) \colon t \geqslant 0\}$ by

$$\tilde{W}(t) = -W(t) \qquad \text{for all } t \geqslant 0,$$

and let $\{\tilde{M}(t) \colon t \geqslant 0\}$ be the associated maximum process. We show that

$$\tilde{M}(t) = L^0(t) \qquad \text{for all } t \geqslant 0,$$

which implies that $\{(|B(t)|, L^0(t)) \colon t \geqslant 0\}$ and $\{(\tilde{M}(t) - \tilde{W}(t), \tilde{M}(t)) \colon t \geqslant 0\}$ agree pointwise, and the result follows as the latter process agrees in distribution with

$$\{(M(t) - B(t), M(t)) \colon t \geqslant 0\}.$$

To show that $\tilde{M}(t) = L^0(t)$ we first note that

$$\tilde{W}(s) = L^0(s) - |B(s)| \leqslant L^0(s),$$

and hence, taking the maximum over all $s \leqslant t$, we get $\tilde{M}(t) \leqslant L^0(t)$. On the other hand, the process $\{L^0(t) \colon t \geqslant 0\}$ increases only on $\{t \colon B(t) = 0\}$, and on this set we have $L^0(t) = \tilde{W}(t) \leqslant \tilde{M}(t)$. Hence the proof is complete, as $\{\tilde{M}(t) \colon t \geqslant 0\}$ is increasing. ∎

7.4 Feynman–Kac formulas and applications

In this section we answer some natural questions about Brownian motion that involve time. For example, we find the probability that linear Brownian motion exits a given interval by a fixed time. Our main tool is the close relationship between the expectation of certain functionals of the Brownian path and the heat equation with dissipation term. This goes under the name of *Feynman–Kac formula*, and the theorems that make up this theory establish a strong link between parabolic partial differential equations and Brownian motion.

Definition 7.41. Let $U \subset \mathbb{R}^d$ be either open and bounded, or $U = \mathbb{R}^d$. A twice differentiable function $u \colon (0, \infty) \times U \to [0, \infty)$ is said to solve the **heat equation with heat dissipation rate** $V \colon U \to \mathbb{R}$ and initial condition $f \colon U \to [0, \infty)$ on U if we have

- $\displaystyle\lim_{\substack{x \to x_0 \\ t \downarrow 0}} u(t,x) = f(x_0)$, whenever $x_0 \in U$,

- $\displaystyle\lim_{\substack{x \to x_0 \\ t \to t_0}} u(t,x) = 0$, whenever $x_0 \in \partial U$,

- $\partial_t u(t,x) = \frac{1}{2}\Delta_x u(t,x) + V(x)u(t,x)$ on $(0,\infty) \times U$,

where the Laplacian Δ_x acts on the space variables x.   ◇

Remark 7.42 The solution $u(t,x)$ describes the temperature at time t at x for a heat flow with *cooling* with rate $-V(x)$ on the set $\{x \in U : V(x) < 0\}$, and *heating* with rate $V(x)$ on the set $\{x \in U : V(x) > 0\}$, where the initial temperature distribution is given by $f(x)$ and the boundary of U is kept at zero temperature.   ◇

Instead of going for the most general results linking the heat equation to Brownian motion, we give some of the more basic forms of the Feynman–Kac formula together with applications. Our first theorem in this spirit, an existence result for the heat equation in the case $U = \mathbb{R}^d$, will lead to a new, more analytic proof of the second arcsine law, Theorem 5.28.

Theorem 7.43 *Suppose* $V \colon \mathbb{R}^d \to \mathbb{R}$ *is bounded. Then* $u \colon [0,\infty) \times \mathbb{R}^d \to \mathbb{R}$ *defined by*

$$u(t,x) = \mathbb{E}_x\left\{ \exp\left(\int_0^t V(B(r))\,dr \right) \right\},$$

solves the heat equation on $\mathbb{R}^d$ *with dissipation rate* V *and initial condition one.*

Proof.   The easiest proof is by a direct calculation. Expand the exponential in a power series, then the terms in the expansion are $a_0(x,t) := 1$ and, for $n \geqslant 1$,

$$
\begin{aligned}
a_n(x,t) &:= \frac{1}{n!}\mathbb{E}_x\left[\int_0^t \cdots \int_0^t V(B(t_1)) \cdots V(B(t_n))dt_1 \ldots dt_n \right]\\
&= \mathbb{E}_x\left[\int_0^t dt_1 \int_{t_1}^t dt_2 \cdots \int_{t_{n-1}}^t dt_n\, V(B(t_1)) \cdots V(B(t_n)) \right]\\
&= \int dx_1 \cdots \int dx_n \int_0^t dt_1 \cdots \int_{t_{n-1}}^t dt_n \prod_{i=1}^n V(x_i) \prod_{i=1}^n \mathfrak{p}(t_i - t_{i-1}, x_{i-1}, x_i),
\end{aligned}
$$

with the conventions $x_0 = x$ and $t_0 = 0$. Using $\frac{1}{2}\Delta_x\mathfrak{p}(t_1,x,x_1) = \partial_{t_1}\mathfrak{p}(t_1,x,x_1)$ and then integration by parts we get

$$
\begin{aligned}
\frac{1}{2}\Delta_x a_n(x) &= \int dx_1 V(x_1) \int_0^t dt_1 \partial_{t_1}\mathfrak{p}(t_1,x,x_1) a_{n-1}(x_1, t-t_1)\\
&= -\int dx_1 V(x_1) \int_0^t dt_1 \mathfrak{p}(t_1,x,x_1)\partial_{t_1} a_{n-1}(x, t-t_1) - V(x)a_{n-1}(x,t)\\
&= \partial_t a_n(x,t) - V(x)a_{n-1}(x,t).
\end{aligned}
$$

Adding up all these terms, and noting that differentiation under the summation sign is allowed, verifies the validity of the differential equation. The requirement on the initial condition follows easily from the boundedness of V.   ∎

As an application we give a proof of the second arcsine law, Theorem 5.28, which does not rely on the first arcsine law. We use Theorem 7.43 with $V(x) = \lambda \mathbf{1}_{[0,\infty)}(x)$. Then

$$u(t,x) := \mathbb{E}_x\left[\exp\left(-\lambda \int_0^t \mathbf{1}_{[0,\infty)}(B(s))\, ds\right)\right]$$

solves

$$\partial_t u(t,x) = \tfrac{1}{2}\partial_{xx} u(t,x) - \lambda \mathbf{1}_{[0,\infty)}(x)\, u(t,x), \qquad u(0,x) = 1 \text{ for all } x \in \mathbb{R}.$$

To turn this partial differential equation into ordinary differential equations, we take the Laplace transform

$$g(x) = \int_0^\infty u(t,x)\, e^{-\rho t}\, dt,$$

which satisfies the equation

$$\rho\, g(x) + \lambda V(x)\, g(x) - \tfrac{1}{2} g''(x) = 1.$$

This can be rewritten as

$$(\rho + \lambda)\, g(x) - \tfrac{1}{2} g''(x) = 1 \text{ if } x > 0,$$
$$\rho\, g(x) - \tfrac{1}{2} g''(x) = 1 \text{ if } x < 0.$$

Solving these two linear ordinary differential equations gives

$$g(x) = \tfrac{1}{\lambda + \rho} + A\, e^{\sqrt{2\rho}\, x} + B\, e^{-\sqrt{2\rho}\, x} \text{ if } x > 0,$$

$$g(x) = \tfrac{1}{\rho} + C\, e^{\sqrt{2\rho}\, x} + D\, e^{-\sqrt{2\rho}\, x} \text{ if } x < 0.$$

As g must remain bounded as $\rho \uparrow \infty$, we must have $A = D = 0$. Moreover, g must be continuously differentiable in zero, hence C and B can be calculated from matching conditions. After an elementary calculation we obtain

$$g(0) = \frac{1}{\sqrt{\rho(\rho + \lambda)}}.$$

On the other hand, with

$$X(t) = \frac{1}{t}\int_0^t \mathbf{1}_{[0,\infty)}(B(s))\, ds$$

we have, using Brownian scaling in the second step,

$$g(0) = \mathbb{E}_0\left[\int_0^\infty \exp\left(-\rho t - \lambda t X(t)\right) dt\right]$$
$$= \mathbb{E}_0\left[\int_0^\infty \exp\left(-\rho t - \lambda t X(1)\right) dt\right] = \mathbb{E}_0\left[\frac{1}{\rho + \lambda X(1)}\right].$$

Now we let $\rho = 1$ and from

$$\mathbb{E}_0\left[\frac{1}{1 + \lambda X(1)}\right] = \frac{1}{\sqrt{1 + \lambda}}$$

and the expansions

$$\frac{1}{\sqrt{1 + \lambda}} = \sum_{n=0}^\infty (-\lambda)^n \frac{\frac{1}{2}\frac{3}{2}\cdots\frac{2n-1}{2}}{n!},$$

and

$$\int_0^1 x^{n-\frac{1}{2}} (1-x)^{-\frac{1}{2}} dx = \frac{\Gamma(\frac{2n+1}{2})\Gamma(\frac{1}{2})}{\Gamma(n+1)} = \pi \frac{\frac{1}{2}\frac{3}{2}\cdots\frac{2n-1}{2}}{n!},$$

we get for the moments of $X(1)$, by a comparison of coefficients,

$$\mathbb{E}\big[X(1)^n\big] = \frac{1}{\pi} \int_0^1 x^n \frac{1}{\sqrt{x(1-x)}} dx,$$

which by (3.11) in Chapter 2 of [Du95] implies that $X(1)$ is arcsine distributed.

Our second version of the Feynman–Kac formula is a uniqueness result for the case of zero dissipation rate, which will allow us to express the probability that linear Brownian motion exits an interval before a fixed time t in two different ways.

Theorem 7.44 *If u is a bounded, twice continuously differentiable solution of the heat equation on the domain U, with zero dissipation rate and continuous initial condition g, then*

$$u(t,x) = \mathbb{E}_x\big[g\big(B(t)\big) 1\{t < \tau\}\big], \tag{7.13}$$

where τ is the first exit time from the domain U.

Proof.   The proof is based on Itô's formula, Theorem 7.15, and Remark 7.16. We let $K \subset U$ be compact and denote by σ the first exit time from K. Fixing $t > 0$ and applying Itô's formula with $f(x,y) = u(t-y,x)$ and $\zeta(s) = s$ gives, for all $s < t$,

$$u(t - s \wedge \sigma, B(s \wedge \sigma)) - u(t, B(0))$$
$$= \int_0^{s \wedge \sigma} \nabla_x u(t - v, B(v)) \cdot dB(v)$$
$$- \int_0^{s \wedge \sigma} \partial_t u(t - v, B(v)) \, dv + \frac{1}{2} \int_0^{s \wedge \sigma} \Delta_x u(t - v, B(v)) \, dv.$$

As u solves the heat equation, the last two terms on the right cancel. Hence, taking expectations,

$$\mathbb{E}_x\big[u(t - s \wedge \sigma, B(s \wedge \sigma))\big] = \mathbb{E}_x\big[u(t, B(0))\big] = u(t,x),$$

using that the stochastic integral has vanishing expectation. Exhausting U by compact sets, i.e. letting $\sigma \uparrow \tau$, and distinguishing the events $s < \sigma$ and $s \geqslant \sigma$ leads to $\mathbb{E}_x[u(t - s, B(s)) 1\{s < \tau\}] = u(t,x)$. Taking a limit $s \uparrow t$ gives the required result. ∎

As an application of Theorem 7.44 we calculate the probability that a linear Brownian motion stays, up to time t, within an interval. As a warm-up we suggest to look at Exercise 7.10 where the easy case of a halfline is treated. Here we focus on intervals $[a,b]$, for $a < 0 < b$, and give two different formulas for the probability of staying in $[a,b]$ up to time t. To motivate the first formula, we start with a heuristic approach, which gives the correct result, and then base the rigorous proof on the Feynman–Kac formula.

For our heuristics we think of the transition (sub-)density, $q_t \colon [0,a] \times [0,a] \to [0,1]$ of a Brownian motion, which is killed upon leaving the interval $[0,a]$. In a first approximation we subtract from the transition density $\mathfrak{p}(t,x,y)$ of an unkilled Brownian motion the transition density for all the paths that reach level 0. By the reflection principle (applied to the first hitting time of level 0) the latter is equal to $\mathfrak{p}(t,x,-y)$.

We then subtract the transition density of all the paths that reach level a, which, again by the reflection principle, equals $\mathfrak{p}(t,x,2a-y)$, then add again the density of all the paths that reach level 0 after hitting a, as these have already been subtracted in the first step. This gives the approximation term $\mathfrak{p}(t,x,y) - \mathfrak{p}(t,x,-y) - \mathfrak{p}(t,x,2a-y) + \mathfrak{p}(t,x,2a+y)$.

Of course the iteration does not stop here (for example we have double-counted some paths that reach level 0 after hitting a). Eventually we have to consider an infinite series of alternating reflections at levels 0 and a to obtain the density

$$q_t(x,y) = \sum_{k=-\infty}^{\infty} \left\{ \mathfrak{p}(t,x,2ka+y) - \mathfrak{p}(t,x,2ka-y) \right\}.$$

Integrating this over $y \in [0,a]$ makes the following theorem plausible.

Theorem 7.45 *Let $0 < x < a$. Then*

$$\mathbb{P}_x\left\{ B(s) \in (0,a) \text{ for all } 0 \leqslant s \leqslant t \right\}$$

$$= \sum_{k=-\infty}^{\infty} \left\{ \Phi\left(\tfrac{2ka+a-x}{\sqrt{t}}\right) - \Phi\left(\tfrac{2ka-x}{\sqrt{t}}\right) - \Phi\left(\tfrac{2ka+a+x}{\sqrt{t}}\right) + \Phi\left(\tfrac{2ka+x}{\sqrt{t}}\right) \right\}, \qquad (7.14)$$

where $\Phi(x)$ is the distribution function of a standard normal distribution.

Proof.   The left hand side in (7.14) agrees with the right hand side in Theorem 7.44 for $U = (0,a)$ and $f = 1$. The series on the right hand side is absolutely convergent, and hence satisfies the boundary conditions at $x = 0$ and $x = a$. It is also not difficult to verify that it is bounded. Elementary calculus gives

$$\partial_t \Phi\left(\tfrac{2ka+a-x}{\sqrt{t}}\right) = -\tfrac{2ka+a-x}{2t^{3/2}}\, \mathfrak{p}(t,x,2ka+a) = \tfrac{1}{2}\, \partial_{xx} \Phi\left(\tfrac{2ka+a-x}{\sqrt{t}}\right),$$

and similarly for the other summands. Hence termwise differentiation shows that the right hand side satisfies the heat equation. To see that the initial condition is fulfilled, note that (as $t \downarrow 0$) the sums over all $k > 0$ and $k < 0$ converge to zero. Among the four terms belonging to $k = 0$, two terms with positive sign and one term with negative sign converge to one, whereas one term converges to zero. ∎

The solution of the heat equation is not in the form one would get by a naïve separation of variables approach. This approach yields a different, equally valuable, expression for the probability of interest. Indeed, writing $u(t,x) = v(t)w(x)$ one expects w to be an eigenfunction of $\tfrac{1}{2}\partial_{xx}$ on $(0,a)$ with zero boundary conditions. These eigenfunctions are

$$\sin\left(\tfrac{k\pi(2x-a)}{2a}\right), \qquad \text{for } k \text{ even}, \qquad \cos\left(\tfrac{k\pi(2x-a)}{2a}\right), \qquad \text{for } k \text{ odd},$$

with eigenvalues $-k^2\pi^2/(2a^2)$. As we are only interested in solutions symmetric about $a/2$ only the cosine terms will contribute. For v we are looking for the eigenfunctions of

∂_t with the same eigenvalues, which are

$$\exp\left(-\frac{k^2\pi^2}{2a^2}t\right), \qquad \text{for } k \text{ odd,}$$

and considering the initial condition (and shifting the cosine by $\pi/2$) leads to the solution

$$u(t,x) = \frac{4}{\pi} \sum_{n=0}^{\infty} \frac{1}{2n+1} \exp\left(-\frac{(2n+1)^2\pi^2}{2a^2}t\right) \sin\left(\frac{(2n+1)\pi x}{a}\right). \tag{7.15}$$

Therefore (7.15) is an alternative representation of the probability in (7.14). For practical purposes this series is more useful when t is large, as the convergence is faster, whereas the series in the theorem converges fast only for small values of $t > 0$.

We now prove an elliptic, or time-stationary, version of the Feynman–Kac formula. This will enable us to describe the distribution of the total time spent by a transient Brownian motion in unit ball in terms of a Laplace transform.

Theorem 7.46 *Let $d \geqslant 3$ and $V\colon \mathbb{R}^d \to [0,\infty)$ be bounded. Define*

$$h(x) := \mathbb{E}_x\left[\exp\left(-\int_0^\infty V(B(t))\,dt\right)\right].$$

Then $h\colon \mathbb{R}^d \to [0,\infty)$ satisfies the equation

$$h(x) = 1 - \int G(x,y)V(y)\,h(y)\,dy \text{ for all } x \in \mathbb{R}^d.$$

Remark 7.47 Informally, the integral equation in Theorem 7.46 implies $\frac{1}{2}\Delta h = Vh$, which is also what one gets from letting $t \uparrow \infty$ in Theorem 7.43. See also Exercise 2.20 for a converse result in a similar spirit. $\diamond$

Proof. Define the 'resolvent operator'

$$R_\lambda^V f(x) := \int_0^\infty e^{-\lambda t}\mathbb{E}_x\left[f(B(t))\,e^{-\int_0^t V(B(s))\,ds}\right]dt.$$

Using the fundamental theorem of calculus in the second step we obtain

$$R_\lambda^0 f(x) - R_\lambda^V f(x) = \mathbb{E}_x \int_0^\infty dt\, e^{-\lambda t - \int_0^t V(B(s))\,ds} f(B(t)) \left(e^{\int_0^t V(B(s))\,ds} - 1\right)$$

$$= \mathbb{E}_x \int_0^\infty dt\, e^{-\lambda t - \int_0^t V(B(s))\,ds} f(B(t)) \int_0^t V(B(s))\, e^{\int_0^s V(B(r))\,dr}\,ds.$$

Using Fubini's theorem and the Markov property, we may continue with

$$= \mathbb{E}_x \int_0^\infty ds\, e^{-\lambda s} V(B(s)) \int_0^\infty dt\, \exp\left(-\lambda t - \int_0^t V(B(s+u))\,du\right) f(B(s+t))$$

$$= \mathbb{E}_x \int_0^\infty ds\, e^{-\lambda s} V(B(s)) R_\lambda^V f(B(s)) = R_\lambda^0\left(V R_\lambda^V f\right)(x).$$

The function h is related to the resolvent operator by the equation

$$h(x) = \lim_{\lambda \downarrow 0} \lambda R_\lambda^V 1(x).$$

Letting $f \equiv 1$ we obtain $1 - \lambda R_\lambda^V 1 = \lambda R_\lambda^0 \left(V R_\lambda^V 1 \right)$, and as $\lambda \downarrow 0$ we get

$$1 - h(x) = R_0^0 \left(V h \right)(x) = \int G(x,y) V(y) h(y) \, dy \,. \qquad \blacksquare$$

We use Theorem 7.46 to give an independent proof of the three-dimensional case of the Ciesielski–Taylor identity, which we have obtained from random walk considerations in Theorem 5.35. Key to this is the following proposition.

Proposition 7.48 *For a standard Brownian motion $\{B(t) \colon t \geqslant 0\}$ in dimension three let $T = \int_0^\infty 1\{B(t) \in \mathcal{B}(0,1)\} \, dt$ be the total occupation time of the unit ball. Then*

$$\mathbb{E}\big[e^{-\lambda T}\big] = \operatorname{sech}(\sqrt{2\lambda})\,.$$

Proof. Let $V(x) = \lambda 1_{\mathcal{B}(0,1)}$ and define $h(x) = \mathbb{E}_x[e^{-\lambda T}]$ as in Theorem 7.46. Then

$$h(x) = 1 - \lambda \int_{\mathcal{B}(0,1)} G(x,y) \, h(y) \, dy \text{ for all } x \in \mathbb{R}^d \,.$$

Clearly, we are looking for a rotationally symmetric function h. The integral on the right can therefore be split into two parts: First, the integral over $\mathcal{B}(0,|x|)$, which is the Newtonian potential due to a symmetric mass distribution on $\mathcal{B}(0,|x|)$ and therefore remains unchanged if the same mass is concentrated at the origin. Second, the integral over $\mathcal{B}(0,1) \setminus \mathcal{B}(0,|x|)$, which is harmonic on the open ball $\mathcal{B}(0,|x|)$ with constant value on the boundary, so itself is constant. Hence, for $x \in \mathcal{B}(0,1)$, to

$$1 - h(x) = \frac{\lambda}{2\pi|x|} \int_{\mathcal{B}(0,|x|)} h(y) \, dy + \lambda \int_{\mathcal{B}(0,1) \setminus \mathcal{B}(0,|x|)} \frac{h(y)}{2\pi|y|} \, dy.$$

With $u(r) = r h(x)$ for $|x| = r$ we have, for $0 < r < 1$,

$$r - u(r) = 2\lambda \int_0^r s u(s) \, ds + 2\lambda r \int_r^1 u(s) \, ds \,,$$

and by differentiation $u'' = 2\lambda u$ on $(0,1)$. Hence

$$u(r) = A e^{\sqrt{2\lambda}\, r} + B e^{-\sqrt{2\lambda}\, r} \,.$$

The boundary conditions $u(0) = 0$ and $u'(1) = 1$ give $A = 1/(\sqrt{2\lambda}(e^{\sqrt{2\lambda}} + e^{-\sqrt{2\lambda}}))$ and $B = -A$. Then

$$h(0) = \lim_{r \downarrow 0} \frac{u(r)}{r} = 1 - 2\lambda \int_0^1 u(r) \, dr$$

$$= 1 - A\sqrt{2\lambda} \left(e^{\sqrt{2\lambda}} + e^{-\sqrt{2\lambda}} - 2 \right) = \frac{2}{e^{\sqrt{2\lambda}} + e^{-\sqrt{2\lambda}}} = \operatorname{sech}(\sqrt{2\lambda}) \,,$$

as required to complete the proof. $\qquad \blacksquare$

Recall that the Ciesielski–Taylor identity, stated in Theorem 5.35, states that the first exit time from the unit ball by a standard Brownian motion in dimension one and the total occupation time of the unit ball by a standard Brownian motion in dimension $d = 3$ have the same distribution.

Proof of the Ciesielski–Taylor identity.   The Laplace transform of the first exit time
from the unit interval $(-1, 1)$ is given in Exercise 2.18. It coincides with the Laplace transform of T given in Proposition 7.48. Hence the two distributions coincide.   ∎

Exercises

Exercise 7.1. Ⓢ Show that for $F \in \mathbf{D}[0, 1]$ the Paley–Wiener integral $\int_0^1 F' \, dB$ of Lemma 1.41
almost surely agrees with the stochastic integral of Definition 7.8.

Exercise 7.2. Ⓢ Suppose $\{H(s, \omega) \colon s \geqslant 0, \omega \in \Omega\}$ is progressively measurable and
$\{B(t) \colon t \geqslant 0\}$ a linear Brownian motion. Show that for any stopping time T with

$$\mathbb{E}\left[\int_0^T H(s)^2 \, ds \right] < \infty,$$

we have

(a) $\mathbb{E}\left[\displaystyle\int_0^T H(s) \, dB(s) \right] = 0,$

(b) $\mathbb{E}\left[\left(\displaystyle\int_0^T H(s) \, dB(s) \right)^2 \right] = \mathbb{E}\left[\displaystyle\int_0^T H(s)^2 \, ds \right].$

Exercise 7.3. Suppose that $f \colon [0, 1] \to \mathbb{R}$ is in the Dirichlet space, i.e. $f(t) = \int_0^t f'(s) \, ds$
for all $t \in [0, 1]$ and $f' \in \mathbf{L}^2(0, 1)$. Then, almost surely,

$$\int_0^1 f'(s) \, dB(s) = \lim_{n \to \infty} n \sum_{j=0}^n \left(f\left(\tfrac{j+1}{n}\right) - f\left(\tfrac{j}{n}\right) \right) \left(B\left(\tfrac{j+1}{n}\right) - B\left(\tfrac{j}{n}\right) \right).$$

Exercise 7.4. Ⓢ Give a detailed proof of the multidimensional Itô formula, Theorem 7.15.

Exercise 7.5. Ⓢ Give an alternative proof of Theorem 2.37 based on a conformal mapping
of the halfplanes $\{(x, y) \colon x > t\}$ onto the unit disk, which exploits our knowledge of harmonic measure on spheres.

Exercise 7.6. Ⓢ Let $\{B(t) \colon t \geqslant 0\}$ be a planar Brownian motion. Show that, if $\theta(t)$ is the
continuous determination of the angle of $B(t)$, we have, almost surely,

$$\liminf_{t \uparrow \infty} \theta(t) = -\infty \qquad \text{and} \qquad \limsup_{t \uparrow \infty} \theta(t) = \infty.$$

Exercise 7.7. Formalise and prove the statement that, for every $\varepsilon > 0$, a planar Brownian
motion winds around its starting point infinitely often in any time interval $[0, \varepsilon]$.

Exercise 7.8. Show that, for simple random walk $\{S_n : n = 0, 1, \ldots\}$ on the integers we have the following analogue of Tanaka's formula: For every $a \in \mathbb{Z}$, almost surely,

$$|S_n - a| - |S_0 - a| = \sum_{j=0}^{n-1} \text{sign}(S_j - a)[S_{j+1} - S_j] + L_a(n),$$

where $L_a(n) = \sum_{j=0}^{n-1} 1\{S_j = a\}$ is the number of visits to a before time n.

Exercise 7.9. ⑤ Show that under suitable conditions, stochastic integrals and ordinary integrals can be interchanged: Suppose $h \colon \mathbb{R} \to [0, \infty)$ is a continuous function with compact support. Then, almost surely,

$$\int_{-\infty}^{\infty} h(a) \left(\int_0^t \text{sign}(B(s) - a) \, dB(s) \right) da = \int_0^t \left(\int_{-\infty}^{\infty} h(a) \, \text{sign}(B(s) - a) \, da \right) dB(s).$$

Hint. Write the outer integral on the left hand side as a limit of Riemann sums and use that the integrand has a continuous modification.

Exercise 7.10.

(a) Show that the function $u \colon (0, \infty) \times (0, \infty) \to \mathbb{R}$ given by

$$u(t, x) = \sqrt{\frac{2}{\pi t}} \int_0^x e^{-\frac{z^2}{2t}} \, dz$$

solves the heat equation on the domain $(0, \infty)$ with zero dissipation rate and constant initial condition $f = 1$.

(b) Infer from this that, for $x > 0$,

$$\mathbb{P}_x\{B(s) > 0 \text{ for all } s \leqslant t\} = \sqrt{\frac{2}{\pi t}} \int_0^x e^{-\frac{z^2}{2t}} \, dz.$$

(c) Explain how the result of (b) could have been obtained from the reflection principle.

Exercise 7.11. Prove the Erdős–Kac theorem: Let $X_1, X_2, \ldots$ be a sequence of independent identically distributed random variables with mean zero and variance one. Let $S_n = X_1 + \cdots + X_n$ and $T_n = \max\{|S_1|, \ldots, |S_n|\}$. Then

$$\lim_{n \to \infty} \mathbb{P}\{T_n < x\} = \frac{4}{\pi} \sum_{n=0}^{\infty} \frac{(-1)^n}{2n+1} \exp\left(-\frac{(2n+1)^2 \pi^2}{8x^2}\right).$$

Exercise 7.12. ⑤ Let T be the total occupation time in the unit ball by a standard Brownian motion in $\mathbb{R}^3$. Show that

(a) $\lim_{x \downarrow 0} \sqrt{\frac{1}{x}} e^{\frac{1}{2x}} \mathbb{P}\{T < x\} = \sqrt{\frac{8}{\pi}}$,

(b) $\lim_{x \uparrow \infty} e^{\frac{\pi^2}{8} x} \mathbb{P}\{T > x\} = \frac{4}{\pi}$.

Notes and comments

The first stochastic integral with a random integrand was defined by Itô [It44] but stochastic integrals with respect to Brownian motion with deterministic integrands were known to Paley, Wiener and Zygmund already in 1933, see [PWZ33] and Section 1.4. Our stochastic integral is by far not the most general construction possible, the complete theory of Itô integration is one of the cornerstones of modern probability. Interesting further material can be found, for example, in the books of McKean [McK69], Chung and Williams [CW90], Rogers and Williams [RW00a, RW00b] or Durrett [Du96]. Itô's formula, first proved in [It51], plays a central rôle in stochastic analysis, quite like the fundamental theorem of calculus does in real analysis. The version we give is designed to minimise the technical effort to get to the desired applications, but a lot more can be said if the discussion is extended to the concept of semimartingales, the references above provide the background for this. The formula is also at the heart of the theory of stochastic differential equations, a recommended introduction into this theory is Øksendal [Ok03] and a standard reference is Ikeda and Watanabe [IW89].

Conformal invariance was known to Lévy and a sketch of a proof is given in the book [Le48]. This fact does not extend to higher dimensions $d \geqslant 3$. There are not many interesting conformally invariant maps anyway, but essentially the only one, inversion on a sphere, fails. This is easy to see, as the image of Brownian motion stopped on the boundary of the punctured domain $\mathcal{B}(0,1) \setminus \{0\}$ has zero probability of not hitting $\mathcal{B}(0,1)$.

There is rich interaction between complex analysis and Brownian motion, which relies on conformal invariance. The conformal invariance of harmonic measure, which we proved in Theorem 7.23, is not easy to obtain by purely analytical means. Another result from complex analysis, which can be proved effectively using Brownian motion is Picard's theorem, see Davis [Da75] for the original paper or Durrett [Du84] for an exposition. The theorem states that a nonconstant entire function has a range which omits at most one point from the complex plane. Only very recently a completely new perspective on conformal invariance has opened up through the theory of conformally invariant random curves developed by Lawler, Schramm, and Werner, see e.g. [We04].

The skew-product representation has many nice applications, for more examples see Le Gall [LG92], which also served as the backbone of our exposition. The first result about the windings of Brownian motion is Spitzer's law, due to F. Spitzer in [Sp58]. There are plenty of extensions of it, including pathwise laws, see Shi [Sh98] or [Mö02], windings around more than one point, and joint laws of windings and additive functionals, see Pitman and Yor [PY86]. A discussion of some problems related to this can be found in Yor [Yo92].

Spitzer's paper [Sp58] also initiated substantial research on Brownian motion in a cone. He shows that, if τ is the first exit time of a planar Brownian motion from a cone with opening angle α, then $\mathbb{E}\tau^p < \infty$ if and only if $p < \frac{\pi}{2\alpha}$. This has been extended to higher dimensions by Burkholder [Bu77] and to more general cones, for example, by Bañuelos and Smits [BS97]. The skew-product representation plays an important rôle in the latter paper, which also contains a formula for the last time before one, when a Brownian motion was in a cone, having started at its vertex. This is a natural generalisation of the first arcsine law to more than one dimension.

Tanaka's formula offers many fruitful openings, among them the Meyer–Tanaka formula, which generalises Itô's formula to general convex functions, see the original work of Meyer [Me76] or the book by Durrett [Du96], and the theory of local times for semimartingales, which is presented in [RY94]. The formula goes back to the paper by Tanaka [Ta63]. Alternative to our approach, Tanaka's formula can be taken as a definition of Brownian local time. Then continuity can be obtained from the Kolmogorov-Čentsov theorem and moment estimates based on the Burkholder–Davis–Gundy inequalities, see for example the book by Karatzas and Shreve [KS91].

The Feynman–Kac formula is a classical application in stochastic calculus, which is discussed in more detail in [KS91]. It can be exploited to obtain an enormous variety of distributional properties of Brownian motion, see Borodin and Salminen [BS02] for (literally!) thousands of examples. The converse, application of Brownian motion to study equations, is of course equally natural. Del Moral [DM04] gives an impressive account of the wide applicability of this formula and its variants.

The identity between the two formulas describing the probability that a Brownian motion stays between two barriers serves as a standard example for the Poisson summation formula, see X.5 and XIX.5 in Feller [Fe66]. According to Feller it was discovered originally in connection with Jacobi's theory of transformations of theta functions, see Satz 277 in Landau [La09]. The 'iterated reflection' argument, which we have used to determine the transition density of a Brownian motion with absorbing barriers may also be used to determine transition densities for a Brownian motion which is reflected at the barriers, see X.5 in [Fe66]. In higher dimensions Brownian motion reflected at the boundaries of a domain is an interesting subject, not least because of its connections to partial differential equations with Neumann boundary conditions, see, for example, Brosamler [Br76].

The Erdős–Kac law plays an important rôle for the Kolmogorov–Smirnov test known from non-parametric statistics, see e.g. [Fe68]. Plenty of proofs of the arcsine law are known: Besides the two provided in this book, there is also an approach of Kac [Ka51] based on the Meyer–Tanaka formula, and Rogers and Williams, see III.24 in [RW00a], provide a proof based on local time theory.

The Ciesielski–Taylor identity was found by Ciesielski and Taylor in 1962 by an explicit calculation, see [CT62]. It extends to arbitrary dimensions d, stating that the law of the exit times from the unit ball by a standard Brownian motion in dimension d equals the law of the total occupation time in the unit ball by the standard Brownian motion in dimension $d + 2$. The argument given here is taken from Spitzer [Sp76], see also III.20 in Rogers and Williams [RW00a]. Many proofs of this fact are known, see for example Yor [Yo92], but none provides a geometrically intuitive explanation and it may well be that none exists. The tail estimates in Exercise 7.12 are crucial ingredients for the Hausdorff dimension results of Dembo et al. [DPRZ00a, DPRZ00b].

8

Potential theory of Brownian motion

In this chapter we develop the key facts of the potential theory of Brownian motion. This theory is centred around the notions of a harmonic function, the energy of a measure, and the capacity of a set. The probabilistic problem at the heart of this chapter is to find the probability that Brownian motion visits a given set.

8.1 The Dirichlet problem revisited

We now take up the study of the Dirichlet problem again and ask for sharp conditions on the domain which ensure the existence of solutions, which allow us to understand the problem for domains with very irregular boundaries, like for example connected components of the complement of a planar Brownian curve. For this task, stochastic integrals and Itô's formula will be a helpful tool. As a warm-up, we suggest to use these tools to give a probabilistic proof of the mean value property of harmonic functions, see Exercise 8.1.

Recall from Example 3.15 that the existence of a solution of the Dirichlet problem may be in doubt by the fact that Brownian motion started at the boundary ∂U may not leave the domain U immediately. Indeed, we show here that this is the only problem that can arise.

Definition 8.1. A point $x \in A$ is called **regular** for the closed set $A \subset \mathbb{R}^d$ if the first hitting time $T_A = \inf\{t > 0 \colon B(t) \in A\}$ satisfies $\mathbb{P}_x\{T_A = 0\} = 1$. A point which is not regular is called **irregular**. ◇

Remark 8.2 In the case $d = 1$ we have already seen that for any starting point $x \in \mathbb{R}$, almost surely a Brownian motion started in x returns to x in every interval $[0, \varepsilon)$ with $\varepsilon > 0$. Hence every point is regular for any set containing it. ◇

We already know a condition which implies that a point is regular, namely the Poincaré cone condition introduced in Chapter 3.

Theorem 8.3 *If the domain $U \subset \mathbb{R}^d$ satisfies the Poincaré cone condition at $x \in \partial U$, then x is regular for the complement of U.*

Proof. Suppose $x \in \partial U$ satisfies the condition, then there is an open cone V with base x and angle $\alpha > 0$, such that $V \cap B(x, r) \subset U^c$ for a suitable $r > 0$. Then the first exit time τ_U for the domain satisfies

$$\mathbb{P}_x\{\tau_U \leqslant t\} \geqslant \mathbb{P}_x\{B(t) \in V \cap \mathcal{B}(x,r)\}$$
$$\geqslant \mathbb{P}_x\{B(t) \in V\} - \mathbb{P}_x\{B(t) \notin \mathcal{B}(x,r)\}$$
$$= \mathbb{P}_x\{B(1) \in V\} - \mathbb{P}_x\{B(1) \notin \mathcal{B}(x,r/\sqrt{t})\},$$

where Brownian scaling was used in the last step. For $t \downarrow 0$ the subtracted term goes to zero, and hence $\mathbb{P}_x\{\tau_U = 0\} = \lim_{t\downarrow 0} \mathbb{P}_x\{\tau_U \leqslant t\} \geqslant \mathbb{P}\{B(1) \in V\} > 0$. By Blumenthal's zero-one law we have $\mathbb{P}_x\{\tau_U = 0\} = 1$, in other words x is regular for U^c. ∎

Remark 8.4 An alternative criterion for regularity, with a similar proof, will be given in Exercise 8.2. At the end of the present chapter we will give a *sharp* condition for a point to be regular, namely *Wiener's test* of regularity. ◇

Theorem 8.5 (Dirichlet Problem) *Suppose $U \subset \mathbb{R}^d$ is a bounded domain and let φ be a continuous function on ∂U. Define $\tau = \min\{t \geqslant 0: B(t) \in \partial U\}$, and define $u: \overline{U} \to \mathbb{R}$ by*

$$u(x) = \mathbb{E}_x\big[\varphi(B(\tau))\big].$$

   (a) *A solution to the Dirichlet problem exists if and only if the function u is a solution to the Dirichlet problem with boundary condition φ.*
   (b) *u is a harmonic function on U with $u(x) = \varphi(x)$ for all $x \in \partial U$ and is continuous at every point $x \in \partial U$ that is regular for the complement of U.*
   (c) *If every $x \in \partial U$ is regular for the complement of U, then u is the unique continuous function $u: \overline{U} \to \mathbb{R}$ which is harmonic on U such that $u(x) = \varphi(x)$ for all $x \in \partial U$.*

Proof.   For the proof of (a) let v be any solution of the Dirichlet problem on U with boundary condition φ. Define open sets $U_n \uparrow U$ by

$$U_n = \big\{x \in U: |x - y| > \tfrac{1}{n} \text{ for all } y \in \partial U\big\}.$$

Let τ_n be the first exit time of U_n and τ the first exit time from U, which are stopping times. By the multidimensional version of Itô's formula, we obtain

$$v(B(t\wedge\tau_n)) = v(B(0)) + \sum_{i=1}^d \int_0^{t\wedge\tau_n} \partial_{x_i} v(B(s))\, dB_i(s) + \frac{1}{2}\sum_{i=1}^d \int_0^{t\wedge\tau_n} \partial_{x_i x_i} v(B(s))\, ds.$$

Note that $\partial_{x_i} v$ is bounded on the closure of U_n, and thus everything is well-defined. The last term vanishes as $\Delta v(x) = 0$ for all $x \in U$. Taking expectations the second term on the right also vanishes, by Exercise 7.2, and we get that

$$\mathbb{E}_x\big[v(B(t \wedge \tau_n))\big] = \mathbb{E}_x\big[v(B(0))\big] = v(x), \qquad \text{for } x \in U_n.$$

Note that v, and hence the integrand on the left hand side, are bounded. Moreover, it is easy to check using boundedness of U and a reduction to the one-dimensional case, that τ is almost surely finite. Hence, as $t \uparrow \infty$ and $n \to \infty$, bounded convergence yields that the left hand side converges to $\mathbb{E}_x[v(B(\tau))] = \mathbb{E}_x[\varphi(B(\tau))]$. The result follows, as the right hand side depends neither on t nor on n.

The harmonicity statement of (b) is included in Theorem 3.8, and $u = \varphi$ on ∂U is obvious from the definition. It remains to show the continuity claim. For a regular $x \in \partial U$ we now show that if Brownian motion is started at a point in $\overline{U}$, which is sufficiently close to x, then with high probability the Brownian motion hits U^c, before leaving a given ball $\mathcal{B}(x, \delta)$. We start by noting that, for every $t > 0$ and $\eta > 0$ the set

$$O(t, \eta) := \{z \in U : \mathbb{P}_z\{\tau \leqslant t\} > \eta\}$$

is open. Indeed, if $z \in O(t, \eta)$, then for some small $s > 0$ and $\delta > 0$ and large $M > 0$, we have

$$\mathbb{P}_z\{|B(s) - z| \leqslant M, \, B(u) \in U^c \text{ for some } s \leqslant u \leqslant t\} > \eta + \delta.$$

By the Markov property the left hand side above can be written as

$$\int_{\mathcal{B}(z, M)} \mathbb{P}_\xi\{B(u) \in U^c \text{ for some } 0 \leqslant u \leqslant t - s\} \, \mathfrak{p}(s, z, \xi) \, d\xi.$$

Now let $\varepsilon > 0$ be sufficiently small, so that $|\mathfrak{p}(s, z, \xi) - \mathfrak{p}(s, y, \xi)| < \delta / \mathcal{L}(\mathcal{B}(0, M))$ for all $|z - y| < \varepsilon$ and $\xi \in \mathbb{R}^d$. Then we have

$$\mathbb{P}_y\{\tau \leqslant t\} \geqslant \mathbb{P}_y\{B(u) \in U^c \text{ for some } s \leqslant u \leqslant t\} > \eta,$$

hence the ball $\mathcal{B}(z, \varepsilon)$ is in $O(t, \eta)$, which therefore must be open. Given $\varepsilon > 0$ and $\delta > 0$ we now choose $t > 0$ small enough, such that for $\tau' = \inf\{s > 0 : B(s) \notin \mathcal{B}(x, \delta)\}$ we have

$$\mathbb{P}_z\{\tau' < t\} < \varepsilon/2 \quad \text{ for all } |x - z| < \delta/2.$$

By regularity we have $x \in O(t, 1 - \varepsilon/2)$, and hence we can choose $0 < \theta < \delta/2$ to achieve $\mathcal{B}(x, \theta) \subset O(t, 1 - \varepsilon/2)$. We have thus shown that,

$$|x - z| < \theta \Rightarrow \mathbb{P}_z\{\tau < \tau'\} > 1 - \varepsilon. \tag{8.1}$$

To complete the proof, let $\varepsilon > 0$ be arbitrary. Then there is a $\delta > 0$ such that $|\varphi(x) - \varphi(y)| < \varepsilon$ for all $y \in \partial U$ with $|x - y| < \delta$. Choose θ as in (8.1). For all $z \in \overline{U}$ with $|z - x| < \delta \wedge \theta$ we get

$$|u(x) - u(z)| = \left| \mathbb{E}_z[\varphi(x) - \varphi(B(\tau))] \right| \leqslant 2\|\varphi\|_\infty \, \mathbb{P}_z\{\tau' < \tau\} + \varepsilon \leqslant \varepsilon \, (2\|\varphi\|_\infty + 1).$$

As $\varepsilon > 0$ can be arbitrarily small, u is continuous at $x \in \partial U$. Finally, part (c) follows easily from (b) and the maximum principle. ∎

A further classical problem of partial differential equations, the Poisson problem, is related to Brownian motion in a way quite similar to the Dirichlet problem.

Definition 8.6. Let $U \subset \mathbb{R}^d$ be a bounded domain and $g : U \to \mathbb{R}$ be continuous. A continuous function $u : \overline{U} \to \mathbb{R}$, which is twice continuously differentiable on U is said to be the **solution of Poisson's problem for** g if

- $u(x) = 0$ for all $x \in \partial U$, and

- $-\frac{1}{2}\Delta u(x) = g(x)$ for all $x \in U$. ◇

A probabilistic approach to the Poisson problem will be developed in Exercises 8.3 and 8.4.

Remark 8.7

(a) For g bounded, any solution u of Poisson's problem for g equals

$$u(x) = \mathbb{E}_x \left[\int_0^T g(B(t)) \, dt \right] \quad \text{for } x \in U,$$

where $T := \inf\{t > 0 \colon B(t) \notin U\}$. Conversely, if g is Hölder continuous and every $x \in \partial U$ is regular for the complement of U, then the function u defined by the displayed equation solves the Poisson problem for g.

(b) If u solves Poisson's problem for $g \equiv 1$ in a domain $U \subset \mathbb{R}^d$, then $u(x) = \mathbb{E}_x[T]$ is the average time it takes a Brownian motion started in x to leave the set U. ◇

8.2 The equilibrium measure

In Chapter 3 we have studied the distribution of the location of the *first entry* of a Brownian motion into a closed set Λ, the harmonic measure. In the case of a transient (or killed) Brownian motion there is a natural counterpart to this by looking at the distribution of the position of the *last exit* from a closed set. This leads to the notion of the *equilibrium measure*, which we discuss and apply in this section.

To motivate the next steps we first look at a simple random walk $\{X_n \colon n \in \mathbb{N}\}$ in $d \geqslant 3$. Let $A \subset \mathbb{Z}^d$ be a bounded set, then by transience the last exit time $\gamma = \max\{n \in \mathbb{N} \colon X_n \in A\}$ is finite on the event that the random walk ever hits A. Note that γ is *not* a stopping time. Then, for any $x \in \mathbb{Z}^d$ and $y \in A$,

$$\mathbb{P}_x\{X \text{ hits } A \text{ and } X_\gamma = y\} = \sum_{k=0}^{\infty} \mathbb{P}_x\{X_k = y, X_j \notin A \text{ for all } j > k\}$$

$$= \sum_{k=0}^{\infty} \mathbb{P}_x\{X_k = y\} \mathbb{P}_y\{\gamma = 0\},$$

and introducing the Green's function $G(x, y) = \sum_{k=0}^{\infty} \mathbb{P}_x\{X_k = y\}$ we get, for all $y \in A$,

$$\mathbb{P}_x\{X \text{ hits } A \text{ and } X_\gamma = y\} = G(x, y) \, \mathbb{P}_y\{\gamma = 0\}.$$

This holds also, obviously, for all $y \in \mathbb{Z}^d \setminus A$. Summing over all $y \in \mathbb{Z}^d$ gives

$$\mathbb{P}_x\{X \text{ ever hits } A\} = \sum_{y \in \mathbb{Z}^d} G(x, y) \mathbb{P}_y\{\gamma = 0\}.$$

The formula allows us to describe the probability of ever hitting a set as a potential with respect to the measure $y \mapsto \mathbb{P}_y\{\gamma = 0\}$, which is supported on A. Our aim in this section is to extend this to Brownian motion.

Note that the argument above relied heavily on the transience of the random walk. This is no different in the case of Brownian motion. In order to include the two-dimensional case we 'kill' the Brownian motion, either when it exits a large domain or at an independent exponential stopping time. Note that both possibilities preserve the strong Markov property, in the case of exponential killing this is due to the lack-of-memory property of the exponential distribution.

To formally explain our setup we now suppose that $\{B(t): 0 \leqslant t \leqslant T\}$ is a transient Brownian motion in the sense of Chapter 3. Recall that this means that $\{B(t): 0 \leqslant t \leqslant T\}$ is a d-dimensional Brownian motion killed at time T, and one of the following holds:

(1) $d \geqslant 3$ and $T = \infty$,
(2) $d \geqslant 2$ and T is an independent exponential time,
(3) $d \geqslant 2$ and T is the first exit time from a bounded domain D containing 0.

We use the convention that $D = \mathbb{R}^d$ in cases (1) and (2). In all cases, transient Brownian motion is a Markov process and, by Theorem 3.30 its transition kernel has a density, which we denote by $\mathfrak{p}^*(t, x, y)$. Note that in case (2,3) the function $\mathfrak{p}^*(t, x, y)$ is only a subprobability density because of the killing, indeed it is strictly smaller than the corresponding density without killing. The associated Green's function

$$G(x, y) = \int_0^\infty \mathfrak{p}^*(t, x, y)\, dt,$$

is always well-defined and finite for all $x \neq y$.

Theorem 8.8 (Last exit formula) *Suppose $\{B(t): 0 \leqslant t \leqslant T\}$ is a transient Brownian motion and $\Lambda \subset \mathbb{R}^d$ a compact set. Let*

$$\gamma = \sup \{t \in (0, T]: B(t) \in \Lambda\}$$

be the last exit time *from Λ, using the convention $\gamma = 0$ if the path does not hit Λ. Then there exists a finite measure ν on Λ called the **equilibrium measure**, such that, for any Borel set $A \subset \Lambda$ and $x \in D$,*

$$\mathbb{P}_x\{B(\gamma) \in A, 0 < \gamma \leqslant T\} = \int_A G(x, y)\, d\nu(y).$$

Remark 8.9 Observe that, given Λ, the equilibrium measure is uniquely determined by the last exit formula. The proof of Theorem 8.8 is similar to the simple calculation in the discrete case, the equilibrium measure is constructed as limit of the measure $\varepsilon^{-1}\mathbb{P}_y\{0 < \gamma \leqslant \varepsilon\}\, dy$. ◇

Proof of Theorem 8.8. Let U_ε be a uniform random variable on $[0, \varepsilon]$, independent of the Brownian motion and the killing time. Then, for any bounded and continuous $f: D \to \mathbb{R}$,

$$\mathbb{E}_x\left[f(B(\gamma - U_\varepsilon))\, 1\{U_\varepsilon < \gamma\}\right]$$

$$= \varepsilon^{-1}\int_0^\infty \mathbb{E}_x\left[f(B(t))1\{t < \gamma \leqslant t + \varepsilon\}\right] dt$$

$$= \varepsilon^{-1}\int_0^\infty \mathbb{E}_x\left[f(B(t))1\{t \leqslant T\}\mathbb{P}_{B(t)}\{0 < \gamma \leqslant \varepsilon\}\right] dt.$$

Using the notation $\psi_\varepsilon(x) = \varepsilon^{-1}\mathbb{P}_x\{0 < \gamma \leqslant \varepsilon\}$ this equals

$$\int_0^\infty \mathbb{E}_x\left[f \cdot \psi_\varepsilon(B(t))1\{t \leqslant T\}\right] dt = \int_0^\infty \int_D \mathfrak{p}^*(t, x, y)\, f \cdot \psi_\varepsilon(y)\, dy\, dt$$

$$= \int_D f(y)\, G(x, y)\, \psi_\varepsilon(y)\, dy.$$

This means that the subprobability measure η_ε defined by

$$\eta_\varepsilon(A) = \mathbb{P}_x\big\{B(\gamma - U_\varepsilon) \in A, \, U_\varepsilon < \gamma \leqslant T\big\}$$

has the density $G(x,y)\,\psi_\varepsilon(y)$. Therefore also,

$$G(x,y)^{-1}\,d\eta_\varepsilon(y) = \psi_\varepsilon(y)\,dy. \tag{8.2}$$

Observe now that, by continuity of the Brownian path, $\lim_{\varepsilon \downarrow 0} \eta_\varepsilon = \eta_0$ in the sense of weak convergence, where the measure η_0 on Λ is defined by

$$\eta_0(A) = \mathbb{P}_x\big\{B(\gamma) \in A, \, 0 < \gamma \leqslant T\big\},$$

for all Borel sets $A \subset \Lambda$. As, for fixed $x \in D$, the function $y \mapsto G(x,y)^{-1}$ is continuous and bounded on Λ, we infer that, in the sense of weak convergence

$$\lim_{\varepsilon \downarrow 0} G(x,y)^{-1}\,d\eta_\varepsilon = G(x,y)^{-1}\,d\eta_0.$$

By (8.2) the measure $\psi_\varepsilon(y)\,dy$ therefore converges weakly to a limit measure ν, which does not depend on x, and satisfies $G(x,y)^{-1}\,d\eta_0(y) = d\nu(y)$ for all $x \in D$. As η_0 has no atom in x we therefore obtain that $d\eta_0(y) = G(x,y)\,d\nu(y)$ for all $x \in D$. Integrating over any Borel set A gives the statement. ∎

A direct representation of the equilibrium measure as a last exit distribution can be obtained in cases (1) and (3) when the Brownian motion is started at a random point.

Theorem 8.10 *Suppose Λ is a compact nonpolar set and*

$$\Lambda \subset \mathcal{B}(z,r).$$

Let $\{B(t)\colon 0 \leqslant t \leqslant T\}$ be a transient Brownian motion started uniformly on $\partial\mathcal{B}(z,r)$ and stopped as in case (1) or as in (3) with $D = \mathcal{B}(z,R)$ for $R > r$. Let γ be the last exit time from Λ, as in Theorem 8.8. Then the equilibrium measure ν satisfies, for any Borel set $A \subset \Lambda$,

$$\frac{\nu(A)}{\nu(\Lambda)} = \mathbb{P}\big\{B(\gamma) \in A \,\big|\, 0 < \gamma \leqslant T\big\}.$$

The proof follows from the following interesting lemma.

Lemma 8.11 *In the setup of Theorem 8.10, the value of the integral*

$$\int G(x,y)\,d\sigma_{z,r}(x)$$

is independent of the choice of $y \in \mathcal{B}(z,r)$.

Proof. By Theorem 3.35 the mapping $y \mapsto I(y) = \int G(x,y)\,d\sigma_{z,r}(x)$ is harmonic on $\mathcal{B}(z,r)$. Fix a point $y \in \mathcal{B}(z,r)$ and let $s = |y - z|$, so that $s < r$. By rotational invariance we have $I(w) = I(y)$ for all $w \in \partial\mathcal{B}(z,s)$. Hence, $I(z) = \int I(w)\,d\varpi_{z,s}(w) = I(y)$. ∎

Proof of Theorem 8.10.   Take the last exit formula from Theorem 8.8 and integrate the variable x with respect to $\sigma_{z,r}$. Using Fubini's theorem, we obtain

$$\mathbb{P}\{B(\gamma) \in A, 0 < \gamma \leqslant T\} = \int \mathbb{P}_x\{B(\gamma) \in A, 0 < \gamma \leqslant T\}\, d\sigma_{z,r}(x)$$

$$= \int_A \int G(x,y)\, d\sigma_{z,r}(x)\, d\nu(y) = c\,\nu(A),$$

where c is the joint value of the integrals in Lemma 8.11. Dividing both sides by $\mathbb{P}\{0 < \gamma \leqslant T\} = c\,\nu(\Lambda)$ gives the result.   ∎

As a first application we give an estimate for the probability that Brownian motion in $\mathbb{R}^d$, for $d \geqslant 3$, hits a set contained in an annulus around x.

Corollary 8.12 *Suppose $\{B(t)\colon t \geqslant 0\}$ is Brownian motion in $\mathbb{R}^d$, with $d \geqslant 3$, and $\Lambda \subset \mathcal{B}(x,R) \setminus \mathcal{B}(x,r)$ is compact. Then*

$$R^{2-d}\nu(\Lambda) \leqslant \mathbb{P}_x\{\{B(t)\colon t \geqslant 0\} \text{ ever hits } \Lambda\} \leqslant r^{2-d}\nu(\Lambda),$$

where ν is the equilibrium measure on Λ.

Proof.   By Theorem 8.8 in the case $A = \Lambda$ we have

$$\mathbb{P}_x\{\{B(t)\colon t \geqslant 0\} \text{ ever hits } \Lambda\} = \int_\Lambda G(x,y)\, d\nu(y).$$

Recall that $G(x,y) = |x-y|^{2-d}$ and use that $R^{2-d} \leqslant G(x,y) \leqslant r^{2-d}$.   ∎

Theorem 8.5 makes us interested in statements claiming that the set of irregular points of a set A is small. The following fundamental result will play an important rôle in the next chapter.

Theorem 8.13 *Suppose $A \subset \mathbb{R}^d$, $d \geqslant 2$, is a closed set and let A^{r} be the set of regular points for A. Then, for all $x \in \mathbb{R}^d$,*

$$\mathbb{P}_x\{B(t) \in A \setminus A^{\mathrm{r}} \text{ for some } t > 0\} = 0,$$

in other words, the set of irregular points is polar for Brownian motion.

For the proof of Theorem 8.13 we have to develop a tool of independent interest, the strong maximum principle. A special case of this is the following statement, from which Theorem 8.13 follows without too much effort.

Theorem 8.14 *Let $\{B(t)\colon t \geqslant 0\}$ be a d-dimensional Brownian motion, and T an independent exponential time. Let $\Lambda \subset \mathbb{R}^d$ be a compact set and define $\tau = \inf\{t > 0\colon B(t) \in \Lambda\}$. If for some $\vartheta < 1$, we have $\mathbb{P}_x\{\tau < T\} \leqslant \vartheta$ for all $x \in \Lambda$, then $\mathbb{P}_x\{\tau < T\} \leqslant \vartheta$ for all $x \in \mathbb{R}^d$.*

Proof of Theorem 8.13. We can write the set of irregular points of A as a countable union of compact sets

$$A \setminus A^{\mathrm{r}} = \bigcup_{\ell=1}^{\infty} \bigcup_{m=1}^{\infty} \bigcup_{n=1}^{\infty} \left\{ x \in A \cap \mathcal{B}(0,m) \colon \mathbb{P}_x\{\tau(A) \leqslant T(n)\} \leqslant 1 - \tfrac{1}{\ell} \right\},$$

where $T(n)$ is an independent exponential time with mean $1/n$ and $\tau(A)$ is the first hitting time of A. It suffices to prove that Brownian motion does not hit any fixed set in the union, so let ℓ, m, n be fixed and take $T = T(n)$, $\vartheta = 1 - 1/\ell$ and a compact set

$$\Lambda = \left\{ x \in A \cap \mathcal{B}(0,m) \colon \mathbb{P}_x\{\tau(A) \leqslant T\} \leqslant \vartheta \right\}.$$

If $x \in \Lambda$, then, writing τ for the first hitting time of $\Lambda \subset A$,

$$\mathbb{P}_x\{\tau \leqslant T\} \leqslant \mathbb{P}_x\{\tau(A) \leqslant T\} \leqslant \vartheta,$$

for all $x \in \Lambda$ and therefore by Theorem 8.14 for all $x \in \mathbb{R}^d$.

Now suppose $x \in \mathbb{R}^d$ is the arbitrary starting point of a Brownian motion $\{B(t) \colon t \geqslant 0\}$ and $\Lambda(\varepsilon) = \{y \in \mathbb{R}^d \colon |y - z| \leqslant \varepsilon \text{ for some } z \in \Lambda\}$. Define τ_ε as the first hitting time of $\Lambda(\varepsilon)$. Clearly, as Λ is closed,

$$\lim_{\varepsilon \downarrow 0} \mathbb{P}_x\{\tau_\varepsilon \leqslant T\} = \mathbb{P}_x\{\tau \leqslant T\}.$$

Moreover, by the strong Markov property applied at the stopping time τ_ε and the lack of memory property of exponential random variables,

$$\mathbb{P}_x\{\tau \leqslant T\} \leqslant \mathbb{P}_x\{\tau_\varepsilon \leqslant T\} \max_{z \in \Lambda_\varepsilon} \mathbb{P}_z\{\tau \leqslant T\} \leqslant \mathbb{P}_x\{\tau_\varepsilon \leqslant T\}\,\vartheta,$$

and letting $\varepsilon \downarrow 0$ we obtain

$$\mathbb{P}_x\{\tau \leqslant T\} \leqslant \mathbb{P}_x\{\tau \leqslant T\}\,\vartheta.$$

As $\vartheta < 1$ this implies $\mathbb{P}_x\{\tau \leqslant T\} = 0$, and as T is independent of the Brownian motion and can take arbitrarily large values with positive probability, we infer that the Brownian motion started in x never hits Λ. ∎

The idea in the proof of Theorem 8.14 is to use the equilibrium measure ν to express $\mathbb{P}_x\{\tau < T\}$ as a potential, which means that, denoting the parameter of the exponential by $\lambda > 0$,

$$\mathbb{P}_x\{\tau < T\} = \int G_\lambda(x,y)\,d\nu(y),$$

where G_λ is the Green's function for the Brownian motion stopped at time T, i.e.

$$G_\lambda(x,y) = \int_0^{\infty} e^{-\lambda t}\,\mathfrak{p}(t,x,y)\,dt.$$

Recall that for any fixed y the function $x \mapsto G_\lambda(x,y)$ is subharmonic on $\mathbb{R}^d \setminus \{y\}$, by Theorem 3.35 (iii), and this implies that

$$U_\lambda \nu(x) = \int G_\lambda(x,y)\,d\nu(y)$$

is subharmonic on Λ^c. If $U_\lambda \nu$ was also continuous on the closure of Λ^c, then the maximum principle in Theorem 3.5 would tell us that $U_\lambda \nu$ has its maxima on the boundary $\partial \Lambda$ and this would prove Theorem 8.14. However we do not know the continuity of $U_\lambda \nu$ on the closure of Λ^c, so we need a strengthening of the maximum principle.

We now let K be a **kernel**, i.e. a measurable function $K \colon \mathbb{R}^d \times \mathbb{R}^d \to [0, \infty]$. Suppose that $x \mapsto K(x, y)$ is subharmonic outside $\{y\}$, and that $K(x, y)$ is a continuous and decreasing function of the distance $|x - y|$. For any finite measure μ without atoms let

$$U_\mu(x) = \int K(x, y) \, d\mu(y)$$

be the potential of μ at x with respect to the kernel K.

Theorem 8.15 (Strong maximum principle) *If μ is supported by the compact set Λ, then, for any $\vartheta > 0$, we have the equivalence*

$$U_\mu(x) \leqslant \vartheta \text{ for all } x \in \Lambda \qquad \Leftrightarrow \qquad U_\mu(x) \leqslant \vartheta \text{ for all } x \in \mathbb{R}^d.$$

Remark 8.16 Note that this completes the proof of Theorem 8.14 and hence of Theorem 8.13 by applying it to the special case of the kernel $K = G_\lambda$ and the equilibrium measure.   ◇

The proof we present relies on a beautiful geometric lemma.

Lemma 8.17 *There is a number N depending only on the dimension d such that the following holds: For every $x \in \mathbb{R}^d$ and every closed set Λ there are N nonoverlapping closed cones $V_1, \ldots, V_N$ with vertex x such that, if ξ_i is a point of $\Lambda \cap V_i$ closest to x, then any point $y \in \Lambda$ with $y \neq x$ is no further to some ξ_i than to x.*

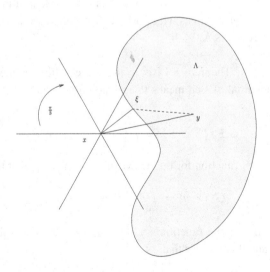

Fig. 8.1. The geometric argument in Lemma 8.17.

Proof. The proof is elementary by looking at Figure 8.1: Let N be the number of closed cones with circular base, vertex in the origin and opening angle $\pi/3$ needed to cover $\mathbb{R}^d$. Replace each of the cones in this collection by a subcone (not necessarily with circular base) such that the collection still covers $\mathbb{R}^d$ but the cones are non-overlapping. Let V be a shift of such a cone with vertex in x, ξ be a point in $V \cap \Lambda$ which is closest to x, and $y \in \Lambda \cap V$ be arbitrary. The triangle with vertices in x, ξ and y has at most angle $\pi/3$ at the vertex x, and hence by the geometry of triangles, the distance of y and ξ is no larger than the distance of y and x. ∎

Proof of Theorem 8.15. Of course, only the implication $\Rightarrow$ needs proof. Take μ satisfying $U_\mu(x) \leqslant \vartheta$ for all $x \in \Lambda$. Note that, by monotone convergence,

$$U_\mu(x) = \lim_{\delta \downarrow 0} \int_{|x-y|>\delta} K(x,y)\,d\mu(y). \tag{8.3}$$

Hence, for a given $\eta > 0$, by Egorov's theorem, there exists a compact subset $F \subset \Lambda$ such that, $\mu(F) > \mu(\Lambda) - \eta$ and the convergence in (8.3) is uniform on F. If we define μ_1 to be the restriction of μ to F, then we can find, for every $\varepsilon > 0$ some $\delta > 0$ such that

$$\sup_{x \in F} \int_{|x-y|\leqslant\delta} K(x,y)\,d\mu_1(y) < \varepsilon.$$

Now let $\{x_n\} \subset \mathbb{R}^d$ be a sequence converging to $x_0 \in F$. Then, as the kernel K is bounded on sets bounded away from the diagonal,

$$\limsup_{n\to\infty} U_{\mu_1}(x_n) \leqslant \int K(x_0,y)\,d\mu_1(y) + \limsup_{n\to\infty} \int_{|y-x_n|\leqslant\delta} K(x_n,y)\,d\mu_1(y).$$

We now want to compare $K(x_n,y)$ with $K(\xi,y)$ for $\xi \in F$. Here we use Lemma 8.17 for the point $x = x_n$ and obtain $\xi_1,\dots,\xi_N \in F$ such that

$$K(x_n,y) \leqslant \sum_{i=1}^{N} K(\xi_i,y),$$

where we have used that K depends only on the distance of the arguments and is decreasing in it. We thus have

$$\int_{|y-x_n|\leqslant\delta} K(x_n,y)\,d\mu_1(y) \leqslant \sum_{i=1}^{N} \int_{|y-\xi_i|\leqslant\delta} K(\xi_i,y)\,d\mu_1(y) \leqslant N\varepsilon.$$

As $\varepsilon > 0$ was arbitrary we get

$$\limsup_{n\to\infty} U_{\mu_1}(x_n) \leqslant U_{\mu_1}(x_0).$$

As the converse statement

$$\liminf_{n\to\infty} U_{\mu_1}(x_n) \geqslant U_{\mu_1}(x_0)$$

holds obviously by Fatou's lemma, we obtain the continuity of U_{μ_1} on F. Continuity of U_{μ_1} on F^c is obvious from the properties of the kernel and the fact that μ_1 is supported by F, so that we have continuity of U_{μ_1} on all of $\mathbb{R}^d$. By the maximum principle, Theorem 3.5, we infer that $U_{\mu_1}(x) \leqslant \vartheta$.

To complete the proof let $x \notin \Lambda$ be arbitrary, and denote its distance to Λ by ϱ. Then $K(x, y) \leqslant C(\varrho)$ for all $y \in \Lambda$. Therefore

$$U_\mu(x) \leqslant U_{\mu_1}(x) + \eta C(\varrho) \leqslant \vartheta + \eta C(\varrho),$$

and the result follows by letting $\eta \downarrow 0$.   ∎

8.3 Polar sets and capacities

One of our ideas to measure the size of sets in Chapter 4 was based on the notion of capacity. While this notion appeared to be useful, but maybe a bit artificial at the time, we can now understand its true meaning. This is linked to the notion of polarity, namely whether a set has a positive probability of being hit by a suitably defined random set.

More precisely, we ask, which sets are polar for the range of a d-dimensional Brownian motion $\{B(t) \colon t \geqslant 0\}$. Recall that a Borel set $A \subset \mathbb{R}^d$ is *polar* for Brownian motion if, for all x,

$$\mathbb{P}_x\{B(t) \in A \text{ for some } t > 0\} = 0.$$

In the case $d = 1$ we already know that only the empty set is polar, whereas by Corollary 2.26 points are polar for Brownian motion in all dimensions $d \geqslant 2$. The general characterisation of polar sets requires an extension of the notion of capacities to a bigger class of kernels.

Definition 8.18. Suppose $A \subset \mathbb{R}^d$ is a Borel set and $K \colon \mathbb{R}^d \times \mathbb{R}^d \to [0, \infty]$ is a kernel. Then the K-energy of a measure μ is defined to be

$$I_K(\mu) = \iint K(x, y) \, d\mu(x) \, d\mu(y),$$

and the K-capacity of A is defined as

$$\mathrm{Cap}_K(A) = \left[\inf\left\{I_K(\mu) \colon \mu \text{ a probability measure on } A\right\}\right]^{-1}.$$

Recall that the α-energy of a measure and the Riesz α-capacity Cap_α of a set defined in Chapter 4 correspond to the kernel $K(x, y) = |x - y|^{-\alpha}$.   ◇

Remark 8.19 In most of our applications the kernels are of the form $K(x, y) = f(|x - y|)$ for some decreasing function $f \colon [0, \infty) \to [0, \infty]$. In this case we simply write I_f instead of I_K and call this the f-energy. We also write Cap_f instead of Cap_K and call this the f-capacity.   ◇

Theorem 8.20 (Kakutani's theorem) *A closed set Λ is polar for d-dimensional Brownian motion if and only if it has zero f-capacity for the **radial potential** f defined by*

$$f(\varepsilon) := \begin{cases} |\log(1/\varepsilon)| & \text{if } d = 2, \\ \varepsilon^{2-d} & \text{if } d \geqslant 3. \end{cases}$$

Remark 8.21 We call the kernel $K(x, y) = f(|x - y|)$, where f is the radial potential, the **potential kernel**. Up to constants, it agrees with the Green kernel in $d \geqslant 3$. ◇

Instead of proving Kakutani's theorem directly, we aim for a stronger, quantitative result in the framework of transient Brownian motions given in Definition 3.28. Recall that this means that $\{B(t): 0 \leqslant t \leqslant T\}$ is a d-dimensional Brownian motion killed at time T, and either (1) $d \geqslant 3$ and $T = \infty$, (2) $d \geqslant 2$ and T is an independent exponential time, or (3) $d \geqslant 2$ and T is the first exit time from a bounded domain D containing the origin. This result gives, for compact sets $\Lambda \subset \mathbb{R}^d$, a quantitative estimate of

$$\mathbb{P}_0\{\exists 0 < t < T \text{ such that } B(t) \in \Lambda\}$$

in terms of capacities. However, even if $d = 3$ and $T = \infty$, one cannot expect that

$$\mathbb{P}_0\{\exists t > 0 \text{ such that } B(t) \in \Lambda\} \asymp \mathrm{Cap}_f(\Lambda)$$

for the radial potential f in Theorem 8.20. Observe, for example, that the left hand side depends strongly on the starting point of Brownian motion, whereas the right hand side is translation invariant. Similarly, if Brownian motion is starting at the origin, the left hand side is invariant under scaling, i.e. remains the same when Λ is replaced by $\lambda\Lambda$ for any $\lambda > 0$, whereas the right hand side is not. For a direct comparison of hitting probabilities and capacities, it is therefore necessary to use a capacity function with respect to a scale-invariant modification of the Green kernel G, called the *Martin kernel*, which we now introduce.

Definition 8.22. We define the Martin kernel $M: D \times D \to [0, \infty]$ by

$$M(x, y) := \frac{G(x, y)}{G(0, y)} \qquad \text{for } x \neq y,$$

and otherwise by $M(x, x) = \infty$. ◇

We need the following technical proposition, which is easy to verify directly from the form of the Green's function G in case (1). For the other two cases we give a conceptual proof.

Proposition 8.23 *For every compact set $\Lambda \subset D \subset \mathbb{R}^d$ there exists a constant C depending only on Λ such that, for all $x, y \in \Lambda$ and sufficiently small $\varepsilon > 0$,*

$$\sup_{|x-z|<\varepsilon} \varepsilon^{-d} \int_{\mathcal{B}(y,\varepsilon)} \frac{G(z, \xi)}{G(x, y)} \, d\xi \leqslant C.$$

Proof. Fix a compact set $\Lambda \subset D$ and $\varepsilon > 0$ smaller than one tenth of the distance of Λ and D^c and let $x, y \in \Lambda$. We abbreviate

$$h_\varepsilon(z, y) = \int_{\mathcal{B}(y,\varepsilon)} G(z, \xi) \, d\xi \qquad \text{for } z \in D.$$

We first assume that $|x - y| > 4\varepsilon$ and show that in this case

$$\sup_{|x-\widetilde{x}|<\varepsilon} \sup_{|y-\widetilde{y}|<\varepsilon} G(\widetilde{x}, \widetilde{y}) \leqslant CG(x, y). \tag{8.4}$$

With $\tau = \inf\{0 < t \leqslant T : B(t) \notin \mathcal{B}(x, 2\varepsilon)\}$ we note that, for all $\widetilde{x} \in \mathcal{B}(x, \varepsilon)$,

$$G(\widetilde{x}, y) = \mathbb{E}_{\widetilde{x}}\big[G(B(\tau), y), \tau \leqslant T\big].$$

This is the average of $G(\cdot, y)$ with respect to the harmonic measure $\mu_{\partial \mathcal{B}(x, 2\varepsilon)}(\widetilde{x}, \cdot)$. This measure has a density with respect to the uniform measure on the sphere $\partial \mathcal{B}(x, 2\varepsilon)$, which is bounded from above by an absolute constant. In the cases (1) and (3) this can be seen directly from Poisson's formula. Therefore $G(\widetilde{x}, y) \leqslant CG(x, y)$ and repetition of this argument, introducing now $\widetilde{y} \in \mathcal{B}(y, \varepsilon)$ and fixing $\widetilde{x}$ gives the claim.

Now look at the case $|x - y| \leqslant 4\varepsilon$. We first observe that, for some constant $c > 0$, $G(x, y) \geqslant c\varepsilon^{2-d}$, which is obvious in all cases. Now let $z \in \mathcal{B}(x, \varepsilon)$. Decomposing the Brownian path on its first exit time τ from $\mathcal{B}(x, 8\varepsilon)$ and denoting the uniform distribution on $\partial \mathcal{B}(x, 8\varepsilon)$ by ϖ we obtain for constants $C_1, C_2 > 0$,

$$
\begin{aligned}
h_\varepsilon(z, y) &\leqslant \mathbb{E}_z[\tau \wedge T] + \mathbb{E}_z\big[h_\varepsilon(B(\tau), y), \tau \leqslant T\big] \\
&\leqslant C_1 \varepsilon^2 + C_2 \varepsilon^d \int G(w, y) \, d\varpi(w),
\end{aligned}
$$

where we have used (8.4). As $\int G(w, y) \, d\varpi(w) \leqslant C_3 G(x, y)$ putting all facts together gives $h_\varepsilon(z, y) \leqslant C_4 \varepsilon^d G(x, y)$, as required. ∎

The following theorem shows that (in all three cases of transient Brownian motions) Martin capacity is indeed a good estimate of the hitting probability.

Theorem 8.24 *Let $\{B(t) : 0 \leqslant t \leqslant T\}$ be a transient Brownian motion and $A \subset D$ closed. Then*

$$\tfrac{1}{2} \operatorname{Cap}_M(A) \leqslant \mathbb{P}_0\{\exists 0 < t \leqslant T \text{ such that } B(t) \in A\} \leqslant \operatorname{Cap}_M(A) \tag{8.5}$$

Proof. Let μ be the (possibly sub-probability) distribution of $B(\tau)$ for the stopping time $\tau = \inf\{0 < t \leqslant T : B(t) \in A\}$. Note that the total mass of μ is

$$\mu(A) = \mathbb{P}_0\{\tau \leqslant T\} = \mathbb{P}_0\{B(t) \in A \text{ for some } 0 < t \leqslant T\}. \tag{8.6}$$

The idea for the upper bound is that if the harmonic measure μ is nonzero, it is an obvious candidate for a measure of finite M-energy. Recall from the definition of the Green's function, for any $y \in D$,

$$\mathbb{E}_0 \int_0^T 1\{|B(t) - y| < \varepsilon\} \, dt = \int_{\mathcal{B}(y, \varepsilon)} G(0, z) \, dz. \tag{8.7}$$

By the strong Markov property applied to the first hitting time τ of A,

$$
\begin{aligned}
\mathbb{P}_0\big\{|B(t) - y| < \varepsilon \text{ and } t \leqslant T\big\} &\geqslant \mathbb{P}_0\big\{|B(t) - y| < \varepsilon \text{ and } \tau \leqslant t \leqslant T\big\} \\
&= \mathbb{E}\mathbb{P}\big\{|B(t - \tau) - y| < \varepsilon \mid \mathcal{F}(\tau)\big\}.
\end{aligned}
$$

Integrating over t and using Fubini's theorem yields

$$\mathbb{E}_0 \int_0^T 1\{|B(t) - y| < \varepsilon\} \, dt \geqslant \int_A \int_{\mathcal{B}(y, \varepsilon)} G(x, z) \, dz \, d\mu(x).$$

Combining this with (8.7) we infer that

$$\int_{\mathcal{B}(y,\varepsilon)} \int_A G(x,z)\,d\mu(x)\,dz \leqslant \int_{\mathcal{B}(y,\varepsilon)} G(0,z)\,dz\,.$$

Dividing by $\mathcal{L}(\mathcal{B}(0,\varepsilon))$ and letting $\varepsilon \downarrow 0$ we obtain

$$\int_A G(x,y)\,d\mu(x) \leqslant G(0,y),$$

i.e. $\int_A M(x,y)\,d\mu(x) \leqslant 1$ for all $y \in D$. Therefore, $I_M(\mu) \leqslant \mu(A)$ and thus if we use $\mu/\mu(A)$ as a probability measure we get

$$\mathrm{Cap}_M(A) \geqslant [I_M(\mu/\mu(A))]^{-1} \geqslant \mu(A),$$

which by (8.6) yields the upper bound on the probability of hitting A.

To obtain a lower bound for this probability, a second moment estimate is used. It is easily seen that the Martin capacity of A is the supremum of the capacities of its compact subsets, so we may assume that A is a compact subset of the domain $D\backslash\{0\}$. We take $\varepsilon > 0$ smaller than half the distance of A to $D^c \cup \{0\}$. For $x, y \in A$ let

$$h_\varepsilon(x,y) = \int_{\mathcal{B}(y,\varepsilon)} G(x,\xi)\,d\xi$$

denote the expected time which a Brownian motion started in x spends in the ball $\mathcal{B}(y,\varepsilon)$. Also define

$$h_\varepsilon^*(x,y) = \sup_{|x-z|<\varepsilon} \int_{\mathcal{B}(y,\varepsilon)} G(z,\xi)\,d\xi.$$

Given a probability measure ν on A, and $\varepsilon > 0$, consider the random variable

$$Z_\varepsilon = \int_A \int_0^T \frac{1\{B(t) \in \mathcal{B}(y,\varepsilon)\}}{h_\varepsilon(0,y)}\,dt\,d\nu(y)\,.$$

Clearly $\mathbb{E}_0 Z_\varepsilon = 1$. By symmetry, the second moment of Z_ε can be written as

$$\mathbb{E}_0 Z_\varepsilon^2 = 2\mathbb{E}_0 \int_0^T ds \int_s^T dt \iint \frac{1\{B(s) \in \mathcal{B}(x,\varepsilon),\, B(t) \in \mathcal{B}(y,\varepsilon)\}}{h_\varepsilon(0,x)h_\varepsilon(0,y)}\,d\nu(x)\,d\nu(y)$$

$$\leqslant 2\mathbb{E}_0 \iint \int_0^T ds\, 1\{B(s) \in \mathcal{B}(x,\varepsilon)\} \frac{h_\varepsilon^*(x,y)}{h_\varepsilon(0,x)h_\varepsilon(0,y)}\,d\nu(x)\,d\nu(y) \qquad (8.8)$$

$$= 2\iint \frac{h_\varepsilon^*(x,y)}{h_\varepsilon(0,y)}\,d\nu(x)\,d\nu(y).$$

Observe that, for all fixed $x, y \in A$ we have $\lim_{\varepsilon \downarrow 0} \mathcal{L}(\mathcal{B}(0,\varepsilon))^{-1} h_\varepsilon^*(x,y) = G(x,y)$ and $\lim_{\varepsilon \downarrow 0} \mathcal{L}(\mathcal{B}(0,\varepsilon))^{-1} h_\varepsilon(0,y) = G(0,y)$. Moreover, by Proposition 8.23 and the fact that $G(0,y)$ is bounded away from zero and infinity for all $y \in A$, for $0 < \varepsilon < 1$ and some constant C,

$$\frac{h_\varepsilon^*(x,y)}{h_\varepsilon(0,y)} \leqslant C\,\frac{G(x,y)}{G(0,y)} = C\,M(x,y).$$

Hence, if ν is a measure of finite energy, we can use dominated convergence and obtain,

$$\lim_{\varepsilon \downarrow 0} \mathbb{E}Z_\varepsilon^2 \leqslant 2\iint \frac{G(x,y)}{G(0,y)}\,d\nu(x)\,d\nu(y) = 2I_M(\nu). \qquad (8.9)$$

Clearly, the hitting probability $\mathbb{P}\{\exists t > 0, y \in A \text{ such that } B(t) \in \mathcal{B}(y, \varepsilon)\}$ is at least

$$\mathbb{P}\{Z_\varepsilon > 0\} \geqslant \frac{(\mathbb{E}Z_\varepsilon)^2}{\mathbb{E}Z_\varepsilon^2} = (\mathbb{E}Z_\varepsilon^2)^{-1},$$

where we have used the Paley–Zygmund inequality in the second step. Compactness of A, together with transience and continuity of Brownian motion, imply that if the Brownian path visits every ε-neighbourhood of the compact set A then it intersects A itself. Therefore, by (8.9),

$$\mathbb{P}\{\exists t > 0 \text{ such that } B(t) \in A\} \geqslant \lim_{\varepsilon \downarrow 0} (\mathbb{E}Z_\varepsilon^2)^{-1} \geqslant \frac{1}{2I_M(\nu)}.$$

Since this is true for all probability measures ν on A, we get the desired conclusion. ∎

Remark 8.25 The right hand inequality in (8.5) can be an equality: look at the case $d = 3$, $T = \infty$, our case (1), and take a sphere in $\mathbb{R}^d$ centred at the origin, which has hitting probability and capacity both equal to one. Exercise 8.7 shows that the constant $1/2$ on the left cannot be increased.   ◇

Proof of Theorem 8.20.   It suffices, by taking countable unions, to consider compact sets Λ which have positive distance from the origin. First consider the case $d \geqslant 3$. Then $G(0, x)$ is bounded away from zero and infinity. Hence the set Λ is polar if and only if its f-capacity vanishes, where $f(\varepsilon) = \varepsilon^{2-d}$.

In the case $d = 2$ we choose a large ball $\mathcal{B}(0, R)$ containing Λ. By Lemma 3.37 the Green's function for the Brownian motion stopped upon leaving $\mathcal{B}(0, R)$ satisfies

$$G(x, y) = -\tfrac{1}{\pi} \log|x - y| + \mathbb{E}_x\left[\tfrac{1}{\pi} \log|B(T) - y|\right].$$

The second summand of $G(x, y)$ is bounded from above if $x, y \in \Lambda$, and $G(0, y)$ is bounded from zero. Hence only the contribution from $-\log|x - y|$ decides about finiteness of the Martin energy of a probability measure. Therefore, any probability measure on Λ with finite Martin energy has finite f-energy for $f(\varepsilon) = -\log \varepsilon$, and vice versa. This completes the proof.   ∎

The estimates in Theorem 8.24 are valid beyond the Brownian motion case. The following proposition, which has a very similar proof to Theorem 8.24, shows that one has an analogous result in a discrete setup. We will see a surprising application of this in Chapter 9.

Proposition 8.26 *Let $\{X_n : n \in \mathbb{N}\}$ be a transient Markov chain on a countable state space S, and, for any initial state ρ, set*

$$G(x, y) = \mathbb{E}_x\left[\sum_{n=0}^{\infty} 1_{\{y\}}(X_n)\right] \quad \text{and} \quad M(x, y) = \frac{G(x, y)}{G(\rho, y)}.$$

Then, for any subset Λ of S,

$$\tfrac{1}{2}\operatorname{Cap}_M(\Lambda) \leqslant \mathbb{P}_\rho\{\{X_n : n \in \mathbb{N}\} \text{ hits } \Lambda\} \leqslant \operatorname{Cap}_M(\Lambda).$$

Proof.   To prove the right hand inequality, we may assume that the hitting probability is positive. Let $\tau = \inf\{n\colon X_n \in \Lambda\}$ and let ν be the measure $\nu(A) = \mathbb{P}_\rho\{\tau < \infty \text{ and } X_\tau \in A\}$. In general, ν is a sub-probability measure, as τ may be infinite. By the Markov property, for $y \in \Lambda$,

$$\int_\Lambda G(x,y)\, d\nu(x) = \sum_{x \in \Lambda} \mathbb{P}_\rho\{X_\tau = x\}\, G(x,y) = G(\rho,y)\,,$$

whence $\int_\Lambda M(x,y)\, d\nu(x) = 1$. Therefore $I_M(\nu) = \nu(\Lambda)$, $I_M\big(\nu/\nu(\Lambda)\big) = [\nu(\Lambda)]^{-1}$; consequently, since $\nu/\nu(\Lambda)$ is a probability measure,

$$\mathrm{Cap}_M(\Lambda) \geqslant \nu(\Lambda) = \mathbb{P}_\rho\{\{X_n\} \text{ hits } \Lambda\}\,.$$

This yields one inequality. Note that the Markov property was used here.

For the reverse inequality, we use the second moment method. Given a probability measure μ on Λ, set

$$Z = \int_\Lambda \sum_{n=0}^\infty 1_{\{y\}}(X_n) \frac{d\mu(y)}{G(\rho,y)}\,.$$

$\mathbb{E}_\rho[Z] = 1$, and the second moment satisfies

$$\begin{aligned}
\mathbb{E}_\rho[Z^2] &= \mathbb{E}_\rho \int_\Lambda \int_\Lambda \sum_{m=0}^\infty \sum_{n=0}^\infty 1_{\{x\}}(X_m) 1_{\{y\}}(X_n) \frac{d\mu(x)\,d\mu(y)}{G(\rho,x)G(\rho,y)} \\
&\leqslant 2\mathbb{E}_\rho \int_\Lambda \int_\Lambda \sum_{m \leqslant n} 1_{\{x\}}(X_m) 1_{\{y\}}(X_n) \frac{d\mu(x)\,d\mu(y)}{G(\rho,x)G(\rho,y)}\,.
\end{aligned}$$

Observe that

$$\sum_{m=0}^\infty \mathbb{E}_\rho \sum_{n=m}^\infty 1_{\{x\}}(X_m) 1_{\{y\}}(X_n) = \sum_{m=0}^\infty \mathbb{P}_\rho\{X_m = x\}\, G(x,y) = G(\rho,x)G(x,y)\,.$$

Hence

$$\mathbb{E}_\rho[Z^2] \leqslant 2 \int_\Lambda \int_\Lambda \frac{G(x,y)}{G(\rho,y)}\, d\mu(x)\, d\mu(y) = 2I_M(\mu)\,,$$

and therefore

$$\mathbb{P}_\rho\{\{X_n\} \text{ hits } \Lambda\} \geqslant \mathbb{P}_\rho\{Z > 0\} \geqslant \frac{\big(\mathbb{E}_\rho[Z]\big)^2}{\mathbb{E}_\rho[Z^2]} \geqslant \frac{1}{2I_M(\mu)}\,.$$

We conclude that $\mathbb{P}_\rho\{\{X_n\} \text{ hits } \Lambda\} \geqslant \frac{1}{2}\mathrm{Cap}_M(\Lambda)$. ∎

Recall from Corollary 8.12 that we have already seen estimates for the probability that Brownian motion hits a set, which were given in terms of the total mass of the equilibrium measure. The following theorem reveals the relationship between the equilibrium measure and capacities.

Theorem 8.27 *Let* $\Lambda \subset \mathbb{R}^d$ *be a nonpolar, compact set and* $G \colon \mathbb{R}^d \times \mathbb{R}^d \to [0, \infty]$ *the Green's function of a transient Brownian motion. Then*

$$\mathrm{Cap}_G(\Lambda) = \left\{ I_G\left(\frac{\nu}{\nu(\Lambda)}\right) \right\}^{-1} = \nu(\Lambda).$$

where ν *is the equilibrium measure of* Λ. *Moreover, the probability measure* $\nu/\nu(\Lambda)$ *is the unique minimiser of the* G*-energy over the set of all probability measures on* Λ.

Remark 8.28 If Λ is polar, we have $\mathrm{Cap}_G(\Lambda) = 0 = \nu(\Lambda)$. ◇

For the proof we first note that, for the Green's function G of a transient Brownian motion, the G-energy of a *signed* measure is always nonnegative.

Lemma 8.29 *Let* μ, ν *be finite measures on* $\mathbb{R}^d$ *and* G *the Green's function* G *of a transient Brownian motion. Then, for* $\sigma = \mu - \nu$, *we have*

$$\iint G(x, y) \, d\sigma(x) \, d\sigma(y) \geqslant 0.$$

Equality holds if and only if $\nu = \mu$.

Proof. From the Chapman–Kolmogorov equation we have

$$\mathfrak{p}^*(t, x, y) = \int \mathfrak{p}^*(t/2, x, z) \, \mathfrak{p}^*(t/2, z, y) \, dz.$$

Integrating with respect to $d\sigma(x) \, d\sigma(y)$ and using the symmetry of $\mathfrak{p}^*(t, \cdot, \cdot)$ gives

$$\iint \mathfrak{p}^*(t, x, y) \, d\sigma(x) \, d\sigma(y) = \int \left(\int \mathfrak{p}^*(t/2, x, z) \, d\sigma(x) \right)^2 dz \geqslant 0.$$

Integrating over time shows that $\iint G(x, y) \, d\sigma(x) \, d\sigma(y) \geqslant 0$.
Equality in the last formula implies that

$$\int \mathfrak{p}^*(t/2, x, z) \, d\sigma(x) = 0 \qquad \text{for } \mathcal{L}\text{-almost every } z \text{ and } t.$$

Now fix a continuous function $f \colon \mathbb{R}^d \to [0, \infty)$ with compact support. We have

$$f(x) = \lim_{t \downarrow 0} \int f(z) \, \mathfrak{p}^*(t/2, x, z) \, dz,$$

and therefore

$$\int f(x) \, d\sigma(x) = \lim_{t \downarrow 0} \iint f(z) \, \mathfrak{p}^*(t/2, x, z) \, dz \, d\sigma(x) = 0,$$

and therefore $\sigma = 0$ as required. ∎

Proof of Theorem 8.27. Let ν be the equilibrium measure and define $\varphi(x) = \int G(x, y) \, d\nu(y)$. By the last exit formula, Theorem 8.8, $\varphi(x)$ is the probability that a Brownian motion started at x hits Λ before time T. Hence $\varphi(x) \leqslant 1$ for every x and, if x is a regular point for Λ, then $\varphi(x) = 1$. Also by the last exit formula, because irregular points are never hit by a Brownian motion, see Theorem 8.13, we have $\varphi(x) = 1$ for ν-almost every point. This implies that

$$I_G(\nu) = \int_\Lambda \varphi(x) \, d\nu(x) = \nu(\Lambda).$$

Suppose now that μ is an arbitrary measure on Λ with $\mu(\Lambda) = \nu(\Lambda)$ and assume that μ has finite energy. Note that μ does not charge the set of irregular points, as otherwise this set would have positive capacity with respect to the Green and hence also the Martin kernel and so would be nonpolar by Theorem 8.24. Then, starting with Lemma 8.29 and using also the symmetry of G,

$$0 \leqslant \iint G(x, y) \, d(\nu - \mu)(x) \, d(\nu - \mu)(y)$$

$$= I_G(\mu) + I_G(\nu) - 2 \iint G(x, y) \, d\nu(x) \, d\mu(y)$$

$$= I_G(\mu) + \nu(\Lambda) - 2 \int_\Lambda \varphi(y) \, d\mu(y) \leqslant I_G(\mu) - \nu(\Lambda),$$

using in the last step that $\varphi(y) = 1$ on the set of regular points, and thus μ-almost everywhere. This implies that $I_G(\mu) \geqslant \nu(\Lambda) = I_G(\nu)$, so that $\nu/\nu(\Lambda)$ is a minimiser in the definition of Cap_G. Conversely, if $I_G(\mu) = I_G(\nu)$ and $\mu(\Lambda) = \nu(\Lambda)$, the same calculation shows that

$$\iint G(x, y) \, d(\nu - \mu)(x) \, d(\nu - \mu)(y) = 0,$$

and hence, by Lemma 8.29, we have $\mu = \nu$. This completes the proof. ∎

If $d \geqslant 3$, Theorem 8.27 shows that the normalised equilibrium measure $\nu_\Lambda := \frac{\nu}{\nu(\Lambda)}$ minimises the energy with respect to the potential kernel, which is

$$\iint f(|x - y|) \, d\mu(x) \, d\mu(y)$$

for the radial potential $f(r) = r^{2-d}$, over the set of all probability measures μ on Λ. We now show an analogous statement in $d = 2$, recall that in this case the radial potential equals $f(r) = -\log(r)$ for $r < 1$.

Theorem 8.30 *Let $\Lambda \subset \mathbb{R}^2$ be a nonpolar, compact set and ν_R be the equilibrium measure of Λ for planar Brownian motion stopped at $\partial \mathcal{B}(0, R)$. Then the limit*

$$\nu_\Lambda = \lim_{R \uparrow \infty} \frac{\nu_R}{\nu_R(\Lambda)}$$

exists and minimises the energy

$$-\iint \log |x - y| \, d\mu(x) \, d\mu(y)$$

over the set of all probability measures μ on Λ.

Remark 8.31 For a compact, nonpolar set $\Lambda \subset \mathbb{R}^d$ the probability measure ν_Λ is defined in the case $d = 2$ by Theorem 8.30 and in the case $d \geqslant 3$ as the normalised equilibrium measure on Λ. We have shown that it minimises the energy

$$\iint f(|x - y|) \, d\mu(x) \, d\mu(y)$$

for the radial potential f, over the set of all probability measures μ on Λ. We therefore call ν_Λ the *energy–minimising measure* on Λ. Only in the case $d \geqslant 3$ we have proved that this measure is the *unique* minimiser of the energy with respect to the radial potential, but in $d = 2$ this will follow from Theorem 8.33 below. ◇

We postpone the proof of the *existence* of the limit of $\nu_R / \nu_R(\Lambda)$ until the proof of Theorem 8.33, and first show the energy–minimisation property for arbitrary sequential limits.

First fix $R > 0$ and recall from Theorem 8.27 that $\nu_R / \nu_R(\Lambda)$ minimises the energy

$$\iint G^{(R)}(x, y) \, d\mu(x) \, d\mu(y)$$

over all probability measure μ on Λ, where $G^{(R)}$ is the Green's function associated with the Brownian motion stopped upon leaving $\mathcal{B}(0, R)$. Our first step shows convergence of these Green's functions to the potential kernel.

Lemma 8.32 *For $x, y \in \mathbb{R}^2$ we have*

$$\lim_{R \uparrow \infty} G^{(R)}(x, y) - \frac{1}{\pi} \log R = -\frac{1}{\pi} \log |x - y|,$$

and the convergence is uniform on compact subsets on $\mathbb{R}^2 \times \mathbb{R}^2$.

Proof. Recall from Lemma 3.37 that

$$G^{(R)}(x, y) = \frac{-1}{\pi} \log |x - y| + \frac{1}{\pi} \mathbb{E}_x \left[\log \left| B(T^{(R)}) - y \right| \right],$$

where $T^{(R)}$ is the first exit time from $\mathcal{B}(0, R)$. Note that, for any compact set $K \subset \mathbb{R}^2$,

$$\log \left| z - \frac{y}{R} \right| \longrightarrow 0, \quad \text{as } R \uparrow \infty,$$

uniformly in $z \in \partial \mathcal{B}(0, 1)$ and $y \in K$. Using this, we see that

$$G^{(R)}(x, y) - \frac{1}{\pi} \log R = -\frac{1}{\pi} \log |x - y| + \frac{1}{\pi} \mathbb{E}_x \left[\log \left| \frac{B(T^{(R)})}{R} - \frac{y}{R} \right| \right]$$

$$\longrightarrow -\frac{1}{\pi} \log |x - y|,$$

uniformly in $x, y \in K$. ∎

Proof of Theorem 8.30.　　　Let μ be an arbitrary probability measure on Λ. For a radius $R > 0$ and threshold $M > 0$ we define

$$G_M^{(R)}(x,y) = \left(G^{(R)}(x,y) - \tfrac{1}{\pi}\log R\right) \wedge M.$$

Then

$$\iint G_M^{(R)}(x,y)\,\frac{d\nu_R(x)\,d\nu_R(y)}{\nu_R(\Lambda)^2} \leqslant \iint G^{(R)}(x,y)\,\frac{d\nu_R(x)\,d\nu_R(y)}{\nu_R(\Lambda)^2} - \tfrac{1}{\pi}\log R$$

$$\leqslant \iint G^{(R)}(x,y)\,d\mu(x)\,d\mu(y) - \tfrac{1}{\pi}\log R.$$

Hence, for any $M > 0$, using Lemma 8.32,

$$\limsup_{R\uparrow\infty} \iint G_M^{(R)}(x,y)\,\frac{d\nu_R(x)\,d\nu_R(y)}{\nu_R(\Lambda)^2} \leqslant -\tfrac{1}{\pi}\iint \log|x-y|\,d\mu(x)\,d\mu(y). \qquad (8.10)$$

To analyse the limsup first note that, by Lemma 8.32,

$$\lim_{R\uparrow\infty} \iint \left[\tfrac{-1}{\pi}\log|x-y| \wedge M\right]\frac{d\nu_R(x)\,d\nu_R(y)}{\nu_R(\Lambda)^2} - \iint G_M^{(R)}(x,y)\,\frac{d\nu_R(x)\,d\nu_R(y)}{\nu_R(\Lambda)^2} = 0.$$

If a sequence $R_n \uparrow \infty$ is chosen such that, in the sense of weak convergence,

$$\lim_{n\uparrow\infty} \frac{\nu_{R_n}}{\nu_{R_n}(\Lambda)} = \nu_\Lambda,$$

then, by Exercise 8.10, we have

$$\lim_{n\uparrow\infty} \iint \left[\tfrac{-1}{\pi}\log|x-y| \wedge M\right]\frac{d\nu_{R_n}(x)\,d\nu_{R_n}(y)}{\nu_{R_n}(\Lambda)^2} = \iint \left[\tfrac{-1}{\pi}\log|x-y| \wedge M\right]d\nu_\Lambda(x)\,d\nu_\Lambda(y).$$

Combining this and inserting the limit in (8.10), we obtain

$$\iint \left[\tfrac{-1}{\pi}\log|x-y| \wedge M\right]d\nu_\Lambda(x)\,d\nu_\Lambda(y) \leqslant -\tfrac{1}{\pi}\iint \log|x-y|\,d\mu(x)\,d\mu(y).$$

Now let $M \uparrow \infty$ and use monotone convergence to obtain

$$-\iint \log|x-y|\,d\nu_\Lambda(x)\,d\nu_\Lambda(y) \leqslant -\iint \log|x-y|\,d\mu(x)\,d\mu(y).$$

As μ was arbitrary, this proves the minimality property of ν_Λ.　　　∎

We conclude this section by showing that the energy–minimising measure agrees with the harmonic measure from infinity, which was introduced in Chapter 3. In the course of the proof we also add the missing part to Theorem 8.30, the existence of the limit in the case $d = 2$.

Theorem 8.33 *Let $\Lambda \subset \mathbb{R}^d$, for $d \geqslant 2$, be a compact, nonpolar set. Then*

$$\nu_\Lambda = \mu_\Lambda,$$

i.e. the energy–minimising measure for the radial potential agrees with the harmonic measure from infinity.

We first give a proof of Theorem 8.33 for the case $d = 2$, which uses the skew-product representation, see Theorem 7.26. Fix $0 < r < R$ and let $\{B(t): t \geqslant 0\}$ be a Brownian motion started uniformly on the sphere $\partial \mathcal{B}(0, R)$. Let $\tau_r := \tau(\mathcal{B}(0, r))$ be the first hitting time of the small ball inside, which is finite almost surely, and $\tau_{r,R} := \tau(\mathcal{B}(0, r), \partial \mathcal{B}(0, R))$ the time of first return to the starting sphere afterwards. Moreover, let

$$\sigma_R := \sup \left\{ 0 \leqslant t < \tau_r \colon B(t) \in \partial \mathcal{B}(0, R) \right\},$$

the time of the last visit to $\partial \mathcal{B}(0, R)$ before the smaller ball is visited. We call the path $\{e(t): 0 \leqslant t \leqslant \tau_{r,R} - \sigma_R\}$ given by

$$e(t) := B(\sigma_R + t)$$

a *Brownian excursion* in $\mathcal{B}(0, R)$ conditioned to hit $\mathcal{B}(0, r)$. We denote by $\tau^e := \tau_{r,R} - \sigma_R$ the *lifetime* of the excursion. Note that this is also the first positive time when the excursion returns to its starting sphere. The main ingredient of the proof is the following time-reversal property of the excursions.

Lemma 8.34 *The laws of the paths $\{e(t): 0 \leqslant t \leqslant \tau^e\}$ and $\{e(\tau^e - t): 0 \leqslant t \leqslant \tau^e\}$ coincide.*

Proof.   We invoke the skew-product representation of $\{B(t): t \geqslant 0\}$ established in Theorem 7.26. This allows us to write

$$B(t) = \exp \left(W_1(H(t)) + \mathrm{i}\, W_2(H(t)) \right), \text{ for all } t \geqslant 0,$$

where $\{W_1(t): t \geqslant 0\}$, with $W_1(0) = \log R$, and $\{W_2(t): t \geqslant 0\}$, with $W_2(0)$ uniformly distributed on $[0, 2\pi)$, are two independent linear Brownian motions. We further have

$$H^{-1}(u) = \int_0^u e^{2W_1(s)} \, ds,$$

so that $\{H^{-1}(t): t \geqslant 0\}$ is a continuous, strictly increasing process adapted to the natural filtration of $\{W_1(t): t \geqslant 0\}$. Hence, $H(\tau_r) = \inf\{u \geqslant 0: W_1(u) = \log r\}$, $H(\sigma_R) = \sup\{0 \leqslant u < H(\tau_r): W_1(u) = \log R\}$ and $H(\tau_{r,R}) = \inf\{u > H(\tau_r): W_1(u) = \log R\}$. By Exercise 5.12 (b) the one-dimensional excursions $\{e_1(s): 0 \leqslant s \leqslant \tau_1^e\}$ defined by

$$e_1(s) = W_1(H(\sigma_R) + s), \quad \tau_1^e = H(\tau_{r,R}) - H(\sigma_R),$$

are time-reversible in law. Marking quantities defined with respect to the time-reversed excursion by $\tilde{\ }$, we obtain for all $0 \leqslant s \leqslant \tau_1^e$,

$$H^{-1}\big(H(\sigma_R) + s\big) - \sigma_R = \int_{H(\sigma_R)}^{H(\sigma_R)+s} e^{2W_1(u)} \, du = \int_0^s e^{2e_1(u)} \, du$$

$$\stackrel{\mathrm{d}}{=} \int_0^s e^{2e_1(\tau_1^e - u)} \, du = \tilde{H}^{-1}(s).$$

For any $0 \leqslant t \leqslant \tau_{r,R} - \sigma_R$ we write $s = H(\sigma_R + t) - H(\sigma_R)$, or equivalently $t = \tilde{H}^{-1}(s)$. Hence

$$\big| B(\sigma_R + t) \big| = \exp \big(W_1(H(\sigma_R + t)) \big) = \exp \big(e_1(s) \big)$$

$$\stackrel{\mathrm{d}}{=} \exp \big(e_1(\tau_1^e - s) \big) = \exp \big(e_1(\tau_1^e - \tilde{H}(t)) \big) = \big| \tilde{B}(t) \big|.$$

As $H(\sigma_r + t) = \int_0^t |B(\sigma_r + u)|^{-2}\, du$, this implies that $H(\sigma_R + t) \overset{\mathrm{d}}{=} \tilde{H}(t)$ and therefore,

$$B(\sigma_R + t) = \exp\left(W_1(H(\sigma_R + t)) + iW_2(H(\sigma_R + t))\right)$$
$$\overset{\mathrm{d}}{=} \exp\left(W_1(\tilde{H}(t)) + iW_2(\tilde{H}(t))\right) = \tilde{B}(t),$$

as required. ∎

Proof of Theorem 8.33 for $d = 2$. Let $\nu_\Lambda = \lim \nu_{R_n}/\nu_{R_n}(\Lambda)$ be any subsequential limit taken along a sequence $R_n \uparrow \infty$. Fix $r > 0$ so that $\Lambda \subset \mathcal{B}(0, r)$. For $R > r$, let $\gamma_R = 0$ if the Brownian motion $\{B(t) : t \geqslant 0\}$ does not hit Λ before time $\tau_{r,R}$, and

$$\gamma_R := \sup\left\{0 \leqslant t \leqslant \tau_{r,R} : B(t) \in \Lambda\right\},$$

otherwise. By Theorem 8.10, for any Borel set $A \subset \Lambda$,

$$\begin{aligned}
\nu_\Lambda(A) &= \lim_{n \to \infty} \frac{\nu_{R_n}(A)}{\nu_{R_n}(\Lambda)} \\
&= \lim_{n \to \infty} \mathbb{P}\big\{B(\gamma_{R_n}) \in A \,\big|\, \gamma_{R_n} > 0\big\} \\
&= \lim_{n \to \infty} \mathbb{P}\big\{e(\gamma_{R_n} - \sigma_{R_n}) \in A \,\big|\, \{e(t) : 0 \leqslant t \leqslant \tau^e\} \text{ hits } \Lambda\big\} \\
&= \lim_{n \to \infty} \mathbb{P}\big\{e(\tau_{r,R_n} - \gamma_{R_n}) \in A \,\big|\, \{e(t) : 0 \leqslant t \leqslant \tau^e\} \text{ hits } \Lambda\big\},
\end{aligned}$$

where we have used Lemma 8.34 in the last step. Now, fixing R_n, let $\{B^*(t) : t \geqslant 0\}$ be a Brownian motion started uniformly on $\partial\mathcal{B}(0, R_n)$ whose associated excursion in $\mathcal{B}(0, R_n)$ conditioned to hit $\mathcal{B}(0, r)$ is $\{e(\tau^e - t) : 0 \leqslant t \leqslant \tau^e\}$. Note that $e(\tau_{r,R_n} - \gamma_{R_n}) = B^*(\tau^*(\Lambda))$, where $\tau^*(\Lambda)$ is the first hitting time of Λ by $\{B^*(t) : t \geqslant 0\}$. Hence the last line in the previous display equals

$$\lim_{n \to \infty} \mathbb{P}\big\{B^*(\tau^*(\Lambda)) \in A \,\big|\, \{B^*(t) : 0 \leqslant t \leqslant \tau^*_{r,R_n}\} \text{ hits } \Lambda\big\},$$

where τ^*_{r,R_n} is the time of first return of $\{B^*(t) : t \geqslant 0\}$ after hitting $\mathcal{B}(0, r)$. As $n \to \infty$, the probability of the conditioning event goes to one, so that we can conclude that

$$\nu_\Lambda(A) = \lim_{n \to \infty} \mathbb{P}\big\{B^*(\tau^*(\Lambda)) \in A\big\} = \mu_\Lambda(A),$$

where we used the definition of the harmonic measure from infinity in the final step. ∎

We now give a proof of Theorem 8.33 for the case $d \geqslant 3$. Again a 'time-reversal' argument is crucial. We start by constructing a family of probability measures μ_t, for $t > 0$, on the space $\mathbf{C}(\mathbb{R}, \mathbb{R}^d)$ of continuous functions from the reals to $\mathbb{R}^d$ by

$$\mu_t(A) = \frac{1}{c_t} \int \mathbb{P}_x\big\{\{B(s) : s \in \mathbb{R}\} \in A, \tau_{\mathcal{B}(0,r)} < t\big\}\, dx, \qquad \text{for } A \subset \mathbf{C}(\mathbb{R}, \mathbb{R}^d) \text{ Borel},$$

where $\{B(s) : s \in \mathbb{R}\}$ under $\mathbb{P}_x$ is a two-sided Brownian motion with $B(0) = x$,

$$\tau_{\mathcal{B}(0,r)} = \inf\{s > 0 : B(s) \in \mathcal{B}(0, r)\}$$

is the first hitting time of the fixed ball $\mathcal{B}(0, r)$ after time zero, and $c_t = \int \mathbb{P}_x\{\tau_{\mathcal{B}(0,r)} <$

$t\}\,dx$ is the normalising constant. Observe that $c_t < \infty$ for any $t > 0$, see Exercise 8.11. The following lemma contains the required time-reversal property.

Lemma 8.35

    (a) *The laws of* $\{B(s)\colon s \geqslant 0\}$ *and* $\{B(t-s)\colon s \geqslant 0\}$ *under* μ_t *agree;*

    (b) *as* $t \uparrow \infty$, *the law of* $\{B(\tau_{\mathcal{B}(0,r)} + s)\colon s \geqslant 0\}$ *under* μ_t *converges in the total variation distance to the law of a Brownian motion started uniformly on the sphere* $\partial\mathcal{B}(0,r)$.

Proof.   **(a)** From Fubini's theorem, we obtain that

$$\int \mathbb{P}_x\big\{\{B(s)\colon s \in \mathbb{R}\} \in A, \tau_{\mathcal{B}(0,r)} < t\big\}\,dx$$

$$= \mathbb{E}_0 \int \mathbb{1}\{\{x + B(s)\colon s \in \mathbb{R}\} \in A, \tau_{\mathcal{B}(-x,r)} < t\}\,dx.$$

Abbreviate $\sigma_{\mathcal{B}(x,r)} = \inf\{s \geqslant 0\colon B(t-s) \in \mathcal{B}(x,r)\}$. Using first the Markov property and then the shift-invariance of the Lebesgue measure, we continue

$$= \mathbb{E}_0 \int \mathbb{1}\{\{x + B(t-s) - B(t)\colon s \in \mathbb{R}\} \in A, \sigma_{\mathcal{B}(B(t)-x,r)} < t\}\,dx$$

$$= \mathbb{E}_0 \int \mathbb{1}\{\{x + B(t-s)\colon s \in \mathbb{R}\} \in A, \sigma_{\mathcal{B}(-x,r)} < t\}\,dx.$$

Finally, using Fubini's theorem again and then observing that $\sigma_{\mathcal{B}(0,r)} < t$ if and only if $\tau_{\mathcal{B}(0,r)} < t$, we can continue,

$$= \int \mathbb{P}_x\big\{\{B(t-s)\colon s \in \mathbb{R}\} \in A, \sigma_{\mathcal{B}(0,r)} < t\big\}\,dx$$

$$= \int \mathbb{P}_x\big\{\{B(t-s)\colon s \in \mathbb{R}\} \in A, \tau_{\mathcal{B}(0,r)} < t\big\}\,dx.$$

(b) It is clear from the symmetry of the Lebesgue measure, that the law of $\{B(\tau_{\mathcal{B}(0,r)} + s)\colon s \geqslant 0\}$ under the probability measure μ_t^*, given by

$$\mu_t^*(A) = \frac{1}{c_t^*} \int_{\mathcal{B}(0,r)^c} \mathbb{P}_x\big\{\{B(s)\colon s \in \mathbb{R}\} \in A, \tau_{\mathcal{B}(0,r)} < t\big\}\,dx, \quad \text{for } A \subset \mathbf{C}(\mathbb{R}, \mathbb{R}^d) \text{ Borel,}$$

is the law of a Brownian motion started uniformly on the sphere $\partial\mathcal{B}(0,r)$. Here the normalising constant is $c_t^* = \int_{\mathcal{B}(0,r)^c} \mathbb{P}_x\{\tau_{\mathcal{B}(0,r)} < t\}\,dx$. The total variation distance of μ_t and μ_t^* is

$$\sup_A \big|\mu_t^*(A) - \mu_t(A)\big|$$

$$\leqslant \left|\frac{1}{c_t^*} - \frac{1}{c_t}\right| \int_{\mathcal{B}(0,r)^c} \mathbb{P}_x\{\tau_{\mathcal{B}(0,r)} < t\}\,dx + \frac{1}{c_t} \int_{\mathcal{B}(0,r)} \mathbb{P}_x\{\tau_{\mathcal{B}(0,r)} < t\}\,dx$$

$$\leqslant \left|\frac{1}{c_t^*} - \frac{1}{c_t}\right| c_t^* + \frac{1}{c_t}\mathcal{L}(\mathcal{B}(0,r)).$$

As $c_t = c_t^* + \mathcal{L}(\mathcal{B}(0,r))$, it suffices to show that $c_t^* \to \infty$. This follows from the hitting estimate of Corollary 3.19 as

$$\lim_{t\uparrow\infty} c_t^* = \int_{\mathcal{B}(0,r)^c} \mathbb{P}_x\{\tau_{\mathcal{B}(0,r)} < \infty\}\, dx = \int_{\mathcal{B}(0,r)^c} \frac{r^{d-2}}{|x|^{d-2}}\, dx$$

$$= r^d \int_{\mathcal{B}(0,1)^c} |x|^{2-d}\, dx = \infty,$$

and this completes the proof. ∎

Proof of Theorem 8.33 for d > 2. Let $r > 0$ such that $\Lambda \subset \mathcal{B}(0,r)$, and look at a Brownian motion started uniformly on $\partial\mathcal{B}(0,r)$. Define $\gamma \geqslant 0$ as $\gamma = 0$, if the Brownian motion never hits Λ, and $\gamma = \sup\{t > 0 : B(t) \in \Lambda\}$ otherwise. By Theorem 8.10 and Lemma 8.35(b), for any Borel set $A \subset \Lambda$,

$$\nu_\Lambda(A) = \mathbb{P}\{B(\gamma) \in A \,|\, \gamma > 0\} = \lim_{t\uparrow\infty} \mu_t\{B(\gamma_t) \in A \,|\, \gamma_t > 0\},$$

where $\gamma_t = 0$ if $\{B(t) : t \geqslant 0\}$ does not hit Λ during the time $[0,t]$ and otherwise is the last time $s \in [0,t]$ with $B(s) \in \Lambda$. We now express all the events in terms of the time reversed Brownian motion $\{B^*(s) : s \geqslant 0\}$ defined by $B^*(s) = B(t-s)$. Recall from Lemma 8.35(a) that, under μ_t, this process has the same law as $\{B(t) : t \geqslant 0\}$. Let τ^* be the first hitting time of Λ by $\{B^*(s) : s \geqslant 0\}$ and note that $\tau^* < t$ if and only if $\gamma_t > 0$. If this is the case, then $\tau^* = t - \gamma_t$. Hence

$$\nu_\Lambda(A) = \lim_{t\uparrow\infty} \mu_t\{B^*(\tau^*) \in A \,|\, \tau^* < t\}.$$

Define $\tau^*_{\mathcal{B}(0,r)} = \inf\{s > 0 : B^*(s) \in \mathcal{B}(0,r)\}$ and look at the embedded Brownian motion $\{B^{**}(s) : s \geqslant 0\}$ defined by $B^{**}(s) = B^*(\tau^*_{\mathcal{B}(0,r)} + s)$. If $B(t) \notin \mathcal{B}(0,r)$ its first hitting time of Λ equals

$$\tau^{**} := \inf\{s : B^{**}(s) \in \Lambda\} = \tau^* - \tau^*_{\mathcal{B}(0,r)}.$$

Hence, we obtain

$$\nu_\Lambda(A) = \lim_{t\uparrow\infty} \mu_t\{B^{**}(\tau^{**}) \in A \,|\, \tau^* < t\} = \mathbb{P}\{B(\tau) \in A \,|\, \tau < \infty\} = \mu_\Lambda(A),$$

where τ is the first hitting time of Λ by the Brownian motion $\{B(t) : t \geqslant 0\}$, which is started uniformly on $\partial\mathcal{B}(0,r)$, and we have used Theorem 3.50 in the last step. ∎

Example 8.36 Recall from Example 7.24 that the Beta$(\frac{1}{2}, \frac{1}{2})$ distribution on $[0,1]$ given by the density

$$g(x) = \frac{1}{\pi} \frac{1}{\sqrt{x(1-x)}}\, dx,$$

is the harmonic measure of the unit interval embedded in the plane. By Theorem 8.33 the function g therefore maximises the expression

$$\int_0^1 \int_0^1 f(x) \log|x - y|\, f(y)\, dx\, dy$$

over all probability densities f on $[0,1]$. ◇

8.4 Wiener's test of regularity

In this section we concentrate on $d \geqslant 3$ and find a sharp criterion for a point to be regular for a closed set $\Lambda \subset \mathbb{R}^d$. This criterion is given in terms of the capacity of the intersection of Λ with annuli, or shells, concentric about x.

To fix some notation let $k > \ell$ be integers and $x \in \mathbb{R}^d$, and define the annulus

$$A_x(k, \ell) := \{ y \in \mathbb{R}^d : 2^{-k} \leqslant |y - x| \leqslant 2^{-\ell} \}.$$

Abbreviate $A_x(k) := A_x(k+1, k)$ and let

$$\Lambda_x^k := \Lambda \cap A_x(k).$$

We aim to prove the following result.

Theorem 8.37 (Wiener's test) *A point $x \in \mathbb{R}^d$ is regular for the closed set $\Lambda \subset \mathbb{R}^d$, $d \geqslant 3$, if and only if*

$$\sum_{k=1}^{\infty} 2^{k(d-2)} C_{d-2}(\Lambda_x^k) = \infty,$$

where C_{d-2} is the Newtonian capacity introduced in Definition 4.31.

In the proof, we may assume, without loss of generality, that $x = 0$. We start the proof with an easy observation.

Lemma 8.38 *There exists a constant $c > 0$, which depends only on the dimension d, such that, for all k, we have*

$$c\, 2^{k(d-2)} C_{d-2}(\Lambda_0^k) \leqslant \mathrm{Cap}_M(\Lambda_0^k) \leqslant c\, 2^{(k+1)(d-2)} C_{d-2}(\Lambda_0^k).$$

Proof. Observe that, as $z \in \Lambda_0^k$ implies $2^{-k-1} \leqslant |z| \leqslant 2^{-k}$, we obtain the statement by estimating the denominator in the Martin kernel M. ∎

The crucial step in the proof is a quantitative estimate, from which Wiener's test follows quickly.

Lemma 8.39 *There exists a constant $c > 0$, depending only on the dimension d, such that*

$$1 - \exp\left(-c \sum_{j=\ell}^{k-1} \mathrm{Cap}_M(\Lambda_0^j)\right) \leqslant \mathbb{P}_0\{\{B(t) : t \geqslant 0\} \text{ hits } \Lambda \cap A_0(k, \ell)\} \leqslant \sum_{j=\ell}^{k-1} \mathrm{Cap}_M(\Lambda_0^j).$$

Proof. For the *upper* bound we look at the event $D(j)$ that a Brownian motion started in 0 hits Λ_0^j. Then, using Theorem 8.24, we get $\mathbb{P}_0(D(j)) \leqslant \mathrm{Cap}_M(\Lambda_0^j)$. Therefore

$$\mathbb{P}_0\{\{B(t) : t \geqslant 0\} \text{ hits } \Lambda \cap A_0(k, \ell)\} \leqslant \mathbb{P}_0\left(\bigcup_{j=\ell}^{k-1} D(j)\right) \leqslant \sum_{j=\ell}^{k-1} \mathrm{Cap}_M(\Lambda_0^j),$$

and this completes the proof of the upper bound.

For the *lower* bound we look at the event $E(z, j)$ that a Brownian motion started in some point $z \in \partial \mathcal{B}(0, 2^{-j})$ and stopped upon hitting $\partial \mathcal{B}(0, 2^{-j+4})$ hits Λ_0^{j-2}. Again we use either Theorem 8.24, or Corollary 8.12 in conjunction with Theorem 8.27, to get, for constants $c_1, c_2 > 0$ depending only on the dimension d,

$$\mathbb{P}_z\{\{B(t) : t \geqslant 0\} \text{ ever hits } \Lambda_0^{j-2}\} \geqslant c_1 \left(2^{-(j-1)} - 2^{-j}\right)^{2-d} C_{d-2}(\Lambda_0^{j-2}),$$

and, for any $y \in \partial \mathcal{B}(0, 2^{-j+4})$,

$$\mathbb{P}_y\{\{B(t) : t \geqslant 0\} \text{ ever hits } \Lambda_0^{j-2}\} \leqslant c_2 \left(2^{-(j-4)} - 2^{-(j-2)}\right)^{2-d} C_{d-2}(\Lambda_0^{j-2}).$$

Therefore, for a constant $c > 0$ depending only on the dimension d,

$$\mathbb{P}(E(z,j)) \geqslant \mathbb{P}_z\{\{B(t)\} \text{ ever hits } \Lambda_0^{j-2}\} - \max_{y \in \partial \mathcal{B}(0,2^{-j+4})} \mathbb{P}_y\{\{B(t)\} \text{ ever hits } \Lambda_0^{j-2}\}$$

$$\geqslant c\, 2^{j(d-2)} C_{d-2}(\Lambda_0^{j-2}).$$

Now divide $\{\ell+2, \ldots, k+1\}$ into (at most) four subsets such that each subset I satisfies $|i-j| \geqslant 4$ for all $i \neq j \in I$. Choose a subset I which satisfies

$$\sum_{j \in I} 2^{(j-2)(d-2)} C_{d-2}(\Lambda_0^{j-2}) \geqslant \tfrac{1}{4} \sum_{j=\ell}^{k-1} 2^{j(d-2)} C_{d-2}(\Lambda_0^j). \tag{8.11}$$

Now we have with $\tau_j = \inf\{t \geqslant 0 \colon |B(t)| = 2^{-j}\}$,

$$\mathbb{P}_0\{\{B(t) : t \geqslant 0\} \text{ avoids } \Lambda \cap A_0(k,\ell)\} \leqslant \mathbb{P}_0\Big(\bigcap_{j \in I} E\big(B(\tau_j), j\big)^c\Big)$$

$$\leqslant \prod_{j \in I} \sup_{z \in \partial \mathcal{B}(0,2^{-j})} \mathbb{P}(E(z,j)^c) \leqslant \prod_{j \in I} \left(1 - c\, 2^{j(d-2)} C_{d-2}(\Lambda_0^{j-2})\right)$$

$$\leqslant \exp\Big(-c \sum_{j \in I} 2^{j(d-2)} C_{d-2}(\Lambda_0^{j-2})\Big),$$

using the estimate $\log(1-x) \leqslant -x$ in the last step. The lower bound now follows from (8.11) and Lemma 8.38 when we pass to the complement. ∎

Proof of Wiener's test. Suppose $\sum_{k=1}^\infty 2^{k(d-2)} C_{d-2}(\Lambda_0^k) = \infty$. Therefore, by Lemma 8.39 and Lemma 8.38, for all $k \in \mathbb{N}$,

$$\mathbb{P}_0\{\{B(t) : t \geqslant 0\} \text{ hits } \Lambda \cap \mathcal{B}(0, 2^{-k})\} \geqslant 1 - \exp\Big(-c \sum_{j=k}^\infty \mathrm{Cap}_M(\Lambda_0^j)\Big) = 1.$$

Since points are polar, for any $\varepsilon, \delta > 0$ there exists a large k such that

$$\mathbb{P}_0\{\{B(t) : t \geqslant \varepsilon\} \text{ hits } \mathcal{B}(0, 2^{-k})\} < \delta.$$

Combining these two facts we get for the first hitting time $\tau = \tau(\Lambda)$ of the set Λ,

$$\mathbb{P}_0\{\tau < \varepsilon\} \geqslant \mathbb{P}_0\{\{B(t) : t \geqslant 0\} \text{ hits } \Lambda \cap \mathcal{B}(0, 2^{-k})\}$$
$$- \mathbb{P}_0\{\{B(t) : t \geqslant \varepsilon\} \text{ hits } \mathcal{B}(0, 2^{-k})\} \geqslant 1 - \delta.$$

As $\varepsilon, \delta > 0$ were arbitrary, the point 0 must be regular.

Now suppose that $\sum_{k=1}^{\infty} 2^{k(d-2)} C_{d-2}\left(\Lambda_0^k\right) < \infty$. Then

$$\sum_{k=1}^{\infty} \mathbb{P}_0\{\{B(t) : t \geqslant 0\} \text{ hits } \Lambda \cap A_0(k)\} \leqslant \sum_{k=1}^{\infty} \mathrm{Cap}_M\left(\Lambda_0^k\right) < \infty.$$

Hence, by the Borel Cantelli lemma, almost surely there exists a ball $\mathcal{B}(0, \varepsilon)$ such that $\{B(t) : t \geqslant 0\}$ does not hit $\mathcal{B}(0, \varepsilon) \cap \Lambda$. By continuity we therefore must have $\inf\{t > 0 : B(t) \in \Lambda\} > 0$ almost surely, hence the point 0 is irregular.   ∎

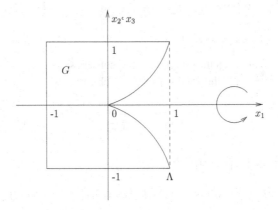

Fig. 8.2. Lebesgue's thorn.

Example 8.40 The following example is due to Lebesgue [Le24], and is usually called **Lebesgue's thorn**. For any $\alpha > 0$ we define an open subset $G \subset (-1, 1)^3$ with a *cusp* at zero by

$$G := \left\{(x_1, x_2, x_3) \in (-1, 1)^3 : \sqrt{x_2^2 + x_3^2} > x_1^\alpha \text{ if } x_1 \geqslant 0\right\},$$

see Figure 8.2. Now the origin is an *irregular* point for $\Lambda = G^c$ if $\alpha > 1$. For the proof it suffices, by Wiener's test, to check that

$$\sum_{k=1}^{\infty} 2^k C_1\left(\Lambda_0^k\right) < \infty.$$

Note that, for any probability measure μ on Λ_0^k, we have $I_1(\mu) \geqslant 2^{\alpha k}$ and, hence,

$$\sum_{k=1}^{\infty} 2^k C_1\left(\Lambda_0^k\right) \leqslant \sum_{k=1}^{\infty} 2^{k(1-\alpha)} < \infty,$$

verifying Wiener's test of irregularity. Conversely, the Poincaré cone condition, see Theorem 8.3, shows that for $\alpha \leqslant 1$ the origin is *regular* for $\Lambda = G^c$.   ◇

Exercises

Exercise 8.1. Ⓢ Let $U \subset \mathbb{R}^d$ be a domain and $u \colon U \to \mathbb{R}$ subharmonic. Use Itô's formula to show that, for any ball $\mathcal{B}(x, r) \subset U$,

$$u(x) \leqslant \frac{1}{\mathcal{L}(\mathcal{B}(x, r))} \int_{\mathcal{B}(x, r)} u(y) \, dy.$$

Exercise 8.2. Let $x \in U \subset \mathbb{R}^d$ be a domain and suppose that

$$\liminf_{r \downarrow 0} \frac{\mathcal{L}(\mathcal{B}(x, r) \cap U^c)}{r^d} > 0.$$

Show that x is regular for the complement of U.

Exercise 8.3. Ⓢ Suppose g is bounded and u a solution of Poisson's problem for g. Show that this solution has the form

$$u(x) = \mathbb{E}_x \left[\int_0^T g(B(t)) \, dt \right], \qquad \text{for } x \in U,$$

where $T := \inf\{t > 0 \colon B(t) \notin U\}$. Observe that this implies that the solution, if it exists, is always uniquely determined.

Exercise 8.4. Let

$$u(x) = \mathbb{E}_x \left[\int_0^T g(B(t)) \, dt \right], \qquad \text{for } x \in U,$$

where $T := \inf\{t > 0 \colon B(t) \notin U\}$. Show that,

(a) If g is Hölder continuous, then the function $u \colon U \to \mathbb{R}$ solves $-\frac{1}{2} \Delta u = g$.

(b) If every point $x \in \partial U$ is regular for the complement of U, then $u(x) = 0$ for all $x \in \partial U$.

Exercise 8.5. Ⓢ Let $a > 0$ and τ a standard exponential random variable independent of the standard Brownian motion $\{B(t) \colon t \geqslant 0\}$ in $\mathbb{R}^d$. Show that there exist constants $0 < c < C$ depending only on a and d, such that for any compact set $A \subset \mathcal{B}(0, a)$, we have

$$c \, \mathbb{P}_0 \{B[0, 1] \cap A \neq \emptyset\} \leqslant \mathbb{P}_0 \{B[0, \tau] \cap A \neq \emptyset\} \leqslant C \, \mathbb{P}_0 \{B[0, 1] \cap A \neq \emptyset\}.$$

Exercise 8.6. Suppose $\Lambda \subset \mathbb{R}^d$, for $d \geqslant 3$, is compact and γ the last exit time from Λ defined as in Theorem 8.8. Show that

$$\lim_{x \to \infty} \mathbb{P}_x \{B(\gamma) \in A \mid \gamma > 0\} = \frac{\nu(A)}{\nu(\Lambda)}.$$

Exercise 8.7. For $d \geqslant 3$ consider the spherical shell

$$\Lambda_R = \{x \in \mathbb{R}^d \ : \ 1 \leqslant |x| \leqslant R\}.$$

Show that $\lim_{R \to \infty} \mathrm{Cap}_M(\Lambda_R) = 2$.

Exercise 8.8. Let $\{X(a) \colon a \geqslant 0\}$ be a stable subordinator of index $\frac{1}{2}$ as defined in Theorem 2.35, and

$$K(s,t) := \begin{cases} (t-s)^{-1/2} & 0 \leqslant s \leqslant t, \\ 0 & s > t \geqslant 0. \end{cases}$$

Let $M(s,t) = K(s,t)/K(0,t)$, then for any subset Λ of $(0,\infty)$,

$$\tfrac{1}{2} \mathrm{Cap}_M(\Lambda) \leqslant \mathbb{P}_0\big\{\{X(a) \colon a \geqslant 0\} \text{ hits } \Lambda\big\} \leqslant \mathrm{Cap}_M(\Lambda).$$

Exercise 8.9. Let $\{B(t) \colon t \geqslant 0\}$ be a standard linear Brownian motion.

(a) For the kernel M of Exercise 8.8, show that $\mathrm{Cap}_M(\text{Zeros}) = 0$ almost surely.

(b) Let $A \subset (0,\infty)$. Show that

$$\mathbb{P}_0\big\{\exists t \in A \text{ with } B(t) = 0\big\} \begin{cases} > 0 & \text{if } \dim A > \tfrac{1}{2}, \\ = 0 & \text{if } \dim A < \tfrac{1}{2}. \end{cases}$$

Exercise 8.10. Ⓢ Let μ_n, μ be Borel probability measures on a compact metric space X. Suppose $\mu_n \to \mu$ in the sense of weak convergence, as defined in Section 12.1 of the appendix. Show that $\mu_n \otimes \mu_n \to \mu \otimes \mu$ in the sense of weak convergence of probability measures on $X \times X$.

Exercise 8.11. Ⓢ Let $\{B(s) \colon s \geqslant 0\}$ under $\mathbb{P}_x$ be a Brownian motion in $\mathbb{R}^d$, $d \geqslant 3$, with $B(0) = x$, and denote by

$$\tau_{\mathcal{B}(0,1)} = \inf\{s > 0 \colon B(s) \in \mathcal{B}(0,1)\}$$

the first hitting time of the unit ball after time zero. Show that there exist constants $0 < c < C < \infty$ such that, for $t \geqslant 1$,

$$ct \leqslant \int \mathbb{P}_x\{\tau_{\mathcal{B}(0,1)} < t\}\, dx \leqslant Ct.$$

Exercise 8.12. Show that exactly one of the probability measures μ on the closed unit disc in the plane that minimise the energies

$$\iint \log \frac{1}{|x-y|}\, d\mu(x)\, d\mu(y) \quad \text{and} \quad \iint \frac{1}{|x-y|}\, d\mu(x)\, d\mu(y)$$

is concentrated on the boundary of the disc.

Exercise 8.13. Let $\Lambda \subset \mathbb{R}^2$ be a nonpolar, compact set and ν^λ be the equilibrium measure for planar Brownian motion stopped at an independent exponential time with parameter λ. Then the limit

$$\lim_{\lambda \downarrow 0} \frac{\nu^\lambda}{\nu^\lambda(\Lambda)}$$

exists and is equal to the energy-minimising measure ν_Λ.
Hint. Use ideas from Theorem 3.34 and Theorem 8.30.

Notes and comments

The proof of the last exit formula is taken from Chung's beautiful paper [Ch73], but the existence of an energy-minimising measure is a much older fact. For the case of the Newtonian potential ($d = 3$) it was determined by Gauss as the charge distribution on the surface of a conductor which minimises the electrostatic energy. Classically, the equilibrium measure is defined as the measure ν on Λ that maximises $\nu(\Lambda)$ among those with potential bounded by one. Then $\nu/\nu(\Lambda)$ is the energy-minimising probability measure, see Carleson [Ca67]. Rigorous results and extensions to general Riesz-potentials are due to Frostman in his ground-breaking thesis [Fr35]. Our discussion of the strong maximum principle follows Carleson [Ca67], Bass [Ba95] describes an alternative approach. The classical proof of Lemma 8.29 uses Fourier transform and Plancherel's theorem, see [Ca67].

Characterising the polar sets for Brownian motion is related to the following question: for which sets $A \subset \mathbb{R}^d$ are there nonconstant bounded harmonic functions on $\mathbb{R}^d \setminus A$? Such sets are called *removable* for bounded harmonic functions. Consider the simplest case first. When A is the empty set, it is obviously polar, and by Liouville's theorem there is no bounded harmonic function on its complement. Nevanlinna [Ne70] proved in the 1920s that for $d \geqslant 3$ there exist nonconstant bounded harmonic functions on $\mathbb{R}^d \setminus A$ if and only if $\mathrm{Cap}_G(A) > 0$, where $G(x, y) = f(|x - y|)$ for the radial potential f as before. Just to make this result more plausible, note that the function $h(x) = \int G(x, y)\mu(dy)$, where μ is a measure on A of finite G-energy, would make a good candidate for such a function, see Theorem 3.35.

Loosely speaking, G-capacity measures whether a set A is big enough to hide a pole of a harmonic function inside. Recall from Theorem 4.32 that $\dim A > d - 2$ implies existence of such functions, and $\dim A < d - 2$ implies nonexistence. Kakutani [Ka44b] showed that there exist bounded harmonic functions on $\mathbb{R}^d \setminus A$ if and only if A is polar for Brownian motion. The precise hitting estimates we give are fairly recent, our proof is a variant of the original proof by Benjamini et al. in [BPP95]. Proposition 8.26 goes back to the same paper.

An interesting question is, which subsets of compact sets are charged by the harmonic measure μ_A. Clearly μ_A does not charge polar sets, and in particular, in $d \geqslant 3$, we have $\mu_A(B) = 0$ for all Borel sets with $\dim(B) < d - 2$. In the plane, by a famous theorem of Makarov, see [Ma85], we have that

- any set B of dimension < 1 has $\mu_A(B) = 0$,
- there is a set $S \subset A$ with $\dim S = 1$ such that $\mu_A(S^c) = 0$.

However, the outer boundary, which supports the harmonic measure, may have a dimension much bigger than one. An interesting question arising in the context of self-avoiding curves asks for the dimension of the outer boundary of the image $B[0,1]$ of a Brownian motion. Based on scaling arguments from polymer physics, Benoit Mandelbrot conjectured in 1982 that this set should have fractal dimension $4/3$. Bishop et al. [BJPP97] showed that the outer boundary has dimension > 1. In 2001 Mandelbrot's conjecture was finally proved by Lawler, Schramm and Werner [LSW01c], see Chapter 11 for more information.

There are some fine results about the hitting probabilities of small balls within a given time in the literature. Le Gall [LG86b] shows, using a classical diffusion argument, that for $d \geqslant 3$ we have, as $\varepsilon \downarrow 0$,

$$\mathbb{P}_0\{\tau(\mathcal{B}(x, \varepsilon)) \leqslant t\} \sim \left(\tfrac{d}{2} - 1\right) \mathcal{L}(\mathcal{B}(0, 1))\, \varepsilon^{d-2} \int_0^t \mathfrak{p}_s(0, y)\, ds.$$

This should be compared to the result of Exercise 8.11. The analogous result for the planar case is due to Spitzer [Sp58]. Further fine results from [LG86b] refer to the hitting of several small balls in a given time, and some asymptotic results for the volume of Wiener sausages, the neighbourhoods of the Brownian path.

9

Intersections and self-intersections of Brownian paths

In this chapter we study multiple points of d-dimensional Brownian motion. We shall see, for example, in which dimensions the Brownian path has double points and explore how many double points there are. This chapter also contains some of the highlights of the book: a proof that planar Brownian motion has points of infinite multiplicity, the intersection equivalence of Brownian motion and percolation limit sets, and the surprising dimension-doubling theorem of Kaufman.

9.1 Intersection of paths: Existence and Hausdorff dimension

9.1.1 Existence of intersections

Suppose that $\{B_1(t) : t \geqslant 0\}$ and $\{B_2(t) : t \geqslant 0\}$ are two independent d-dimensional Brownian motions started in arbitrary points. The question we ask in this section is, in which dimensions the ranges, or paths, of the two motions have a nontrivial intersection, in other words whether there exist times $t_1, t_2 > 0$ such that $B_1(t_1) = B_2(t_2)$. As this question is easy if $d = 1$ we assume $d \geqslant 2$ throughout this section.

We have developed the tools to decide this question in Chapter 4 and Chapter 8. Keeping the path $\{B_1(t) : t \geqslant 0\}$ fixed, we have to decide whether it is a polar set for the second Brownian motion. By Kakutani's theorem, Theorem 8.20, this question depends on its capacity with respect to the potential kernel. As the capacity is again related to Hausdorff measure and dimension, the results of Chapter 4 are crucial in the proof of the following result.

Theorem 9.1

(a) *For $d \geqslant 4$, almost surely, two independent Brownian paths in $\mathbb{R}^d$ have an empty intersection, except for a possible common starting point.*

(b) *For $d \leqslant 3$, almost surely, the intersection of two independent Brownian paths in $\mathbb{R}^d$ is nontrivial, i.e. contains points other than a possible common starting point.*

Remark 9.2 In the case $d \leqslant 3$, if the Brownian paths are started at the same point, then almost surely, the paths intersect before any positive time $t > 0$, see Exercise 9.1 (a). ◇

Proof of (a). Note that it suffices to look at one Brownian motion and show that its path is, almost surely, a set of capacity zero with respect to the potential kernel. If $d \geqslant 4$, the capacity with respect to the potential kernel is a multiple of the Riesz $(d - 2)$-capacity. By Theorem 4.27 this capacity is zero for sets of finite $(d-2)$-dimensional Hausdorff measure. Now note that if $d \geqslant 5$ the dimension of a Brownian path is two, and hence strictly smaller than $d - 2$, so that the $(d - 2)$-dimensional Hausdorff measure is zero, which shows that the capacity must be zero.

If $d = 4$ the situation is only marginally more complicated, although the dimension of the Brownian path is $2 = d - 2$ and the simple argument above does not apply. However, we know from (4.2) in Chapter 4 that $\mathcal{H}^2(B[0,1]) < \infty$ almost surely, which implies that $\mathrm{Cap}_2(B[0,1]) = 0$ by Theorem 4.27. This implies that an independent Brownian motion almost surely does not hit any of the segments $B[n, n + 1]$, and therefore avoids the path entirely. ∎

Proof of (b). If $d = 3$, the capacity with respect to the potential kernel is a multiple of the Riesz 1-capacity. As the Hausdorff dimension of a path is two, this capacity is positive by Theorem 4.32. Therefore two Brownian paths in $d = 3$ intersect with positive probability.

Suppose now the two Brownian motions start at different points. We may assume that one is the origin and the other one is denoted x. By rotational invariance, the probability that the paths do not intersect depends only on $|x|$, and by Brownian scaling we see that it is completely independent of the choice of $x \neq 0$. Denote this probability by q and, given any $\varepsilon > 0$, choose a large time t such that

$$\mathbb{P}\{B_1(t_1) \neq B_2(t_2) \text{ for all } 0 < t_1, t_2 \leqslant t\} \leqslant q + \varepsilon.$$

Then, using the Markov property,

$$q \leqslant \mathbb{P}\{B_1(t_1) \neq B_2(t_2) \text{ for all } t_1, t_2 \leqslant t\} \, \mathbb{P}\{B_1(t_1) \neq B_2(t_2) \text{ for all } t_1, t_2 > t\}$$
$$\leqslant q(q + \varepsilon).$$

As $\varepsilon > 0$ was arbitrary, we get $q \leqslant q^2$, and as we know that $q < 1$ we obtain that $q = 0$. This shows that two Brownian paths started in different points intersect almost surely. If they start in the same point, by the Markov property,

$$\mathbb{P}\{B_1(t_1) \neq B_2(t_2) \text{ for all } t_1, t_2 > 0\} = \lim_{\substack{t \downarrow 0 \\ t > 0}} \mathbb{P}\{B_1(t_1) \neq B_2(t_2) \text{ for all } t_1, t_2 > t\} = 0,$$

as required to complete the argument in the case $d = 3$. A path in $d \leqslant 2$ is the projection of a three dimensional path on a lower dimensional subspace, hence if two paths in $d = 3$ intersect almost surely, then so do two paths in $d = 2$. ∎

It is equally natural to ask, for integers $p > 2$ and $d \leqslant 3$, whether a collection of p independent d-dimensional Brownian motions

$$\{B_1(t): t \geqslant 0\}, \ldots, \{B_p(t): t \geqslant 0\}$$

intersect, i.e. whether there exist times $t_1, \ldots, t_p > 0$ such that $B_1(t_1) = \cdots = B_p(t_p)$.

Theorem 9.3

(a) *For $d \geqslant 3$, almost surely, three independent Brownian paths in $\mathbb{R}^d$ have an empty intersection, except for a possible common starting point.*

(b) *For $d = 2$, almost surely, the intersection of any finite number p of independent Brownian paths in $\mathbb{R}^d$ is nontrivial, i.e. contains points other than a possible common starting point.*

In the light of our discussion of the case $p = 2$, it is natural to approach the question about the existence of intersections of p paths, by asking for the Hausdorff dimension and measure of the intersection of $p - 1$ paths. This leads to an easy proof of (a).

Lemma 9.4 *Suppose $\{B_i(t): t \geqslant 0\}$, for $i = 1, 2$, are two independent Brownian motions in $d = 3$. Then, almost surely, for every compact set $\Lambda \subset \mathbb{R}^3$ not containing the starting points of the Brownian motions, we have $\mathcal{H}^1(B_1[0, \infty) \cap B_2[0, \infty) \cap \Lambda) < \infty$.*

Proof. Fix a cube Cube $\subset \mathbb{R}^3$ of unit side length not containing the starting points. It suffices to show that, almost surely, $\mathcal{H}^1(B_1[0, \infty) \cap B_2[0, \infty) \cap \text{Cube}) < \infty$. For this purpose let $\mathfrak{C}_n$ be the collection of dyadic subcubes of Cube of side length 2^{-n}, and $\mathfrak{I}_n$ be the collection of cubes in $\mathfrak{C}_n$ which are hit by both motions. By our hitting estimates, Corollary 3.19, there exists $C > 0$ such that, for any cube $E \in \mathfrak{C}_n$,

$$\mathbb{P}\{E \in \mathfrak{I}_n\} = \mathbb{P}\{\exists s > 0 \text{ with } B(s) \in E\}^2 \leqslant C2^{-2n}.$$

Now, for every n, the collection $\mathfrak{I}_n$ is a covering of $B_1[0, \infty) \cap B_2[0, \infty) \cap \text{Cube}$, and

$$\mathbb{E}\left[\sum_{E \in \mathfrak{I}_n} |E|\right] = 2^{3n} \, \mathbb{P}\{E \in \mathfrak{I}_n\} \sqrt{3} 2^{-n} \leqslant C\sqrt{3}.$$

Therefore, by Fatou's lemma, we obtain

$$\mathbb{E}\left[\liminf_{n \to \infty} \sum_{E \in \mathfrak{I}_n} |E|\right] \leqslant \liminf_{n \to \infty} \mathbb{E}\left[\sum_{E \in \mathfrak{I}_n} |E|\right] \leqslant C\sqrt{3}.$$

Hence the liminf is finite almost surely, and we infer from this that $\mathcal{H}^1(B_1[0, \infty) \cap B_2[0, \infty) \cap \text{Cube})$ is finite almost surely. ∎

Proof of Theorem 9.3 (a). It suffices to show that, for any cube Cube of unit side length which does not contain the origin, we have $\text{Cap}_1(B_1[0, \infty) \cap B_2[0, \infty) \cap \text{Cube}) = 0$. This follows directly from Lemma 9.4 and the energy method, Theorem 4.27. ∎

For Theorem 9.3 (b) it would suffice to show that the Hausdorff dimension of the set $B_1(0, \infty) \cap \ldots \cap B_{p-1}(0, \infty)$ is positive in the case $d = 2$. In fact, it is a natural question to ask for the Hausdorff dimension of the intersection of Brownian paths in any case when the set is nonempty. The problem was raised by Itô and McKean in the first edition of their influential book [IM74], and has since been resolved by Taylor [Ta66] and Fristedt [Fr67]. The substantial problem of finding lower bounds for the Hausdorff dimension of the intersection sets is best approached using the technique of *stochastic co-dimension*, which we discuss now.

9.1.2 Stochastic co-dimension and percolation limit sets

Given a set A, the idea behind the stochastic co-dimension approach is to take a suitable random test set Θ, and check whether $\mathbb{P}\{\Theta \cap A \neq \emptyset\}$ is zero or positive. In the latter case this indicates that the set is large, and we should therefore get a lower bound on the dimension of A. A natural choice of such a random test set would be the range of Brownian motion. Recall that, for example in the case $d = 3$, if $\mathbb{P}\{B[0, \infty) \cap A \neq \emptyset\} > 0$, this implies that $\dim A \geqslant 1$.

Of course, in order to turn this idea into a systematic technique for finding lower bounds for the Hausdorff dimension, an entire family of test sets is needed to tune the size of the test set in order to give sharp bounds. For this purpose, Taylor [Ta66] used stable processes instead of Brownian motion. This is not the easiest way and also limited, because stable processes only exist across a limited range of parameters. The approach we use in this book is based on using the family of percolation limit sets as test sets.

Suppose that $C \subset \mathbb{R}^d$ is a fixed compact unit cube. We denote by $\mathfrak{C}_n$ the collection of compact dyadic subcubes (relative to C) of side length 2^{-n}. We also let

$$\mathfrak{C} = \bigcup_{n=0}^{\infty} \mathfrak{C}_n.$$

Given $\gamma \in [0, d]$ we construct a random compact set $\Gamma[\gamma] \subset C$ inductively as follows: We keep each of the 2^d compact cubes in $\mathfrak{C}_1$ independently with probability $p = 2^{-\gamma}$. Let $\mathfrak{S}_1$ be the collection of cubes kept in this procedure and $\mathsf{S}(1)$ their union. Pass from $\mathfrak{S}_n$ to $\mathfrak{S}_{n+1}$ by keeping each cube of $\mathfrak{C}_{n+1}$, which is not contained in a previously rejected cube, independently with probability p. Denote by $\mathfrak{S} = \bigcup_{n=1}^{\infty} \mathfrak{S}_n$ and let $\mathsf{S}(n+1)$ be the union of the cubes in $\mathfrak{S}_{n+1}$. Then the random set

$$\Gamma[\gamma] := \bigcap_{n=1}^{\infty} \mathsf{S}(n)$$

is called a **percolation limit set**. The usefulness of percolation limit sets in fractal geometry comes from the following theorem.

Theorem 9.5 (Hawkes 1981) *For every* $\gamma \in [0, d]$ *and every closed set* $A \subset C$ *the following properties hold*

(i) *if* $\dim A < \gamma$, *then almost surely,* $A \cap \Gamma[\gamma] = \emptyset$,

(ii) *if* $\dim A > \gamma$, *then* $A \cap \Gamma[\gamma] \neq \emptyset$ *with positive probability,*

(iii) *if* $\dim A > \gamma$, *then*

(a) *almost surely* $\dim \left(A \cap \Gamma[\gamma] \right) \leqslant \dim A - \gamma$ *and,*

(b) *for all* $\varepsilon > 0$, *with positive probability* $\dim \left(A \cap \Gamma[\gamma] \right) \geqslant \dim A - \gamma - \varepsilon$.

Remark 9.6 Observe that the first part of the theorem gives a *lower* bound γ for the Hausdorff dimension of a set A, if we can show that $A \cap \Gamma[\gamma] \neq \emptyset$ with positive probability. As with so many ideas in fractal geometry one of the roots of this method lies in the study of trees, more precisely percolation on trees, see [Ly90]. ◇

Remark 9.7

(a) The stochastic co-dimension technique and the energy method are closely related: A set A is called *polar for the percolation limit set*, if

$$\mathbb{P}\{A \cap \Gamma[\gamma] \neq \emptyset\} = 0.$$

We shall see in Theorem 9.18 that a set is polar for the percolation limit set if and only if it has γ-capacity zero.

(b) For $d \geqslant 3$, the criterion for polarity of a percolation limit set with $\gamma = d - 2$ therefore agrees with the criterion for the polarity for Brownian motion, recall Theorem 8.20. This 'equivalence' between percolation limit sets and Brownian motion has a quantitative strengthening which is discussed in Section 9.2 of this chapter. ◇

Proof of (i) in Hawkes' theorem.   The proof of part (i) is based on the *first moment method*, which means that we essentially only have to calculate an expectation. Because $\dim A < \gamma$ there exists, for every $\varepsilon > 0$, a covering of A by countably many sets $D_1, D_2, \ldots$ with $\sum_{i=1}^{\infty} |D_i|^{\gamma} < \varepsilon$. As each set is contained in no more than a constant number of dyadic cubes of smaller diameter, we may even assume that $D_1, D_2, \ldots \in \mathcal{C}$. Suppose that the side length of D_i is 2^{-n}, then the probability that $D_i \in \mathcal{S}_n$ is $2^{-n\gamma}$. By picking from $D_1, D_2, \ldots$ those cubes which are in $\mathcal{S}$ we get a covering of $A \cap \Gamma[\gamma]$. Let N be the number of cubes picked in this procedure, then

$$\mathbb{P}\{A \cap \Gamma[\gamma] \neq \emptyset\} \leqslant \mathbb{P}\{N > 0\} \leqslant \mathbb{E}N = \sum_{i=1}^{\infty} \mathbb{P}\{D_i \in \mathcal{S}\} = \sum_{i=1}^{\infty} |D_i|^{\gamma} < \varepsilon.$$

As this holds for all $\varepsilon > 0$ we infer that, almost surely, we have $A \cap \Gamma[\gamma] = \emptyset$. ∎

Proof of (ii) in Hawkes' theorem. The proof of part (ii) is based on the *second moment method*, which means that a variance has to be calculated. We also use the easy part of Frostman's lemma in the form of Theorem 4.32, which states that, as $\dim A > \gamma$, there exists a probability measure μ on A such that $I_\gamma(\mu) < \infty$.

Now let n be a positive integer and define the random variables

$$Y_n = \sum_{C \in \mathfrak{S}_n} \frac{\mu(C)}{|C|^\gamma} = \sum_{C \in \mathfrak{C}_n} \mu(C) 2^{n\gamma} \, 1_{\{C \in \mathfrak{S}_n\}}.$$

Note that $Y_n > 0$ implies $S(n) \cap A \neq \emptyset$ and, by compactness, if $Y_n > 0$ for all n we even have $A \cap \Gamma[\gamma] \neq \emptyset$. As $Y_{n+1} > 0$ implies $Y_n > 0$, we get that

$$\mathbb{P}\{A \cap \Gamma[\gamma] \neq \emptyset\} \geqslant \mathbb{P}\{Y_n > 0 \text{ for all } n\} = \lim_{n \to \infty} \mathbb{P}\{Y_n > 0\}.$$

It therefore suffices to give a positive lower bound for $\mathbb{P}\{Y_n > 0\}$ independent of n. A straightforward calculation gives for the first moment $\mathbb{E}[Y_n] = \sum_{C \in \mathfrak{C}_n} \mu(C) = 1$. For the second moment we find

$$\mathbb{E}[Y_n^2] = \sum_{C \in \mathfrak{C}_n} \sum_{D \in \mathfrak{C}_n} \mu(C)\mu(D) \, 2^{2n\gamma} \, \mathbb{P}\{C \in \mathfrak{S}_n \text{ and } D \in \mathfrak{S}_n\}.$$

The latter probability depends on the dyadic distance of the cubes C and D: if 2^{-m} is the side length of the smallest dyadic cube which contains both C and D, then the probability in question is $2^{-2\gamma(n-m)}2^{-\gamma m}$. The value m can be estimated in terms of the Euclidean distance of the cubes, indeed if $x \in C$ and $y \in D$ then

$$|x - y| \leqslant \sqrt{d} 2^{-m}.$$

This gives a handle to estimate the second moment in terms of the energy of μ. We find that

$$\mathbb{E}[Y_n^2] = \sum_{C \in \mathfrak{C}_n} \sum_{D \in \mathfrak{C}_n} \mu(C)\mu(D) 2^{\gamma m} \leqslant d^{\gamma/2} \iint \frac{d\mu(x)\, d\mu(y)}{|x - y|^\gamma} = d^{\gamma/2} I_\gamma(\mu).$$

Plugging these moment estimates into the easy form of the Paley–Zygmund inequality, Lemma 3.23, gives $\mathbb{P}\{Y_n > 0\} \geqslant d^{-\gamma/2} I_\gamma(\mu)^{-1}$, as required. ∎

Proof of (iii) in Hawkes' theorem. For part (iii) note that the intersection $\Gamma[\gamma] \cap \Gamma[\delta]$ of two independent percolation limit sets has the same distribution as $\Gamma[\gamma + \delta]$. Suppose first that $\delta > \dim A - \gamma$. Then, by part (i), $A \cap \Gamma[\gamma] \cap \Gamma[\delta] = \emptyset$ almost surely, and hence, by part (ii), $\dim A \cap \Gamma[\gamma] \leqslant \delta$ almost surely. Letting $\delta \downarrow \dim A - \gamma$ completes the proof of part (a). Now suppose that $\delta < \dim A - \gamma$. Then, with positive probability, $(A \cap \Gamma[\gamma]) \cap \Gamma[\delta] \neq \emptyset$, by part (ii). And using again part (i) we get that $\dim A \cap \Gamma[\gamma] \geqslant \delta$ with positive probability, completing the proof of part (b). ∎

9.1.3 Hausdorff dimension of intersections

We can now use the stochastic codimension approach to find the Hausdorff dimension of the intersection of two Brownian paths, whenever it is nonempty. Note that the following theorem also implies Theorem 9.3 (b).

Theorem 9.8 *Suppose $d \geqslant 2$ and $p \geqslant 2$ are integers such that $p(d-2) < d$. Suppose that*

$$\{B_1(t): t \geqslant 0\}, \ldots, \{B_p(t): t \geqslant 0\}$$

are p independent d-dimensional Brownian motions. Let $\text{Range}_i = B_i[0, \infty)$ be the range of the process $\{B_i(t): t \geqslant 0\}$, for $1 \leqslant i \leqslant p$. Then, almost surely,

$$\dim\left(\text{Range}_1 \cap \ldots \cap \text{Range}_p\right) = d - p(d-2).$$

Remark 9.9 A good way to make this result plausible is by recalling the situation for the intersection of linear subspaces of $\mathbb{R}^d$: If the spaces are in general position, then the co-dimension of the intersection is the sum of the co-dimensions of the subspaces. As the Hausdorff dimension of a Brownian path is two, the plausible codimension of the intersection of p paths is $p(d-2)$, and hence the dimension is $d - p(d-2)$. ◇

Remark 9.10 Assuming the theorem, if the Brownian paths are started in the same point, then almost surely, $\dim(B_1[0, t_1] \cap \cdots \cap B_p[0, t_p]) = d - p(d-2)$, for any $t_1, \ldots, t_p > 0$, see Exercise 9.1 (b). ◇

For the proofs of the lower bounds in Theorem 9.8 we use the stochastic codimension method, but first we provide a useful zero-one law.

Lemma 9.11 *For any $\gamma > 0$ the probability of the event*

$$\left\{ \dim\left(\text{Range}_1 \cap \ldots \cap \text{Range}_p\right) \geqslant \gamma \right\}$$

is either zero or one, and independent of the starting points of the Brownian motions.

Proof. For $t \in (0, \infty]$ denote $S(t) = B_1(0, t) \cap \cdots \cap B_p(0, t)$ and let

$$p(t) = \mathbb{P}\{\dim S(t) \geqslant \gamma\}.$$

We start by considering the case that all Brownian motions start at the origin. Then, by monotonicity of the events,

$$\mathbb{P}\{ \dim S(t) \geqslant \gamma \text{ for all } t > 0 \} = \lim_{t \downarrow 0} p(t).$$

The event on the left hand side is in the germ-σ-algebra and hence, by Blumenthal's zero-one law, has probability zero or one. By scaling, however, $p(t)$ does not depend on t at all, so we have either $p(t) = 0$ for all $t > 0$ or $p(t) = 1$ for all $t > 0$.

In the first case we note that, by the Markov property applied at time t,

$$0 = \mathbb{P}\{ \dim S(\infty) \geqslant \gamma \}$$
$$= \int \mathbb{P}\{ \dim S(\infty) \geqslant \gamma \mid B_1(t) = x_1, \ldots, B_p(t) = x_p \} \, d\mu(x_1, \ldots, x_p),$$

where μ is the product of p independent centred, normally distributed random variables with variances t. As $\mu \ll \mathcal{L}_{pd}$, we have $\mathbb{P}\{\dim S(\infty) \geqslant \gamma\} = 0$ for $\mathcal{L}_{pd}$-almost every vector of starting points. Finally, for an arbitrary configuration of starting points,

$$\mathbb{P}\{\dim S(\infty) \geqslant \gamma\}$$
$$= \lim_{t\downarrow 0} \mathbb{P}\{\dim\{x \in \mathbb{R}^d : \exists t_i \geqslant t \text{ such that } x = B_1(t_1) = \cdots = B_p(t_p)\} \geqslant \gamma\} = 0.$$

A completely analogous argument can be carried out for the second case. $\blacksquare$

Proof of Theorem 9.8.   First we look at $d = 3$ (and hence $p = 2$) and note that, by Lemma 9.4, we have $\dim(\text{Range}_1 \cap \text{Range}_2) \leqslant 1$, and hence only the lower bound remains to be proved. Suppose $\gamma < 1$ is arbitrary, and pick $\beta > 1$ such that $\gamma + \beta < 2$. Let $\Gamma[\gamma]$ and $\Gamma[\beta]$ be two independent percolation limit sets, independent of the Brownian motions. Note that $\Gamma[\gamma] \cap \Gamma[\beta]$ is a percolation limit set with parameter $\gamma + \beta$. Hence, by Theorem 9.5 (ii) and the fact that $\dim(\text{Range}_1) = 2 > \gamma + \beta$, we have

$$\mathbb{P}\{\text{Range}_1 \cap \Gamma[\gamma] \cap \Gamma[\beta] \neq \emptyset\} > 0.$$

Interpreting $\Gamma[\beta]$ as the test set and using Theorem 9.5 (i) we obtain

$$\dim\left(\text{Range}_1 \cap \Gamma[\gamma]\right) \geqslant \beta \qquad \text{with positive probability.}$$

As $\beta > 1$, given this event, the set $\text{Range}_1 \cap \Gamma[\gamma]$ has positive capacity with respect to the potential kernel in $\mathbb{R}^3$ and is therefore nonpolar with respect to the independent Brownian motion $\{B_2(t) : t \geqslant 0\}$. We therefore have

$$\mathbb{P}\{\text{Range}_1 \cap \text{Range}_2 \cap \Gamma[\gamma] \neq \emptyset\} > 0.$$

Using Theorem 9.5 (i) we infer that $\dim(\text{Range}_1 \cap \text{Range}_2) \geqslant \gamma$ with positive probability. Lemma 9.11 shows that this must in fact hold almost surely, and the result follows as $\gamma < 1$ was arbitrary.

Next, we look at $d = 2$ and any $p \geqslant 2$. Note that the upper bounds are trivial. For the lower bounds, suppose $\gamma < 2$ is arbitrary, and pick $\beta_1, \ldots, \beta_p > 0$ such that $\gamma + \beta_1 + \cdots + \beta_p < 2$. Let $\Gamma[\gamma]$ and $\Gamma[\beta_1], \ldots, \Gamma[\beta_p]$ be independent percolation limit sets, independent of the p Brownian motions. Then

$$\Gamma[\gamma] \cap \bigcap_{i=1}^{p} \Gamma[\beta_i]$$

is a percolation limit set with parameter $\gamma + \beta_1 + \cdots + \beta_p$. Hence, by Theorem 9.5 (ii) and the fact that $\dim(\text{Range}_1) = 2 > \gamma + \beta_1 + \cdots + \beta_p$, we have

$$\mathbb{P}\left\{\text{Range}_1 \cap \Gamma[\gamma] \cap \bigcap_{i=1}^{p} \Gamma[\beta_i] \neq \emptyset\right\} > 0.$$

Interpreting $\Gamma[\beta_p]$ as the test set and using Theorem 9.5 (i) we obtain

$$\dim\left(\text{Range}_1 \cap \Gamma[\gamma] \cap \bigcap_{i=1}^{p-1} \Gamma[\beta_i]\right) \geqslant \beta_p \qquad \text{with positive probability.}$$

As $\beta_p > 0$, given this event, the set

$$\mathrm{Range}_1 \cap \Gamma[\gamma] \cap \bigcap_{i=1}^{p-1} \Gamma[\beta_i]$$

has positive capacity with respect to the potential kernel in $\mathbb{R}^2$ and is therefore nonpolar with respect to the independent Brownian motion $\{B_2(t) \colon t \geqslant 0\}$. We therefore have

$$\mathbb{P}\Big\{\mathrm{Range}_1 \cap \mathrm{Range}_2 \cap \Gamma[\gamma] \cap \bigcap_{i=1}^{p-1} \Gamma[\beta_i] \neq \emptyset\Big\} > 0.$$

Iterating this procedure $p - 1$ times we obtain

$$\mathbb{P}\Big\{\bigcap_{i=1}^{p} \mathrm{Range}_i \cap \Gamma[\gamma] \neq \emptyset\Big\} > 0.$$

Using Theorem 9.5 (i) we infer that $\dim(\bigcap_{i=1}^{p} \mathrm{Range}_i) \geqslant \gamma$ with positive probability. Lemma 9.11 shows that this must in fact hold almost surely, and the result follows as $\gamma < 2$ was arbitrary. ∎

9.2 Intersection equivalence of Brownian motion and percolation limit sets

The idea of quantitative estimates of hitting probabilities has a natural extension: two random sets may be called *intersection-equivalent* if their hitting probabilities for a large class of test sets are comparable. This concept of equivalence allows surprising relationships between random sets which, at first sight, might not have much in common. In this section we establish intersection equivalence between Brownian motion and suitably defined percolation limit sets, and use this to characterise the polar sets for the intersection of Brownian paths. We start the discussion by formalising the idea of intersection equivalence.

Definition 9.12. Two random closed sets A and B in $\mathbb{R}^d$ are **intersection-equivalent** in the compact set U if there exist two positive constants c, C such that, for any closed set $\Lambda \subset U$,

$$c\,\mathbb{P}\{A \cap \Lambda \neq \emptyset\} \leqslant \mathbb{P}\{B \cap \Lambda \neq \emptyset\} \leqslant C\,\mathbb{P}\{A \cap \Lambda \neq \emptyset\}. \tag{9.1}$$

Using the symbol $a \asymp b$ to indicate that the ratio of a and b is bounded from above and below by positive constants which do not depend on Λ we can write this as

$$\mathbb{P}\{A \cap \Lambda \neq \emptyset\} \asymp \mathbb{P}\{B \cap \Lambda \neq \emptyset\}.$$

◇

Remark 9.13 Let $\mathcal{G}$ be the collection of all closed subsets of $\mathbb{R}^d$. Formally, we define a random closed set as a mapping $A \colon \Omega \to \mathcal{G}$ such that, for every compact $\Lambda \subset \mathbb{R}^d$, the set $\{\omega \colon A(\omega) \cap \Lambda = \emptyset\}$ is measurable. ◇

The philosophy of the main result of this section is that we would like to find a class of particularly simple sets which are intersection-equivalent to the paths of transient Brownian motion. If these sets are easier to study, we can 'translate' easy results about the simple sets into hard ones for Brownian motion. A good candidate for these simple sets are percolation limit sets: they have excellent features of *self-similarity* and *independence* between the fine structures in different parts. Many of their properties can be obtained from classical facts about Galton–Watson branching processes.

We introduce percolation limit sets with *generation dependent* retention probabilities. Denote by $\mathfrak{C}_n$ the compact dyadic cubes of side length 2^{-n}. For any sequence $p_1, p_2, \dots$ in $(0, 1)$ we define families $\mathfrak{S}_n$ of compact dyadic cubes inductively by including any cube in $\mathfrak{C}_n$ which is not contained in a previously rejected cube, independently with probability p_n. Define

$$\Gamma = \bigcap_{n=1}^{\infty} \bigcup_{S \in \mathfrak{S}_n} S,$$

to be the **percolation limit set** for the sequence $p_1, p_2, \dots$.

To find a suitable sequence of retention probabilities we compare the hitting probabilities of dyadic cubes by a percolation limit set on the one hand and a transient Brownian on the other. (This is obviously necessary to establish intersection equivalence). We assume that percolation is performed in a cube Cube at positive distance from the origin, at which a transient Brownian motion is started. Supposing for the moment that the retention probabilities are such that the survival probability of any retained cube is bounded from below, for any cube $Q \in \mathfrak{C}_n$, the hitting estimates for the percolation limit set are

$$\mathbb{P}\{\Gamma \cap Q \neq \emptyset\} \asymp p_1 \cdots p_n.$$

By Theorem 8.24, on the other hand,

$$\mathbb{P}\{B[0,T] \cap Q \neq \emptyset\} \asymp \mathrm{Cap}_M(Q) \asymp 1/f(2^{-n}),$$

for the radial potential

$$f(\varepsilon) = \begin{cases} \log_2(1/\varepsilon) & \text{for } d = 2, \\ \varepsilon^{2-d} & \text{for } d \geqslant 3, \end{cases}$$

where we have chosen basis 2 for the logarithm for convenience of this argument. Then we choose the sequence $p_1, p_2, \dots$ of retention probabilities such that $p_1 \cdots p_n = 1/f(2^{-n})$. More explicitly, we choose $p_1 = 2^{2-d}$ and, for $n \geqslant 2$,

$$p_n = \frac{f(2^{-n+1})}{f(2^{-n})} = \begin{cases} \frac{n-1}{n} & \text{for } d = 2, \\ 2^{2-d} & \text{for } d \geqslant 3. \end{cases} \tag{9.2}$$

The retention probabilities are constant for $d \geqslant 3$, but generation dependent for $d = 2$.

Theorem 9.14 Let $\{B(t) \colon 0 \leqslant t \leqslant T\}$ *denote transient Brownian motion started at the origin and* Cube $\subset \mathbb{R}^d$ *a compact cube of unit side length not containing the origin. Let* Γ *be a percolation limit set in* Cube *with retention probabilities chosen as in* (9.2). *Then the range of the Brownian motion is intersection-equivalent to the percolation limit set* Γ *in the cube* Cube.

Before discussing the proof, we look at an application of Theorem 9.14 to our understanding of Brownian motion. We first make two easy observations.

Lemma 9.15 *Suppose that* $A_1, \ldots, A_k, F_1, \ldots, F_k$ *are independent random closed sets, with* A_i *intersection-equivalent to* F_i *for* $1 \leqslant i \leqslant k$. *Then* $A_1 \cap A_2 \cap \ldots \cap A_k$ *is intersection-equivalent to* $F_1 \cap F_2 \cap \ldots \cap F_k$.

Proof. By induction, we can reduce this to the case $k = 2$. It then clearly suffices to show that $A_1 \cap A_2$ is intersection-equivalent to $F_1 \cap A_2$. This is done by conditioning on A_2,

$$
\begin{aligned}
\mathbb{P}\{A_1 \cap A_2 \cap \Lambda \neq \emptyset\} &= \mathbb{E}\big[\mathbb{P}\{A_1 \cap A_2 \cap \Lambda \neq \emptyset \,|\, A_2\}\big] \\
&\asymp \mathbb{E}\big[\mathbb{P}\{F_1 \cap A_2 \cap \Lambda \neq \emptyset \,|\, A_2\}\big] \\
&= \mathbb{P}\{F_1 \cap A_2 \cap \Lambda \neq \emptyset\}.
\end{aligned}
$$
∎

Lemma 9.16 *For independent percolation limit sets* Γ_1 *and* Γ_2 *with retention probabilities* $p_1, p_2, \ldots$ *and* $q_1, q_2, \ldots$, *respectively, their intersection* $\Gamma_1 \cap \Gamma_2$ *is a percolation limit set with retention probabilities* $p_1 q_1, p_2 q_2, \ldots$.

Proof. This is obvious from the definition of percolation limit sets and independence. ∎

These results enable us to recover the results about existence of nontrivial intersections of Brownian paths from the survival criteria of Galton–Watson trees, see Section 12.4 of the appendix.

As an example, we take a look at the intersection of two Brownian paths in $\mathbb{R}^d$, $d \geqslant 3$. By Theorem 9.14 and Lemma 9.15, the intersection of these paths is intersection-equivalent (in any unit cube not containing the starting points) to the intersection of two independent percolation limit sets with constant retention parameters $p = 2^{2-d}$. This intersection, by Lemma 9.16, is another percolation limit set, but now with parameter $p^2 = 2^{4-2d}$. Now observe that this set has a positive probability of being nonempty if and only if a Galton–Watson process with binomial offspring distribution with parameters $n = 2^d$ and $p = 2^{4-2d}$ has a positive survival probability. Recalling the criterion for survival of Galton–Watson trees from Proposition 12.37 in the appendix, we see that this is the case if and only if the mean offspring number np strictly exceeds 1, i.e. if $4 - d > 0$. In other words, in $d = 3$ the two paths intersect with positive probability, in all higher dimensions they almost surely do not intersect.

We now give the proof of Theorem 9.14. A key rôle in the proof is played by a fundamental result of Russell Lyons concerning survival probabilities of general trees under the percolation process, which has great formal similarity with the quantitative hitting estimates for Brownian paths of Theorem 8.24.

Recall the notation for trees from Section 12.4 in the appendix. As usual we define, for any kernel $K \colon \partial T \times \partial T \to [0, \infty]$, the K-energy of the measure μ on ∂T as

$$
I_K(\mu) = \iint K(x, y) \, d\mu(x) \, d\mu(y),
$$

and the K-capacity of the boundary of the tree by

$$\mathrm{Cap}_K(\partial T) = \left[\inf\left\{I_K(\mu)\colon \mu \text{ a probability measure on } \partial T\right\}\right]^{-1}.$$

Given a sequence $p_1, p_2, \ldots$ of probabilities, *percolation* on T is obtained by removing each edge of T of order n independently with probability $1 - p_n$ and retaining it otherwise, with mutual independence among edges. Say that a ray ξ **survives the percolation** if all the edges on ξ are retained, and say that the tree boundary ∂T survives if some ray of T survives.

Theorem 9.17 (Lyons) *If percolation with retention probabilities $p_1, p_2, \ldots$ is performed on a rooted tree T, then*

$$\mathrm{Cap}_K(\partial T) \leqslant \mathbb{P}\left\{\, \partial T \text{ survives the percolation }\right\} \leqslant 2\mathrm{Cap}_K(\partial T),\qquad(9.3)$$

where the kernel K is defined by $K(x,y) = \prod_{i=1}^{|x \wedge y|} p_i^{-1}$.

Proof. For two vertices v, w we write $v \leftrightarrow w$ if the shortest path between the vertices is retained in the percolation. We also write $v \leftrightarrow \partial T$ if a ray through vertex v survives the percolation and $v \leftrightarrow T_n$ if there is a self-avoiding path of retained edges connecting v to a vertex of generation n. Note that, with ρ denoting the root of the tree, $K(x,y) = \mathbb{P}\{\rho \leftrightarrow x \wedge y\}^{-1}$ by definition of the kernel K. By the finiteness of the degrees,

$$\{\rho \leftrightarrow \partial T\} = \bigcap_n \{\rho \leftrightarrow T_n\}.$$

We start with the left inequality in (9.3) and consider the case of a finite tree T first. We extend the definition of the boundary ∂T to finite trees by letting ∂T be the set of leaves, i.e., the vertices with no offspring. Let μ be a probability measure on ∂T and set

$$Y = \sum_{x \in \partial T} \mu(x)\frac{1\{\rho \leftrightarrow x\}}{\mathbb{P}\{\rho \leftrightarrow x\}}.$$

Then $\mathbb{E}[Y] = \sum_{x \in \partial T} \mu(x) = 1$, and

$$\mathbb{E}[Y^2] = \mathbb{E}\left[\sum_{x \in \partial T}\sum_{y \in \partial T} \mu(x)\mu(y)\frac{1\{\rho \leftrightarrow x, \rho \leftrightarrow y\}}{\mathbb{P}\{\rho \leftrightarrow x\}\,\mathbb{P}\{\rho \leftrightarrow y\}}\right]$$

$$= \sum_{x \in \partial T}\sum_{y \in \partial T} \mu(x)\mu(y)\frac{\mathbb{P}\{\rho \leftrightarrow x \text{ and } \rho \leftrightarrow y\}}{\mathbb{P}\{\rho \leftrightarrow x\}\mathbb{P}\{\rho \leftrightarrow y\}}.$$

Thus,

$$\mathbb{E}[Y^2] = \sum_{x,y \in \partial T} \mu(x)\mu(y)\,\frac{1}{\mathbb{P}\{\rho \leftrightarrow x \wedge y\}} = I_K(\mu).$$

Using the Paley–Zygmund inequality in the second step, we obtain

$$\mathbb{P}\{\rho \leftrightarrow \partial T\} \geqslant \mathbb{P}\{Y > 0\} \geqslant \frac{\left(\mathbb{E}[Y]\right)^2}{\mathbb{E}[Y^2]} = \frac{1}{I_K(\mu)}.$$

The left hand side does not depend on μ, so optimising the right hand side over μ yields

$$\mathbb{P}\{\rho \leftrightarrow \partial T\} \geqslant \sup_{\mu} \frac{1}{I_K(\mu)} = \mathrm{Cap}_K(\partial T)\,,$$

which proves the lower bound for finite trees. For T infinite, let μ be any probability measure on ∂T. This induces a probability measure $\widetilde{\mu}$ on the set T_n, consisting of those vertices which become leaves when the tree T is cut off after the nth generation, by letting

$$\widetilde{\mu}(v) = \mu\{\xi \in \partial T \colon v \in \xi\}, \qquad \text{for any vertex } v \in T_n\,.$$

By the finite case considered above,

$$\mathbb{P}\{\rho \leftrightarrow T_n\} \geqslant \Big(\sum_{x,y \in T_n} K(x,y)\widetilde{\mu}(x)\widetilde{\mu}(y) \Big)^{-1}.$$

Each ray ξ must pass through some vertex $x \in T_n$. This implies that $K(x,y) \leqslant K(\xi,\eta)$ for $x \in \xi$ and $y \in \eta$. Therefore,

$$\int_{\partial T} \int_{\partial T} K(\xi,\eta)\,d\mu(\xi)\,d\mu(\eta) \geqslant \sum_{x,y \in T_n} K(x,y)\widetilde{\mu}(x)\widetilde{\mu}(y) \geqslant \frac{1}{\mathbb{P}\{\rho \leftrightarrow T_n\}}\,.$$

Hence $\mathbb{P}\{\rho \leftrightarrow T_n\} \geqslant I_K(\mu)^{-1}$ for any probability measure μ on ∂T. Optimising over μ and passing to the limit as $n \to \infty$, we get $\mathbb{P}\{\rho \leftrightarrow \partial T\} \geqslant \mathrm{Cap}_K(\partial T)\,.$

It remains to prove the right hand inequality in (9.3). Assume first that T is finite. There is a Markov chain $\{V_k : k \in \mathbb{N}\}$ hiding here: Suppose the offspring of each individual is ordered from left to right, and note that this imposes a natural order on all vertices of the tree by saying that x is to the left of y if there are siblings v, w with v to the left of w, such that x is a descendant of v and y is a descendant of w. The random set of leaves that survive the percolation may thus be enumerated from left to right as $V_1, V_2, \ldots, V_r$. The key observation is that the random sequence $\rho, V_1, V_2, \ldots, V_r, \Delta, \Delta, \ldots$ is a Markov chain on the state space $\partial T \cup \{\rho, \Delta\}$, where ρ is the root and Δ is a formal absorbing cemetery. Indeed, given that $V_k = x$, all the edges on the unique path from ρ to x are retained, so that survival of leaves to the right of x is determined by the edges strictly to the right of the path from ρ to x, and is thus conditionally independent of $V_1, \ldots, V_{k-1}$, see Figure 9.1.

This verifies the Markov property, so Proposition 8.26 may be applied. The transition probabilities for the Markov chain above are complicated, but it is easy to write down the Green kernel G. For any vertex x let $\mathtt{path}(x)$ be the set of edges on the shortest path from ρ to x. Clearly, $G(\rho,y)$ equals the probability that y survives percolation, so

$$G(\rho,y) = \prod_{n=1}^{|y|} p_n\,.$$

If x is to the left of y, then $G(x,y)$ is equal to the probability that the range of the Markov chain contains y given that it contains x, which is just the probability of y surviving given that x survives. Therefore,

$$G(x,y) = \prod_{n=|x \wedge y|+1}^{|y|} p_n\,,$$

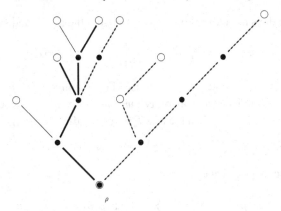

Fig. 9.1. The Markov chain embedded in the tree.

and hence

$$M(x,y) = \frac{G(x,y)}{G(\rho,y)} = \prod_{n=1}^{|x \wedge y|} p_n^{-1} \, .$$

Now $G(x,y) = 0$ for x on the right of y; thus (keeping the diagonal in mind)

$$K(x,y) \leqslant M(x,y) + M(y,x)$$

for all $x, y \in \partial T$, and therefore $I_K(\mu) \leqslant 2 I_M(\mu)$. Now apply Proposition 8.26 to $\Lambda = \partial T$:

$$\mathrm{Cap}_K(\partial T) \geqslant \tfrac{1}{2}\mathrm{Cap}_M(\partial T) \geqslant \tfrac{1}{2}\mathbb{P}\{\{V_k : k \in \mathbb{N}\} \text{ hits } \partial T\} = \tfrac{1}{2}\mathbb{P}\{\rho \leftrightarrow \partial T\}\,.$$

This establishes the upper bound for finite T. The inequality for general T follows from the finite case by taking limits. ∎

The main remaining task is to translate Lyons' theorem, Theorem 9.17, into hitting estimates for percolation limit sets using a 'tree representation' as in Figure 9.2, and relating the capacity of the tree boundary to the capacity of the percolation limit set.

Theorem 9.18 *Let Γ be a percolation limit set in the unit cube* Cube *with retention parameters $p_1, p_2, \ldots$. Then, for any closed set $\Lambda \subset$* Cube *we have*

$$\mathbb{P}\{\Gamma \cap \Lambda \neq \emptyset\} \asymp \mathrm{Cap}_f(\Lambda)\,,$$

for any decreasing f satisfying $f(2^{-k}) = p_1^{-1} \cdots p_k^{-1}$.

Remark 9.19 This result extends parts (i) and (ii) of Hawkes' theorem, Theorem 9.5, in two ways: It includes generation dependent retention and gives a quantitative estimate. ⋄

The key to this theorem is the following representation for the f-energy of a measure. Recall that $\mathfrak{D}_n$ denotes the collection of half-open dyadic cubes of side length 2^{-n}.

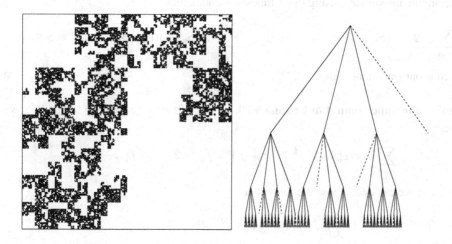

Fig. 9.2. Percolation limit set and associated tree

Lemma 9.20 *Suppose* $f\colon (0,\infty) \to (0,\infty)$ *is a decreasing function, and denote* $h(n) = f(2^{-n}) - f(2^{1-n})$ *for* $n \geqslant 1$, *and* $h(0) = f(1)$. *Then, for any measure* μ *on the unit cube* $[0,1)^d$,

$$I_f(\mu) \asymp \sum_{n=0}^{\infty} h(n) \Big(\sum_{Q \in \mathfrak{D}_n} \mu(Q)^2 \Big),$$

where the implied constants depend only on d.

Proof of the lower bound in Lemma 9.20. Fix an integer ℓ such that $\sqrt{d} \leqslant 2^{\ell}$. For any $x,y \in [0,1]^d$ we write $n(x,y) = \max\{n\colon x,y \in Q \text{ for some } Q \in \mathfrak{D}_n\}$. Note that $n(x,y) = n+\ell$ implies $|x - y| \leqslant \sqrt{d}2^{-n-\ell} \leqslant 2^{-n}$ and hence $f(|x - y|) \geqslant f(2^{-n})$. We thus get

$$I_f(\mu) = \iint f(|x - y|)\, d\mu(x)\, d\mu(y)$$
$$\geqslant \sum_{n=0}^{\infty} f(2^{-n})\, \mu \otimes \mu\{(x,y)\colon n(x,y) = n+\ell\}$$
$$= \sum_{n=0}^{\infty} f(2^{-n}) \left[S_{n+\ell}(\mu) - S_{n+\ell+1}(\mu) \right],$$

where $S_n(\mu) = \sum_{Q \in \mathfrak{D}_n} \mu(Q)^2$. Note that, by the Cauchy–Schwarz inequality,

$$S_n(\mu) = \sum_{Q \in \mathfrak{D}_n} \mu(Q)^2 = \sum_{Q \in \mathfrak{D}_n} \Big(\sum_{\substack{V \in \mathfrak{D}_{n+1} \\ V \subset Q}} \mu(V) \Big)^2$$

$$\leqslant 2^d \sum_{V \in \mathfrak{D}_{n+1}} \mu(V)^2 = 2^d\, S_{n+1}(\mu). \tag{9.4}$$

Rearranging the sum and using this ℓ times, we obtain that

$$\sum_{n=0}^{\infty} f(2^{-n}) \left[S_{n+\ell}(\mu) - S_{n+\ell+1}(\mu) \right] = \sum_{n=0}^{\infty} h(n) \, S_{n+\ell}(\mu) \geqslant \sum_{n=0}^{\infty} h(n) \, 2^{-d\ell} \, S_n(\mu) \,,$$

which is our statement with $c = 2^{-d\ell}$. ∎

Proof of the upper bound in Lemma 9.20. For $\sqrt{d} \, 2^{1-n} \geqslant |x - y| > \sqrt{d} \, 2^{-n}$, we have

$$\sum_{k=0}^{\infty} h(k) \mathbf{1}\{\sqrt{d} \, 2^{1-k} \geqslant |x - y|\} = f(\sqrt{d} \, 2^{-n}) \geqslant f(|x - y|) \,,$$

and hence we can decompose the integral as

$$I_f(\mu) = \iint f(|x - y|) \, d\mu(x) \, d\mu(y)$$

$$\leqslant \iint \sum_{k=0}^{\infty} h(k) \mathbf{1}\{\sqrt{d} \, 2^{1-k} \geqslant |x - y|\} \, d\mu(x) \, d\mu(y) \,.$$

For cubes $Q_1, Q_2 \in \mathfrak{D}_n$ we write $Q_1 \sim Q_2$ if there exist $q_1 \in Q_1$ and $q_2 \in Q_2$ with $|q_1 - q_2| \leqslant \sqrt{d} 2^{-n}$ (though note that $\sim$ is not an equivalence relation). Then

$$\iint \mathbf{1}\{\sqrt{d} \, 2^{1-k} \geqslant |x - y|\} \, d\mu(x) \, d\mu(y) = \mu \otimes \mu\{(x, y) \colon |x - y| \leqslant \sqrt{d} \, 2^{1-k}\}$$

$$\leqslant \sum_{\substack{Q_1, Q_2 \in \mathfrak{D}_{k-1} \\ Q_1 \sim Q_2}} \mu(Q_1) \mu(Q_2) \leqslant \tfrac{1}{2} \sum_{\substack{Q_1, Q_2 \in \mathfrak{D}_{k-1} \\ Q_1 \sim Q_2}} \left(\mu(Q_1)^2 + \mu(Q_2)^2 \right) ,$$

using the inequality of the geometric and arithmetic mean in the last step. As, for each cube Q_1, the number of cubes Q_2 with $Q_1 \sim Q_2$ is bounded by some constant $C_d > 0$, we obtain that

$$I_f(\mu) \leqslant \tfrac{C_d+1}{2} \sum_{k=0}^{\infty} h(k) \sum_{Q \in \mathfrak{D}_{k-1}} \mu(Q)^2 \leqslant (C_d + 1) \, 2^{d-1} \sum_{k=0}^{\infty} h(k) \sum_{Q \in \mathfrak{D}_k} \mu(Q)^2 \,,$$

using (9.4) from above. This completes the proof of the upper bound. ∎

Proof of Theorem 9.18. Denote the coordinatewise minimum of Cube by x_0. We employ the canonical mapping $\mathcal{R}$ from the boundary of a 2^d-ary tree Υ, where every vertex has 2^d children, to the cube Cube. Formally, label the edges from each vertex to its children in a one-to-one manner with the vectors in $\Theta = \{0, 1\}^d$. Then the boundary $\partial \Upsilon$ is identified with the sequence space $\Theta^{\mathbb{Z}^+}$ and we define $\mathcal{R} \colon \partial \Upsilon = \Theta^{\mathbb{Z}^+} \to$ Cube by

$$\mathcal{R}(\omega_1, \omega_2, \ldots) = x_0 + \sum_{n=1}^{\infty} 2^{-n} \omega_n \,.$$

We now use the representation given in Lemma 9.20 to relate the K-energy of a measure μ on $\partial \Upsilon$ (with K as in Theorem 9.17) to the f-energy of its image measure $\mu \circ \mathcal{R}^{-1}$ on Cube, showing that

$$I_K(\mu) \asymp I_f(\mu \circ \mathcal{R}^{-1}) \tag{9.5}$$

where the implied constants depend only on the dimension d. Indeed the K-energy of a measure μ on $\partial\Upsilon$ satisfies, by definition,

$$I_K(\mu) = \iint \prod_{i=1}^{|x \wedge y|} p_i^{-1}\, d\mu(x)\, d\mu(y) = \iint \sum_{v \leqslant x \wedge y} \left(\prod_{i=1}^{|v|} p_i^{-1} - \prod_{i=1}^{|v|-1} p_i^{-1} \right) d\mu(x)\, d\mu(y),$$

where we count all ancestors v of $x \wedge y$ and we interpret the contribution of the root $v = \rho$ as one. Interchanging summation and integration we obtain

$$
\begin{aligned}
I_K(\mu) &= \sum_{v \in \Upsilon} \left(\prod_{i=1}^{|v|} p_i^{-1} - \prod_{i=1}^{|v|-1} p_i^{-1} \right) \iint 1\{x \geqslant v, y \geqslant v\}\, d\mu(x)\, d\mu(y) \\
&= \sum_{v \in \Upsilon} \left(\prod_{i=1}^{|v|} p_i^{-1} - \prod_{i=1}^{|v|-1} p_i^{-1} \right) \mu(\{\xi \in \partial T : v \in \xi\})^2,
\end{aligned}
$$

whereas the f-energy of the measure $\mu \circ \mathcal{R}^{-1}$ satisfies, by Lemma 9.20,

$$I_f\left(\mu \circ \mathcal{R}^{-1}\right) \asymp \sum_{k=0}^{\infty} h(k) \sum_{D \in \mathfrak{D}_k} \mu\left(\mathcal{R}^{-1}(D)\right)^2,$$

where

$$h(k) = f(2^{-k}) - f(2^{-k+1}) = p_1^{-1} \cdots p_k^{-1} - p_1^{-1} \cdots p_{k-1}^{-1}$$

by our assumptions on f. Now $\mathcal{R}^{-1}(D)$ is contained in no more than 3^d sets of the form $\{\xi \in \partial T : v \in \xi\}$, for $|v| = k$, in such a way that over all cubes $D \in \mathfrak{D}_k$ no such set is used in more than 3^d covers. Conversely each set $\mathcal{R}^{-1}(D)$ contains an individual set of this form entirely, so that we obtain (9.5).

Any closed set Λ in the unit cube Cube can be written as the image $\mathcal{R}(\partial T)$ of the boundary of some subtree T of the regular 2^d-ary tree. As any measure ν on $\mathcal{R}(\partial T) \subset$ Cube can be written as $\mu \circ \mathcal{R}^{-1}$ for an appropriate measure μ on ∂T it follows from (9.5) that $\mathrm{Cap}_K(\partial T) \asymp \mathrm{Cap}_f(\mathcal{R}(\partial T))$. We perform percolation with retention parameters $p_1, p_2, \ldots$ on the tree T. Then, by Theorem 9.17,

$$
\begin{aligned}
\mathbb{P}\{\Gamma \cap \Lambda \neq \emptyset\} &= \mathbb{P}\{\, \partial T \text{ survives the percolation }\} \\
&\asymp \mathrm{Cap}_K(\partial T) \asymp \mathrm{Cap}_f(\Lambda).
\end{aligned}
$$
∎

Proof of Theorem 9.14.   As the cube Cube has positive distance to the starting point of Brownian motion, we can remove the denominator and smaller order terms from the Martin kernel in Theorem 8.24, as in the proof of Theorem 8.20. We thus obtain

$$\mathbb{P}\{B[0, T] \cap \Lambda \neq \emptyset\} \asymp \mathrm{Cap}_f(\Lambda),$$

where f is the radial potential. For the choice of retention probabilities in (9.2) we can apply Theorem 9.18, which implies

$$\mathrm{Cap}_f(\Lambda) \asymp \mathbb{P}\{\Gamma \cap \Lambda \neq \emptyset\},$$

and combining the two displays gives the result. ∎

The intersection equivalence approach enables us to characterise the polar sets for the intersection of p independent Brownian motions in $\mathbb{R}^d$ and give a quantitative estimate of the hitting probabilities.

Theorem 9.21 *Let $B_1, \ldots, B_p$ be independent Brownian motions in $\mathbb{R}^d$ starting in arbitrary fixed points and suppose $p(d-2) < d$. Let*

$$S = \{x \in \mathbb{R}^d \colon \exists\, t_1, \ldots, t_p > 0 \text{ with } x = B_1(t_1) = \cdots = B_p(t_p)\}.$$

Then, for any closed set Λ, we have

$$\mathbb{P}\{S \cap \Lambda \neq \emptyset\} > 0 \qquad \textit{if and only if} \qquad \mathrm{Cap}_{f^p}(\Lambda) > 0,$$

where f is the radial potential.

Proof.   We may assume that Λ is contained in a unit cube at positive distance from the starting points. Let Γ be a percolation limit set in that cube, with retention probabilities $p_1, p_2, \ldots$ satisfying $p_1 \cdots p_n = 1/f^p(2^{-n})$. By Theorem 9.14, Lemma 9.15 and Lemma 9.16, the random set S is intersection-equivalent to Γ in that cube. Theorem 9.18 characterises the polar sets for Γ, completing the argument.   ∎

9.3 Multiple points of Brownian paths

A point $x \in \mathbb{R}^d$ has **multiplicity** p, or is a p-**fold multiple point**, for a Brownian motion $\{B(t) \colon t \geqslant 0\}$ in $\mathbb{R}^d$, if there exist times $0 < t_1 < \cdots < t_p$ with

$$x = B(t_1) = \cdots = B(t_p).$$

The results of the previous section also provide the complete answer to the question of the existence of such points.

Theorem 9.22 *Suppose $d \geqslant 2$ and $\{B(t) \colon t \in [0,1]\}$ is a d-dimensional Brownian motion. Then, almost surely,*

- *if $d \geqslant 4$ no double points exist, i.e. Brownian motion is injective,*

- *if $d = 3$ double points exist, but triple points fail to exist,*

- *if $d = 2$ points of any finite multiplicity exist.*

Proof.   To show *nonexistence* of double points in $d \geqslant 4$ it suffices to show that for any rational $\alpha \in (0,1)$, almost surely, there exists no times $0 \leqslant t_1 < \alpha < t_2 \leqslant 1$ with $B(t_1) = B(t_2)$. Fixing such an α, the Brownian motions $\{B_1(t) \colon 0 \leqslant t \leqslant 1 - \alpha\}$ and $\{B_2(t) \colon 0 \leqslant t \leqslant \alpha\}$ given by

$$B_1(t) = B(\alpha + t) - B(\alpha) \quad \text{and} \quad B_2(t) = B(\alpha - t) - B(\alpha)$$

are independent and hence, by Theorem 9.1, they do not intersect, almost surely, proving the statement.

To show *existence* of double points in $d \leqslant 3$ we apply Theorem 9.1 in conjunction with Remark 9.2, to the independent Brownian motions $\{B_1(t) : 0 \leqslant t \leqslant \frac{1}{2}\}$ and $\{B_2(t) : 0 \leqslant t \leqslant \frac{1}{2}\}$ given by

$$B_1(t) = B\left(\tfrac{1}{2} + t\right) - B\left(\tfrac{1}{2}\right) \quad \text{and} \quad B_2(t) = B\left(\tfrac{1}{2} - t\right) - B\left(\tfrac{1}{2}\right),$$

to see that, almost surely, the two ranges intersect.

To show nonexistence of *triple* points in $d = 3$ we observe that it suffices to show that for any three rationals $0 < \alpha_1 < \alpha_2 < \alpha_3$ and $\varepsilon < (\alpha_3 - \alpha_2) \wedge (\alpha_2 - \alpha_1)$, almost surely there are no times $t_i \in (\alpha_i, \alpha_i + \varepsilon)$ such that $B(t_1) = B(t_2) = B(t_3)$. By conditioning the Brownian motion on its values at the times α_i and $\alpha_i + \varepsilon$, for $i \in \{1, 2, 3\}$, we obtain three Brownian bridges $\{B_i(t) : 0 \leqslant t \leqslant \varepsilon\}$ given by

$$B_i(t) = B(\alpha_i + t) - B(\alpha_i), \quad \text{for } i \in \{1, 2, 3\}.$$

By Exercise 9.2 the probability that these three bridges intersect is zero, for any values $B(\alpha_i)$, $B(\alpha_i + \varepsilon)$. Taking an expectation over these values we obtain the result.

To show the existence of *p-multiple* points in $\mathbb{R}^2$ fix $\delta > 0$ and numbers

$$0 < \alpha_1 < \alpha_2 < \cdots < \alpha_p < \alpha_{p+1} = \delta.$$

Let $\varepsilon > 0$ small enough that $\alpha_i + \varepsilon < \alpha_{i+1}$ for $i \in \{1, \ldots, p\}$ and condition the Brownian motion on its values at the times α_i and $\alpha_i + \varepsilon$, for $i \in \{1, \ldots, p\}$. We obtain p Brownian bridges $\{B_i(t) : 0 \leqslant t \leqslant \varepsilon\}$ given by

$$B_i(t) = B(\alpha_i + t) - B(\alpha_i), \quad \text{for } i \in \{1, \ldots, p\}.$$

By Exercise 9.2 these bridges intersect with positive probability, for any values $B(\alpha_i)$, $B(\alpha_i + \varepsilon)$. Taking an expectation over these values we obtain that, for $\delta > 0$, the path $\{B(t) : 0 \leqslant t \leqslant \delta\}$ has a p-multiple point with positive probability. By Brownian scaling this probability is independent of the choice of δ, and letting $\delta \downarrow 0$, we obtain

$$\text{Prob}\{ \text{ for all } \delta > 0 \text{ exist } 0 < t_1 < \cdots < t_p < \delta \text{ with } B(t_1) = \cdots = B(t_p)\} > 0.$$

By Blumenthal's zero–one law this probability must be one, so that we have a p-multiple point almost surely, which completes the proof. ∎

Theorem 9.23 *Let $\{B(t) : 0 \leqslant t \leqslant 1\}$ be a planar Brownian motion. Then, almost surely, for every positive integer p, there exist points $x \in \mathbb{R}^2$ which are visited exactly p times by the Brownian motion.*

Proof. Note first that it suffices to show this with positive probability. Indeed, by Brownian scaling, the probability that the path $\{B(t) : 0 \leqslant t \leqslant r\}$ has points of multiplicity exactly p does not depend on r. By Blumenthal's zero-one law it therefore must be zero or one. The idea of the proof is now to construct a set Λ such that $\text{Cap}_{f^p}(\Lambda) > 0$ but $\text{Cap}_{f^{p+1}}(\Lambda) = 0$ for the radial potential f. By Exercise 9.3 the first condition implies that the probability that Λ contains a p-fold multiple point is positive. The second condition ensures that it almost surely does not contain a $p + 1$-fold multiple point. Hence the p-multiple points found in Λ must be strictly p-multiple.

We construct the set Λ by iteration, starting from a compact unit cube Cube. In the nth construction step we divide each cube retained in the previous step into its four nonoverlapping dyadic subcubes and retain only one of them, say the bottom left cube, except at the steps with number

$$n = \lceil 4^{\frac{k}{p+1}} \rceil, \qquad \text{for } k = p+1, p+2, \ldots,$$

when we retain all four subcubes. The number $k(n)$ of times within the first n steps when we have retained all four cubes satisfies $k(n) \asymp (\log n) \frac{p+1}{\log 4}$. Denoting by $\mathfrak{S}_n$ the set of all dyadic cubes retained in the nth step, we define the compact set

$$\Lambda = \bigcap_{n=1}^{\infty} \bigcup_{S \in \mathfrak{S}_n} S.$$

The calculation of the capacity of Λ will be based on the formula given in Lemma 9.20. Observe that, if $f^p(\varepsilon) = \log^p(1/\varepsilon)$ is the pth power of the 2-dimensional radial potential, then the associated function is

$$h^{(p)}(n) = f^p(2^{-n}) - f^p(2^{1-n}) \asymp n^p - (n-1)^p \asymp n^{p-1}.$$

Note that the number $g(n)$ of cubes kept in the first n steps of the construction satisfies $g(n) \asymp 4^{k(n)} \asymp n^{p+1}$. By our construction $\sum_{n=0}^{\infty} n^{p-1} g(n)^{-1} < \infty$, but $\sum_{n=0}^{\infty} n^p g(n)^{-1} = \infty$. For the measure μ distributing the unit mass equally among the retained cubes of the same side length (hence giving mass $g(n)^{-1}$ to each retained cube), we have

$$I_{f^p}(\mu) \asymp \sum_{n=0}^{\infty} h^{(p)}(n) \Big(\sum_{Q \in \mathfrak{D}_n} \mu(Q)^2 \Big) \asymp \sum_{n=0}^{\infty} n^{p-1} g(n)^{-1} < \infty,$$

and hence $\text{Cap}_{f^p}(\Lambda) > 0$. For the converse statement, note that

$$\Big(\sum_{Q \in \mathfrak{D}_n} \nu(Q)^2 \Big) \Big(\sum_{Q \in \mathfrak{D}_n} 1\{Q \text{ retained }\} \Big) \geqslant 1,$$

for any probability measure ν on Λ, by the Cauchy–Schwarz inequality. Hence,

$$I_{f^{p+1}}(\nu) \asymp \sum_{n=0}^{\infty} h^{(p+1)}(n) \Big(\sum_{Q \in \mathfrak{D}_n} \nu(Q)^2 \Big) \geqslant \sum_{n=0}^{\infty} h^{(p+1)}(n) g(n)^{-1}$$

$$\asymp \sum_{n=0}^{\infty} n^p g(n)^{-1} = \infty,$$

verifying that $\text{Cap}_{f^{p+1}}(\Lambda) = 0$. This completes the proof. ∎

Knowing that planar Brownian motion has points of arbitrarily large *finite* multiplicity, it is an interesting question whether there are points of *infinite* multiplicity.

Theorem* 9.24 *Let $\{B(t): t \geqslant 0\}$ be a planar Brownian motion. Then, almost surely, there exists a point $x \in \mathbb{R}^2$ such that the set $\{t \geqslant 0: B(t) = x\}$ is uncountable.*

The rest of this section is devoted to the proof of this interesting result and will not be used in the remainder of the book. It may be skipped on first reading.

Let us first describe the rough strategy of the proof: We start by finding two disjoint intervals I_1 and I_2 with $B(I_1) \cap B(I_2) \neq \emptyset$. Inside these we find disjoint subintervals $I_{11}, I_{12} \subset I_1$ and $I_{21}, I_{22} \subset I_2$ such that the four Brownian images $B(I_{ij})$ intersect. Continuing this way, we construct a binary tree T of time intervals where rays in T represent sequences of nested intervals and the intersection of each such sequence will be mapped to the same point by the Brownian motion.

Throughout the proof we use the following notation. For any open or closed sets A_1, A_2, ... and a Brownian motion $B \colon [0, \infty) \to \mathbb{R}^2$ define stopping times

$$\tau(A_1) := \inf\{t \geqslant 0 \colon B(t) \in A_1\},$$

$$\tau(A_1, \ldots, A_n) := \inf\{t \geqslant \tau(A_1, \ldots, A_{n-1}) \colon B(t) \in A_n\}, \quad \text{for } n \geqslant 2,$$

where, as usual, the infimum of the empty set is taken to be infinity. We say the Brownian motion *upcrosses the shell* $\mathcal{B}(x, 2r) \setminus \mathcal{B}(x, r)$ *twice* before a stopping time T if,

$$\tau\big(\mathcal{B}(x, r), \mathcal{B}(x, 2r)^c, \mathcal{B}(x, r), \mathcal{B}(x, 2r)^c\big) < T.$$

We call the paths of Brownian motion between $\tau(\mathcal{B}(x, r))$ and $\tau(\mathcal{B}(x, r), \mathcal{B}(x, 2r)^c)$, and between $\tau(\mathcal{B}(x, r), \mathcal{B}(x, 2r)^c, \mathcal{B}(x, r))$ and $\tau(\mathcal{B}(x, r), \mathcal{B}(x, 2r)^c, \mathcal{B}(x, r), \mathcal{B}(x, 2r)^c)$ the *upcrossing excursions*, see Figure 9.3.

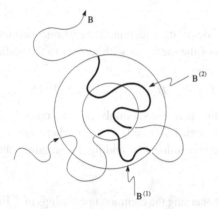

Fig. 9.3. The path $B \colon [0, \infty) \to \mathbb{R}^2$ upcrosses the shell twice; the upcrossing excursions are bold and marked $B^{(1)}$, $B^{(2)}$.

From now on let T be the first exit time of Brownian motion from the unit ball.

Lemma 9.25 *There exist constants $0 < c_0 < C_0$ such that, if $2 < m < n$ are two integers and B a ball of radius 2^{-n} with centre at distance at least 2^{-m} and at most 3×2^{-m} from the origin, we have*

$$c_0 \frac{m}{n} \leqslant \mathbb{P}_0\{\tau(B) < T\} \leqslant C_0 \frac{m}{n}.$$

Proof. For the lower bound we note that the disk of radius $\frac{1}{2}$ around the centre of B is contained in the unit disk, so that the first exit time T' from this disk satisfies $T' \leqslant T$. Theorem 3.18 gives the lower bound for $\mathbb{P}_0\{\tau(B) < T'\}$. Similarly, for the upper bound we look at the disk of radius 2 around the centre of B, which contains the unit disk. ∎

Recall from Theorem 3.44 that the density of $B(T)$ under $\mathbb{P}_z$ is given by the Poisson kernel, which is

$$\mathcal{P}(z, w) = \frac{1 - |z|^2}{|z - w|^2} \qquad \text{for any } z \in \mathcal{B}(0, 1) \text{ and } w \in \partial\mathcal{B}(0, 1).$$

Lemma 9.26 *Consider Brownian motion started at $z \in \mathcal{B}(0, r)$ where $r < 1$, and stopped at time T when it exits the unit ball. Let $\tau \leqslant T$ be a stopping time, and let $A \in \mathcal{F}(\tau)$. Then we have*

(i) $\mathbb{P}_z\big(A \,\big|\, B(T)\big) = \mathbb{P}_z(A) \dfrac{\mathbb{E}_z\big[\mathcal{P}(B(\tau), B(T)) \,\big|\, A\big]}{\mathcal{P}(z, B(T))}.$

(ii) *If $\mathbb{P}_z\big(\{B(\tau) \in \mathcal{B}(0, r)\} \,\big|\, A\big) = 1$, then*

$$\Big(\frac{1 - r}{1 + r}\Big)^2 \mathbb{P}_z(A) \leqslant \mathbb{P}_z\big(A \,\big|\, B(T)\big) \leqslant \Big(\frac{1 + r}{1 - r}\Big)^2 \mathbb{P}_z(A) \qquad \text{almost surely.}$$

Proof. (i) Let $I \subset \partial\mathcal{B}(0, 1)$ be a Borel set. Using the strong Markov property and the assumption $A \in \mathcal{F}(\tau)$ in the second step, we get

$$\mathbb{P}_z\big(A \,\big|\, \{B(T) \in I\}\big) \,\mathbb{P}_z\{B(T) \in I\} = \mathbb{P}_z(A) \,\mathbb{P}_z\big(\{B(T) \in I\} \,\big|\, A\big)$$
$$= \mathbb{P}_z(A) \,\mathbb{E}_z\big[\mathbb{P}_{B(\tau)}\{B(T) \in I\} \,\big|\, A\big].$$

As a function of I, both sides of the equation define a finite measure with total mass $\mathbb{P}_z(A)$. Comparing the densities of the measures with respect to the surface measure on $\partial\mathcal{B}(0, 1)$ gives

$$\mathbb{P}_z\big(A \,\big|\, B(T)\big) \,\mathcal{P}(z, B(T)) = \mathbb{P}_z(A) \,\mathbb{E}_z\big[\mathcal{P}(B(\tau), B(T)) \,\big|\, A\big].$$

(ii) The assumption of this part and (i) imply that the ratio $\mathbb{P}_z(A|B(T))/\mathbb{P}_z(A)$ can be written as an average of ratios $\mathcal{P}(u, w)/\mathcal{P}(z, w)$ where $w = B(T) \in \partial\mathcal{B}(0, 1)$ and $u, z \in \mathcal{B}(0, r)$. The assertion follows by finding the minimum and maximum of $\mathcal{P}(u, w)$ as u ranges over $\mathcal{B}(0, r)$. ∎

The following lemma, concerning the common upcrossings of L Brownian excursions, will be the engine driving the proof of Theorem 9.24.

Lemma 9.27 *Let $n > 5$ and let $\{x_1, \ldots, x_{4^{n-5}}\}$ be points such that the balls $\mathcal{B}(x_i, 2^{1-n})$ are disjoint and contained in the shell $\{z\colon \frac{1}{4} \leqslant |z| \leqslant \frac{3}{4}\}$. Consider L independent Brownian upcrossing excursions $W_1, \ldots, W_L$, started at prescribed points on $\partial\mathcal{B}(0, 1)$ and stopped when they reach $\partial\mathcal{B}(0, 2)$. Let S denote the number of centres x_i, $1 \leqslant i \leqslant 4^{n-5}$ such that the shell $\mathcal{B}(x_i, 2^{-n+1}) \setminus \mathcal{B}(x_i, 2^{-n})$ is upcrossed twice by each of $W_1, \ldots, W_L$. Then there exist constants $c, c_* > 0$ such that*

$$\mathbb{P}\big\{S > 4^n (c/n)^L\big\} \geqslant \frac{c_*^L}{L!}. \tag{9.6}$$

Moreover, the same estimate (with a suitable constant c_*) is valid if we condition on the end points of the excursions $W_1, \ldots, W_L$.

Proof of Lemma 9.27. By Lemma 9.25, for any $z \in \partial \mathcal{B}(0,1)$, the probability of Brownian motion starting at z hitting the ball $\mathcal{B}(x_i, 2^{-n})$ before reaching $\partial \mathcal{B}(0,2)$ is at least $\frac{c_0}{n}$, and the probability of the second upcrossing excursion of $\mathcal{B}(x_i, 2^{-n+1}) \setminus \mathcal{B}(x_i, 2^{-n})$, when starting at $\partial \mathcal{B}(x_i, 2^{1-n})$ is at least $1/2$. Thus

$$\mathbb{E}S \geqslant 4^{n-5} \left(\frac{c_0}{2n} \right)^L. \tag{9.7}$$

We now estimate the second moment of S. Consider a pair of centres x_i, x_j such that $2^{-m} \leqslant |x_i - x_j| \leqslant 2^{1-m}$ for some $m < n-1$. For each $k \leqslant L$, let $V_k = V_k(x_i, x_j)$ denote the event that the balls $\mathcal{B}(x_i, 2^{-n})$ and $\mathcal{B}(x_j, 2^{-n})$ are both visited by W_k. Given that $\mathcal{B}(x_i, 2^{-n})$ is reached first, the conditional probability that W_k will also visit $\mathcal{B}(x_j, 2^{-n})$ is at most $C_0 \frac{m}{n}$, by Lemma 9.25. We conclude that $\mathbb{P}(V_k) \leqslant 2C_0^2 \frac{m}{n^2}$ whence

$$\mathbb{P}\left(\bigcap_{k=1}^{L} V_k \right) \leqslant \left(2C_0^2 \frac{m}{n^2} \right)^L.$$

For each $m < n-1$ and $i \leqslant 4^{n-5}$, the number of centres x_j such that $2^{-m} \leqslant |x_i - x_j| \leqslant 2^{1-m}$ is at most a constant multiple of 4^{n-m}. Using that the diagonal terms are of lower order, we deduce that there exists $C_1 > 0$ such that

$$\mathbb{E}S^2 \leqslant \frac{C_1^L 4^{2n}}{n^{2L}} \sum_{m=1}^{n} m^L 4^{-m} \leqslant \frac{(2C_1)^L 4^{2n} L!}{n^{2L}}, \tag{9.8}$$

where the last inequality follows, e.g., from taking $x = 1/4$ in the binomial identity

$$\sum_{m=0}^{\infty} \binom{m+L}{L} x^m = (1-x)^{-L-1}.$$

Now (9.7), (9.8) and the Paley–Zygmund inequality, see Exercise 3.5, yield (9.6). The final statement of the lemma follows from Lemma 9.26. ∎

Proof of Theorem 9.24. Fix an increasing sequence $\{n_i : i \geqslant 1\}$ to be chosen later, and let $N_\ell = \sum_{i=1}^{\ell} n_i$ with $N_0 = 0$. Denote $q_i = 4^{n_i-5}$ and $Q_i = 4^{N_i-5i}$. We begin by constructing a nested sequence of centres with which we associate a forest, i.e. a collection of trees, in the following manner. The first level of the forest consists of Q_1 centres, $\{x_1^{(1)}, \ldots, x_{Q_1}^{(1)}\}$, chosen such that the balls $\{\mathcal{B}(x_k^{(1)}, 2^{-N_1+1}) : 1 \leqslant k \leqslant Q_1\}$ are disjoint and contained in the annulus $\{z : \frac{1}{4} \leqslant |z| \leqslant \frac{3}{4}\}$.

Continue this construction recursively. For $\ell > 1$ suppose that level $\ell - 1$ of the forest has been constructed. Level ℓ consists of Q_ℓ vertices $\{x_1^{(\ell)}, \ldots, x_{Q_\ell}^{(\ell)}\}$. Each vertex $x_i^{(\ell-1)}$, $1 \leqslant i \leqslant Q_{(\ell-1)}$, at level $\ell - 1$ has q_ℓ children $\{x_j^{(\ell)} : (i-1)q_\ell < j \leqslant iq_\ell\}$ at level ℓ; the balls of radius $2^{-N_\ell+1}$ around these children are disjoint and contained in the annulus

$$\{z : \tfrac{1}{2} 2^{-N_\ell-1} \leqslant |z - x_i^{(\ell-1)}| \leqslant \tfrac{3}{4} 2^{-N_\ell-1}\}.$$

Recall that $T = \inf\{t > 0 : |B(t)| = 1\}$. We say that a level one vertex $x_k^{(1)}$ *survived* if the Brownian motion upcrosses the shell $\mathcal{B}(x_k^{(1)}, 2^{-N_1+1}) \setminus \mathcal{B}(x_k^{(1)}, 2^{-N_1})$ twice before T.

A vertex at the second level $x_k^{(2)}$ is said to have *survived* if its parent vertex survived, and in each upcrossing excursion of its parent, the Brownian motion upcrosses the shell $\mathcal{B}(x_k^{(2)}, 2^{-N_2+1}) \setminus \mathcal{B}(x_k^{(2)}, 2^{-N_2})$ twice. Recursively, we say a vertex $x_k^{(\ell)}$, at level ℓ of the forest, *survived* if its parent vertex survived, and in each of the $2^{\ell-1}$ upcrossing excursions of its parent, the Brownian motion upcrosses the shell

$$\mathcal{B}(x_k^{(\ell)}, 2^{-N_\ell+1}) \setminus \mathcal{B}(x_k^{(\ell)}, 2^{-N_\ell})$$

twice. Note at this point that if there is an infinite ray of surviving vertices

$$x_{k(1)}^{(1)}, x_{k(2)}^{(2)}, x_{k(3)}^{(3)}, \cdots$$

such that $x_{k(\ell+1)}^{(\ell+1)}$ is a child of $x_{k(\ell)}^{(\ell)}$, for $\ell = 1, 2, \ldots$, then the sequence of compact balls centred in $x_{k(\ell)}^{(\ell)}$ with radius 2^{-N_ℓ} is nested. Therefore there exists exactly one point x in the intersection of these balls. For any level ℓ, there are 2^ℓ disjoint upcrossing excursions of the shell $\mathcal{B}(x_{k(\ell)}^{(\ell)}, 2^{-N_\ell+1}) \setminus \mathcal{B}(x_{k(\ell)}^{(\ell)}, 2^{-N_\ell})$. Each of these contains two disjoint excursions at level $\ell+1$. Thus the time intervals corresponding to these excursions form a binary tree, where the children of an interval at level ℓ are the two intervals at level $\ell + 1$ it contains. An infinite ray in this tree is a nested sequence of compact intervals and their intersection is a time t with $B(t) = x$. Since there are uncountably many rays, x has uncountable multiplicity.

Now, for any $\ell \geqslant 1$, let S_ℓ denote the number of vertices at level ℓ of the forest that survived. Using the notation of Lemma 9.27, let

$$\Gamma_\ell = 4^{n_\ell} \left(\frac{c}{n_\ell}\right)^L \quad \text{and} \quad p_\ell = \frac{c_*^L}{L!},$$

where $L = L(\ell) = 2^{\ell-1}$. Lemma 9.27 with $n = n_1$ states that

$$\mathbb{P}\{S_1 > \Gamma_1\} \geqslant p_1 = c_*. \tag{9.9}$$

For $\ell > 1$, the same lemma, and independence of excursions in disjoint shells given their endpoints, yield

$$\mathbb{P}\big(\{S_{\ell+1} \leqslant \Gamma_{\ell+1}\} \,\big|\, \{S_\ell > \Gamma_\ell\}\big) \leqslant (1 - p_{\ell+1})^{\Gamma_\ell} \leqslant \exp(-p_{\ell+1}\Gamma_\ell). \tag{9.10}$$

By picking n_ℓ large enough, we can ensure that $p_{\ell+1}\Gamma_\ell > \ell$, whence the right hand side of (9.10) is summable in ℓ. Consequently

$$\alpha = \mathbb{P}\Big(\bigcap_{\ell=1}^{\infty}\{S_\ell > \Gamma_\ell\}\Big)$$

$$= \mathbb{P}\{S_1 > \Gamma_1\} \prod_{\ell=1}^{\infty} \mathbb{P}\big(\{S_{\ell+1} > \Gamma_{\ell+1}\} \,\big|\, \{S_\ell > \Gamma_\ell\}\big) > 0. \tag{9.11}$$

Thus with probability at least α, there is a ray of surviving vertices $x_{k(\ell)}^{(\ell)}$ and, as seen above, this yields a point visited by Brownian motion uncountably many times before it exits the unit disk.

Let H_r denote the event that Brownian motion, killed on exiting $\mathcal{B}(0, r)$, has a point of uncountable multiplicity. As explained above, (9.11) implies that $\mathbb{P}(H_1) \geq \alpha > 0$. By Brownian scaling, $\mathbb{P}(H_r)$ does not depend on r, whence

$$\mathbb{P}\Big(\bigcap_{n=1}^{\infty} H_{1/n} \Big) \geq \alpha.$$

The Blumenthal zero-one law implies that this intersection has probability 1, so there are points of uncountable multiplicity almost surely. ∎

9.4 Kaufman's dimension doubling theorem

In Theorem 4.33 we have seen that d-dimensional Brownian motion maps any set of dimension α almost surely into a set of dimension 2α. Surprisingly, by a famous result of Kaufman, the dimension doubling property holds almost surely *simultaneously* for all sets.

Theorem 9.28 (Kaufman 1969) *Let* $\{B(t): t \geq 0\}$ *be Brownian motion in dimension* $d \geq 2$. *Almost surely, for any set* $A \subset [0, \infty)$, *we have*

$$\dim B(A) = 2 \dim A.$$

Before discussing the proof, let us look at some consequences of Theorem 9.28. The power of this result lies in the fact that the dimension doubling formula can now be applied to arbitrary random sets.

As a first application we ask, how big the sets

$$T(x) = \{t \geq 0 : B(t) = x\}$$

of times mapped by d-dimensional Brownian motion onto the same point x can possibly be. We have seen so far in this chapter and Theorem 6.40 that, almost surely,

- in dimension $d \geq 4$ all sets $T(x)$ consist of at most one point,
- in dimension $d = 3$ all sets $T(x)$ consist of at most two points,
- in dimension $d = 2$ at least one of the sets $T(x)$ is uncountable,
- in dimension $d = 1$ all sets $T(x)$ have at least Hausdorff dimension $\frac{1}{2}$.

We use Kaufman's theorem to determine the Hausdorff dimension of the sets $T(x)$ in the case of planar and linear Brownian motion.

Corollary 9.29 *Suppose* $\{B(t): t \geq 0\}$ *is a planar Brownian motion. Then, almost surely, for all* $x \in \mathbb{R}^2$, *we have* $\dim T(x) = 0$.

Proof. By Kaufman's theorem, almost surely, $\dim T(x) = \frac{1}{2} \dim\{x\} = 0$ for all x. ∎

Corollary 9.30 *Suppose* $\{B(t): t \geq 0\}$ *is a linear Brownian motion. Then, almost surely, for all* $x \in \mathbb{R}$, *we have* $\dim T(x) = \frac{1}{2}$.

Proof. The lower bound was shown in Theorem 6.40. For the upper bound let $\{W(t): t \geqslant 0\}$ be a Brownian motion independent of $\{B(t): t \geqslant 0\}$. Applying Kaufman's theorem for the planar Brownian motion given by $\widetilde{B}(t) = (B(t), W(t))$ we get, almost surely, for every x,

$$\dim T(x) = \dim \widetilde{B}^{-1}(\{x\} \times \mathbb{R}) \leqslant \tfrac{1}{2} \dim(\{x\} \times \mathbb{R}) = \tfrac{1}{2},$$

which proves the upper bound. ∎

We now prove Kaufman's theorem. Recall that, by Corollary 1.20, almost surely, the function $\{B(t): t \geqslant 0\}$ is α-Hölder continuous for any $\alpha < \tfrac{1}{2}$. Hence, by Proposition 4.14, irrespective of the dimension d, almost surely,

$$\dim B(A) \leqslant 2 \dim A \qquad \text{and for all sets } A \subset [0, \infty).$$

Hence only the lower bound $\dim B(A) \geqslant 2 \dim A$ requires proof. We first focus on the case $d \geqslant 3$. The crucial idea here is that one uses a standardised covering of $B(A)$ by dyadic cubes and ensures that, simultaneously for all possible covering cubes the preimages allow an efficient covering. An upper bound for $\dim A$ follows by selecting from the coverings of all preimages.

Lemma 9.31 *Consider a cube $Q \subset \mathbb{R}^d$ centred at a point x and having diameter $2r$. Let $\{B(t): t \geqslant 0\}$ be d-dimensional Brownian motion, with $d \geqslant 3$. Define recursively*

$$
\begin{aligned}
\tau_1^Q &= \inf\{t \geqslant 0 : B(t) \in Q\}, \\
\tau_{k+1}^Q &= \inf\{t \geqslant \tau_k^Q + r^2 : B(t) \in Q\}, \qquad \text{for } k \geqslant 1,
\end{aligned}
$$

with the usual convention that $\inf \emptyset = \infty$. Then there exists $0 < \theta < 1$ depending only on the dimension d, such that $\mathbb{P}_z\{\tau_{n+1}^Q < \infty\} \leqslant \theta^n$ for all $z \in \mathbb{R}^d$ and $n \in \mathbb{N}$.

Proof. It is sufficient to show that for some θ as above,

$$\mathbb{P}_z\left\{\tau_{k+1}^Q = \infty \,\middle|\, \tau_k^Q < \infty\right\} > 1 - \theta.$$

Observe that the quantity on the left can be bounded from below by

$$\mathbb{P}_z\{\tau_{k+1}^Q = \infty \mid |B(\tau_k^Q + r^2) - x| > 3r, \tau_k^Q < \infty\} \mathbb{P}_z\{|B(\tau_k^Q + r^2) - x| > 3r \mid \tau_k^Q < \infty\}.$$

The second factor is bounded from below by $\inf_{y \in Q} \mathbb{P}_y\{|B(r^2) - x| > 3r\}$, by the strong Markov property. Using transience of Brownian motion in $d \geqslant 3$, the first factor is bounded from below by $\inf_{y \notin \mathcal{B}(x, 3r)} \mathbb{P}_y\{\tau(Q) = \infty\}$, where, as before, $\tau(Q)$ denotes the first hitting time of Q. Both bounds are positive and do not depend on the scaling factor r. ∎

Recall that $\mathfrak{C}_m$ denotes the set of dyadic cubes of side length 2^{-m} inside Cube $= [-\tfrac{1}{2}, \tfrac{1}{2}]^d$.

Lemma 9.32 *In the setup of Lemma 9.31, there exists a random variable $C = C(\omega)$ such that, almost surely, for all m and for all cubes $Q \in \mathfrak{C}_m$ we have $\tau_{\lceil mC+1 \rceil}^Q = \infty$.*

Proof. From Lemma 9.31 we get that

$$\sum_{m=1}^{\infty} \sum_{Q \in \mathfrak{C}_m} \mathbb{P}\{\tau_{\lceil cm+1 \rceil}^Q < \infty\} \leqslant \sum_{m=1}^{\infty} 2^{dm} \theta^{cm}.$$

Now choose c so large that $2^d \theta^c < 1$. Then, by the Borel–Cantelli lemma, for all but finitely many m we have $\tau_{\lceil cm+1 \rceil}^Q = \infty$ for all $Q \in \mathfrak{C}_m$. Finally, we can choose a random $C(\omega) > c$ to handle the finitely many exceptional cubes. ∎

Proof of Theorem 9.28 for d > 2. As mentioned before we can focus on the '$\geqslant$' direction. We fix L and show that, almost surely, for all subsets S of $[-L, L]^d$ we have

$$\dim B^{-1}(S) \leqslant \tfrac{1}{2} \dim S. \tag{9.12}$$

Applying this to $S = B(A) \cap [-L, L]^d$ successively for a countable unbounded set of L we get the desired conclusion. By scaling, it is sufficient to prove (9.12) for $L = 1/2$. The idea now is to verify (9.12) for all paths satisfying Lemma 9.32 using completely deterministic reasoning. As this set of paths has full measure, this verifies the statement.

Hence fix a path $\{B(t)\colon t \geqslant 0\}$ satisfying Lemma 9.32 for a constant $C > 0$. If $\beta > \dim S$ and $\varepsilon > 0$ there exists a covering of S by binary cubes $\{Q_j\colon j \in \mathbb{N}\} \subset \bigcup_{m=1}^{\infty} \mathfrak{C}_m$ such that $\sum |Q_j|^\beta < \varepsilon$. If N_m denotes the number of cubes from $\mathfrak{C}_m$ in such a covering, then

$$\sum_{m=1}^{\infty} N_m \, 2^{-m\beta} < \varepsilon.$$

Consider the inverse image of these cubes under $\{B(t)\colon t \geqslant 0\}$. Since we chose this path so that Lemma 9.32 is satisfied, this yields a covering of $B^{-1}(S)$, which for each $m \geqslant 1$ uses at most CmN_m intervals of length $r^2 = d2^{-2m}$.

For $\gamma > \beta$ we can bound the $\gamma/2$-dimensional Hausdorff content of $B^{-1}(S)$ from above by

$$\sum_{m=1}^{\infty} Cm \, N_m \, (d2^{-2m})^{\gamma/2} = C \, d^{\gamma/2} \sum_{m=1}^{\infty} m \, N_m \, 2^{-m\gamma}.$$

This can be made small by choosing a suitable $\varepsilon > 0$. Thus $B^{-1}(S)$ has Hausdorff dimension at most $\gamma/2$ for all $\gamma > \beta > \dim S$, and therefore $\dim B^{-1}(S) \leqslant \dim S/2$. ∎

In $d = 2$ we cannot rely on transience of Brownian motion. To get around this problem, we can look at the Brownian path up to a stopping time. A convenient choice of stopping time for this purpose is $\tau_R^* = \min \{t : |B(t)| = R\}$. For the two dimensional version of Kaufman's theorem it is sufficient to show that, almost surely,

$$\dim B(A) \geqslant 2 \dim(A \cap [0, \tau_R^*]) \text{ for all } A \subset [0, \infty).$$

Lemma 9.31 has to be changed accordingly.

Lemma 9.33 *Consider a cube $Q \subset \mathbb{R}^2$ centred at a point x and having diameter $2r$, and assume that the cube Q is inside the ball of radius R about the origin. Let $\{B(t)\colon t \geqslant 0\}$*

be planar Brownian motion. Define τ_k^Q as in Lemma 9.31. Then there exists $c = c(R) > 0$ such that, with $2^{-m-1} < r < 2^{-m}$, for any $z \in \mathbb{R}^2$,

$$\mathbb{P}_z\{\tau_k^Q < \tau_R^*\} \leqslant \left(1 - \frac{c}{m}\right)^k \leqslant e^{-ck/m}. \tag{9.13}$$

Proof. It suffices to bound $\mathbb{P}_z\{\tau_{k+1}^Q \geqslant \tau_R^* \mid \tau_k^Q < \tau_R^*\}$ from below by

$$\mathbb{P}_z\{\tau_{k+1}^Q \geqslant \tau_R^* \mid |B(\tau_k^Q + r^2) - x| > 2r, \tau_k^Q < \tau_R^*\} \, \mathbb{P}_z\{|B(\tau_k^Q + r^2) - x| > 2r \mid \tau_k^Q < \tau_R^*\}.$$

The second factor can be bounded from below by a positive constant, which does not depend on r and R. The first factor is bounded from below by the probability that planar Brownian motion started at any point in $\partial \mathcal{B}(0, 2r)$ hits $\partial \mathcal{B}(0, 2R)$ before $\partial \mathcal{B}(0, r)$. Using Theorem 3.18 this probability is given by

$$\frac{\log 2r - \log r}{\log 2R - \log r} \geqslant \frac{1}{\log_2 R + 2 + m}.$$

This is at least c/m for some $c > 0$ which depends on R only. ∎

The bound (9.13) on $\mathbb{P}\{\tau_k^Q < \tau_R^*\}$ in two dimensions is worse by a linear factor than the corresponding bound in higher dimensions. This, however, does not make a significant difference in the proof of the two dimensional version of Theorem 9.28, which can now be completed in the same way, see Exercise 9.9.

There is also a version of Kaufman's theorem for Brownian motion in dimension one.

Theorem 9.34 *Suppose $\{B(t) \colon t \geqslant 0\}$ is a linear Brownian motion. Then, almost surely, for all nonempty closed sets $S \subset \mathbb{R}$, we have*

$$\dim B^{-1}(S) = \tfrac{1}{2} + \tfrac{1}{2} \dim S.$$

Remark 9.35 Note that here it is essential to run Brownian motion on an unbounded time interval. For example, for the point $x = \max_{0 \leqslant t \leqslant 1} B(t)$ the set $\{t \in [0, 1] \colon B(t) = x\}$ is a singleton almost surely. The restriction to closed sets comes from Frostman's lemma, which we have proved for closed sets only, and can be relaxed accordingly. ◇

Proof. For the proof of the upper bound let $\{W(t) \colon t \geqslant 0\}$ be a Brownian motion independent of $\{B(t) \colon t \geqslant 0\}$. Applying Kaufman's theorem for the planar Brownian motion given by $\widetilde{B}(t) = (B(t), W(t))$ we get almost surely, for all $S \subset \mathbb{R}$,

$$\dim B^{-1}(S) = \dim \widetilde{B}^{-1}(S \times \mathbb{R}) \leqslant \tfrac{1}{2} \dim(S \times \mathbb{R}) = \tfrac{1}{2} + \tfrac{1}{2} \dim S,$$

where we have used the straightforward fact that $\dim(S \times \mathbb{R}) = 1 + \dim S$.
The lower bound requires a more complicated argument, based on Frostman's lemma. For this purpose we may suppose that $S \subset (-M, M)$ is closed and $\dim S > \alpha$. Then there exists a measure μ supported by S such that

$$\mu(\mathcal{B}(x, r)) \leqslant r^\alpha \quad \text{for all } x \in S, \, 0 < r < 1.$$

Let ℓ^a be the measure with cumulative distribution function given by the local time at level a. Let ν be the measure on $B^{-1}(S)$ given by

$$\nu(A) = \int \mu(da)\, \ell^a(A), \quad \text{for } A \subset [0, \infty) \text{ Borel.}$$

Then, by Theorem 6.19, for a given $\varepsilon > 0$, one can find a constant $C > 0$ such that

$$\ell^a(\mathcal{B}(x,r)) = L^a(x+r) - L^a(x-r) \leqslant Cr^{\frac{1}{2}-\varepsilon}$$

for all $a \in [-M, M]$ and $0 < r < 1$. By Hölder continuity of Brownian motion there exists, for given $\varepsilon > 0$, a constant $c > 0$ such that, for every $x \in [0, 1]$,

$$|B(x+s) - B(x)| \leqslant cr^{\frac{1}{2}-\varepsilon} \text{ for all } s \in [-r, r].$$

From this we get the estimate

$$\nu(\mathcal{B}(x,r)) = \int \mu(da)\big[L^a(x+r) - L^a(x-r)\big]$$

$$\leqslant \int_{B(x)-cr^{\frac{1}{2}-\varepsilon}}^{B(x)+cr^{\frac{1}{2}-\varepsilon}} \mu(da)\big[L^a(x+r) - L^a(x-r)\big]$$

$$\leqslant c^\alpha r^{\frac{\alpha}{2}-\varepsilon\alpha} Cr^{\frac{1}{2}-\varepsilon} \quad \text{for all } x \in S,\ 0 < r < 1.$$

Hence, by the mass distribution principle, we get the lower bound $\alpha/2 + 1/2 - \varepsilon(1 + \alpha)$ for the dimension and the result follows when $\varepsilon \downarrow 0$ and $\alpha \uparrow \dim S$. $\blacksquare$

As briefly remarked in the discussion following Theorem 4.33, Brownian motion is also 'capacity-doubling'. This fact holds for a very general class of kernels, we give an elegant proof of this fact here.

Theorem 9.36 *Let $\{B(t) : t \in [0,1]\}$ be d-dimensional Brownian motion and $A \subset [0,1]$ a closed set. Suppose f is decreasing and there is a constant $C > 0$ with*

$$\int_0^1 \frac{f(r^2 x)}{f(x)} r^{d-1}\, dr \leqslant C \text{ for all } x \in (0,1), \tag{9.14}$$

and let $\phi(x) = x^2$. Then, almost surely,

$$\mathrm{Cap}_f(A) > 0 \quad \text{if and only if} \quad \mathrm{Cap}_{f \circ \phi}(B(A)) > 0.$$

Remark 9.37 Condition (9.14) is only used in the 'only if' part of the statement. Note that if $f(x) = x^{-\alpha}$ is a power law, then (9.14) holds if and only if $2\alpha < d$. ◇

Proof. We start with the 'only if' direction, which is easier. Suppose $\mathrm{Cap}_f(A) > 0$. This implies that there is a mass distribution μ on A such that the f-energy of μ is finite. Then $\mu \circ B^{-1}$ is a mass distribution on $B(A)$ and we will show that it has finite $f \circ \phi$-energy.

Indeed,

$$I_{f \circ \phi}(\mu \circ B^{-1}) = \iint f \circ \phi(|x - y|) \, \mu \circ B^{-1}(dx) \, \mu \circ B^{-1}(dy)$$

$$= \iint f(|B(s) - B(t)|^2) \, \mu(ds) \, \mu(dt).$$

Hence,

$$\mathbb{E} I_{f \circ \phi}(\mu \circ B^{-1}) = \iint \mathbb{E} f(|X|^2 |s - t|) \, \mu(ds) \, \mu(dt),$$

where X is a d-dimensional standard normal random variable. Using polar coordinates and the monotonicity of f we get, for a constant $\kappa(d)$ depending only on the dimension,

$$\mathbb{E}\big[f(|X|^2 |s - t|)\big] = \kappa(d) \int_0^\infty f(r^2 |s - t|) \, e^{-r^2/2} \, r^{d-1} \, dr$$

$$\leqslant f(|s - t|) \, \kappa(d) \left(\int_0^1 \frac{f(r^2 |s - t|)}{f(|s - t|) \, r^{1-d}} \, dr + \int_1^\infty e^{-r^2/2} \, r^{d-1} \, dr \right).$$

By (9.14) the bracket on the right hand side is bounded by a constant independent of $|s - t|$, and hence $\mathbb{E}[I_{f \circ \phi}(\mu \circ B^{-1})] < \infty$, which in particular implies $I_{f \circ \phi}(\mu \circ B^{-1}) < \infty$ almost surely.

The difficulty in the 'if' direction is that a measure on $B(A)$ with finite $f \circ \phi$-energy cannot easily be transported backwards onto A. To circumvent this problem we use the characterisation of capacity in terms of polarity with respect to percolation limit sets, recall Theorem 9.18. We may assume, without loss of generality, that $f(1/4) = 1$.

Fix a unit cube Cube such that $\mathrm{Cap}_{f \circ \phi}(B(A) \cap \mathrm{Cube}) > 0$ with positive probability, and let Γ be a percolation limit set with retention probabilities associated to the decreasing function $f(x^2/4)$ as in Theorem 9.18, which is independent of Brownian motion. Then, by Theorem 9.18, we have $B(A) \cap \Gamma \neq \emptyset$ with positive probability. Define a random variable

$$T = \inf \{t \in A \colon B(t) \in \Gamma\},$$

which is finite with positive probability. Hence the measure μ given by

$$\mu(B) = \mathbb{P}\{T \in B, T < \infty\}$$

is a mass distribution on A. We shall show that it has finite f-energy, which completes the proof. Again we use the polarity criterion of Theorem 9.18 to do this. Let $\mathsf{S}_n = \bigcup_{S \in \mathfrak{S}_n} S$ be the union of all cubes retained in the construction up to step n. Then, by looking at the retention probability of any fixed point in Cube, we have, for any $s \in A$,

$$\mathbb{P}\{B(s) \in \mathsf{S}_n\} \leqslant p_1 \cdots p_n = \frac{1}{f \circ \phi(2^{-n-1})}. \tag{9.15}$$

Conversely, by a first entrance decomposition,

$$\mathbb{P}\{B(s) \in \mathsf{S}_n\} \geqslant \mathbb{P}\{B(s) \in \mathsf{S}_n, B(T) \in \mathsf{S}_n, T < \infty\}$$

$$= \int_0^s \mu(dt) \, \mathbb{P}\{B(s) \in \mathsf{S}_n \mid B(t) \in \mathsf{S}_n\}$$

Given $B(t) \in \mathsf{S}_n$ and $\sqrt{s-t} \leqslant 2^{-n+k}$ for some $k \in \{0, \ldots, n\}$, the probability that $B(s)$ and $B(t)$ are contained in the same dyadic cube $Q \in \mathfrak{C}_{n-k}$ is bounded from below by a constant. Given this event, we know that Q is retained in the percolation (otherwise we could not have $B(t) \in \mathsf{S}_n$) and the probability that the cube in $\mathfrak{C}_n$, that contains $B(s)$, is retained in the percolation is at least $p_{n-k+1} \cdots p_n$ (interpreted as 1 if $k = 0$). Therefore

$$\int_0^s \mu(dt)\, \mathbb{P}\{B(s) \in \mathsf{S}_n \mid B(t) \in \mathsf{S}_n\}$$

$$\geqslant c \sum_{k=0}^n \mu\big([s - 2^{-2n+2k}, s - 2^{-2n+2k-2})\big)\, p_{n-k+1} \cdots p_n$$

$$\geqslant c \sum_{k=0}^n \mu\big([s - 2^{-2n+2k}, s - 2^{-2n+2k-2})\big)\, \frac{f \circ \phi(2^{-n+k-1})}{f \circ \phi(2^{-n-1})}$$

$$\geqslant c\, \frac{1}{f \circ \phi(2^{-n-1})} \int_0^{s-2^{-2n-2}} \mu(dt)\, f(s-t),$$

using the monotonicity of f in the last step. Finiteness of the f-energy follows by comparing this with (9.15), cancelling the factor $1/f \circ \phi(2^{-n})$, integrating over $\mu(ds)$, and letting $n \to \infty$. This completes the proof. ∎

Exercises

Exercise 9.1.

  (a) Suppose that $\{B_1(t): t \geqslant 0\}, \{B_2(t): t \geqslant 0\}$ are independent standard Brownian motions in $\mathbb{R}^3$. Then, almost surely, $B_1[0,t] \cap B_2[0,t] \neq \{0\}$ for any $t > 0$.

  (b) Suppose that $\{B_1(t): t \geqslant 0\}, \ldots, \{B_p(t): t \geqslant 0\}$ are p independent standard Brownian motions in $\mathbb{R}^d$, and $d > p(d-2)$. Then, almost surely,

$$\dim\big(B_1[0,t_1] \cap \cdots \cap B_p[0,t_p]\big) = d - p(d-2) \qquad \text{for any } t_1, \ldots, t_p > 0.$$

Exercise 9.2. Let $\{X^{(1)}(t): 0 \leqslant t \leqslant 1\}, \ldots, \{X^{(p)}(t): 0 \leqslant t \leqslant 1\}$ be p independent d-dimensional Brownian bridges with $X^{(i)}(0) = x_i \in \mathbb{R}^d$ and $X^{(p)}(1) = y_i \in \mathbb{R}^d$.

  (a) Show that if $d = 3$ and $p = 3$, almost surely, the intersection of the ranges of the Brownian bridges is empty (except possibly at the start and end points).

  (b) Show that if $d = 2$ and p arbitrary, with positive probability, the intersection of the ranges of the independent Brownian bridges is nonempty.

Exercise 9.3. For a d-dimensional Brownian motion $\{B(t): t \geqslant 0\}$ we denote by

$$S(p) = \big\{x \in \mathbb{R}^d : \exists\, 0 < t_1 < \cdots < t_p < 1 \text{ with } x = B(t_1) = \cdots = B(t_p)\big\}$$

the set of p-fold multiple points. Show that, for $d > p(d-2)$,

  (a) $\dim S(p) = d - p(d-2)$, almost surely.

(b) for any closed set Λ, we have

$$\mathbb{P}\{S(p) \cap \Lambda \neq \emptyset\} > 0 \qquad \text{if and only if} \qquad \text{Cap}_{f^p}(\Lambda) > 0 \,,$$

where the decreasing function f is the radial potential.

Exercise 9.4. In the situation of Exercise 9.3, show that the ratio

$$\frac{\mathbb{P}\{S(p) \cap \Lambda \neq \emptyset\}}{\text{Cap}_{f^p}(\Lambda)}$$

may be unbounded.

Exercise 9.5. Let $\{B(t) \colon t \geqslant 0\}$ be a standard linear Brownian motion. Show that its zero set is intersection-equivalent to $\Gamma[\frac{1}{2}]$ in any compact unit interval not containing the origin. **Hint.** Use Exercise 8.8.

Exercise 9.6.

(a) Let A be a set of rooted trees. We say that A is *inherited* if every finite tree is in A, and if $T \in A$ and $v \in V$ is a vertex of the tree then the tree $T(v)$, consisting of all successors of v, is in A.

Prove the *Galton–Watson 0–1 law*: For a Galton–Watson tree, conditional on survival, every inherited set has probability zero or one.

(b) Show that for the percolation limit sets $\Gamma[\gamma] \subset \mathbb{R}^d$ with $0 < \gamma < d$ we have

$$\mathbb{P}\{\dim \Gamma[\gamma] = d - \gamma \mid \Gamma[\gamma] \neq \emptyset\} = 1.$$

Exercise 9.7. Consider a linear Brownian motion $\{B(t) \colon t \geqslant 0\}$ and let $A_1, A_2 \subset [0, \infty)$.

(a) Show that if $\dim(A_1 \times A_2) < 1/2$ then $\mathbb{P}\{B(A_1) \text{ intersects } B(A_2)\} = 0$.

(b) Derive the same conclusion under the weaker assumption that $A_1 \times A_2$ has vanishing $1/2$-dimensional Hausdorff measure.

(c) Show that if $\text{Cap}_{1/2}(A_1 \times A_2) > 0$, then $\mathbb{P}\{B(A_1) \text{ intersects } B(A_2)\} > 0$.

Exercise 9.8. Ⓢ Use Exercise 9.7 to find a set $A \subset [0, \infty)$ such that the probability that a linear Brownian motion $\{B(t) \colon t \geqslant 0\}$ is one-to-one on A is strictly between zero and one.

Exercise 9.9. Complete the proof of Theorem 9.28 in the case $d = 2$.

Exercise 9.10. Ⓢ Let $\{B(t) \colon 0 \leqslant t \leqslant 1\}$ be a planar Brownian motion. For every $a \in \mathbb{R}$ define the sets $S(a) = \{y \in \mathbb{R} \colon (a, y) \in B[0, t]\}$, consisting of the vertical slices of the path. Show that, almost surely, $\dim S(a) = 1$, for every $a \in (\min\{x \colon (x, y) \in B[0, t]\}, \max\{x \colon (x, y) \in B[0, t]\})$.

Notes and comments

The question whether there exist p-multiple points of a d-dimensional Brownian motion was solved in various stages in the early 1950s. First, Lévy showed in [Le40] that almost all paths of a planar Brownian motion have double points, and Kakutani [Ka44a] showed that if $n \geqslant 5$ almost no paths have double points. The cases of $d = 3, 4$ where added by Dvoretzky, Erdős and Kakutani in [DEK50] and the same authors showed in [DEK54] that planar Brownian motion has points of arbitrary multiplicity. Finally, Dvoretzky, Erdős, Kakutani and Taylor showed in [DEKT57] that there are no triple points in $d = 3$. Clearly the existence of p-fold multiple points is essentially equivalent to the problem whether p independent Brownian motions have a common intersection.

The problem of finding the Hausdorff dimension of the set of p-fold multiple points in the plane, and of double points in $\mathbb{R}^3$, was still open when Itô and McKean wrote their influential book on the sample paths of diffusions in 1964, see p.261 in [IM74], but was solved soon after by Taylor [Ta66] and Fristedt [Fr67]. Perkins and Taylor [PT88] provide fine results when Brownian paths in higher dimensions 'come close' to self-intersecting. The method of stochastic codimension, which we use to find these dimensions, is due to Taylor [Ta66], who used the range of stable processes as 'test sets'. The restriction of the stable indices to the range $\alpha \in (0, 2]$ leads to complications, which can be overcome by a projection method of Fristedt [Fr67] or by using multiparameter processes, see Khosh-nevisan [Kh02]. The use of percolation limit sets as test sets is much more recent and due to Khoshnevisan et al. [KPX00], though similar ideas are used in the context of trees at least since the pioneering work of Lyons [Ly90]. The latter paper is also the essential source for our proof of Hawkes' theorem.

Some very elegant proofs of these classical facts were given later: Rosen [Ro83] provides a local time approach, and Kahane [Ka86] proves a general formula for the intersection of independent random sets satisfying suitable conditions. The bottom line of Kahane's approach is that the formula 'codimension of the intersection is equal to the sum of codimensions of the intersected sets' which is well-known from linear subspaces in general position can be extended to the Hausdorff dimension of a large class of random sets, which includes the paths of Brownian motion, see also Falconer [Fa97a] and Mattila [Ma95]. The intersection equivalence approach we describe in Section 9.2 is taken from [Pe96a], [Pe96b]. The proof of Lyons' theorem we give is taken from Benjamini et al. [BPP95]. See Theorem 2.1 in Lyons [Ly92] for the original proof.

Exact Hausdorff gauges allow a distinction of the sizes of the set of p-multiple points of a planar Brownian motion for different values of p. Le Gall [LG87b] showed that, for $d = 2$, the set of p-multiple points has positive and σ-finite Hausdorff measure for the gauge function

$$\psi_p(r) = r^2 \left[\log(1/r) \log \log \log(1/r) \right]^p,$$

and in $d = 3$ the set of double points has positive and finite Hausdorff measure for

$$\psi(r) = r \left[\log \log(1/r) \right]^2.$$

Turning to packing measures, which we will introduce properly in Chapter 10, results for
the packing gauge of the double points were given by Le Gall in [LG87c] in $d = 2$, where
it turns out that the ϕ-packing measure is either zero or infinite, depending whether

$$\int_{0+} \frac{\phi(r)}{r^3 [\log(1/r)]^{p+1}} < \infty.$$

The case of $d = 3$ turned out to be quite different and was only recently resolved in [MS09],
where it turns out that the ϕ-packing measure is either zero or infinite, and the integral test
distinguishing between these cases depends on an *intersection exponent*, see for example
Chapter 11 for a definition. These dimension gauges imply in particular that $\mathcal{H}^2(S_p) =
\mathcal{P}^2(S_p) = 0$ almost surely, if S_p is the set of p-multiple points of a planar Brownian
motion, and that $\mathcal{H}^1(S_2) = 0$, $\mathcal{P}^1(S_2) = \infty$ almost surely, if S_2 is the set of double points
of Brownian motion in $\mathbb{R}^3$.

An interesting line of generalisation is whether almost-sure properties of Brownian
motion also hold quasi-everywhere, a stronger notion due to Fukushima [Fu80]. Roughly
speaking, a property holds quasi-everywhere if an Ornstein-Uhlenbeck process on path
space, whose stationary measure is the Wiener measure, never fails to have the property.
For example, in the context of intersections, Lyons [Ly86] showed that Brownian motion
has no double points quasi-everywhere if and only if $d \geqslant 6$, and Penrose [Pe89] that the
set of double points of quasi-every Brownian motion in dimension three has Hausdorff
dimension one. A similar line of research are the dynamical theories of Brownian motion
initiated by Nelson [Ne67].

Hendricks and Taylor conjectured in 1976 a characterisation of the polar sets for the
multiple points of a Brownian motion or a more general Markov process, which included
the statement of Theorem 9.21. Sufficiency of the capacity criterion in Theorem 9.21 was
proved by Evans [Ev87a, Ev87b] and independently by Tongring [To88], see also Le Gall,
Rosen and Shieh [LRS89]. The full equivalence was later proved in a much more general
setting by Fitzsimmons and Salisbury [FS89]. A quantitative treatment of the question,
which sets contain double points of Brownian motion is given in [PP07].

Points of multiplicity strictly n where identified by Adelman and Dvoretzky [AD85]
and the result is also an immediate consequence of the exact Hausdorff gauge function
identified by Le Gall [LG86a]. The existence of points of infinite multiplicity in the planar
case was first stated in Dvoretzky et al. [DEK58] though their proof seems to have a gap.
Le Gall [LG87a] proves a stronger result: Two sets $A, B \subset \mathbb{R}$ are said to be of the same
order type if there exists an increasing homeomorphism ϕ of $\mathbb{R}$ such that $\phi(A) = B$. Le
Gall shows that, for any totally disconnected, compact $A \subset \mathbb{R}$, almost surely there exists
a point $x \in \mathbb{R}^2$ such that the set $\{t \geqslant 0 : B(t) = x\}$ has the same order type as A. In
particular, there exist points of countably infinite and uncountable multiplicity. Le Gall's
proof is based on the properties of natural measures on the intersection of Brownian paths.
Our proof avoids this and seems to be new.

Substantial generalisations of Exercise 9.7 can be found in papers by Khoshnevisan [Kh99] and Khoshnevisan and Xiao [KX05]. For example, in Theorem 8.2 of [Kh99] it is shown that the condition in part (c) is an equivalence. The question of the Hausdorff dimension of the intersection of a Brownian image $B(A)$ with a given set $F \subset \mathbb{R}^d$ has been open for a while. It seems that at the time of writing a solution has been achieved by Khoshnevisan and Xiao.

Kaufman proved his dimension doubling theorem in [Ka69]. The version for Brownian motion in dimension one is due to Serlet [Se95]. The capacity-doubling result in the given generality is new, but Khoshnevisan and Xiao, see Question 1.1 and Theorem 7.1 in [KX05], prove the special case when f is a power law using a different method. Their argument is based on the investigation of additive Lévy processes and works for a class of processes much more general than Brownian motion. Theorem 9.36 does not hold uniformly for all sets A. Examples can be constructed along the lines in Perkins and Taylor [PT87].

In this book we do not construct a measure on the intersection of p Brownian paths. However this is possible and yields the *intersection local time* first studied by Geman, Horowitz and Rosen [GHR84], see also Rosen [Ro83]. This quantity plays a key rôle in the analysis of Brownian paths and Le Gall [LG92] gives a very accessible account of the state of research in 1991, which is still worth reading. Recent research deals with the Hausdorff dimension of subsets of the intersections with special properties, like thick times, see [DPRZ02] and [KM02], or thin times, see [KM05].

10

Exceptional sets for Brownian motion

The techniques developed in this book so far give a fairly satisfactory picture of the behaviour of a Brownian motion at a typical time, like a fixed time or a stopping time. In this chapter we explore exceptional times, for example times where the path moves slower or faster than in the law of the iterated logarithm, or does not wind as in Spitzer's law. Again Hausdorff dimension is the right tool to describe just how rare an exceptional behaviour is, but we shall see that another notion of dimension, the packing dimension, can provide additional insight.

10.1 The fast times of Brownian motion

In a famous paper from 1974, Orey and Taylor raise the question how often on a Brownian path the law of the iterated logarithm fails. To understand this, recall that, by Corollary 5.3 and the Markov property, for a linear Brownian motion $\{B(t) \colon t \geqslant 0\}$ and for every $t \in [0, 1]$, almost surely,

$$\limsup_{h \downarrow 0} \frac{|B(t+h) - B(t)|}{\sqrt{2h \log \log(1/h)}} = 1.$$

This contrasts sharply with the following result (note the absence of the iterated logarithm!).

Theorem 10.1 *Almost surely, we have*

$$\max_{0 \leqslant t \leqslant 1} \limsup_{h \downarrow 0} \frac{|B(t+h) - B(t)|}{\sqrt{2h \log(1/h)}} = 1.$$

Remark 10.2 At the time $t \in [0, 1]$ where the maximum in Theorem 10.1 is attained, the law of the iterated logarithm fails and it is therefore an *exceptional* time. ◇

Proof. The upper bound follows from Lévy's modulus of continuity, Theorem 1.14, as

$$\sup_{0 \leqslant t < 1} \limsup_{h \downarrow 0} \frac{|B(t+h) - B(t)|}{\sqrt{2h \log(1/h)}} \leqslant \limsup_{h \downarrow 0} \sup_{0 \leqslant t \leqslant 1-h} \frac{|B(t+h) - B(t)|}{\sqrt{2h \log(1/h)}} = 1.$$

Readers who have skipped the proof of Theorem 1.14 given in Chapter 1 will be able to infer the upper bound directly from Remark 10.5 below. It remains to show that there exists a time $t \in [0, 1]$ such that

$$\limsup_{h \downarrow 0} \frac{|B(t+h) - B(t)|}{\sqrt{2h \log(1/h)}} \geq 1.$$

Recall from Theorem 1.13 and scaling that, almost surely, for every constant $c < \sqrt{2}$ and every $\varepsilon > 0$ there exist $0 < h < \varepsilon$ and $t \in [0, 1 - h]$ with

$$|B(t+h) - B(t)| > c\sqrt{h \log(1/h)}.$$

Using the Markov property this implies that, for $c < \sqrt{2}$, the sets

$$M(c, \varepsilon) = \left\{ t \in [0, 1] \colon \text{there is } h \in (0, \varepsilon) \text{ such that } |B(t+h) - B(t)| > c\sqrt{h \log(1/h)} \right\}$$

are almost surely dense in $[0, 1]$. By continuity of Brownian motion they are open, and clearly $M(c, \varepsilon) \subset M(d, \delta)$ whenever $c > d$ and $\varepsilon < \delta$. Hence, by Baire's (category) theorem, the intersection

$$\bigcap_{\substack{c < \sqrt{2}, \varepsilon > 0 \\ c, \varepsilon \in \mathbb{Q}}} M(c, \varepsilon) = \left\{ t \in [0, 1] \colon \limsup_{h \downarrow 0} \frac{|B(t+h) - B(t)|}{\sqrt{2h \log(1/h)}} \geq 1 \right\}$$

is dense and hence nonempty almost surely. ∎

To explore how often we come close to the exceptional behaviour described in Theorem 10.1 we introduce a spectrum of exceptional points. Given $a > 0$ we call a time $t \in [0, 1]$ an a-**fast time** if

$$\limsup_{h \downarrow 0} \frac{|B(t+h) - B(t)|}{\sqrt{2h \log(1/h)}} \geq a,$$

and $t \in [0, 1]$ is a **fast time** if it is a-fast for some $a > 0$. By Theorem 10.1 fast times exist, in fact the proof even shows that the set of fast times is the intersection of countably many open dense sets in $[0, 1]$ and hence is dense and uncountable. Conversely it is immediate from the law of the iterated logarithm that the set has Lebesgue measure zero, recall Remark 1.28. The appropriate notion to measure the quantity of a-fast times is, again, Hausdorff dimension.

Theorem 10.3 (Orey and Taylor 1974) *Suppose $\{B(t) \colon t \geq 0\}$ is a linear Brownian motion. Then, for every $a \in [0, 1]$, we have almost surely,*

$$\dim \left\{ t \in [0, 1] : \limsup_{h \downarrow 0} \frac{|B(t+h) - B(t)|}{\sqrt{2h \log(1/h)}} \geq a \right\} = 1 - a^2.$$

The rest of this section is devoted to the proof of this result. We start with a proof of the *upper bound*, which also shows that there are almost surely no a-fast times for $a > 1$.

So fix an arbitrary $a > 0$. Let $\varepsilon > 0$ and $\eta > 1$, having in mind that we later let $\eta \downarrow 1$ and $\varepsilon \downarrow 0$. The basic idea is to cover the interval $[0, 1)$ by a collection of intervals of the form $[j\eta^{-k}, (j+1)\eta^{-k})$ with $j = 0, \ldots, \lceil \eta^k - 1 \rceil$ and $k \geqslant 1$. Any such interval of length $h := \eta^{-k}$ is included in the covering if, for $h' := k\eta^{-k}$,

$$|B(jh + h') - B(jh)| > a(1 - 4\varepsilon)\sqrt{2h' \log(1/h')}.$$

Let $\mathfrak{I}_k = \mathfrak{I}_k(\eta, \varepsilon)$ be the collection of intervals of length η^{-k} chosen in this procedure.

Lemma 10.4 *Almost surely, for every $\varepsilon > 0$ and $\delta > 0$, there is an $\eta > 1$ and $m \in \mathbb{N}$ such that the collection $\mathfrak{I} = \mathfrak{I}(\varepsilon, \delta) = \{ I \in \mathfrak{I}_k(\eta, \varepsilon) \colon k \geqslant m \}$ is a covering of the set of a-fast times consisting of intervals of diameter no bigger than δ.*

Proof.   We first note that by Theorem 1.12 there exists a constant $C > 0$ such that, almost surely, there exists $\rho > 0$ such that, for all $s, t \in [0, 2]$ with $|s - t| \leqslant \rho$,

$$|B(s) - B(t)| \leqslant C\sqrt{|s - t| \log(1/|s - t|)}. \tag{10.1}$$

Choose $\eta > 1$ such that $\sqrt{\eta - 1} \leqslant a\varepsilon/C$. Let M be the minimal integer with $M\eta^{-M} \leqslant \rho$ and $m \geqslant M$ such that $m\eta^{-m} < \delta$ (to ensure that our covering sets have diameter no bigger than δ) and $k\eta^{-k} < \ell\eta^{-\ell}$ for all $k > \ell \geqslant m$. Now suppose that $t \in [0, 1]$ is an a-fast time. By definition there exists $0 < u < m\eta^{-m}$ such that

$$|B(t + u) - B(t)| \geqslant a(1 - \varepsilon)\sqrt{2u \log(1/u)}.$$

We pick the unique $k \geqslant m$ such that $k\eta^{-k} < u \leqslant (k - 1)\eta^{-k+1}$, and let $h' = k\eta^{-k}$. By (10.1), we have

$$|B(t + h') - B(t)| \geqslant |B(t + u) - B(t)| - |B(t + u) - B(t + h')|$$
$$\geqslant a(1 - \varepsilon)\sqrt{2u \log(1/u)} - C\sqrt{(u - h')\log(1/(u - h'))}.$$

As $0 \leqslant u - h' \leqslant (\eta - 1)k\eta^{-k}$, and by our choice of η and by choosing m sufficiently large, the subtracted term can be made smaller than $a\varepsilon\sqrt{2h' \log(1/h')}$. Hence there exists $k \geqslant m$ such that

$$|B(t + h') - B(t)| \geqslant a(1 - 2\varepsilon)\sqrt{2h' \log(1/h')}.$$

Now let j be such that $t \in [j\eta^{-k}, (j + 1)\eta^{-k})$. As before let $h = \eta^{-k}$. Then, by the triangle inequality and using (10.1) twice, we have

$$|B(jh + h') - B(jh)|$$
$$\geqslant |B(t + h') - B(t)| - |B(t) - B(jh)| - |B(jh + h') - B(t + h')|$$
$$\geqslant a(1 - 2\varepsilon)\sqrt{2h' \log(1/h')} - 2C\sqrt{h \log(1/h)}$$
$$> a(1 - 4\varepsilon)\sqrt{2h' \log(1/h')},$$

using in the last step that, by choosing m sufficiently large, the subtracted term can be made smaller than $2a\varepsilon\sqrt{2h' \log(1/h')}$.   ∎

Proof of the upper bound in Theorem 10.3.   This involves only a first moment calculation. All there is to show is that, for any $\gamma > 1 - a^2$ there exists $\varepsilon > 0$ such that, for any $\delta > 0$ sufficiently small, the random variable $\sum_{I \in \mathcal{J}(\varepsilon, \delta)} |I|^{\gamma}$ is finite, almost surely. For this it suffices to verify that its expectation is finite. Note that

$$\mathbb{E}\left[\sum_{I \in \mathcal{J}(\varepsilon, \delta)} |I|^{\gamma} \right] = \sum_{k=m}^{\infty} \sum_{j=0}^{\lceil \eta^k - 1 \rceil} \eta^{-k\gamma} \, \mathbb{P}\left\{ \frac{|B(j\eta^{-k} + k\eta^{-k}) - B(j\eta^{-k})|}{\sqrt{2 k \eta^{-k} \log(\eta^k/k)}} > a(1 - 4\varepsilon) \right\}.$$

So it all boils down to an estimate of a single probability, which is very simple as it involves just one normal random variable, namely $B(j\eta^{-k} + k\eta^{-k}) - B(j\eta^{-k})$. More precisely, for X standard normal,

$$\mathbb{P}\left\{ \frac{|B(j\eta^{-k} + k\eta^{-k}) - B(j\eta^{-k})|}{\sqrt{2k\eta^{-k} \log(\eta^k/k)}} > a(1 - 4\varepsilon) \right\}$$

$$= \mathbb{P}\left\{ |X| > a(1 - 4\varepsilon) \sqrt{2 \log(\eta^k/k)} \right\} \tag{10.2}$$

$$\leqslant \frac{1}{a(1-4\varepsilon)\sqrt{\log(\eta^k/k)\pi}} \exp\left\{ -a^2 (1 - 4\varepsilon)^2 \log(\eta^k/k) \right\} \leqslant \eta^{-ka^2 (1-4\varepsilon)^3},$$

for all sufficiently large k and all $0 \leqslant j < 2^k$, using the estimate for normal random variables of Lemma 12.9 in the penultimate step. Given $\gamma > 1 - a^2$ we can finally find $\varepsilon > 0$ such that $\gamma + a^2 (1 - 4\varepsilon)^3 > 1$, so that

$$\sum_{k=m}^{\infty} \sum_{j=0}^{\eta^k - 1} \eta^{-k\gamma} \, \mathbb{P}\left\{ \frac{|B(j\eta^{-k} + k\eta^{-k}) - B(j\eta^{-k})|}{\sqrt{2k\eta^{-k} \log(\eta^k/k)}} > a(1 - 4\varepsilon) \right\}$$

$$\leqslant \sum_{k=1}^{\infty} \eta^k \eta^{-k\gamma} \, \eta^{-ka^2 (1-4\varepsilon)^3} < \infty,$$

completing the proof of the upper bound in Theorem 10.3. ∎

Remark 10.5 If $a > 1$ one can choose $\gamma < 0$ in the previous proof, which shows that there are no a-fast times as the empty collection is suitable to cover the set of a-fast times. ◇

For the *lower bound* we have to work harder. We divide, for any positive integer k, the interval $[0,1]$ into nonoverlapping dyadic subintervals $[j2^{-k}, (j+1)2^{-k}]$ for $j = 0, \ldots, 2^k - 1$. As before, we denote this collection of intervals by $\mathfrak{C}_k$ and by $\mathfrak{C}$ the union over all collections $\mathfrak{C}_k$ for $k \geqslant 1$. To each interval $I \in \mathfrak{C}$ we associate a $\{0,1\}$-valued random variable $Z(I)$ and then define sets

$$A(k) := \bigcup_{\substack{I \in \mathfrak{C}_k \\ Z(I) = 1}} I \qquad \text{and} \qquad A := \bigcap_{n=1}^{\infty} \bigcup_{k=n}^{\infty} A(k).$$

Because $1_A = \limsup 1_{A(k)}$ the set A is often called the **limsup fractal** associated with the family $(Z(I) : I \in \mathfrak{C})$. We shall see below that the set of a-fast times contains a large limsup fractal and derive the lower bound from the following general result on limsup fractals.

Theorem 10.6 *Suppose that* $(Z(I): I \in \mathfrak{C})$ *is a collection of random variables with values in* $\{0, 1\}$ *such that* $p_k := \mathbb{P}\{Z(I) = 1\}$ *is the same for all* $I \in \mathfrak{C}_k$. *For* $I \in \mathfrak{C}_m$, *with* $m \leqslant n$, *define*

$$M_n(I) := \sum_{\substack{J \in \mathfrak{C}_n \\ J \subset I}} Z(J).$$

Let $\zeta(n) \geqslant 1$ *and* $0 < \gamma < 1$ *be such that*

(1) $\mathrm{Var}(M_n(I)) \leqslant \zeta(n)\, \mathbb{E}[M_n(I)] = \zeta(n)\, p_n\, 2^{n-m}$,

(2) $\lim\limits_{n \uparrow \infty} 2^{n(\gamma-1)}\, \zeta(n)\, p_n^{-1} = 0$,

then $\dim A \geqslant \gamma$ *almost surely for the limsup fractal* A *associated with* $(Z(I): I \in \mathfrak{C})$.

The *idea of the proof* of Theorem 10.6 is to construct a probability measure μ on A and then use the energy method. To this end, we choose an increasing sequence $\ell_0, \ell_1, \ldots$ such that $M_{\ell_k}(D) > 0$ for all $D \in \mathfrak{C}_{\ell_{k-1}}$. We then define a (random) probability measure μ in the following manner: Assign mass $2^{-\ell_0}$ to each of the intervals $I \in \mathfrak{C}_{\ell_0}$. Proceed inductively: if $J \in \mathfrak{C}_m$ with $\ell_{k-1} < m \leqslant \ell_k$ and $J \subset D$ for $D \in \mathfrak{C}_{\ell_{k-1}}$ define

$$\mu(J) = \frac{M_{\ell_k}(J)\mu(D)}{M_{\ell_k}(D)}. \tag{10.3}$$

Then μ is consistently defined on all intervals in $\mathfrak{C}$ and therefore can be extended to a probability measure on $[0, 1]$ by the measure extension theorem. Note that $\mu(A^c) = 0$, so that μ is supported by A. The crucial part of the proof is then to show that, for a suitable choice of $\ell_0, \ell_1, \ldots$ the measure μ has finite γ-energy.

For the proof of Theorem 10.6 we need two lemmas. The first one is a simple combination of two facts, which have been established at other places in the book: The bounds for the energy of a measure established in Lemma 9.20, and the lower bound of Hausdorff dimension in terms of capacity which follows from the energy method, see Theorem 4.27.

Lemma 10.7 *Suppose* $B \subset [0, 1]$ *is a Borel set and* μ *is a probability measure on* B. *Then*

$$\sum_{m=1}^{\infty} \sum_{J \in \mathfrak{C}_m} \frac{\mu(J)^2}{2^{-\alpha m}} < \infty \quad \text{implies} \quad \dim B \geqslant \alpha.$$

Proof. By Lemma 9.20 with $f(x) = x^{-\alpha}$ and $h(n) = 2^{n\alpha}(1 - 2^{-\alpha})$ we obtain, for a suitable constant $C > 0$ that

$$I_\alpha(\mu) \leqslant C \sum_{m=1}^{\infty} \sum_{J \in \mathfrak{C}_m} \frac{\mu(J)^2}{2^{-\alpha m}}.$$

If the right hand side is finite, then so is the α-energy of the measure μ. We thus obtain $\dim B \geqslant \alpha$ by Theorem 4.27. ∎

For the formulation of the second lemma we use (2) to pick, for any $\ell \in \mathbb{N}$ an integer $n = n(\ell) \geqslant \ell$ such that $2^{n(\gamma-1)}\, \zeta(n) \leqslant p_n\, 2^{-3\ell}$.

Lemma 10.8 *There exists an almost surely finite random variable ℓ_0 such that, for all $\ell \geqslant \ell_0$ and $D \in \mathfrak{C}_\ell$, with $n = n(\ell)$,*

- *for all $D \in \mathfrak{C}_\ell$ we have*

$$\left| M_n(D) - \mathbb{E}M_n(D) \right| < \tfrac{1}{2}\mathbb{E}M_n(D),$$

and, in particular, $M_n(D) > 0$;

- *for a constant C depending only on γ,*

$$\sum_{m=\ell}^{n} 2^{\gamma m} \sum_{\substack{J \in \mathfrak{C}_m \\ J \subset D}} \frac{M_n(J)^2}{(2^{n-\ell}p_n)^2} \leqslant C2^{\gamma\ell}.$$

Remark 10.9 The first statement in the lemma says intuitively that the variance of the random variables $M_n(D)$ is small, i.e. they are always close to their mean. This is essentially what makes this proof work. ◇

Proof of Lemma 10.8. For $m \leqslant n, J \in \mathfrak{C}_m$ we denote $\Delta_n(J) := M_n(J) - \mathbb{E}M_n(J)$ and, for $\ell \leqslant n$ and $D \in \mathfrak{C}_\ell$, set

$$\Upsilon_n(D) := \sum_{m=\ell}^{n} 2^{m\gamma} \sum_{\substack{J \in \mathfrak{C}_m \\ J \subset D}} \Delta_n(J)^2.$$

By assumption (1) in Theorem 10.6 we have $\mathbb{E}\left[\Delta_n(J)^2\right] \leqslant \zeta(n)p_n 2^{n-m}$ and therefore, for all $D \in \mathfrak{C}_\ell$,

$$\mathbb{E}\Upsilon_n(D) \leqslant \sum_{m=\ell}^{n} 2^{m\gamma} \sum_{\substack{J \in \mathfrak{C}_m \\ J \subset D}} \mathbb{E}[\Delta_n(J)^2] \leqslant \sum_{m=\ell}^{n} 2^{m\gamma}\zeta(n)\, p_n\, 2^{n-\ell} \leqslant \frac{2^{(n+1)\gamma}}{2^\gamma - 1}\,\zeta(n)\, p_n\, 2^{n-\ell}.$$

By our choice of $n = n(\ell)$ we thus obtain

$$\mathbb{E}\left[\sum_{D \in \mathfrak{C}_\ell} \frac{\Upsilon_n(D)}{(2^{n-\ell}p_n)^2} \right] \leqslant \frac{2^\gamma}{2^\gamma - 1}\,\zeta(n)\, 2^{2\ell-n+n\gamma}p_n^{-1} \leqslant \frac{2^\gamma}{2^\gamma - 1}\, 2^{-\ell}.$$

Since the right hand side is summable in ℓ we conclude that, almost surely, the summands inside the last expectation converge to zero as $\ell \uparrow \infty$. In particular, there exists $\ell_0 < \infty$ such that, for all $\ell \geqslant \ell_0$ we have $2^{-\ell\gamma} \leqslant 1/4$ and, for $n = n(\ell)$ and $D \in \mathfrak{C}_\ell$,

$$\Upsilon_n(D) \leqslant \left(2^{n-\ell}p_n\right)^2 = \left(\mathbb{E}M_n(D)\right)^2.$$

The first statement follows from this very easily: For any $\ell \geqslant \ell_0$ and $n = n(\ell)$ we have (recalling the definition of $\Upsilon_n(D)$),

$$\Delta_n(D)^2 \leqslant 2^{-\ell\gamma}\Upsilon_n(D) \leqslant 2^{-\ell\gamma}\left(\mathbb{E}M_n(D)\right)^2 \leqslant \tfrac{1}{4}\left(\mathbb{E}M_n(D)\right)^2.$$

In order to get the second statement we calculate,

$$\sum_{\substack{J \in \mathfrak{C}_m \\ J \subset D}} \frac{(\mathbb{E}M_n(J))^2}{(2^{n-\ell}p_n)^2} = \sum_{\substack{J \in \mathfrak{C}_m \\ J \subset D}} 2^{2(\ell-m)} = 2^{\ell-m}.$$

Therefore

$$\sum_{\substack{m=\ell}}^{n} 2^{m\gamma} \sum_{\substack{J \in \mathfrak{C}_m \\ J \subset D}} \frac{(\mathbb{E}M_n(J))^2}{(2^{n-\ell}p_n)^2} = 2^{\ell} \sum_{m=\ell}^{n} 2^{-m(1-\gamma)} \leqslant \frac{2^{\ell\gamma}}{1 - 2^{-(1-\gamma)}}. \tag{10.4}$$

Now, recalling the choice of n,

$$\sum_{m=\ell}^{n} 2^{m\gamma} \sum_{\substack{J \in \mathfrak{C}_m \\ J \subset D}} \frac{\Delta_n(J)^2}{(2^{n-\ell}p_n)^2} = \frac{\Upsilon_n(D)}{(2^{n-\ell}p_n)^2} \leqslant 1. \tag{10.5}$$

Since $M_n(J)^2 = \big(\mathbb{E}M_n(J) + \Delta_n(J)\big)^2 \leqslant 2\big(\mathbb{E}M_n(J)\big)^2 + 2\big(\Delta_n(J)\big)^2$, adding (10.4) and (10.5) and setting $C := 2 + 2/(1 - 2^{-(1-\gamma)})$ proves the second statement. ∎

We now define $\ell_{k+1} = n(\ell_k)$ for all integers $k \geqslant 0$. The first statement of Lemma 10.8 ensures that μ is well defined by (10.3), and together with the second statement will enable us to check that μ has finite γ-energy.

Proof of Theorem 10.6.   We can now use Lemma 10.8 to verify the condition of Lemma 10.7 and finish the proof of Theorem 10.6. Indeed, by definition of μ,

$$\sum_{m=\ell_0+1}^{\infty} \sum_{J \in \mathfrak{C}_m} \frac{\mu(J)^2}{2^{-\gamma m}} = \sum_{k=0}^{\infty} \sum_{m=\ell_k+1}^{\ell_{k+1}} 2^{\gamma m} \sum_{D \in \mathfrak{C}_{\ell_k}} \frac{\mu(D)^2}{M_{\ell_{k+1}}(D)^2} \sum_{\substack{J \in \mathfrak{C}_m \\ J \subset D}} M_{\ell_{k+1}}(J)^2. \tag{10.6}$$

Recall that $q_{k+1} := \mathbb{E}M_{\ell_{k+1}}(D) = 2^{\ell_{k+1}-\ell_k} p_{\ell_{k+1}}$ and, by the first statement of Lemma 10.8, for every $k \in \mathbb{N}$ and $D \in \mathfrak{C}_{\ell_k}$,

$$\tfrac{1}{2} q_{k+1} \leqslant M_{\ell_{k+1}}(D) \leqslant 2q_{k+1}. \tag{10.7}$$

Now, from the definition of the measure μ we get, with $D \subset D' \in \mathfrak{C}_{\ell_{k-1}}$,

$$\mu(D) = \frac{M_{\ell_k}(D)\mu(D')}{M_{\ell_k}(D')} \leqslant 2Z(D)/q_k,$$

and therefore we can continue (10.6) with the upper bound

$$16 \sum_{k=0}^{\infty} \frac{1}{q_k^2} \sum_{D \in \mathfrak{C}_{\ell_k}} Z(D) \sum_{m=\ell_k+1}^{\ell_{k+1}} 2^{\gamma m} \sum_{\substack{J \in \mathfrak{C}_m \\ J \subset D}} \frac{M_{\ell_{k+1}}(J)^2}{q_{k+1}^2} \leqslant 16\,C \sum_{k=0}^{\infty} \frac{1}{q_k^2} \sum_{D \in \mathfrak{C}_{\ell_k}} Z(D)\, 2^{\gamma\ell_k},$$

using the second statement of Lemma 10.8 and the definition of q_{k+1}. Recall that the sum of the indicator variables above is, by definition, equal to $M_{\ell_k}([0,1])$. Finally, using (10.7) and the definition of $\ell_k = n(\ell_{k-1})$ we note that,

$$\sum_{k=1}^{\infty} \frac{1}{q_k^2} M_{\ell_k}([0,1])\, 2^{\gamma\ell_k} \leqslant 2 \sum_{k=1}^{\infty} \frac{2^{\ell_{k-1}}}{q_k}\, 2^{\gamma\ell_k} = 2 \sum_{k=1}^{\infty} 2^{2\ell_{k-1}-\ell_k}\, \frac{2^{\gamma\ell_k}}{p_{\ell_k}}$$

$$\leqslant \sum_{k=1}^{\infty} 2^{-\ell_{k-1}+1} < \infty.$$

This ensures convergence of the sequence (10.6) and thus completes the proof. ∎

Coming back to the lower bound in Theorem 10.3 we fix $\varepsilon > 0$. Given $I = [jh, (j+1)h]$ with $h := 2^{-k}$ we let $Z(I) = 1$ if and only if

$$|B(jh + h') - B(jh)| \geqslant a(1 + \varepsilon)\sqrt{2h'\log(1/h')}, \quad \text{for } h' := k2^{-k}.$$

Lemma 10.10 *Almost surely, the set A associated with this family $(Z(I)\colon I \in \mathfrak{C})$ of random variables is contained in the set of a-fast times.*

Proof. Recall that by Theorem 1.12 there exists a constant $C > 0$ such that, almost surely,

$$|B(s) - B(t)| \leqslant C\sqrt{|t - s|\log(1/|t - s|)}, \quad \text{for all } s, t \in [0, 2].$$

Now assume that k is large enough that $\left(\frac{2C}{a\varepsilon}\right)^2 \log 2 + \log k \leqslant k \log 2$. Let $t \in A$ and suppose that $t \in I \in \mathfrak{C}_k$ with $Z(I) = 1$. Then, by the triangle equality,

$$
\begin{aligned}
|B(t + h') - B(t)| \\
&\geqslant |B(jh + h') - B(jh)| - |B(t + h') - B(jh + h')| - |B(jh) - B(t)| \\
&\geqslant a(1 + \varepsilon)\sqrt{2h'\log(1/h')} - 2C\sqrt{h\log(1/h)} \\
&\geqslant a\sqrt{2h'\log(1/h')}.
\end{aligned}
$$

As this happens for infinitely many k, this proves that t is an a-fast time. ∎

The next lemma singles out the crucial estimates of expectation and variance needed to apply Theorem 10.6. The first is based on the upper tail estimate for a standard normal distribution, the second on the 'short range dependence' of the family $(Z(I)\colon I \in \mathfrak{C})$.

Lemma 10.11 *Define $p_n = \mathbb{E}[Z(I)]$ for $I \in \mathfrak{C}_n$, and $\eta(n) := 2n + 1$. Then,*

(a) *for $I \in \mathfrak{C}_k$ we have $\mathbb{E}[Z(I)] \geqslant 2^{-k a^2 (1+\varepsilon)^3}$;*

(b) *for $m \leqslant n$ and $J \in \mathfrak{C}_m$, we have $\operatorname{Var} M_n(J) \leqslant p_n 2^{n-m} \eta(n)$.*

Proof. For part (a), denoting by X a standard normal random variable,

$$
\begin{aligned}
\mathbb{P}\{|B(jh + h') - B(jh)| &\geqslant a(1 + \varepsilon)\sqrt{2h'\log(1/h')}\} \\
&= \mathbb{P}\{|X| > a(1 + \varepsilon)\sqrt{2\log(1/h')}\} \\
&\geqslant \frac{a(1+\varepsilon)\sqrt{2\log(1/h')}}{1+2a^2(1+\varepsilon)^2\log(1/h')} \frac{1}{\sqrt{2\pi}} \exp\{-a^2(1+\varepsilon)^2\log(1/h')\} \geqslant 2^{-k a^2 (1+\varepsilon)^3},
\end{aligned}
\tag{10.8}
$$

for all sufficiently large k and all $0 \leqslant j < 2^k$, using the lower estimate for normal random variables of Lemma 12.9 in the penultimate step.

For part (b) note that for two intervals $J_1, J_2 \in \mathfrak{C}_n$ the associated random variables $Z(J_1)$ and $Z(J_2)$ are independent if their distance is at least $n2^{-n}$. Using this whenever possible and the trivial estimate $\mathbb{E}[Z(J_1)Z(J_2)] \leqslant \mathbb{E}Z(J_1)$ otherwise, we get

$$
\mathbb{E}M_n(J)^2 = \sum_{\substack{J_1, J_2 \in \mathfrak{C}_n \\ J_1, J_2 \subset J}} \mathbb{E}\big[Z(J_1)Z(J_2)\big]
$$

$$
\leqslant \sum_{\substack{J_1 \in \mathfrak{C}_n \\ J_1 \subset J}} \Big\{ (2n+1)\mathbb{E}Z(J_1) + \mathbb{E}Z(J_1) \sum_{\substack{J_2 \in \mathfrak{C}_n \\ J_2 \subset J}} \mathbb{E}Z(J_2) \Big\}.
$$

Hence we obtain

$$
\mathbb{E}\big[(M_n(J) - \mathbb{E}M_n(J))^2\big] \leqslant \sum_{\substack{J_1 \in \mathfrak{C}_n \\ J_1 \subset J}} (2n+1)p_n = p_n 2^{n-m}(2n+1),
$$

which proves the lemma.   ∎

Proof of the lower bound in Theorem 10.3.   By Lemma 10.11 the conditions of Theorem 10.6 hold for any $\gamma < 1 - a^2(1 + \varepsilon)^3$. As, for $\varepsilon > 0$, the set A associated to $(Z(I) \colon I \in \mathfrak{C})$ is contained in the set of a-fast times, the latter has dimension $\geqslant 1 - a^2$. ∎

10.2 Packing dimension and limsup fractals

In this section we ask for a precise criterion, whether a set E contains a-fast times for various values of a. It turns out that such a criterion depends not on the Hausdorff, but on the packing dimension of the set E. We therefore begin this section by introducing the concept of *packing dimension*, which was briefly mentioned in the beginning of Chapter 4, in some detail. We choose to define packing dimension in a way which indicates its conceptual nature as a *dual* to the notion of Hausdorff dimension. The natural *dual* operation to covering a set with balls, as in the case of Hausdorff dimension, is the operation of *packing* balls disjointly into the set.

Definition 10.12.   Suppose E is a metric space. For every $\delta > 0$, a δ-**packing** of $A \subset E$ is a countable collection of *disjoint* balls

$$
\mathcal{B}(x_1, r_1), \mathcal{B}(x_2, r_2), \mathcal{B}(x_3, r_3), \ldots
$$

with centres $x_i \in A$ and radii $0 \leqslant r_i \leqslant \delta$. For every $s \geqslant 0$ we introduce the **s-value** of the packing as $\sum_{i=1}^{\infty} r_i^s$. The **s-packing number** of A is defined as

$$
P^s(A) = \lim_{\delta \downarrow 0} P^s_\delta(A) \quad \text{for } P^s_\delta(A) = \sup \Big\{ \sum_{i=1}^{\infty} r_i^s \colon (\mathcal{B}(x_i, r_i)) \text{ a } \delta\text{-packing of } A \Big\}. \quad \diamond
$$

Note that the packing number is defined in the same way as the Hausdorff measure with efficient (small) coverings replaced by efficient (large) packings. A difference is that the packing numbers do *not* define a reasonable measure. However a small modification gives the so-called packing measure,

$$
\mathcal{P}^s(A) = \inf \Big\{ \sum_{i=1}^{\infty} P^s(A_i) \colon A = \bigcup_{i=1}^{\infty} A_i \Big\}.
$$

The packing dimension has a definition analogous to the definition of Hausdorff dimension with Hausdorff measures replaced by packing measures.

Definition 10.13. The **packing dimension** of E is $\dim_P E = \inf\{s\colon \mathcal{P}^s(E) = 0\}$. ◇

Remark 10.14 It is not hard to see that

$$\dim_P E = \inf\{s\colon \mathcal{P}^s(E) < \infty\} = \sup\{s\colon \mathcal{P}^s(E) > 0\} = \sup\{s\colon \mathcal{P}^s(E) = \infty\},$$

a proof of this fact is suggested as Exercise 10.1. ◇

An alternative approach to packing dimension is to use a suitable *regularisation* of the upper Minkowski dimension, recall Remark 4.4 where we have hinted at this possibility.

Theorem 10.15 *For every metric space E we have*

$$\dim_P E = \inf\left\{ \sup_{i=1}^{\infty} \overline{\dim}_M E_i \; : \; E = \bigcup_{i=1}^{\infty} E_i \, , E_i \text{ bounded}\right\}.$$

Remark 10.16 This characterisation of the packing dimension shows that $\dim_P E \leqslant \overline{\dim}_M E$ for all bounded sets E, and, of course, strict inequality may hold. Every countable set has packing dimension 0, compare with the example in Exercise 4.2. Moreover, it is not hard to see that the countable stability property is satisfied. ◇

Proof. Define, for every $A \subset E$ and $\varepsilon > 0$,

$$P(A, \varepsilon) = \max\left\{k\colon \text{ there are disjoint balls } \mathcal{B}(x_1, \varepsilon), \ldots, \mathcal{B}(x_k, \varepsilon) \text{ with } x_i \in A\right\}.$$

Recall from (4.1) the definition of the numbers $M(A, \varepsilon)$ giving the number of sets of diameter at most ε needed to cover A. We first show that

$$P(A, 4\varepsilon) \leqslant M(A, 2\varepsilon) \leqslant P(A, \varepsilon).$$

Indeed, if $k = P(A, \varepsilon)$ let $\mathcal{B}(x_1, \varepsilon), \ldots, \mathcal{B}(x_k, \varepsilon)$ be disjoint balls with $x_i \in A$. Suppose $x \in A \setminus \bigcup_{i=1}^{k} \mathcal{B}(x_i, 2\varepsilon)$, then $\mathcal{B}(x, \varepsilon)$ is disjoint from all balls $\mathcal{B}(x_i, \varepsilon)$ contradicting the choice of k. Hence $\mathcal{B}(x_1, 2\varepsilon), \ldots, \mathcal{B}(x_k, 2\varepsilon)$ is a covering of A and we have shown $M(A, 2\varepsilon) \leqslant P(A, \varepsilon)$. For the other inequality let $m = M(A, 2\varepsilon)$ and $k = P(A, 4\varepsilon)$ and choose $x_1, \ldots, x_m \in A$ and $y_1, \ldots, y_k \in A$ such that

$$A \subset \bigcup_{i=1}^{m} \mathcal{B}(x_i, 2\varepsilon) \text{ and } \mathcal{B}(y_1, 4\varepsilon), \ldots, \mathcal{B}(y_k, 4\varepsilon) \text{ disjoint}.$$

Then each y_j belongs to some $\mathcal{B}(x_i, 2\varepsilon)$ and no such ball contains more than one such point. Thus $k \leqslant m$, which proves $P(A, 4\varepsilon) \leqslant M(A, 2\varepsilon)$.

Suppose now that $\inf\{t\colon \mathcal{P}^t(E) = 0\} < s$. Then there is $t < s$ and $E = \bigcup_{i=1}^{\infty} A_i$ such that, for every set $A = A_i$, we have $P^t(A) < 1$. Obviously, $P_\varepsilon^t(A) \geqslant P(A, \varepsilon)\varepsilon^t$. Letting $\varepsilon \downarrow 0$ gives

$$\limsup_{\varepsilon \downarrow 0} M(A, \varepsilon)\varepsilon^t \leqslant \limsup_{\varepsilon \downarrow 0} P(A, \varepsilon/2)\varepsilon^t \leqslant 2^t P^t(A) < 2^t.$$

Hence $\overline{\dim}_M A \leqslant t$ and by definition $\sup_{i=1}^{\infty} \overline{\dim}_M A_i \leqslant t < s$.

To prove the opposite inequality, let $0 < t < s < \dim_P(E)$, and $A_i \subset E$ be bounded with $E = \bigcup_{i=1}^{\infty} A_i$. It suffices to show that $\overline{\dim}_M(A_i) \geqslant t$ for some i. Since $\mathcal{P}^s(E) > 0$ there is i such that $P^s(A_i) > 0$. Let $0 < \alpha < P^s(A_i)$, then for all $\delta \in (0, 1)$ we have $P_\delta^s(A_i) > \alpha$ and there exist disjoint balls $\mathcal{B}(x_1, r_1), \mathcal{B}(x_2, r_2), \mathcal{B}(x_3, r_3), \ldots$ with centres $x_j \in A_i$ and radii r_j smaller than δ with

$$\sum_{j=1}^{\infty} r_j^s \geqslant \alpha.$$

For every m let k_m be the number of balls with radius $2^{-m-1} < r_j \leqslant 2^{-m}$. Then,

$$\sum_{m=0}^{\infty} k_m 2^{-ms} \geqslant \sum_{j=1}^{\infty} r_j^s \geqslant \alpha.$$

This yields, for some integer $N \geqslant 0$, $2^{Nt}(1 - 2^{t-s})\alpha \leqslant k_N$, since otherwise

$$\sum_{m=0}^{\infty} k_m 2^{-ms} < \sum_{m=0}^{\infty} 2^{mt}(1 - 2^{t-s})2^{-ms}\alpha = \alpha.$$

Since $r_j \leqslant \delta$ for all j, we have $2^{-N-1} < \delta$. Moreover,

$$P(A_i, 2^{-N-1}) \geqslant k_N \geqslant 2^{Nt}(1 - 2^{t-s})\alpha,$$

which gives

$$\sup_{0 \leqslant \varepsilon \leqslant \delta} P(A_i, \varepsilon)\varepsilon^t \geqslant P(A_i, 2^{-N-1})2^{-Nt-t} \geqslant 2^{-t}(1 - 2^{t-s})\alpha.$$

Letting $\delta \downarrow 0$, and recalling the relation of $M(A, \varepsilon)$ and $P(A, \varepsilon)$ established at the beginning of the proof, we obtain

$$\limsup_{\varepsilon \downarrow 0} M(A_i, \varepsilon)\varepsilon^t \geqslant \limsup_{\varepsilon \downarrow 0} P(A_i, 2\varepsilon)\varepsilon^t > 0,$$

and thus $\overline{\dim}_M A_i \geqslant t$, as required.   ∎

Remark 10.17 It is easy to see that, for every metric space, $\dim_P E \geqslant \dim E$. This is suggested as Exercise 10.2.   ◇

The following result shows that every closed subset of $\mathbb{R}^d$ has a large subset, which is 'regular' in a suitable sense. It will be used in the proof of Theorem 10.28 below.

Lemma 10.18 *Let $A \subset \mathbb{R}^d$ be closed.*

   (i) *If any open set V intersecting A satisfies $\overline{\dim}_M(A \cap V) \geqslant \alpha$, then $\dim_P(A) \geqslant \alpha$.*
   (ii) *If $\dim_P(A) > \alpha$, then there is a (relatively closed) nonempty subset $\widetilde{A}$ of A, such that, for any open set V which intersects $\widetilde{A}$, we have $\dim_P(\widetilde{A} \cap V) > \alpha$.*

Proof.   Let $A \subset \bigcup_{j=1}^{\infty} A_j$, where the A_j are closed. We are going to show that there exist an open set V and an index j such that $V \cap A \subset A_j$. For this V and j we have,

$$\overline{\dim}_M(A_j) \geqslant \overline{\dim}_M(A_j \cap V) \geqslant \overline{\dim}_M(A \cap V) \geqslant \alpha.$$

This in turn implies that $\dim_P(A) \geqslant \alpha$.

Suppose now that for any V open such that $V \cap A \neq \emptyset$, it holds that $V \cap A \not\subset A_j$. Then A_j^c is a dense open set relative to A. By Baire's (category) theorem $A \cap \bigcap_j A_j^c \neq \emptyset$, which means that $A \not\subset \bigcup_j A_j$, contradicting our assumption and proving (i).

Now choose a countable basis $\mathcal{B}$ of the topology of $\mathbb{R}^d$ and define

$$\widetilde{A} = A \setminus \bigcup \{B \in \mathcal{B}: \dim_P (B \cap A) \leqslant \alpha\}.$$

Then, $\dim_P (A \setminus \widetilde{A}) \leqslant \alpha$ using stability of packing dimension. From this we conclude that

$$\dim_P \widetilde{A} = \dim_P A > \alpha.$$

If for some V open, $V \cap \widetilde{A} \neq \emptyset$ and $\dim_P (\widetilde{A} \cap V) \leqslant \alpha$ then V contains some set $B \in \mathcal{B}$ such that $\widetilde{A} \cap B \neq \emptyset$. For that set we have $\dim_P (A \cap B) \leqslant \dim_P (A \setminus \widetilde{A}) \vee \dim_P (\widetilde{A} \cap B) \leqslant \alpha$, contradicting the construction of $\widetilde{A}$. ∎

Example 10.19 An example of a result demonstrating the duality between Hausdorff and packing dimension is the *product formula*, see [BP96]. In the dimension theory of smooth sets (manifolds, linear spaces) we have the following formula for product sets

$$\dim(E \times F) = \dim E + \dim F.$$

The example discussed in Exercise 10.3 shows that this formula fails for Hausdorff dimension, a reasonable formula for the Hausdorff dimension of product sets necessarily involves information about the packing dimension of one of the factor sets. In [BP96] it is shown that, for every Borel set $A \subset \mathbb{R}^d$,

$$\dim_P (A) = \sup_B \left\{ \dim(A \times B) - \dim(B) \right\}$$

where the supremum is over all compact sets $B \subset \mathbb{R}^d$. One can also show that, if A satisfies $\dim A = \dim_P A$, then the product formula $\dim(A \times B) = \dim A + \dim B$ holds. ◇

Before moving back to our study of Brownian paths we study the packing dimension of the 'test sets' we have used in the stochastic codimension method, see Section 9.9.1.

Theorem 10.20 *Let $\gamma \in [0, d]$ and $\Gamma[\gamma]$ be a percolation limit set in $\mathbb{R}^d$ with retention parameter $2^{-\gamma}$. Then*

- $\dim_P \Gamma[\gamma] \leqslant d - \gamma$ *almost surely,*

- $\dim_P \Gamma[\gamma] = d - \gamma$ *almost surely on $\Gamma[\gamma] \neq \emptyset$.*

Proof. For the first item, as packing dimension is bounded from above by the upper Minkowski dimension, it suffices to show that $\overline{\dim}_M \Gamma[\gamma] \leqslant d - \gamma$ almost surely. For this purpose we use the formula for the upper Minkowski dimension given in Remark 4.2. For a given n, we cover the percolation limit set by $\mathfrak{S}_n$, the collection of cubes retained in the nth construction step. The probability that a given cube of side length 2^{-n} is in $\mathfrak{S}_n$ is $2^{-n\gamma}$ and hence the expected number of cubes in $\mathfrak{S}_n$ is $2^{n(d-\gamma)}$. Hence, for any $\varepsilon > 0$,

$$\mathbb{P}\{2^{n(\gamma-d-\varepsilon)} \#\mathfrak{S}_n > 1\} \leqslant 2^{n(\gamma-d-\varepsilon)} \mathbb{E}\#\mathfrak{S}_n \leqslant 2^{-n\varepsilon},$$

which is summable. Hence, almost surely, $2^{n(\gamma-d-\varepsilon)} \#\mathfrak{S}_n \leqslant 1$ for all but finitely many n. Thus, almost surely,

$$\overline{\dim}_{\mathrm{M}} \leqslant \limsup_{n\uparrow\infty} \frac{\log \#\mathfrak{S}_n}{n \log 2} \leqslant d - \gamma + \varepsilon \qquad \text{for every } \varepsilon > 0.$$

For the second item recall the corresponding statement for Hausdorff dimension from Exercise 9.6. The result follows, as packing dimension is bounded from below by the Hausdorff dimension, see Remark 10.17. ∎

Remark 10.21 Simple modifications of the corresponding proofs for the upper bounds in the case of Hausdorff dimension, see Exercise 10.4, show that

- $\dim_P \mathsf{Range}[0, 1] = 2$, for Brownian motion in $d \geqslant 2$,

- $\dim_P \mathsf{Graph}[0, 1] = \frac{3}{2}$, for Brownian motion in $d = 1$,

- $\dim_P \mathsf{Zeros} = \frac{1}{2}$, for Brownian motion in $d = 1$.

Hence, at a first glance the concept of packing dimension does not seem to add a substantial contribution to the discussion of fine properties of d-dimensional Brownian motion. However, a first sign that something interesting might be going on can be found in Exercise 10.5, where we show that the Hausdorff and packing dimension of the sets of a-fast times differ. This is indicative of the fact that optimal coverings of these sets use covering sets of widely differing size, and that optimal packings use sets of quite different scale. ◇

Given a set $E \subset [0, 1]$ we now ask for the maximal value of a such that E contains an a-fast time with positive probability. This notion of size is most intimately linked to packing dimension as the following theorem shows. We denote by $F(a) \subset [0, 1]$ the set of a-fast times.

Theorem 10.22 (Khoshnevisan, Peres and Xiao) *For any compact set $E \subset [0, 1]$, almost surely,*

$$\sup_{t\in E} \limsup_{h\downarrow 0} \frac{|B(t + h) - B(t)|}{\sqrt{2h \log(1/h)}} = \sqrt{\dim_P(E)}.$$

Moreover, if $\dim_P(E) > a^2$, then $\dim_P(F(a) \cap E) = \dim_P(E)$.

Remark 10.23 The result can be extended from compact sets E to more general classes of sets, more precisely the *analytic* sets, see [KPX00]. ◇

Remark 10.24 An equivalent formulation of the theorem is that, for any compact $E \subset [0, 1]$, almost surely,

$$\mathbb{P}\{F(a) \cap E \neq \emptyset\} = \begin{cases} 1 & \text{if } \dim_P(E) > a^2, \\ 0 & \text{if } \dim_P(E) < a^2. \end{cases}$$

Using the compact percolation limit sets $E = \Gamma[\gamma]$ in this result and Hawkes' theorem, Theorem 9.5, one can obtain an alternative proof of the Orey–Taylor theorem. Indeed, by Theorem 10.20, if $\gamma < 1 - a^2$ we have $\dim_P(E) > a^2$ with positive probability, and therefore, $\mathbb{P}\{F(a) \cap E \neq \emptyset\} > 0$. Hence, by Hawkes' theorem, $\dim F(a) \geqslant \gamma$ with positive probability. Brownian scaling maps a-fast times onto a-fast times. Therefore there exists $\varepsilon > 0$ such that, for any $n \in \mathbb{N}$ and $0 \leqslant j \leqslant n - 1$,

$$\mathbb{P}\{ \dim(F(a) \cap [j/n, (j+1)/n]) \geqslant \gamma\} \geqslant \varepsilon,$$

and hence

$$\mathbb{P}\{ \dim F(a) \geqslant \gamma\} \geqslant 1 - (1 - \varepsilon)^n \to 1.$$

Letting $\gamma \uparrow 1 - a^2$ gives $\dim F(a) \geqslant 1 - a^2$ almost surely. Conversely, by Theorem 10.20, if $\gamma > 1 - a^2$ we have $\dim_P(E) < a^2$ almost surely, and therefore, $\mathbb{P}\{F(a) \cap E \neq \emptyset\} = 0$. Hence, by Hawkes' theorem, we have $\dim F(a) \leqslant 1 - a^2$ almost surely. ◇

Theorem 10.22 can be seen as a probabilistic interpretation of packing dimension. The upper and lower Minkowski dimensions allow a similar definition when the order of sup and lim are interchanged.

Theorem 10.25 *For any compact $E \subset [0, 1]$, almost surely,*

$$\limsup_{h \downarrow 0} \sup_{t \in E} \frac{|B(t + h) - B(t)|}{\sqrt{2h \log(1/h)}} = \sqrt{\overline{\dim}_M(E)}. \qquad (10.9)$$

Proof of the upper bounds in Theorems 10.22 and 10.25. Suppose $E \subset [0, 1]$ is compact. We assume that $\overline{\dim}_M(E) < \lambda < a^2$ and show that

$$\limsup_{h \downarrow 0} \sup_{t \in E} \frac{|B(t + h) - B(t)|}{\sqrt{2h \log(1/h)}} \leqslant a \quad \text{almost surely.} \qquad (10.10)$$

Note that this is the upper bound in Theorem 10.25. Once this is shown it immediately implies

$$\sup_{t \in E} \limsup_{h \downarrow 0} \frac{|B(t + h) - B(t)|}{\sqrt{2h \log(1/h)}} \leqslant \sqrt{\overline{\dim}_M(E)} \quad \text{almost surely.}$$

Now, for any decomposition $E = \bigcup_{i=1}^{\infty} E_i$, we have

$$\sup_{t \in E} \limsup_{h \downarrow 0} \frac{|B(t + h) - B(t)|}{\sqrt{2h \log(1/h)}} = \sup_{i=1}^{\infty} \sup_{t \in \mathrm{cl}(E_i)} \limsup_{h \downarrow 0} \frac{|B(t + h) - B(t)|}{\sqrt{2h \log(1/h)}}$$

$$\leqslant \sup_{i=1}^{\infty} \sqrt{\overline{\dim}_M(E_i)},$$

where we have made use of the fact that the upper Minkowski dimension is insensitive under taking the closure of a set. Theorem 10.15 now implies that

$$\sup_{t \in E} \limsup_{h \downarrow 0} \frac{|B(t + h) - B(t)|}{\sqrt{2h \log(1/h)}} \leqslant \sqrt{\dim_P(E)} \quad \text{almost surely,}$$

which is the upper bound in Theorem 10.22.

For the proof of (10.10) cover E by disjoint subintervals $I = [(j/k)\eta^{-k}, ((j+1)/k)\eta^{-k})$ for $j = 0, \ldots, \lceil k\eta^k - 1 \rceil$, of equal length $h = \eta^{-k}/k$ such that $I \cap E \neq \emptyset$. By definition of the upper Minkowski dimension there exists an m such that, for all $k \geqslant m$, no more than $\eta^{\lambda k}$ different such intervals of length $h = \eta^{-k}/k$ intersect E.

Now fix $\varepsilon > 0$ such that $\lambda < a^2 (1 - 4\varepsilon)^3$, which is possible by our condition on λ. Let $Z(I) = 1$ if, for $h' = \eta^{-k}$,

$$|B(jh + h') - B(jh)| > a(1 - 4\varepsilon)\sqrt{2h' \log(1/h')}.$$

Recall from the proof of Lemma 10.4 that there is an $\eta > 1$ such that, for any $m \in \mathbb{N}$, the collection

$$\{I = [(j/k)\eta^{-k}, ((j+1)/k)\eta^{-k}) \colon Z(I) = 1, I \cap E \neq \emptyset, k \geqslant m\}$$

is a covering of the set

$$M(m) := \left\{ t \in E : \sup_{\eta^{-k} < u \leqslant \eta^{-k+1}} \frac{|B(t + u) - B(t)|}{\sqrt{2u \log(1/u)}} \geqslant a(1 - \varepsilon) \text{ for some } k \geqslant m \right\}.$$

Moreover, we recall from (10.2), that

$$\mathbb{P}\{Z(I) = 1\} \leqslant \eta^{-ka^2 (1-4\varepsilon)^3},$$

and, sticking to our notation $I = [(j/k)\eta^{-k}, ((j+1)/k)\eta^{-k})$ for a little while longer,

$$\sum_{k=0}^{\infty} \sum_{j=0}^{\lceil k\eta^k - 1 \rceil} \mathbb{P}\{Z(I) = 1\} \mathbf{1}\{I \cap E \neq \emptyset\} \leqslant \sum_{k=0}^{\infty} \eta^{\lambda k} \eta^{-ka^2 (1-4\varepsilon)^3} < \infty,$$

and hence by the Borel–Cantelli lemma there exists an m such that $Z(I) = 0$ whenever $I = [(j/k)\eta^{-k}, ((j+1)/k)\eta^{-k})$ for some $k \geqslant m$. This means that the set $M(m)$ can be covered by the empty covering, so it must itself be empty. This shows (10.10) and completes the proof. ∎

We embed the proof of the lower bound into a more general framework, including the discussion of limsup fractals in a d-dimensional cube.

Definition 10.26. Fix an open unit cube $\mathsf{Cube} = x_0 + (0, 1)^d \subset \mathbb{R}^d$. For any nonnegative integer k, denote by $\mathfrak{C}_k$ the collection of dyadic cubes

$$x_0 + \prod_{i=1}^{d} [j_i 2^{-k}, (j_i + 1)2^{-k}] \qquad \text{with } j_i \in \{0, \ldots, 2^k - 1\} \text{ for all } i \in \{1, \ldots, d\},$$

and $\mathfrak{C} = \bigcup_{k \geqslant 0} \mathfrak{C}_k$. Denote by $(Z(I) \colon I \in \mathfrak{C})$ a collection of random variables each taking values in $\{0, 1\}$. The **limsup fractal** associated to this collection is the random set

$$A := \bigcap_{n=1}^{\infty} \bigcup_{k=n}^{\infty} \bigcup_{\substack{I \in \mathfrak{C}_k \\ Z(I)=1}} \mathrm{int}(I),$$

where $\mathrm{int}(I)$ is the interior of the cube I.   ◇

Remark 10.27 Compared with the setup of the previous section we have switched to the use of *open* cubes in the definition of limsup fractals. This choice is more convenient when we prove hitting estimates, whereas in Theorem 10.6 the choice of closed cubes was more convenient when constructing random measures on A. ◇

The key to our result is the hitting probabilities for the discrete limsup fractal A under some conditions on the random variables $(Z(I): I \in \mathfrak{C})$.

Theorem 10.28 *Suppose that*

  (i) *the means $p_n = \mathbb{E}[Z(I)]$ are independent of the choice of $I \in \mathfrak{C}_n$ and satisfy*

$$\liminf_{n \uparrow \infty} \frac{\log p_n}{n \log 2} \geqslant -\gamma, \qquad \textit{for some } \gamma > 0;$$

  (ii) *there exists $c > 0$ such that the random variables $Z(I)$ and $Z(J)$ are independent whenever $I, J \in \mathfrak{C}_n$ and the distance of I and J exceeds $cn2^{-n}$.*

Then, for any compact $E \subset$ Cube with $\dim_P(E) > \gamma$, we have

$$\mathbb{P}\{A \cap E \neq \emptyset\} = 1.$$

Remark 10.29 The second assumption, which gives us the necessary independence for the lower bound, can be weakened, see [KPX00]. Note that no assumption is made concerning the dependence of random variables $Z(I)$ for intervals I of different size. ◇

Proof of Theorem 10.28. Let $E \subset$ Cube be compact with $\dim_P E > \gamma$. Let $\widetilde{E}$ be defined as in Lemma 10.18 for example as

$$\widetilde{E} = E \setminus \bigcup_{\substack{a_i < b_i \\ \text{rational}}} \Big\{ \prod_{i=1}^{d} (a_i, b_i) : \overline{\dim}_M \big(E \cap \prod_{i=1}^{d} (a_i, b_i)\big) < \gamma \Big\}.$$

From the proof of Lemma 10.18 we have $\dim_P E = \dim_P \widetilde{E}$. Define open sets

$$A_n = \bigcup_{I \in \mathfrak{C}_n} \{\text{int}(I) : Z(I) = 1\},$$

and

$$A_n^* = \bigcup_{m \geqslant n} A_m = \bigcup_{m \geqslant n} \bigcup_{I \in \mathfrak{C}_m} \{\text{int}(I) : Z(I) = 1\}.$$

By definition $A_n^* \cap \widetilde{E}$ is open in $\widetilde{E}$. We will show that it is also dense in $\widetilde{E}$ with probability one. This, by Baire's category theorem, will imply that

$$A \cap \widetilde{E} = \bigcap_{n=1}^{\infty} A_n^* \cap \widetilde{E} \neq \emptyset, \qquad \text{almost surely,}$$

as required. To show that $A_n^* \cap \widetilde{E}$ is dense in $\widetilde{E}$, we need to show that for any open binary cube J which intersects $\widetilde{E}$, the set $A_n^* \cap \widetilde{E} \cap J$ is almost surely nonempty.

For the rest of the proof, fix J and recall that $\overline{\dim}_M(\tilde{E} \cap J) \geqslant \dim_P(\tilde{E} \cap J) > \gamma$. Take $\varepsilon > 0$ small and n large enough so that $\tilde{E} \cap J$ intersects more than $2^{n(\gamma + 2\varepsilon)}$ binary cubes of side length 2^{-n}, and so that $(\log p_n)/n > -(\log 2)(\gamma + \varepsilon)$. Let $\mathcal{S}_n$ be the set of cubes in $\mathfrak{C}_n$ that intersect $\tilde{E} \cap J$. Define

$$T_n = \sum_{I \in \mathcal{S}_n} Z(I),$$

so that $\mathbb{P}\{A_n \cap \tilde{E} \cap J = \emptyset\} = \mathbb{P}\{T_n = 0\}$. To show that this probability converges to zero, by the Paley–Zygmund inequality, it suffices to prove that $(\operatorname{Var} T_n)/(\mathbb{E}T_n)^2$ does. The first moment of T_n is given by

$$\mathbb{E}T_n = s_n\, p_n > 2^{(\gamma + 2\varepsilon)n} 2^{-\gamma n - \varepsilon n} = 2^{\varepsilon n},$$

where s_n denotes the cardinality of $\mathcal{S}_n$. The variance can be written as

$$\operatorname{Var} T_n = \operatorname{Var} \sum_{I \in \mathcal{S}_n} Z(I) = \sum_{I \in \mathcal{S}_n} \sum_{J \in \mathcal{S}_n} \operatorname{Cov}(Z(I), Z(J)).$$

Here each summand is at most p_n, and the summands for which I and J have distance at least $cn2^{-n}$ vanish by assumption. Thus

$$\sum_{I \in \mathcal{S}_n} \sum_{J \in \mathcal{S}_n} \operatorname{Cov}(Z(I), Z(J)) \leqslant p_n\, \#\big\{(I, J) \in \mathcal{S}_n \times \mathcal{S}_n : \operatorname{dist}(I, J) \leqslant cn\, 2^{-n}\big\}$$

$$\leqslant p_n s_n\, (2cn + 1)^d = c(2cn + 1)^d\, \mathbb{E}T_n.$$

This implies that $(\operatorname{Var} T_n)/(\mathbb{E}T_n)^2 \to 0$. Hence, almost surely, A_n^* is an open dense set, concluding the proof. ∎

We now show how the main statement of Theorem 10.22 follows from this, and how the ideas in the proof also lead to the lower bound in Theorem 10.25.

Proof of the lower bound in Theorem 10.22 and 10.25. For the lower bound we look at a compact set $E \subset (0, 1)$ with $\dim_P(E) > a^2$ and first go for the result in Theorem 10.22. Choose $\varepsilon > 0$ such that $\dim_P(E) > a^2(1 + \varepsilon)^3$. Associate to every dyadic interval $I = [jh, (j+1)h] \in \mathfrak{C}_k$ with $h = 2^{-k}$ the random variable $Z(I)$, which takes the value one if and only if, for $h' = k2^{-k}$,

$$|B(jh + h') - B(jh)| \geqslant a(1 + \varepsilon)\sqrt{2h' \log(1/h')},$$

and note that by Lemma 10.10 the limsup fractal associated to these random variables is contained in the set of a-fast times. It remains to note that the collection $\{Z(I) : I \in \mathfrak{C}_k, k \geqslant 0\}$ satisfies the condition (i) with $\gamma = a^2(1 + \varepsilon)^3$ by (10.8) and condition (ii) with $c = 1$. Theorem 10.28 now gives that

$$\mathbb{P}\{A \cap E \neq \emptyset\} = 1,$$

and therefore

$$\sup_{t \in E} \limsup_{h \downarrow 0} \frac{|B(t + h) - B(t)|}{\sqrt{2h \log(1/h)}} \geqslant \sqrt{\dim_P(E)}.$$

For the lower bound in Theorem 10.25 we look at a compact set $E \subset (0, 1)$ with $\overline{\dim}_M(E)$ $> a^2$ and fix $\varepsilon > 0$ such that $\overline{\dim}_M(E) \geqslant a^2(1 + \varepsilon)^6$. Hence there exists a sequence $(n_k : k \in \mathbb{N})$ such that

$$\#\{I \in \mathfrak{C}_{n_k} : I \cap E \neq \emptyset\} \geqslant 2^{n_k \, a^2 (1+\varepsilon)^5}.$$

With $Z(I)$ defined as above we obtain, using notation and proof of Theorem 10.28, that

$$\mathbb{P}\{Z(I) = 1\} \geqslant 2^{-n_k \gamma}, \qquad \text{with } \gamma = a^2(1+\varepsilon)^4,$$

and

$$\operatorname{Var} T_{n_k} \leqslant (2n_k + 1)^d \, \mathbb{E} T_{n_k}, \qquad \text{for } T_n = \sum_{I \in \mathfrak{C}_n} Z(I) \mathbb{1}\{I \cap E \neq \emptyset\}.$$

By Chebyshev's inequality we get, for $1/2 < \eta < 1$,

$$\mathbb{P}\{|T_{n_k} - \mathbb{E} T_{n_k}| \geqslant (\mathbb{E} T_{n_k})^\eta\} \leqslant (2n_k + 1)^d \, (\mathbb{E} T_{n_k})^{1-2\eta}.$$

As $\mathbb{E} T_{n_k}$ is exponentially increasing in n_k we can infer, using the Borel–Cantelli lemma, that

$$\lim_{k \uparrow \infty} \frac{T_{n_k}}{\mathbb{E} T_{n_k}} = 1 \qquad \text{almost surely.}$$

This implies that $T_{n_k} \neq 0$ for all sufficiently large k. Hence, as $Z(I) = 1$ and $I \cap E \neq \emptyset$ imply that there exists $t \in I \cap E$ with $|B(t + h') - B(t)| \geqslant a\sqrt{2h' \log(1/h')}$ for $h' = n_k 2^{n_k}$, completing the proof of Theorem 10.25. ∎

10.3 Slow times of Brownian motion

At the fast times Brownian motion has, in infinitely many small scales, unusually large growth. Conversely, one may ask whether there are times where a Brownian path has, at *all* small scales, unusually small growth. The notion of a *slow time* for the Brownian motion is related to the nondifferentiability of the Brownian path. Indeed, in our proof of non-differentiability, we showed that almost surely,

$$\limsup_{h \downarrow 0} \frac{|B(t + h) - B(t)|}{h} = \infty, \qquad \text{for all } t \in [0, 1],$$

and in 1963 Dvoretzky showed that there exists a constant $\delta > 0$ such that almost surely,

$$\limsup_{h \downarrow 0} \frac{|B(t + h) - B(t)|}{\sqrt{h}} > \delta, \qquad \text{for all } t \in [0, 1].$$

In 1983 Davis and, independently, Perkins and Greenwood, found that the optimal constant in this result is equal to one.

Theorem 10.30 (Davis, Perkins and Greenwood) *Almost surely,*

$$\inf_{t \in [0,1]} \limsup_{h \downarrow 0} \frac{|B(t + h) - B(t)|}{\sqrt{h}} = 1.$$

Remark 10.31 We call $t \in [0,1]$ an a-**slow time** if

$$\limsup_{h \downarrow 0} \frac{|B(t+h) - B(t)|}{\sqrt{h}} \leqslant a. \tag{10.11}$$

The result shows that a-slow times exist for $a > 1$ but not for $a < 1$. The Hausdorff dimension of the set of a-slow times is studied in Perkins [Pe83].   ◇

For the proof of Theorem 10.30 we need to investigate the probability that the graph of a Brownian motion stays within a parabola open to the right. The following lemma is what we need for a lower bound.

Lemma 10.32 *Let* $M := \max_{0 \leqslant t \leqslant 1} |B(t)|$ *and, for* $r < 1$, *define the stopping time*

$$T = \inf\{t \geqslant 1 : |B(t)| = M + r\sqrt{t}\}.$$

Then $\mathbb{E}T < \infty$.

Proof. By Theorem 2.48, for every $t \geqslant 1$, we have

$$\mathbb{E}[T \wedge t] = \mathbb{E}[B(T \wedge t)^2] \leqslant \mathbb{E}[(M + r\sqrt{T \wedge t})^2]$$
$$= \mathbb{E}M^2 + 2r\mathbb{E}[M\sqrt{T \wedge t}] + r^2\mathbb{E}[T \wedge t]$$
$$\leqslant \mathbb{E}M^2 + 2r(\mathbb{E}M^2)^{1/2}(\mathbb{E}[T \wedge t])^{1/2} + r^2\mathbb{E}[T \wedge t],$$

where Hölder's inequality was used in the last step. This gives

$$(1 - r^2)\mathbb{E}[T \wedge t] \leqslant \mathbb{E}[M^2] + 2r(\mathbb{E}M^2)^{1/2}(\mathbb{E}[T \wedge t])^{1/2},$$

and as $\mathbb{E}[M^2] < \infty$ we get that $\mathbb{E}[T \wedge t]$ is bounded and hence $\mathbb{E}T < \infty$. ∎

Proof of the lower bound in Theorem 10.30.   It suffices to show that, for any fixed $r < 1$ and $h_0 > 0$, the set

$$A = \{t \in [0, h_0] : |B(t+h) - B(t)| < r\sqrt{h} \text{ for all } 0 < h \leqslant h_0\}$$

is empty almost surely. By Brownian scaling we may further assume that $h_0 = 1$. For any interval $I = [a, b] \subset [0, 1]$, we have, by the triangle inequality and Brownian scaling, for $M = \max\{|B(t) - B(a)| : a \leqslant t \leqslant b\}$, that

$$\mathbb{P}\{\exists t \in I \; : \; |B(t+h) - B(t)| < r\sqrt{h} \text{ for all } 0 < h \leqslant 1\}$$
$$\leqslant \mathbb{P}\{|B(a+h) - B(a)| < M + r\sqrt{h} \text{ for all } b - a < h \leqslant 1\}$$
$$\leqslant \mathbb{P}\{T \geqslant \tfrac{1}{b-a}\},$$

where T is as in Lemma 10.32. Dividing $[0, 1]$ into n intervals of length $1/n$ we get

$$\mathbb{P}\{A \neq \emptyset\} \leqslant \sum_{k=0}^{n-1} \mathbb{P}\{A \cap [k/n, (k+1)/n] \neq \emptyset\} \leqslant n\mathbb{P}\{T \geqslant n\}$$
$$= \mathbb{E}[n\mathbf{1}\{T \geqslant n\}] \to 0,$$

using in the final step that $n\mathbf{1}\{T \geqslant n\}$ is dominated by the integrable random variable T. ∎

We turn to the proof of the upper bound. Again we start by studying exit times from a parabola. For $0 < r < \infty$ and $a > 0$ let

$$T(r, a) := \inf\{t \geqslant 0 \colon |B(t)| = r\sqrt{t + a}\}.$$

For the moment it suffices to note the following property of $T(1, a)$.

Lemma 10.33 *We have* $\mathbb{E}T(1, a) = \infty$.

Proof. Suppose that $\mathbb{E}T(1, a) < \infty$. Then, by Theorem 2.48, we have that $\mathbb{E}T(1, a) = \mathbb{E}B(T(1, a))^2 = \mathbb{E}T(1, a) + a$, which is a contradiction. Hence $\mathbb{E}T(1, a) = \infty$. ∎

For $0 < r < \infty$ and $a > 0$ we now define further stopping times

$$S(r, a) := \inf\{t \geqslant a \colon |B(t)| \geqslant r\sqrt{t}\}.$$

Lemma 10.34 *If* $r > 1$ *there is a* $p = p(r) < 1$ *such that* $\mathbb{E}[S(r, 1)^p] = \infty$. *In particular,*

$$\limsup_{n \uparrow \infty} n \, \frac{\mathbb{P}\{S(r, 1) > n\}}{\mathbb{E}[S(r, 1) \wedge n]} > 0.$$

The proof uses the following general lemma.

Lemma 10.35 *Suppose* X *is a nonnegative random variable and* $\mathbb{E}X^p = \infty$ *for some* $p < 1$. *Then*

$$\limsup_{n \uparrow \infty} n \, \mathbb{P}\{X > n\} / \mathbb{E}[X \wedge n] > 0.$$

Proof. Let $p < 1$ and suppose for contradiction that, for some $\varepsilon < \frac{1-p}{2}$,

$$n \, \mathbb{P}\{X > n\} < \varepsilon \, \mathbb{E}[X \wedge n] \qquad \text{for all integers } n \geqslant y_0 \geqslant 2. \tag{10.12}$$

For all $N \geqslant 1$, using Fubini's theorem in the first and substitution of variables in the second step, we get that

$$\mathbb{E}[(X \wedge N)^p] = \int_0^{N^p} \mathbb{P}\{X^p > x\} \, dx = p \int_0^N y^{p-1} \, \mathbb{P}\{X > y\} \, dy \, ,$$

and hence, using (10.12) for $n = \lfloor y \rfloor$, we obtain

$$\mathbb{E}[(X \wedge N)^p] \leqslant p \int_0^{y_0} y^{p-1} \, dy + \varepsilon \tfrac{y_0}{y_0 - 1} p \int_{\lfloor y_0 \rfloor}^N y^{p-2} \, \mathbb{E}[X \wedge y] \, dy$$

$$\leqslant y_0^p + 2\varepsilon p \int_{\lfloor y_0 \rfloor}^N y^{p-2} \int_0^y \mathbb{P}\{X > z\} \, dz \, dy$$

$$\leqslant y_0^p + 2\varepsilon p \int_0^N \mathbb{P}\{X > z\} \int_z^\infty y^{p-2} \, dy \, dz$$

$$\leqslant y_0^p + \varepsilon \tfrac{2}{1-p} \, \mathbb{E}[(X \wedge N)^p],$$

and hence, by choice of ε,

$$\mathbb{E}[(X \wedge N)^p] \leqslant \frac{y_0^p}{1 - \frac{2\varepsilon}{1-p}}.$$

This implies $\mathbb{E}[X^p] = \sup \mathbb{E}[(X \wedge N)^p] < \infty$, which contradicts to our assumption. ∎

Proof of Lemma 10.34. Define a sequence of stopping times by $\tau_0 = 1$ and, for $k \geqslant 1$,

$$\tau_k = \begin{cases} \inf\{t \geqslant \tau_{k-1} : B(t) = 0 \text{ or } |B(t)| \geqslant r\sqrt{t}\} & \text{if } k \text{ is odd,} \\ \inf\{t \geqslant \tau_{k-1} : |B(t)| \geqslant \sqrt{t}\} & \text{if } k \text{ is even.} \end{cases}$$

For any fixed $\lambda > 0$ let $\varphi(a) = \mathbb{P}\{T(1,a) > \lambda a\}$ and note that, by Brownian scaling, $\varphi(a) = \varphi(1)$ for all $a > 0$. Hence, by the strong Markov property,

$$\mathbb{P}\{\tau_{2k} - \tau_{2k-1} > \lambda\tau_{2k-1} \mid B(\tau_{2k-1}) = 0\} = \mathbb{E}[\varphi(\tau_{2k-1}) \mid B(\tau_{2k-1}) = 0]$$
$$= \mathbb{P}\{T(1,1) > \lambda\}.$$

Define $c := \mathbb{P}\{S(0,1) < S(r,1)\}$. Now, for $k \geqslant 2$ and $\lambda > 0$, on $\{\tau_{2k-2} < S(r,1)\}$,

$$\mathbb{P}\{\tau_{2k} - \tau_{2k-1} > \lambda\tau_{2k-2} \mid \mathcal{F}(\tau_{2k-2})\}$$
$$\geqslant \mathbb{P}\{\tau_{2k} - \tau_{2k-1} > \lambda\tau_{2k-1} \mid \mathcal{F}(\tau_{2k-2}), B(\tau_{2k-1}) = 0\}\,\mathbb{P}\{B(\tau_{2k-1}) = 0 \mid \mathcal{F}(\tau_{2k-2})\}$$
$$= c\,\mathbb{P}\{T(1,1) > \lambda\}.$$

To pass from this estimate to the pth moments we use that, for any nonnegative random variable X, we have $\mathbb{E}X^p = \int_0^\infty \mathbb{P}\{X^p > \lambda\}\,d\lambda$. This gives

$$\mathbb{E}\big[(\tau_{2k} - \tau_{2k-1})^p\big] = \mathbb{E}\int_0^\infty \tau_{2k-2}^p\,\mathbb{P}\big\{(\tau_{2k} - \tau_{2k-1})^p > \lambda\tau_{2k-2}^p \mid \mathcal{F}(\tau_{2k-2})\big\}\,d\lambda$$
$$\geqslant \mathbb{E}\int_0^\infty \tau_{2k-2}^p\,\mathbb{P}\big\{\tau_{2k} - \tau_{2k-1} > \lambda^{1/p}\tau_{2k-2} \mid \mathcal{F}(\tau_{2k-2})\big\}\mathbf{1}\{\tau_{2k-2} < S(r,1)\}\,d\lambda$$
$$\geqslant c\,\mathbb{E}\int_0^\infty \tau_{2k-2}^p\,\mathbb{P}\{T(1,1) > \lambda^{1/p}\}\mathbf{1}\{\tau_{2k-2} < S(r,1)\}\,d\lambda.$$

Now, using the formula for $\mathbb{E}X^p$ again, but for $X = T(1,1)$ and noting that $\{\tau_{2k-2} < S(r,1)\} = \{\tau_{2k-3} < \tau_{2k-2}\}$, we obtain

$$\mathbb{E}\big[(\tau_{2k} - \tau_{2k-1})^p\big] \geqslant c\,\mathbb{E}[T(1,1)^p]\,\mathbb{E}\big[\tau_{2k-2}^p\mathbf{1}\{\tau_{2k-2} < S(r,1)\}\big]$$
$$\geqslant c\,\mathbb{E}[T(1,1)^p]\,\mathbb{E}\big[(\tau_{2k-2} - \tau_{2k-3})^p\big],$$

and by iterating this,

$$\mathbb{E}\big[(\tau_{2k} - \tau_{2k-1})^p\big] \geqslant \big(c\,\mathbb{E}[T(1,1)^p]\big)^{k-1}\,\mathbb{E}\big[(\tau_2 - \tau_1)^p\big].$$

Note that, by Fatou's lemma and by Lemma 10.33, $\liminf_{p\uparrow 1} \mathbb{E}[T(1,1)^p] \geqslant \mathbb{E}T(1,1) = \infty$. Hence we may pick $p < 1$ such that $\mathbb{E}[T(1,1)^p] > 1/c$. Then

$$\mathbb{E}[S(r,1)^p] \geqslant \mathbb{E}[\tau_{2k}^p] \geqslant \mathbb{E}\big[(\tau_{2k} - \tau_{2k-1})^p\big] \longrightarrow \infty,$$

as $k \uparrow \infty$, which is the first statement we wanted to prove. The second statement follows directly from the general fact stated as Lemma 10.35. ∎

Proof of the upper bound in Theorem 10.30.   Fix $r > 1$ and let

$$A(n) = \{t \in [0,1]: |B(t+h) - B(t)| < r\sqrt{h}, \text{ for all } \tfrac{1}{n} \leqslant h \leqslant 1\}.$$

Note that $n \geqslant m$ implies $A(n) \subset A(m)$. We show that

$$\mathbb{P}\Big\{ \bigcap_{n=1}^{\infty} A(n) \neq \emptyset \Big\} = \lim_{n \to \infty} \mathbb{P}\{A(n) \neq \emptyset\} > 0. \tag{10.13}$$

Fix $n \in \mathbb{N}$ and let $v(0,n) = 0$ and, for $i \geqslant 1$,

$$v(i,n) := (v(i-1,n)+1)$$

$$\wedge \inf\big\{t \geqslant v(i-1,n) + \tfrac{1}{n}: |B(t) - B(v(i-1,n))| \geqslant r\sqrt{t - v(i-1,n)}\big\}.$$

Then $\mathbb{P}\{v(i+1,n) - v(i,n) = 1 \mid \mathcal{F}(v(i,n))\} = \mathbb{P}\{S(r,1) \geqslant n\}$, and by Brownian scaling,

$$\mathbb{E}[v(i+1,n) - v(i,n) \mid \mathcal{F}(v(i,n))] = \tfrac{1}{n}\mathbb{E}[S(r,1) \wedge n]. \tag{10.14}$$

Of course $v(k,n) \geqslant 1$ if $v(i,n) - v(i-1,n) = 1$ for some $i \leqslant k$. Thus, for any m,

$$\mathbb{P}\{v(i+1,n) - v(i,n) = 1 \text{ for some } i \leqslant m \text{ such that } v(i,n) \leqslant 1\}$$

$$= \sum_{i=1}^{m} \mathbb{P}\{S(r,1) \geqslant n\}\mathbb{P}\{v(i,n) \leqslant 1\}$$

$$\geqslant m\mathbb{P}\{S(r,1) \geqslant n\}\mathbb{P}\{v(m,n) \leqslant 1\}.$$

Let $(n_k : k \in \mathbb{N})$ be an increasing sequence of integers such that

$$n_k \frac{\mathbb{P}\{S(r,1) \geqslant n_k\}}{\mathbb{E}[S(r,1) \wedge n_k]} \geqslant \varepsilon > 0,$$

and $\mathbb{E}[S(r,1) \wedge n_k] \leqslant n_k/6$ for all k, which is possible by Lemma 10.34. Choose the integers m_k so that they satisfy

$$\frac{1}{3} \leqslant \frac{m_k}{n_k} \mathbb{E}[S(r,1) \wedge n_k] \leqslant \frac{1}{2}.$$

Summing (10.14) over all $i = 0, \ldots, m_k - 1$ and taking the expectation,

$$\mathbb{E}v(m_k, n_k) = \frac{m_k}{n_k} \mathbb{E}[S(r,1) \wedge n_k],$$

hence $\mathbb{P}\{v(m_k,n_k) \geqslant 1\} \leqslant 1/2$. Now we get, putting all our ingredients together,

$\mathbb{P}\{A(n_k) \neq \emptyset\}$

$$\geqslant \mathbb{P}\{v(i+1,n_k) - v(i,n_k) = 1 \text{ for some } i \leqslant m_k \text{ such that } v(i,n_k) \leqslant 1\}$$

$$\geqslant m_k\mathbb{P}\{S(r,1) \geqslant n_k\}\mathbb{P}\{v(m_k,n_k) \leqslant 1\}$$

$$\geqslant m_k\mathbb{P}\{S(r,1) \geqslant n_k\}/2 \geqslant \frac{m_k}{2n_k}\varepsilon\mathbb{E}[S(r,1) \wedge n_k] \geqslant \frac{\varepsilon}{6}.$$

This proves (10.13). It remains to observe that, by Brownian scaling, there exists $\delta > 0$ such that, for all $n \in \mathbb{N}$,

$$\mathbb{P}\{\exists t \in [0,1/n]: \limsup_{h \downarrow 0} |B(t+h) - B(t)|/\sqrt{h} \leqslant r\} \geqslant \delta.$$

Hence, by independence,

$$\mathbb{P}\{\exists t \in [0,1]: \limsup_{h\downarrow 0} |B(t+h) - B(t)|/\sqrt{h} \leqslant r\} \geqslant 1 - (1-\delta)^n \longrightarrow 1.$$

This completes the proof of the upper bound, and hence the proof of Theorem 10.30. ∎

10.4 Cone points of planar Brownian motion

We now focus on a planar Brownian motion $\{B(t): t \geqslant 0\}$. Recall from Section 7.2 that around a *typical point* on the path this motion performs an infinite number of windings in both directions. It is easy to see that there are exceptional points to this behaviour: Let

$$x_0 = \min\{x: (x,0) \in B[0,1]\}.$$

Then the Brownian motion does not perform *any* windings around $(x_0, 0)$, as this would necessarily imply that it crosses the half-line $\{(x,0): x < x_0\}$ contradicting the minimality of x_0. More generally, each point $(x_0, y_0) \in \mathbb{R}^2$ with $x_0 = \min\{x: (x, y_0) \in B[0,1]\}$ has this property, if the set is nonempty. Hence, the set of such points has dimension at least one, as the projection onto the y-axis gives a nondegenerate interval. We shall see below that this set has indeed Hausdorff dimension one.

We now look at points where a cone-shaped area with the tip of the cone placed in the point is avoided by the Brownian motion. These points are called cone points.

Definition 10.36.   Let $\{B(t): t \geqslant 0\}$ be a planar Brownian motion. For any angle $\alpha \in (0, 2\pi)$ and direction $\xi \in [0, 2\pi)$, define the closed **cone**

$$W[\alpha, \xi] := \{re^{i(\theta - \xi)}: |\theta| \leqslant \alpha/2, r \geqslant 0\} \subset \mathbb{R}^2.$$

Given a cone $x + W[\alpha, \xi]$ we call its **dual** the reflection of its complement about the tip, i.e. the cone $x + W[2\pi - \alpha, \xi + \pi]$. A point $x = B(t)$, $0 < t < 1$, is an α-**cone point** if there exists $\varepsilon > 0$ and $\xi \in [0, 2\pi)$ such that

$$B(0,1) \cap \mathcal{B}(x, \varepsilon) \subset x + W[\alpha, \xi]. \qquad \diamond$$

Remark 10.37 Clearly, if $x = B(t)$ is a cone point, then there exists a small $\delta > 0$ such that $B(t - \delta, t + \delta) \subset x + W[\alpha, \xi]$. Hence the path $\{B(t): 0 \leqslant t \leqslant 1\}$ performs only a finite number of windings around x. $\qquad \diamond$

We now identify the opening angles α for which there exist α-cone points. In the cases where they exist, we determine the Hausdorff dimension of the set of α-cone points.

Theorem 10.38 (Evans 1985) *Let* $\{B(t): 0 \leqslant t \leqslant 1\}$ *be a planar Brownian motion. Then, almost surely, α-cone points exist for any $\alpha \geqslant \pi$ but not for $\alpha < \pi$. Moreover, if $\alpha \in [\pi, 2\pi)$, then*

$$\dim \left\{ x \in \mathbb{R}^2 : x \text{ is an } \alpha\text{-cone point} \right\} = 2 - \frac{2\pi}{\alpha}.$$

In the proof of Theorem 10.38 we identify $\mathbb{R}^2$ with the complex plane and use complex notation wherever convenient. Suppose that $\{B(t): t \geqslant 0\}$ is a planar Brownian motion defined for all positive times. We first fix an angle $\alpha \in (0, 2\pi)$ and a direction $\xi \in [0, 2\pi)$ and define the notion of an approximate cone point as follows: For any $0 < \delta < \varepsilon$ we let

$$T_\delta(z) := \inf \left\{ s \geqslant 0 : B(s) \in \mathcal{B}(z, \delta) \right\}$$

and

$$S_{\delta,\varepsilon}(z) := \inf \left\{ s \geqslant T_{\delta/2}(z) : B(s) \notin \mathcal{B}(z, \varepsilon) \right\}.$$

We say that $z \in \mathbb{R}^2$ is a (δ, ε)-*approximate cone point* if

$$B(0, T_\delta(z)) \subset z + W[\alpha, \xi], \quad \text{and} \quad B(T_{\delta/2}(z), S_{\delta,\varepsilon}(z)) \subset z + W[\alpha, \xi].$$

Note that we do not require (δ, ε)-approximate cone points to belong to the Brownian path. The relation between cone points and approximate cone points will become clear later, we first collect the necessary information about the probability that a given point is a (δ, ε)-approximate cone point. The strong Markov property allows us to consider the events happening during the intervals $[0, T_\delta(z)]$ and $[T_{\delta/2}(z), 1]$ separately.

Lemma 10.39 *There exist $0 < c < C$ (depending on α) such that, for every $\delta > 0$,*

(a) *for all $z \in \mathbb{R}^2$,*

$$\mathbb{P}\{B(0, T_\delta(z)) \subset z + W[\alpha, \xi]\} \leqslant C \left(\tfrac{\delta}{|z|}\right)^{\frac{\pi}{\alpha}},$$

(b) *for all $z \in \mathbb{R}^2$ with $0 \in z + W[\alpha/2, \xi]$,*

$$\mathbb{P}\{B(0, T_\delta(z)) \subset z + W[\alpha, \xi]\} \geqslant c \left(\tfrac{\delta}{|z|}\right)^{\frac{\pi}{\alpha}}.$$

Proof. We write $z = |z| \, e^{i\theta}$ and apply the skew-product representation, Theorem 7.26, to the Brownian motion $\{z - B(t): t \geqslant 0\}$ and obtain

$$B(t) = z - R(t) \exp(i\,\theta(t)), \quad \text{for all } t \geqslant 0,$$

for $R(t) = \exp(W_1(H(t))$ and $\theta(t) = W_2(H(t))$, where $\{W_1(t): t \geqslant 0\}$ and $\{W_2(t): t \geqslant 0\}$ are independent linear Brownian motions started in $\log |z|$, resp. in θ, and a strictly increasing time-change $\{H(t): t \geqslant 0\}$ which depends only on the first of these motions. This implies that $T_\delta(z) = \inf\{s \geqslant 0 : R(s) \leqslant \delta\}$ and therefore

$$H(T_\delta(z)) = \inf \left\{ u \geqslant 0 : W_1(u) \leqslant \log \delta \right\} =: \tau_{\log \delta}.$$

We infer that

$$\{B(0,T_\delta(z)) \subset z + W[\alpha,\xi]\} = \{|W_2(u) + \pi - \xi| \leqslant \tfrac{\alpha}{2} \text{ for all } u \in [0,\tau_{\log \delta}]\}.$$

The latter event means that a linear Brownian motion started in θ stays inside the interval $[\xi - \pi - \alpha/2, \xi - \pi + \alpha/2]$ up to the independent random time $\tau_{\log \delta}$. For the probability of such events we have found two formulas, (7.14) and (7.15) in Chapter 7. The latter formula gives

$$\mathbb{P}\{|W_2(u) + \pi - \xi| \leqslant \tfrac{\alpha}{2} \text{ for all } u \in [0,\tau_{\log \delta}]\}$$

$$= \sum_{k=0}^{\infty} \tfrac{4}{(2k+1)\,\pi} \sin\left(\tfrac{(2k+1)\pi(\alpha/2+\xi-\pi-\theta)}{\alpha}\right) \mathbb{E}\left[\exp\left(-\tfrac{(2k+1)^2\pi^2}{2\alpha^2}\,\tau_{\log \delta}\right)\right]$$

$$= \sum_{k=0}^{\infty} \tfrac{4}{(2k+1)\,\pi} \sin\left(\tfrac{(2k+1)\pi(\alpha/2+\xi-\pi-\theta)}{\alpha}\right)\left(\tfrac{\delta}{|z|}\right)^{(2k+1)\frac{\pi}{\alpha}},$$

using Exercise 2.18 (a) to evaluate the Laplace transform of the first hitting times of a point by linear Brownian motion. Now note that the upper bound, part (a) of the lemma, is easy if $|z| \leqslant 2\delta$, and otherwise one can bound the exact formula from above by

$$\left(\tfrac{\delta}{|z|}\right)^{\frac{\pi}{\alpha}} \sum_{k=0}^{\infty} \tfrac{4}{(2k+1)\,\pi}\, 2^{-2k\frac{\pi}{\alpha}}.$$

The lower bound, part (b) of the lemma, follows from Brownian scaling if $\delta/|z|$ is bounded from below. Otherwise note that, under our assumption $0 \in z + W[\alpha/2,\xi]$, we have $|\theta + \pi - \xi| \leqslant \tfrac{\alpha}{4}$ and thus the sine term corresponding to $k = 0$ is bounded from below by $\sin(\pi/4) > 0$. Thus we get a lower bound of

$$\left(\tfrac{\delta}{|z|}\right)^{\frac{\pi}{\alpha}}\left[\tfrac{4}{\pi}\sin(\pi/4) - \sum_{k=1}^{\infty} \tfrac{4}{(2k+1)\,\pi}\left(\tfrac{\delta}{|z|}\right)^{2k\frac{\pi}{\alpha}}\right],$$

and the term in the square bracket is bounded from zero, if $\delta/|z|$ is sufficiently small. ∎

An entirely analogous argument also provides the estimates needed for the events imposed *after* the Brownian motion has hit the ball $\mathcal{B}(z,\delta/2)$. Define, for later reference,

$$S_\varepsilon^{(t)}(z) := \inf\{s > t\colon B(s) \notin \mathcal{B}(z,\varepsilon)\}.$$

Lemma 10.40 *There exist constants $C > c > 0$ such that, for every $0 < \delta < \varepsilon$,*

(a) *for all $x, z \in \mathbb{R}^2$ with $|x - z| = \delta/2$,*

$$\mathbb{P}_x\{B(0,S_\varepsilon^{(0)}(z)) \subset z + W[\alpha,\xi]\} \leqslant C\left(\tfrac{\delta}{\varepsilon}\right)^{\frac{\pi}{\alpha}}.$$

(b) *for all $x, z \in \mathbb{R}^2$ with $|x - z| = \delta/2$ and $x - z \in W[\alpha/2,\xi]$,*

$$\mathbb{P}_x\{B(0,S_\varepsilon^{(0)}(z)) \subset z + W[\alpha,\xi]\} \geqslant c\left(\tfrac{\delta}{\varepsilon}\right)^{\frac{\pi}{\alpha}}.$$

We now focus on the *upper bound* in Theorem 10.38. Using the strong Markov property we may combine Lemmas 10.39 (a) and 10.40 (a) to obtain the following lemma.

Lemma 10.41 *There exists a constant $C_0 > 0$ such that, for any $z \in \mathbb{R}^2$,*

$$\mathbb{P}\{z \text{ is a } (\delta, \varepsilon)\text{-approximate cone point }\} \leqslant C_0 |z|^{-\frac{\pi}{\alpha}} \varepsilon^{-\frac{\pi}{\alpha}} \delta^{\frac{2\pi}{\alpha}}.$$

Proof. By the strong Markov property applied at the stopping time $T_{\delta/2}(z)$ we get

$$\mathbb{P}\{z \text{ is a } (\delta, \varepsilon)\text{-approximate cone point }\}$$
$$\leqslant \mathbb{E}\left[1\{B(0, T_\delta(z)) \subset z + W[\alpha, \xi]\} \, \mathbb{P}_{B(T_{\delta/2}(z))}\{B(0, S_\varepsilon^{(0)}(z)) \subset z + W[\alpha, \xi]\}\right]$$
$$\leqslant C^2 \left(\tfrac{\delta}{|z|}\right)^{\frac{\pi}{\alpha}} \left(\tfrac{\delta}{\varepsilon}\right)^{\frac{\pi}{\alpha}},$$

where we have used Lemmas 10.39 (a) and 10.40 (a). The result follows with $C_0 := C^2$. ∎

Let $M(\alpha, \xi, \varepsilon)$ be the set of all points in the plane which are (δ, ε)-approximate cone points for all $\delta > 0$. Obviously $z \in M(\alpha, \xi, \varepsilon)$ if and only if there exists $t > 0$ such that $z = B(t)$ and $B(0, t) \subset z + W[\alpha, \xi]$, and $B(t, S_\varepsilon^{(t)}(z)) \subset z + W[\alpha, \xi]$.

Lemma 10.42 *Almost surely,*

   (a) *if $\alpha \in (0, \pi)$ then $M(\alpha, \xi, \varepsilon) = \emptyset$,*

   (b) *if $\alpha \in [\pi, 2\pi)$ then $\dim M(\alpha, \xi, \varepsilon) \leqslant 2 - \frac{2\pi}{\alpha}$.*

Proof. Take a compact cube Cube of unit side length not containing the origin. It suffices to show that $M(\alpha, \xi, \varepsilon) \cap \text{Cube} = \emptyset$ if $\alpha \in (0, \pi)$ and $\dim M(\alpha, \xi, \varepsilon) \cap \text{Cube} \leqslant 2 - \frac{2\pi}{\alpha}$ if $\alpha \in (\pi, 2\pi)$.

Given a dyadic subcube $D \in \mathfrak{D}_k$ of Cube of side length 2^{-k} let $D^* \supset D$ be a concentric ball around D with radius $(1 + \sqrt{2})2^{-k}$. Define the *focal point* $x = x(D)$ of D to be

- if $\alpha < \pi$ the tip of the cone $x + W[\alpha, \xi]$ whose boundary halflines are tangent to D^*,
- if $\alpha > \pi$ the tip of the cone whose dual has boundary halflines tangent to D^*.

The following properties are easy to check: For every $\varepsilon > 0$ and $\alpha \in [0, 2\pi)$, there exists $k_0 \in \mathbb{N}$ such that for all $k \geqslant k_0$ and $D \in \mathfrak{D}_k$ and $y \in D$, we have $\mathcal{B}(y, \varepsilon) \supset \mathcal{B}(x, \varepsilon/2)$, and $y + W[\alpha, \xi] \subset x + W[\alpha, \xi]$. Moreover there exist constants $C_1 > c_1 > 0$ depending only on α, such that

- $\mathcal{B}(y, C_1 2^{-k}) \subset \mathcal{B}(x, C_1^2 2^{-k})$,
- $\mathcal{B}(y, \tfrac{1}{2} C_1 2^{-k}) \supset \mathcal{B}(x, c_1 C_1 2^{-k})$, and
- $|x - y| < c_1 C_1 2^{-k}$.

Altogether, these properties imply that, for k large enough, if the cube $D \in \mathfrak{D}_k$ contains a $(C_1 2^{-k}, \varepsilon)$-approximate cone point, then its focal point x satisfies

- $B(0, T_{C_1^2 2^{-k}}(x)) \subset x + W[\alpha, \xi]$, and
- $B(T_{c_1 C_1 2^{-k}}(x), S_{c_1 C_1 2^{-k}, \varepsilon/2}(x)) \subset x + W[\alpha, \xi]$.

See Figure 10.1 for an illustration.

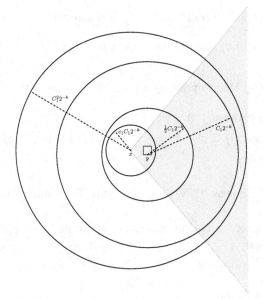

Fig. 10.1. Position of the points in Lemma 10.42.

Hence, by combining Lemma 10.39 (a) and Lemma 10.40 (a) as in Lemma 10.41, we find a constant $C_2 > 0$,

$$\mathbb{P}\{D \text{ contains a } (C_1\,2^{-k}, \varepsilon)\text{-approximate cone point}\} \leqslant C_2\,|x(D)|^{-\frac{\pi}{\alpha}}\,\varepsilon^{-\frac{\pi}{\alpha}}\,2^{-k\frac{2\pi}{\alpha}}.$$

Note that, given Cube and $\varepsilon > 0$ we can find $k_1 \geqslant k_0$ such that $|x(D)|$ is bounded away from zero over all $D \in \mathfrak{D}_k$ and $k \geqslant k_1$. Hence we obtain $C_3 > 0$ such that, for all $k \geqslant k_1$,

$$\mathbb{P}\{D \text{ contains a } (C_1\,2^{-k}, \varepsilon)\text{-approximate cone point}\} \leqslant C_3\,2^{-k\frac{2\pi}{\alpha}}.$$

Then, if $\alpha \in (0, \pi)$,

$$\mathbb{P}\{M(\alpha, \xi, \varepsilon) \neq \emptyset\} \leqslant \sum_{D \in \mathfrak{D}_k} \mathbb{P}\{D \text{ contains a } (C_1\,2^{-k}, \varepsilon)\text{-approximate cone point}\}$$

$$\leqslant C_3\,2^{2k}\,2^{-k\frac{2\pi}{\alpha}} \quad \overset{k \to \infty}{\longrightarrow} \quad 0,$$

proving part (a). Moreover, if $\alpha \in (\pi, 2\pi)$ and $k \geqslant k_1$, we may cover $M(\alpha, \xi, \varepsilon) \cap$ Cube by the collection of cubes $D \in \mathfrak{D}_k$ which contain a $(C_1 2^{-k}, \varepsilon)$-approximate cone point. Then, for any $\gamma > 2 - \frac{2\pi}{\alpha}$ the expected γ-value of this covering is

$$\mathbb{E} \sum_{D \in \mathfrak{D}_k} 2^{-k\gamma + \frac{1}{2}\gamma} \mathbf{1}\{D \text{ contains a } (C_1\,2^{-k}, \varepsilon)\text{-approximate cone point}\}$$

$$\leqslant 2^{\frac{1}{2}\gamma} \sum_{D \in \mathfrak{D}_k} 2^{-k\gamma}\,\mathbb{P}\{D \text{ contains a } (C_1\,2^{-k}, \varepsilon)\text{-approximate cone point}\}$$

$$\leqslant C_3\,2^{k(2 - \frac{2\pi}{\alpha} - \gamma)} \quad \overset{k \to \infty}{\longrightarrow} \quad 0,$$

and this proves that, almost surely, $\dim M(\alpha, \xi, \varepsilon) \leqslant \gamma$. ∎

Proof of the upper bound in Theorem 10.38. Suppose $\delta > 0$ is arbitrary and $z \in \mathbb{R}^2$ is an α-cone point. Then there exist a rational number $q \in [0, 1)$, a rational direction $\xi \in [0, 2\pi)$, and a rational $\varepsilon > 0$, such that $z = B(t)$ for some $t \in (q, 1)$ and

$$B(q, t) \subset z + W[\alpha + \delta, \xi], \quad \text{and} \quad B(t, S_\varepsilon^{(t)}(z)) \subset z + W[\alpha + \delta, \xi].$$

By Lemma 10.42 for every fixed choice of rational parameters this set is empty almost surely if $\alpha + \delta < \pi$. For any $\alpha < \pi$ we can pick $\delta > 0$ with $\alpha + \delta < \pi$ and hence there are no α-cone points almost surely. Similarly, if $\alpha \geqslant \pi$, we use Lemma 10.42 and the countable stability of Hausdorff dimension to obtain an almost sure upper bound of $2 - 2\pi/(\alpha + \delta)$ for the set of α-cone points. The result follows as $\delta > 0$ was arbitrary. ∎

We now establish the framework to prove the lower bound in Theorem 10.38. Again we fix $x_0 \in \mathbb{R}^d$ and a cube Cube $= x_0 + [0, 1)^d$. Recall the definition of the collection $\mathfrak{D}_k$ of dyadic half-open subcubes of side length 2^{-k} and let $\mathfrak{D} = \bigcup_{k=1}^\infty \mathfrak{D}_k$. Suppose that $\{Z(I) \colon I \in \mathfrak{D}\}$ is a collection of random variables each taking values in $\{0, 1\}$. With this collection we associate the random set

$$A := \bigcap_{k=1}^\infty \bigcup_{\substack{I \in \mathfrak{D}_k \\ Z(I)=1}} I.$$

Theorem 10.43 *Suppose that the random variables $\{Z(I) \colon I \in \mathfrak{D}\}$ satisfy the monotonicity condition*

$$I \subset J \text{ and } Z(I) = 1 \quad \Rightarrow \quad Z(J) = 1.$$

Assume that, for some positive constants γ, c_1 and C_1,

   (i) $c_1 |I|^\gamma \leqslant \mathbb{E}Z(I) \leqslant C_1 |I|^\gamma$ *for all $I \in \mathfrak{D}$,*

   (ii) $\mathbb{E}[Z(I)Z(J)] \leqslant C_1 |I|^{2\gamma} \operatorname{dist}(I, J)^{-\gamma}$ *for all $I, J \in \mathfrak{D}_k$, $\operatorname{dist}(I, J) > 0$, $k \geqslant 1$.*

Then, for $\lambda > \gamma$ and $\Lambda \subset$ Cube closed with $\mathcal{H}^\lambda(\Lambda) > 0$, there exists a $p > 0$, such that

$$\mathbb{P}\{\dim(A \cap \Lambda) \geqslant \lambda - \gamma\} \geqslant p.$$

Remark 10.44 Though formally, if the monotonicity condition holds, A is a limsup fractal, the monotonicity establishes a strong dependence of the random variables $\{Z(I) \colon I \in \mathfrak{D}_k\}$ which in general invalidates the second assumption of Theorem 10.28. We therefore need a result which deals specifically with this situation. ◊

We prepare the proof with a little lemma, based on Fubini's theorem.

Lemma 10.45 *Suppose ν is a probability measure on $\mathbb{R}^d$ such that $\nu\mathcal{B}(x, r) \leqslant Cr^\lambda$ for all $x \in \mathbb{R}^d, r > 0$. Then, for all $0 < \beta < \lambda$ there exists $C_2 > 0$ such that,*

$$\int_{\mathcal{B}(x,r)} |x - y|^{-\beta} \nu(dy) \leqslant C_2 \, r^{\lambda - \beta}, \quad \text{for every } x \in \mathbb{R}^d \text{ and } r > 0.$$

This implies, in particular, that

$$\iint |x - y|^{-\beta} \, d\nu(x) \, d\nu(y) < \infty.$$

Proof.   Fubini's theorem gives

$$\int_{\mathcal{B}(x,r)} |x - y|^{-\beta} \, \nu(dy) = \int_0^\infty \nu\{y \in \mathcal{B}(x,r) : |x - y|^{-\beta} > s\} \, ds$$

$$\leqslant \int_{r^{-\beta}}^\infty \nu \mathcal{B}(x, s^{-1/\beta}) \, ds + C \, r^{\lambda - \beta}$$

$$\leqslant C \int_{r^{-\beta}}^\infty s^{-\lambda/\beta} \, ds + C \, r^{\lambda - \beta},$$

which implies the first statement. Moreover,

$$\iint |x - y|^{-\beta} \, d\nu(x) \, d\nu(y) \leqslant \int d\nu(x) \int_{\mathcal{B}(x,1)} |x - y|^{-\beta} \, d\nu(y) + 1 \leqslant C_2 + 1. \quad \blacksquare$$

Proof of Theorem 10.43.   We show that there exists $p > 0$ such that, for every $0 < \beta < \lambda - \gamma$, with probability at least p, there exists a positive measure μ on $\Lambda \cap A$ such that its β-energy $I_\beta(\mu)$ is finite. This implies $\dim(A \cap \Lambda) \geqslant \beta$ by the energy method, see Theorem 4.27.

First, given $\Lambda \subset$ Cube with $\mathcal{H}^\lambda(\Lambda) > 0$, we use Frostman's lemma to find a Borel probability measure ν on Λ and a positive constant C such that $\nu(D) \leqslant C|D|^\lambda$ for all Borel sets $D \subset \mathbb{R}^d$. Writing

$$A_n := \bigcup_{\substack{I \in \mathfrak{D}_n \\ Z(I) = 1}} I,$$

we define μ_n to be the measure supported on Λ given by

$$\mu_n(B) = 2^{n\gamma} \, \nu(B \cap A_n) \qquad \text{for any Borel set } B \subset \mathbb{R}^d.$$

Then, using (i), we get

$$\mathbb{E}\big[\mu_n(A_n)\big] = 2^{n\gamma} \sum_{I \in \mathfrak{D}_n} \nu(I) \, \mathbb{E}Z(I) \geqslant c_1 \, d^{\gamma/2} \sum_{I \in \mathfrak{D}_n} \nu(I) = c_1 \, d^{\gamma/2}.$$

Moreover, using (ii), we obtain

$$\mathbb{E}\big[\mu_n(A_n)^2\big] = 2^{2n\gamma} \sum_{I \in \mathfrak{D}_n} \sum_{J \in \mathfrak{D}_n} \mathbb{E}[Z(I)Z(J)]\nu(I)\nu(J)$$

$$\leqslant C_1 d^\gamma \sum_{I \in \mathfrak{D}_n} \sum_{\substack{J \in \mathfrak{D}_n \\ \mathrm{dist}(I,J) > 0}} \mathrm{dist}(I, J)^{-\gamma} \nu(I)\nu(J)$$

$$+ C 3^d \sqrt{d}^\lambda \, 2^{2n\gamma} \, 2^{-n\lambda} \sum_{I \in \mathfrak{D}_n} \mathbb{E}[Z(I)]\nu(I),$$

since for every cube I there are 3^d cubes J with $\mathrm{dist}(I, J) = 0$. Hence

$$\mathbb{E}\big[\mu_n(A_n)^2\big] \leqslant C_1\big((1 + 2\sqrt{d})d\big)^\gamma \iint |x - y|^{-\gamma} \, d\nu(x) \, d\nu(y) + C \, 3^d \sqrt{d}^\lambda \, d^{\gamma/2},$$

where we use that for $x \in I$, $y \in J$ with $\mathrm{dist}(I, J) > 0$ we have $|x - y| \leqslant (1 + 2\sqrt{d}) \, \mathrm{dist}(I, J)$. Finiteness of the right hand side, denoted C_3, follows from the second statement of Lemma 10.45. We now show that, for $\beta < \lambda - \gamma$ we can find $k(\beta)$ such that $\mathbb{E}I_\beta(\mu_n) \leqslant k(\beta)$. Indeed,

$$
\mathbb{E}I_\beta(\mu_n) = 2^{2n\gamma} \sum_{I, J \in \mathfrak{D}_n} \mathbb{E}[Z(I)Z(J)] \int_I d\nu(x) \int_J d\nu(y) \, |x - y|^{-\beta}
$$

$$
\leqslant C_1 \, d^\gamma \sum_{I \in \mathfrak{D}_n} \sum_{\substack{J \in \mathfrak{D}_n \\ \mathrm{dist}(I,J) > 0}} \mathrm{dist}(I, J)^{-\gamma} \int_I d\nu(x) \int_J d\nu(y) \, |x - y|^{-\beta}
$$

$$
+ C_1 \, d^{\gamma/2} \, 2^{n\gamma} \sum_{I \in \mathfrak{D}_n} \sum_{\substack{J \in \mathfrak{D}_n \\ \mathrm{dist}(I,J) = 0}} \int_I d\nu(x) \int_J d\nu(y) \, |x - y|^{-\beta} \, .
$$

For the first summand, we use that $\mathrm{dist}(I, J)^{-\gamma} \leqslant (3\sqrt{d})^\gamma \, |x - y|^{-\gamma}$ whenever $x \in I$ and $y \in J$, and infer boundedness from the second statement of Lemma 10.45. For the second summand, the first statement of Lemma 10.45 gives a bound of

$$
C_1 C_2 \, d^{\gamma/2} \, 2^{n\gamma} \, (3\sqrt{d}2^{-n})^{\lambda - \beta} \sum_{I \in \mathfrak{D}_n} \nu(I) \leqslant C_1 \, C_2 \, d^{\gamma/2} \, (3\sqrt{d})^{\lambda - \beta} \, .
$$

Hence, $\mathbb{E}I_\beta(\mu_n)$ is bounded uniformly in n, as claimed. We thus find $\ell(\beta) > 0$ such that

$$
\mathbb{P}\{I_\beta(\mu_n) \geqslant \ell(\beta)\} \leqslant \frac{k(\beta)}{\ell(\beta)} \leqslant \frac{c_1^2}{8C_3} \, .
$$

Now, by the Paley–Zygmund inequality, see Lemma 3.23,

$$
\mathbb{P}\{\mu_n(A_n) > \tfrac{c_1}{2}\} \geqslant \mathbb{P}\{\mu_n(A_n) > \tfrac{1}{2}\mathbb{E}[\mu_n(A_n)]\} \geqslant \frac{1}{4} \frac{\mathbb{E}[\mu_n(A_n)]^2}{\mathbb{E}[\mu_n(A_n)^2]} \geqslant \frac{c_1^2}{4C_3} \, .
$$

Hence we obtain that

$$
\mathbb{P}\{\mu_n(A_n) > \tfrac{c_1}{2}, \, I_\beta(\mu_n) < \ell(\beta)\} \geqslant p := \frac{c_1^2}{8C_3} \, .
$$

Using Fatou's lemma we infer that

$$
\mathbb{P}\{\mu_n(A_n) > \tfrac{c_1}{2}, \, I_\beta(\mu_n) < \ell(\beta) \text{ infinitely often}\}
$$
$$
\geqslant \liminf_{n \to \infty} \mathbb{P}\{\mu_n(A_n) > \tfrac{c_1}{2}, \, I_\beta(\mu_n) < \ell(\beta)\} \geqslant p.
$$

On this event we can pick a subsequence along which μ_n converges to some measure μ. Then μ is supported by A and $\mu(\mathrm{Cube}) \geqslant \liminf \mu_n(\mathrm{Cube}) = \liminf \mu_n(A_n) \geqslant c_1/2$. Finally, for each $\varepsilon > 0$, where the limit is taken along the chosen subsequence,

$$
\iint_{|x-y| > \varepsilon} |x - y|^{-\beta} d\mu(x) \, d\mu(y) = \lim \iint_{|x-y| > \varepsilon} |x - y|^{-\beta} d\mu_n(x) \, d\mu_n(y)
$$
$$
\leqslant \lim I_\beta(\mu_n) \leqslant \ell(\beta),
$$

and for $\varepsilon \downarrow 0$ we get $I_\beta(\mu) \leqslant \ell(\beta)$. ∎

We now use Theorem 10.43 to give a *lower bound* for the dimension of the set of cone points. Fix $\alpha \in (\pi, 2\pi)$ and a unit cube

$$\mathsf{Cube} = x_0 + [0,1]^2 \subset W[\alpha/2, 0],$$

and recall the definition of the classes $\mathfrak{C}$ and $\mathfrak{C}_k$ of compact dyadic subcubes. Choose a large radius $R > 2$ such that $\mathsf{Cube} \subset \mathcal{B}(0, R/2)$ and define

$$r_k := R - \sum_{j=1}^{k} 2^{-j} > R/2.$$

Given a cube $I \in \mathfrak{C}_k$ we denote by z its centre and let $Z(I) = 1$ if z is a $(2^{-k}, r_k)$-approximate cone point with direction $\xi = \pi$, i.e. if

$$B(0, T_{2^{-k}}(z)) \subset z + W[\alpha, \pi], \quad \text{and} \quad B(T_{2^{-k-1}}(z), S_{2^{-k}, r_k}(z)) \subset z + W[\alpha, \pi],$$

and otherwise let $Z(I) = 0$. By our choice of the sequence (r_k) we have

$$I \subset J \text{ and } Z(I) = 1 \quad \Rightarrow \quad Z(J) = 1.$$

Lemma 10.46 *There are constants $0 < c_1 < C_1 < \infty$ such that, for any cube $I \in \mathfrak{C}$, we have*

$$c_1 \, |I|^{\frac{2\pi}{\alpha}} \leqslant \mathbb{P}\{Z(I) = 1\} \leqslant C_1 \, |I|^{\frac{2\pi}{\alpha}}.$$

Proof. The upper bound is immediate from Lemma 10.41. For the lower bound we use that, for any $z \in \mathsf{Cube}$ and $\delta > 0$,

$$\inf_{|x-z|=\delta} \mathbb{P}_x\big\{B(T_{\delta/2}(z)) \in z + W[\alpha/2, \pi]\big\}$$
$$= \inf_{|x|=1} \mathbb{P}_x\big\{B(T_{1/2}(0)) \in W[\alpha/2, \pi]\big\} =: c_0 > 0,$$

and hence, if z is the centre of $I \in \mathfrak{C}_k$ and $\delta = 2^{-k}$, using Lemmas 10.39 (b) and 10.40 (b),

$$\mathbb{P}\{Z(I) = 1\}$$
$$\geqslant \mathbb{E}\Big[1\{B(0, T_\delta(z)) \subset z + W[\alpha, \pi]\} \, \mathbb{E}_{B(T_\delta(z))}\big[1\{B(T_{\delta/2}(z)) \in z + W[\alpha/2, \pi]\}$$
$$\times \mathbb{P}_{B(T_{\delta/2}(z))}\{B(0, S_{r_k}^{(0)}(z)) \subset z + W[\alpha, \pi]\}\big]\Big]$$
$$\geqslant c_0 \, c^2 \, \delta^{\frac{2\pi}{\alpha}} \, (R|z|)^{-\frac{\pi}{\alpha}},$$

which gives the desired statement, as $|z|$ is bounded away from infinity. ∎

Lemma 10.47 *There is a constant $0 < C_1 < \infty$ such that, for any cubes $I, J \in \mathfrak{C}_k$, $k \geqslant 1$, we have*

$$\mathbb{E}\big[Z(I)Z(J)\big] \leqslant C_1 \, |I|^{\frac{4\pi}{\alpha}} \, \mathrm{dist}(I, J)^{-\frac{2\pi}{\alpha}}.$$

Proof.   Let z_I, z_J be the centres of I, resp. J, and abbreviate $\eta := |z_I - z_J|$ and $\delta := 2^{-k}$. Then, for $\eta > 2\delta$, using the strong Markov property and Lemmas 10.39 (a) and 10.40 (a),

$$\mathbb{E}\big[Z(I)\, Z(J)\, 1\{T_{\delta/2}(z_I) < T_{\delta/2}(z_J)\}v\big]$$
$$\leqslant \mathbb{E}\Big[1\{B(0, T_\delta(z_I)) \subset z_I + W[\alpha, \pi]\}$$
$$\times \mathbb{E}_{B(T_{\delta/2}(z_I))}\Big[1\{B(0, S^{(0)}_{\eta/2}(z_I)) \subset z_I + W[\alpha, \pi]\}$$
$$\times \mathbb{E}_{B(T_{\eta/2}(z_J))}\Big[1\{B(0, T_\delta(z_J)) \subset z_J + W[\alpha, \pi]\}$$
$$\times \mathbb{P}_{B(T_{\delta/2}(z_J))}\{B(0, S^{(0)}_{r_k}(z_J)) \subset z_J + W[\alpha, \pi]\}\Big]\Big]\Big]$$
$$\leqslant C^4 \Big(\tfrac{\delta}{|z_I|}\Big)^{\frac{\pi}{\alpha}} \Big(\tfrac{\delta}{\eta}\Big)^{\frac{2\pi}{\alpha}} \Big(\tfrac{2\delta}{R}\Big)^{\frac{\pi}{\alpha}} \leqslant C_2\, |I|^{\frac{4\pi}{\alpha}}\, \mathrm{dist}(I, J)^{-\frac{2\pi}{\alpha}},$$

where we recall that Cube does not contain the origin and let $C_2 > 0$ be an appropriate constant. Suppose now that $\eta \leqslant 2\delta$. Then, by a simpler argument,

$$\mathbb{E}\big[Z(I)\, Z(J)\, 1\{T_{\delta/2}(z_I) < T_{\delta/2}(z_J)\}\big]$$
$$\leqslant \mathbb{E}\big[1\{B(0, T_\delta(z_I)) \subset z_I + W[\alpha, \pi]\}$$
$$\times \mathbb{P}_{B(T_{\delta/2}(z_J))}\{B(0, S^{(0)}_{r_k}(z_J)) \subset z_J + W[\alpha, \pi]\}\big]$$
$$\leqslant C^2 \Big(\tfrac{\delta}{|z_I|}\Big)^{\frac{\pi}{\alpha}} \Big(\tfrac{2\delta}{R}\Big)^{\frac{\pi}{\alpha}} \leqslant C_3\, |I|^{\frac{4\pi}{\alpha}}\, \mathrm{dist}(I, J)^{-\frac{2\pi}{\alpha}}.$$

Exchanging the rôle of I and J gives the corresponding estimate

$$\mathbb{E}[Z(I)Z(J)1\{T_{\delta/2}(z_I) > T_{\delta/2}(z_J)\}] \leqslant C_3\, |I|^{\frac{4\pi}{\alpha}}\, \mathrm{dist}(I, J)^{-\frac{2\pi}{\alpha}},$$

and the proof is completed by adding the two estimates.   ■

Proof of the lower bound in Theorem 10.38.   The set A which we obtain from our choice of $\{Z(I): I \in \mathfrak{C}\}$ is contained in the set

$$\tilde{A} := \big\{B(t) : t > 0 \text{ and } B(0, S^{(t)}_{R/2}(B(t))) \subset B(t) + W[\alpha, \pi]\big\}.$$

Therefore, by Theorem 10.43, we have $\dim \tilde{A} \geqslant 2 - 2\pi/\alpha$ with positive probability. Given any $0 < \delta < 1/2$ and $r > 0$, we define a sequence $\tau_1^{(\delta)} \leqslant \tau_2^{(\delta)} \leqslant \ldots$ of stopping times by $\tau_1^{(\delta)} = 0$ and, for $k \geqslant 1$,

$$\tau_k^{(\delta)} := S^{(\tau_{k-1}^{(\delta)})}_{\delta r}(B(\tau_{k-1}^{(\delta)})).$$

Denoting $\eta = R/(2r)$ and

$$A_k^{(\delta)} := \big\{B(t) : \tau_{k-1}^{(\delta)} \leqslant t \leqslant \tau_k^{(\delta)} \text{ and } B(\tau_{k-1}^{(\delta)}, S^{(t)}_{\eta r}(B(t))) \subset B(t) + W[\alpha, \pi]\big\}$$

we have that

$$\tilde{A} \subset \bigcup_{k=1}^{\infty} A_k^{(\delta)}.$$

Now fix $\beta < 2 - 2\pi/\alpha$. The events $\{\dim A_k^{(\delta)} \geqslant \beta\}$ all have the same probability, which cannot be zero as this would contradict the lower bound on the dimension of $\tilde{A}$. In particular, there exists $p_R^{(\delta)} > 0$ such that

$$\mathbb{P}\big\{ \dim \big\{ B(t) \colon 0 \leqslant t \leqslant S_{\delta r}^{(0)}(0) \text{ and } B(0, S_{\eta r}^{(t)}(B(t))) \subset B(t) + W[\alpha, \pi] \big\} \geqslant \beta \big\} \geqslant p_R^{(\delta)}.$$

By scaling we get that $p_R^{(\delta)}$ does not depend on r. Hence, by Blumenthal's zero-one law, we have that $p_R^{(\delta)} = 1$ for all $\delta > 0, R > 0$. Letting $\beta \uparrow 2 - 2\pi/\alpha$ we get, almost surely,

$$\dim \big\{ B(t) \colon 0 \leqslant t \leqslant S_\delta^{(0)}(0), \ B(0, S_\eta^{(t)}(B(t))) \subset B(t) + W[\alpha, \pi] \big\} \geqslant 2 - \tfrac{2\pi}{\alpha}$$

for every $\delta > 0, \eta > 0$.

Given $\varepsilon > 0$, we may choose $\delta, \eta > 0$ such that, with probability $> 1 - \varepsilon$, we have $S_\delta^{(0)}(0) < 1$ and $S_\eta^{(t)}(B(t)) > 1$ for all $0 \leqslant t \leqslant 1$. This implies that

$$\dim \big\{ B(t) \colon 0 \leqslant t \leqslant 1, \ B(0,1) \subset B(t) + W[\alpha, \pi] \big\} \geqslant 2 - \tfrac{2\pi}{\alpha}$$

with probability $> 1 - \varepsilon$, and the result follows as $\varepsilon > 0$ was arbitrary. ∎

A surprising consequence of the non-existence of cone points for angles smaller then π is that the convex hull of the planar Brownian curve is a fairly smooth set.

Theorem 10.48 (Adelman) *Almost surely, the convex hull of $\{B(s) \colon 0 \leqslant s \leqslant 1\}$ has a differentiable boundary.*

Proof.   A compact, convex subset $H \subset \mathbb{R}^2$ is said to have a *corner* at $x \in \partial H$ if there exists a cone with vertex x and opening angle $\alpha > \pi$ which avoids $H \setminus \{x\}$. If H does not have corners, the supporting hyperplanes are unique at each point $x \in \partial H$ and thus ∂H is a differentiable boundary. So all we have to show is that the convex hull H of $\{B(s) \colon 0 \leqslant s \leqslant 1\}$ has no corners. Clearly, by Spitzer's theorem, $B(0)$ and $B(1)$ are not corners almost surely. Suppose any other point $x \in \partial H$ is a corner, then obviously it is contained in the path, and therefore it is a $(2\pi - \alpha)$-cone point for some $\alpha > \pi$. By Theorem 10.38, almost surely, such points do not exist and this is a contradiction. ∎

Exercises

Exercise 10.1. Show that, for every metric space E,

$$\dim_P E = \inf\{s \colon \mathcal{P}^s(E) < \infty\} = \sup\{s \colon \mathcal{P}^s(E) > 0\} = \sup\{s \colon \mathcal{P}^s(E) = \infty\}.$$

Exercise 10.2. ⑤ Show that, for every metric space E, we have

$$\dim_P E \geqslant \dim E.$$

Exercise 10.3. Let $\{m_k \colon k \geqslant 1\}$ be a rapidly increasing sequence of positive integers such that

$$\lim_{k \to \infty} \frac{m_k}{m_{k+1}} = 0.$$

Define two subsets of $[0, 1]$ by

$$E = \left\{ \sum_{i=1}^{\infty} \frac{x_i}{2^i} \, : \, x_i \in \{0, 1\} \text{ and } x_i = 0 \text{ if } m_k + 1 \leqslant i \leqslant m_{k+1} \text{ for some even } k \right\}$$

and

$$F = \left\{ \sum_{i=1}^{\infty} \frac{x_i}{2^i} \, : \, x_i \in \{0, 1\} \text{ and } x_i = 0 \text{ if } m_k + 1 \leqslant i \leqslant m_{k+1} \text{ for some odd } k \right\}.$$

Show that

(a) $\dim E = \underline{\dim}_M E = 0$ and $\dim F = \underline{\dim}_M F = 0$,

(b) $\dim_P E = \overline{\dim}_M E = 1$ and $\dim_P F = \overline{\dim}_M F = 1$,

(c) $\dim(E \times F) \geqslant 1$.

Exercise 10.4. Show that, almost surely,

(a) $\dim_P \text{Range}[0, 1] = 2$, for Brownian motion in $d \geqslant 2$,

(b) $\dim_P \text{Graph}[0, 1] = \frac{3}{2}$, for Brownian motion in $d = 1$,

(c) $\dim_P \text{Zeros} = \frac{1}{2}$, for Brownian motion in $d = 1$.

Exercise 10.5. Show that, for every $a \in [0, 1]$, we have almost surely,

$$\dim_P \left\{ t \in [0, 1] \, : \, \limsup_{h \downarrow 0} \frac{|B(t+h) - B(t)|}{\sqrt{2h \log(1/h)}} \geqslant a \right\} = 1.$$

Hint. This can be done directly, but it can also be derived from more general ideas, as formulated for example in Exercise 10.9.

Exercise 10.6. Show that

$$\liminf_{h \downarrow 0} \sup_{t \in E} \frac{|B(t+h) - B(t)|}{\sqrt{2h \log(1/h)}} = \sqrt{\underline{\dim}_M (E)}.$$

Exercise 10.7. ⑤ Use Theorem 10.43 to prove once more that the zero set of linear Brownian motion has Hausdorff dimension $\frac{1}{2}$ almost surely.

Exercise 10.8. Show that, if

$$\limsup_{n \uparrow \infty} \frac{\log p_n}{n \log 2} \leqslant -\gamma, \qquad \text{for some } \gamma > 0,$$

then, for any compact $E \subset [0, 1]$ with $\dim_P(E) < \gamma$, we have

$$\mathbb{P}\{A \cap E \neq \emptyset\} = 0.$$

Note that no independence assumption is needed for this statement.

Exercise 10.9. §

(a) Suppose A is a discrete limsup fractal associated to random variables $\{Z(I)\colon I \in \mathfrak{C}_k, k \geqslant 1\}$ satisfying the conditions of Theorem 10.28. Then, if $\dim_P(E) > \gamma$, we have almost surely, $\dim_P(A \cap E) = \dim_P(E)$.

(b) Show that, if $\dim_P(E) > a^2$, then almost surely

$$\dim_P(F(a) \cap E) = \dim_P(E),$$

where $F(a)$ is the set of a-fast times.

Exercise 10.10. § Give a proof of Lemma 10.40 (a) based on Theorem 7.25.

Exercise 10.11. Suppose $K \subset \mathbb{R}^2$ is a compact set and $x \in \mathbb{R}^2 \setminus K$ a point outside the set. Imagine K as a solid body, and x as the position of an observer. This observer can only see a part of the body, which can be formally described as

$$K(x) = \{y \in K \colon [x, y] \cap K = \{y\}\},$$

where $[x, y]$ denotes the compact line segment connecting x and y. It is natural to ask for the Hausdorff dimension of the visible part of a set K. Assuming that $\dim K \geqslant 1$, an unresolved conjecture in geometric measure theory claims that, for Lebesgue-almost every $x \notin K$, the Hausdorff dimension of $K(x)$ is one.
Show that this conjecture holds for the path of planar Brownian motion, $K = B[0, 1]$, in other words, almost surely, for Lebesgue-almost every $x \in \mathbb{R}^2$, the Hausdorff dimension of the visible part $B[0, 1](x)$ is one.

Exercise 10.12. Let $\{B(t)\colon t \geqslant 0\}$ be a planar Brownian motion and $\alpha \in [\pi, 2\pi)$. Show that, almost surely, no double points are α-cone points.

Exercise 10.13. § Let $\{B(t)\colon t \geqslant 0\}$ be a planar Brownian motion and $\alpha \in (0, \pi]$. A point $x = B(t), 0 < t < 1$, is a *one-sided* α-cone point if there exists $\xi \in [0, 2\pi)$ such that

$$B(0, t) \subset x + W[\alpha, \xi].$$

(a) Show that for $\alpha \leqslant \frac{\pi}{2}$, almost surely, there are no one-sided α-cone points.

(b) Show that for $\alpha \in (\frac{\pi}{2}, \pi]$, almost surely, the set of one-sided α-cone points has Hausdorff dimension $2 - \frac{\pi}{\alpha}$.

Notes and comments

The paper [OT74] by Orey and Taylor is a seminal work in the study of dimension spectra for exceptional points of Brownian motion. It contains a proof of Theorem 10.3 using the mass distribution principle and direct construction of the Frostman measure. This approach can be extended to other limsup fractals, but this method requires quite strong independence assumptions which make this method difficult in many more general situations. In

[OT74] the question how often on a Brownian path the law of the iterated logarithm fails is also answered in the sense that, for $\theta > 1$, almost surely, the set

$$\left\{ t > 0 \colon \limsup_{h \downarrow 0} \frac{B(t+h) - B(t)}{\sqrt{2h \log \log(1/h)}} \geqslant \theta \right\}$$

has zero or infinite Hausdorff measure for the gauge function $\phi(r) = r \log(1/r)^\gamma$ depending whether $\gamma < \theta^2 - 1$ or $\gamma > \theta^2 - 1$. Finer results do not seem to be known at the moment.

Our proof of Theorem 10.3 is based on estimates of energy integrals. This method was used by Hu and Taylor [HT97] and Shieh and Taylor [ST99], and our exposition follows Dembo et al. [DPRZ00a] closely. In the latter paper an interesting class of exceptional times for the Brownian motion is treated, the *thick times* of Brownian motion in dimension $d \geqslant 3$. For any time $t \in (0, 1)$ we let $U(t, \varepsilon) = \mathcal{L}\{s \in (0, 1) \colon |B(s) - B(t)| \leqslant \varepsilon\}$ the set of times where the Brownian is up to ε near to its position at time t. It is shown that, for all $0 \leqslant a \leqslant \frac{16}{\pi^2}$, almost surely,

$$\dim \left\{ t \in [0, 1] \colon \limsup_{\varepsilon \downarrow 0} \frac{U(t, \varepsilon)}{\varepsilon^2 \log(1/\varepsilon)} \geqslant a \right\} = 1 - a \frac{\pi^2}{16} \,.$$

This paper should be very accessible to anyone who followed the arguments of Section 10.1. The method of Dembo et al. [DPRZ00a] can be extended to limsup fractals with somewhat weaker independence properties and also extends to the study of dimension spectra with strict equality.

A third way to prove Theorem 10.3 is the method of stochastic codimension explored in Section 10.10.2. An early reference for this method is Taylor [Ta66] who suggested to use the range of stable processes as test sets, and made use of the potential theory of stable processes to obtain lower bounds for Hausdorff dimension. This class of test sets is not big enough for all problems: the Hausdorff dimension of a stable process is bounded from above by its index, hence cannot exceed 2, and therefore these test sets can only test dimensions in the range $[d - 2, d]$. A possible remedy is to pass to multiparameter processes, see the recent book of Khoshnevisan [Kh02] for a survey. Later, initiated by seminal papers of Hawkes [Ha81] and R. Lyons [Ly90], it was discovered that percolation limit sets are a very suitable class of test functions, see Khoshnevisan et al. [KPX00]. Our exposition closely follows the latter reference.

The result about the thick times of Brownian motion stated above can be interpreted as a multifractal analysis of the occupation measure. Such an analysis can also be performed in two dimensions, but the result and techniques are entirely different, see Dembo et al. [DPRZ01]. Times at which $U(t, \varepsilon)$ is exceptionally small for infinitely many scales $\varepsilon > 0$, the *thin times*, are investigated in Dembo et al. [DPRZ00b]. Other measures associated with Brownian paths that have been studied from a multifractal point of view are the local times, see Hu and Taylor [HT97] and Shieh and Taylor [ST99], and the intersection local times of several Brownian paths, see [KM02] and [KM05].

Kaufman [Ka75] showed that every compact set $E \subset [0,1]$ with $\dim(E) > a^2$ almost surely contains an a-fast time, but the more precise result involving the packing dimension is due to Khoshnevisan et al. [KPX00]. The concept of packing dimension was introduced surprisingly late by Tricot in [Tr82] and in [TT85] it was investigated together with the packing measure and applied to the Brownian path by Taylor and Tricot. Lemma 10.18(i) is from [Tr82], Lemma 10.18(ii) for trees can be found in [BP94], see Proposition 4.2(b), the general version given is in Falconer and Howroyd [FH96] and in Mattila and Mauldin [MM97].

Several people contributed to the investigation of slow points, for example Dvoretzky [Dv63], Kahane [Ka76], Davis [Da83], Greenwood and Perkins [GP83] and Perkins [Pe83]. There are a number of variants, for example one can allow $h < 0$ in (10.11) or omit the modulus signs. The Hausdorff dimension of a-slow points is discussed in [Pe83], this class of exceptional sets is not tractable with the limsup-method: note that an exceptional behaviour is required at all small scales. The crucial ingredient, the finiteness criterion for moments of the stopping times $T(r, a)$ is due to Shepp [Sh67].

Cone points were discussed by Evans in [Ev85], an alternative discussion can be found in Lawler's survey paper [La99]. Our argument essentially follows the latter paper. The correlation condition in Theorem 10.43 appears in the strongly related context of quasi-Bernoulli percolation on trees, see Lyons [Ly92]. An alternative notion of *global* cone points requires that the entire path of the Brownian motion $\{B(t): t \geqslant 0\}$ stays inside the cone with tip in the cone point. The same dimension formula holds for this concept. The upper bound follows of course from our consideration of local cone points, and our proof gives the lower bound with positive probability. The difficult part is to show that the lower bound holds with probability one. A solution to this problem is contained in Burdzy and San Martín [BSM89], and this technique has also been successfully used in the study of the outer boundary, or frontier, of Brownian motion, see Lawler [La96b] and Bishop et al. [BJPP97].

A discussion of the smoothness of the boundary of the convex hull can be found in Cranston, Hsu and March [CHM89], but our Theorem 10.48 is older. The result was stated by Lévy [Le48] and was probably first proved by Adelman in 1982, though this does not seem to be published.

It is conjectured in geometric measure theory that for any set of Hausdorff dimension $\dim K \geqslant 1$, for Lebesgue-almost every $x \notin K$, the Hausdorff dimension of the visible part $K(x)$ is one. For upper bounds on the dimension and the state of the art on this conjecture, see O'Neil [ON07]. It is natural to compare this to Makarov's theorem on the support of harmonic measure: if the rays of light were following Brownian paths rather than straight lines, the conjecture would hold by Makarov's theorem, see [Ma85].

11

Stochastic Loewner evolution and planar Brownian motion

by Oded Schramm and Wendelin Werner

This chapter presents an overview over some aspects of the recent development of the stochastic Loewner evolution from the point of view of Brownian motion. Stochastic Loewner evolution allows to address a variety of important questions on the geometry of planar Brownian motion that cannot be answered otherwise. This chapter is intended as an invitation to further study, and therefore does not intend to provide the same level of detail as the chapters in the main body of the book.

11.1 Some subsets of planar Brownian paths

11.1.1 The questions

The conformal invariance of planar Brownian motion and the powerful tools of one-dimensional complex analysis open the way to a deep understanding of some aspects of the geometry of the Brownian curve. For the sake of concreteness, let us begin by presenting a couple of motivating questions.

Question 11.1 (Intersection exponent) *Let $\{Z^1(t): t \geqslant 0\}$ and $\{Z^2(t): t \geqslant 0\}$ be two independent planar Brownian motions started at distinct points. What is the asymptotic decay rate as $t \to \infty$ of $\mathbb{P}\{Z^1[0,t] \cap Z^2[0,t] = \emptyset\}$?*

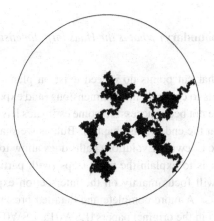

Fig. 11.1. Non-intersecting Brownian motions

Let us insist on the fact that we are looking at the probability that (simultaneously) for all $t_1, t_2 \leqslant t$ we have $Z^1(t_1) \neq Z^2(t_2)$, and that this is quite different from questions about the process $\{Z^1(t) - Z^2(t) : t \geqslant 0\}$.

Clearly, recurrence of planar Brownian motion implies that this probability goes to 0 as $t \to \infty$. In fact, one can easily deduce from a subadditivity argument that the answer to the question is $t^{-\xi + o(1)}$ for some positive constant ξ, and the problem is really about the identification of ξ. The exponent ξ is often called the **intersection exponent** of planar Brownian motion. As we will later discuss, knowing its value is instrumental in studying the set of **cut points** in the Brownian path $Z[0, 1]$, i.e. the set of points $x \in \mathbb{R}^2$ such that $Z[0, 1] \setminus \{x\}$ is disconnected:

Question 11.2 (Cut points) *Are there cut points on a planar Brownian path? If so, what is the Hausdorff dimension of the set of cut points?*

Another interesting subset of the planar Brownian path $Z[0, t]$ is its **outer boundary**, defined as the boundary of the unbounded connected component of $\mathbb{R}^2 \setminus Z[0, t]$; see Figure 11.2.

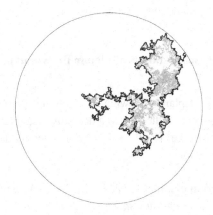

Fig. 11.2. A Brownian path and its outer boundary.

Question 11.3 (Outer boundary) *What is the Hausdorff dimension of the outer boundary of the Brownian path?*

Chris Burdzy showed that cut points do indeed exist on planar Brownian paths [Bu89, Bu95], but direct attempts to compute these dimensions (and exponents) through the study of Brownian motion have not been successful (some estimates have however been obtained, see the historical notes at the end of this chapter). But, as we shall now try to explain, the study of SLE (Stochastic Loewner Evolution) paths does allow to determine these values. The goal of this chapter is to explain the main steps (with partial proofs only) that lead to these answers. We will focus mainly on the intersection exponent ξ and the related question about cut points. A more complete and detailed presentation of the results and their proofs can be found in the original papers [LSW01a, LSW01b, LSW03], and are also discussed in [We04, La05, La09].

11.1.2 Reformulation in terms of Brownian hulls

We recall from Theorem 7.20 that conformal invariance is nicely expressed for Brownian paths that are stopped at their exit times from given domains. For example, consider a planar Brownian motion $\{Z(t): t \geqslant 0\}$ started from the origin, and stopped at its first exit time $T = T_D$ of a given bounded simply connected domain D that contains the origin. Consider the conformal mapping $\Phi = \Phi_D$ from D onto the unit disc $\mathbb{U} = \mathcal{B}(0, 1)$ such that $\Phi(0) = 0$ and $\Phi'(0)$ is a positive real (this map exists and is unique by Riemann's mapping theorem, see for instance [Ah78] for basic background in complex analysis). Then, the law of $\{\Phi(Z(t)): 0 \leqslant t \leqslant T\}$ is that of a time-changed Brownian motion started at the origin and stopped at its first exit time from the unit disc. In other words, if we forget about the time parametrisation and worry only about the 'trace' of the paths (i.e. the set of points that the Brownian motion has visited), we get an identity in law between $\{\Phi(Z(t)): 0 \leqslant t \leqslant T\}$ and $\{Z(t): 0 \leqslant t \leqslant \sigma\}$, where $\sigma = T_U$ is the exit time from the unit disc.

As we shall see, it is useful to consider the random set K defined as follows: We look at the trace $Z[0, \sigma]$ and we fill in its 'holes'. In other words, we say that K is the complement of the unbounded connected component of the complement of $Z[0, \sigma]$ in the plane. We call K the **hull** of $Z[0, \sigma]$.

Let us now explain why the previous two questions can be reformulated in terms of the law of the hull K.

- Let us first focus on the question about the outer boundary of the Brownian motion. We can expect that if we can determine the Hausdorff dimension of the outer boundary of $Z[0, \sigma]$ and prove that it is almost surely equal to some value d, then the Hausdorff dimension of the outer boundary of $Z[0, 1]$ will also be equal to d almost surely. But the boundary of the hull K is exactly the outer boundary of $Z[0, \sigma]$. Hence, Question 11.3 reduces to: 'What is the Hausdorff dimension of the boundary of K?'

- A similar and slightly more involved argument applies to the set of cut points. The goal is therefore first to determine the Hausdorff dimensions of the set of cut points of K, i.e. of the set of points p in K such that $K \setminus \{p\}$ is disconnected.

- Consider now two independent Brownian paths $\{Z^1(t): t \geqslant 0\}$ and $\{Z^2(t): t \geqslant 0\}$ started at the two points $Z^1(0) = 0$ and $Z^2(0) = 1$ (here 0 and 1 are viewed as elements of the complex plane). Define for each $R > 1$, the respective exit times T_R^1 and T_R^2 of Z^1 and Z^2 from the disc $\mathcal{B}(0, R)$. It is easy to see that for each $\epsilon > 0$, the probability that T_R^1 does not belong to $[R^{2-\epsilon}, R^{2+\epsilon}]$ does decay rapidly as $R \to \infty$. More precisely, we get that for some positive β and all sufficiently large R,

$$\mathbb{P}\{T_R^1 \notin [R^{2-\varepsilon}, R^{2+\varepsilon}] \text{ or } T_R^2 \notin [R^{2-\varepsilon}, R^{2+\varepsilon}]\} \leqslant e^{-\beta R^\varepsilon}.$$

Indeed, because of scaling, the left hand side is equal to $\mathbb{P}\{\sigma \notin [R^{-\varepsilon}, R^\varepsilon]\}$, and on the one hand,

$$\mathbb{P}\{\sigma < R^{-\varepsilon}\} \leqslant \mathbb{P}\{\max_{s \leqslant R^{-\varepsilon}} |\mathfrak{Re}(Z(s))| \geqslant 1/\sqrt{2}\} + \mathbb{P}\{\max_{s \leqslant R^{-\varepsilon}} |\mathfrak{Im}(Z(s))| \geqslant 1/\sqrt{2}\}$$

$$\leqslant 8\,\mathbb{P}\{\mathfrak{Re}(Z(R^{-\varepsilon})) \geqslant 1/\sqrt{2}\},$$

while, on the other hand,

$$\mathbb{P}\{\sigma > N\} \leqslant \mathbb{P}\{|Z(j+1) - Z(j)| \leqslant 2 \forall j = 0, 1, \ldots, N-1\} \leqslant \mathbb{P}\{|Z(1)| \leqslant 2\}^N.$$

Hence, up to a small error, estimating the probability that

$$Z^1[0, t] \cap Z^2[0, t] = \emptyset$$

boils down to estimating the probability that

$$Z^1[0, T_R^1] \cap Z^2[0, T_R^2] = \emptyset$$

for $R = \sqrt{t}$ when $R \to \infty$. More precisely, if we can show that the second one behaves like $R^{-2\xi + o(1)}$ as $R \to \infty$, then it will follow that the first one is equivalent to $t^{-\xi + o(1)}$ as $t \to \infty$.

Let us now define the hulls K_R^1 and K_R^2 of $Z^1[0, T_R^1]$ and $Z^2[0, T_R^2]$. We can note that $Z^1[0, T_R^1] \cap Z^2[0, T_R^2] = \emptyset$ if and only if $K_R^1 \cap K_R^2 = \emptyset$. Furthermore, conformal invariance of planar Brownian motion shows readily that the law of K_R^2 is just the image of the law of K_R^1 under the conformal transformation from $\mathcal{B}(0, R)$ onto itself that sends the starting point of Z^2 onto the origin. Hence, we have also reformulated Question 11.1 in terms of the law of K.

11.1.3 An alternative characterisation of Brownian hulls

We now explain why conformal invariance makes it possible to give a simple description of the law of K that does seemingly not involve Brownian motion.

Let $\mathcal{U}$ denote the set of simply connected open subsets U' of the unit disc $\mathbb{U}$ such that $0 \in U'$. For any two such U' and U'' in $\mathcal{U}$, we define $U' \wedge U''$ to be the connected component of $U' \cap U''$ that contains the origin. Clearly, $U' \wedge U'' \in \mathcal{U}$.

For any $U' \in \mathcal{U}$, we denote by $m(U')$ the harmonic measure of $\partial \mathbb{U} \cap \partial U'$ in U' at the origin. This is just the probability that a Brownian path started at the origin does exit U' via a point on the unit circle. Because U' is simply connected, this happens exactly if the hull of this Brownian motion stays in U' up to the time σ. Hence, if we define K as before and set $K^* = K \setminus \{Z_\sigma\}$, we get immediately that

$$\mathbb{P}\{K^* \subset U'\} = m(U'). \tag{11.1}$$

Now suppose that $\mathcal{K}$ is the hull of some other continuous random path $(\eta_t, t \leqslant \tau)$ stopped at its first hitting of the unit disc; this second random path is not necessarily a Brownian motion, but we suppose that it also satisfies

$$\mathbb{P}\{\mathcal{K}^* \subset U'\} = m(U'), \tag{11.2}$$

for all $U' \in \mathcal{U}$, where $\mathcal{K}^* = \mathcal{K} \setminus \{\eta_\tau\}$.

We can note that the set of events of the type $\{K \colon K \subset U'\}$ (for such hulls) when U' spans $\mathcal{U}$ is stable under finite intersections, and in fact generate the σ-algebra on which we can define the measure on hulls. Hence, it follows from standard measure-theoretical arguments that the laws of K and of $\mathcal{K}$ are identical. In other words, K (or rather its law) is the only 'random hull' that has the property that for any $U' \in \mathcal{U}$, (11.1) holds.

Note that (11.2) can be also expressed in terms of the path η directly. Suppose that for any U', the exit point of U' by η is distributed according to harmonic measure from 0 in U', then (11.2) follows.

Let us sum up our analysis so far: We have first reformulated our questions in terms of the random hull K, and we have now given a simple characterisation of the law of K. The plan will now be the following:

- Construct a random curve η that exits every domain U' in $\mathcal{U}$ according to harmonic measure. We have just argued that this implies that the laws of K and $\mathcal{K}$ are identical.
- Using the construction of this other random curve η, compute the exponents and dimensions that we are looking for.

It turns out that such a random path η indeed exists, and that it is one of the stochastic Loewner evolutions, more precisely SLE(6).

11.2 Paths of stochastic Loewner evolution

11.2.1 Heuristic description

Suppose that one wishes to describe a 'continuously growing' curve $\{\eta_t : t \geqslant 0\}$ that is always 'growing towards infinity'. More precisely, let us first suppose that $\{\eta_t : t \geqslant 0\}$ is a simple random curve starting at the origin.

At each time $t > 0$, we define the conformal map ψ_t from $\mathbb{R}^2 \setminus \eta[0, t]$ into the complement of the unit disc, such that $\psi_t(\infty) = \infty$ and $\psi_t(\eta_t) = 1$. By Riemann's mapping theorem, this map ψ_t is unique.

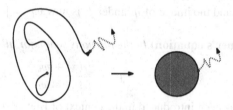

Fig. 11.3. The conformal map ψ_t.

The crucial assumption that we will make is that for each $t > 0$, the random path

$$\{\psi_t(\eta_{t+s}) : s \geqslant 0\}$$

(or rather its trace) is independent of $\eta[0, t]$, and that its law is independent of t. In other words, the curve is growing towards infinity from η_t in the set $\mathbb{R}^2 \setminus \eta[0, t]$ in a 'conformally invariant way'.

This suggests that it is possible to define the curve $\{\eta_t : t \geqslant 0\}$ progressively, by iterating independent identically distributed pieces. Suppose for instance that we have already defined $\eta[0, 1]$ and that we wish to define what happens after time 1. The curve $\psi_1(\eta[1, u])$ is independent of $\eta[0, 1]$. It is a piece of curve in $\mathbb{R}^2 \setminus \mathbb{U}$ that starts at $\psi_1(\eta_1) = 1$. The conformal map $\psi_u \circ \psi_1^{-1}$ maps $\mathbb{R}^2 \setminus (\mathbb{U} \cup \psi_1(\eta[1, u]))$ onto $\mathbb{R}^2 \setminus \mathbb{U}$. It therefore characterises the set $\eta[1, u]$ and is characterised by it. Hence, it is independent from $\eta[0, 1]$.

It follows that for $u_1 < u_2 < \ldots < u_n$, the conformal maps

$$\psi_{u_n} \circ \psi_{u_{n-1}}^{-1}, \ldots, \psi_{u_2} \circ \psi_{u_1}^{-1}$$

are independent. Furthermore, if we choose the time-parametrisation correctly, then they will be identically distributed. This leads to the idea that ψ_t are obtained via iterations of i.i.d. random conformal maps.

11.2.2 Loewner's equation

Suppose now that $\{\eta_t : t \geqslant 0\}$ is a given continuous simple curve (with no double points) in the plane starting at the origin such that $\lim_{t \to \infty} \eta_t = \infty$. We define as in the previous paragraph the conformal map ψ_t from $\mathbb{R}^2 \setminus \eta[0, t]$ onto $\mathbb{R}^2 \setminus \mathbb{U}$ such that $\psi_t(\infty) = \infty$ and $\psi_t(\eta_t) = 1$. Recall that $z \mapsto 1/\psi_t(1/z)$ extends analytically to the origin (one can for instance first define this analytic map via Riemann's mapping theorem and then define ψ_t), so that ψ_t can be expanded as a power series in the neighbourhood of infinity. In particular,

$$\psi_t(z) \sim a(t)z$$

when $z \to \infty$ for some $a(t)$. It is not difficult to see that $t \mapsto a(t)$ is a continuous function, and that $t \mapsto |a(t)|$ is decreasing (because the set $\mathbb{R}^2 \setminus \eta[0, t]$ is decreasing). Furthermore, simple estimates imply that $\lim_{t \to 0} |a(t)| = \infty$ and $\lim_{t \to \infty} |a(t)| = 0$.

It is therefore possible (and natural) to reparametrise the curve $\{\eta_t : t \geqslant 0\}$ in such a way that the parameter t now lives in $\mathbb{R}$, that $\lim_{t \to -\infty} \eta_t = 0$ and that $|a(t)| = \exp(-t)$. We then define the conformal map f_t from $\mathbb{R}^2 \setminus \eta[-\infty, t]$ onto $\mathbb{R}^2 \setminus \mathbb{U}$, but this time, we normalise it in such a way that $f_t(z) \sim e^{-t}z$ as $z \to \infty$. In other words, f_t is just obtained from ψ_t by a rotation, and the image of η_t under f_t is now $w_t := |a(t)|/a(t)$.

Theorem 11.4 (Loewner's equation) *In the previous setup, for all $t \geqslant 0$, one has*

$$\frac{\partial}{\partial t} f_t(z) = -f_t(z) \frac{f_t(z) + w_t}{f_t(z) - w_t}. \tag{11.3}$$

Loewner's equation has been introduced in the context of Bieberbach's conjecture for harmonic functions, see for instance [Du83] for a derivation of this equation.

Let us give a brief indication of where this ordinary differential equation comes from. Recall first that the Poisson representation theorem shows that the only harmonic function in the unit disc such that $G(0) = 1$ and $G(z) \to 0$ on $\partial \mathbb{U} \setminus \{1\}$ is the function $z \mapsto \mathfrak{Re}((1 + z)/(1 - z))$.

A first step is to prove directly (using harmonic measure estimates) that the map $t \mapsto w_t$ is continuous. Then, one notes that (for instance because of scaling) it suffices to consider the case where $t = 1$. When $s > 1$, the function $f_s \circ f_1^{-1}$ is analytic from $\mathbb{R}^2 \setminus (\mathbb{U} \cup \eta[1, s])$ onto $\mathbb{R}^2 \setminus \mathbb{U}$ (where we view these sets as subsets of the Riemann sphere). If we define

$$h_s(z) = -\log\big(f_s \circ f_1^{-1}(z)/z\big),$$

we get a bounded analytic function on $\mathbb{R}^2 \setminus (\mathbb{U} \cup \eta[1, s])$. The boundary values of $\mathfrak{Re}(h_s)$ are zero on $\partial \mathbb{U}$ and $\log|z|$ on $f_1(\eta(1, s])$, and moreover, $h_s(\infty) = s - 1$. Hence it follows (for instance from the maximum principle) that $\mathfrak{Re}(h_s)$ is nonnegative on $\mathbb{R}^2 \setminus (\mathbb{U} \setminus \eta[1, s])$.

Consider the limit $\lim_{s \downarrow 1} h_s/(s-1)$. Existence of this limit can be justified as follows: First, standard compactness properties of analytic functions imply that subsequential limits exist. Let $h\colon \mathbb{R}^2 \setminus \mathbb{U} \to \mathbb{R}$ denote one such subsequential limit. Clearly, $h(\infty) = 1$ and $\mathfrak{Re}(h) \geqslant 0$. It is then not too hard to verify that $\mathfrak{Re}(h)$ is continuous up to the boundary except near $w_1 = f_1(\eta_1)$ and that $\mathfrak{Re}(h) = 0$ on $\partial \mathbb{U} \setminus \{w_1\}$. Hence, the Poisson representation theorem (applied to $z \mapsto h(w_1/z)$) implies that

$$\mathfrak{Re}\big(h(z)\big) = -\mathfrak{Re}\left(\frac{z+w_1}{z-w_1}\right),$$

and since $\mathfrak{Im}(h(\infty)) = 0$, we conclude that

$$h(z) = -\frac{z+w_1}{z-w_1}.$$

As this limit does not depend on the choice of subsequence, it follows that as $s \downarrow 1$,

$$\frac{f_s \circ f_1^{-1}(z)}{z} - 1 \sim \log\left(\frac{f_s \circ f_1^{-1}(z)}{z}\right) \sim (s-1) \times \frac{w_1+z}{w_1-z},$$

which implies that

$$\partial_s^+\big|_{s=1} f_s(z) = f_1(z) \times \frac{w_1 + f_1(z)}{w_1 - f_1(z)},$$

where ∂_s^+ denotes the one sided derivative from the right.

The reader can now probably already guess how to define SLEs: Just choose w_t to be a Brownian motion on the unit circle. The conformal maps f_t are then defined via (11.3), and the SLE curve can then be deduced from it.

11.2.3 The loop-erased random walk

Even if it is not really necessary in order to define stochastic Loewner evolutions and to study consequences of their study to Brownian paths, we believe that it is useful at this point to explain some background and motivation using discrete models. In the next subsections, we will therefore describe two particular lattice models and their relation to SLE paths. In those settings it can be more useful to consider random curves that grow 'towards the inside' of domains. This is of course almost identical to the previous case (just use the $z \mapsto 1/z$ transformation to transform outside into inside i.e. look for instance at the conformal map $g_t(z) = 1/f_t(1/z)$ instead of f_t).

One such discrete example is the loop-erased random walk. In fact, this model is the one for which the SLE model was first introduced, see [Sc00]. If G is a recurrent connected graph containing a vertex v and a nonempty set of vertices A, then the loop-erased random walk from v to A is the random path obtained from the simple random walk started at v and stopped when hitting A by erasing the loops as they are created.

Let us now give a more precise definition. First, define the simple random walk S on the graph G, which we assume to have more than one vertex. Let $S(0) = v$ and for each positive integer n, let the conditional distribution of $S(n)$ given $(S(0), S(1), \ldots, S(n-1))$ be uniform among the neighbours of $S(n-1)$. Let $T := \inf\{n \in \mathbb{N}\colon S(n) \in A\}$. Since G is recurrent, we know that T is almost surely finite.

We now define the **loop-erasure** $(\beta(0), \beta(1), \ldots, \beta(\tau))$ of $S[0, T]$ by induction: We set $\beta(0) = v$ and then for each $k > 1$, if $\beta(k-1) = S(T)$, we set $\tau = k - 1$ and finish the procedure, whereas if $\beta(k-1) \neq S(T)$, we set

$$m = \max\{j < T \ : \ S(j) = \beta(k-1)\} \text{ and } \beta(k) = S(m+1).$$

An important property of the loop-erased random walk is given by the following lemma.

Lemma 11.5 *Let G be a recurrent connected graph, v a vertex in G, and A a nonempty set of vertices in G. The conditional law of $(\beta(0), \ldots, \beta(\tau - j))$ given $\beta(\tau) = x_0, \ldots, \beta(\tau - j) = x_j$ is that of the loop-erasure of a random walk S' started at v, stopped at it first hitting time T' of $A' = A \cup \{x_0, \ldots, x_k\}$ and conditioned to first hit this set at $S'(T') = x_k$.*

This lemma can be interpreted as some sort of Markov property of the time-reversal of β.

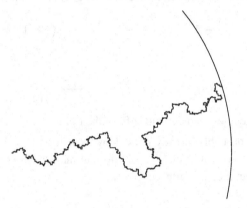

Fig. 11.4. A loop-erased random walk

Consider a simply connected domain D in the plane $\mathbb{R}^2 = \mathbb{C}$, with $D \neq \mathbb{C}$. Suppose that $0 \in D$. Take $\delta > 0$ small, and consider the square lattice $\delta \mathbb{Z}^2$ of mesh δ. We are interested in the loop-erasure β of the random walk on $\delta \mathbb{Z}^2$ started at 0 and stopped when it first uses an edge intersecting ∂D. More specifically, we are interested in the scaling limit of β, which is the limit of the law of β as $\delta \downarrow 0$. (For the sake of brevity, we will not specify the precise topology in which the limit is taken. This discussion is meant as a motivation, and hence we allow ourselves not to be completely rigorous.) Let μ_D denote the limit law. Figure 11.4 shows a sample of the loop-erasure of simple random walk on $\delta \mathbb{Z}^2$ started at 0 and stopped on exiting the unit disk $\mathbb{U}$.

In order to use Lemma 11.5, it turns out to be better to parametrise time 'backwards', i.e. to define $\gamma(0)$ to be the 'end-point' on ∂D, and $\gamma(\infty) = 0$ (we will discuss the precise time-parametrisation later). Suppose that $\gamma \colon [0, \infty] \to D$ is a sample from μ_D. Since Brownian motion is conformally invariant up to a time change and it is the limit of simple random walk, it is somewhat reasonable to expect that γ is also conformally invariant, i.e., that if $G \colon D \to \mathbb{U}$ is a conformal homeomorphism from D to the unit disk, then $\mu_D = \mu_{\mathbb{U}} \circ G$ (in fact, this result has now been proved using SLE, see [LSW04].) Moreover, Lemma 11.5 suggests that if $t < \infty$ is fixed, then the conditional law of $\gamma[t, \infty]$ given $\gamma[0, t]$ is $\mu_{D \setminus \gamma[0, t]}$

where the path is conditioned to exit this domain through $\gamma(t)$. Now, the latter domain $D \setminus \gamma[0, t]$ is probably geometrically rather complicated, because γ is a fractal curve. But we may simplify this domain using a conformal map.

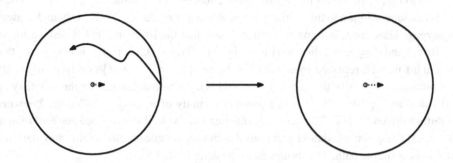

Fig. 11.5. The conformal map g_t

Let us now consider the special case where $D = \mathbb{U}$. Note that $\gamma(0)$ is then distributed uniformly on the unit circle. Let $g_t \colon \mathbb{U} \setminus \gamma[0, t] \to \mathbb{U}$ denote the conformal homeomorphism normalised such that $g_t(0) = 0$ and $g'_t(0)$ is real and positive. It can be shown that $g'_t(0)$ is continuous and increasing in t and that $\lim_{t \to \infty} g'_t(0) = \infty$. Consequently, we may, and will, choose the time parameter t so that $g'_t(0) = e^t$. This is sometimes called the parametrisation by capacity.

Loewner's theorem allows to reconstruct γ from the function $W \colon t \mapsto g_t(\gamma(t))$. In the present setting, γ defined on $[0, \infty]$ is a simple path in $\mathbb{U} \cup \{\gamma(0)\}$ satisfying $\gamma(0) \in \partial \mathbb{U}$ and $\gamma(\infty) = 0$, and which is parametrised by capacity in D. (The assumptions that γ is a simple path and that ∂D is a simple closed path may be relaxed, but it is best at this point to keep the setting simple.)

Applying Theorem 11.4 to the functions $z \mapsto 1/g_t(1/z)$, we get that the conformal homeomorphisms g_t satisfy the differential equation

$$\frac{\partial}{\partial t} g_t(z) = -g_t(z) \frac{g_t(z) + W_t}{g_t(z) - W_t} \tag{11.4}$$

at every pair of points (z, t) such that $t \geq 0$ and $z \in \mathbb{U} \setminus \gamma[0, t]$. If $z \in \mathbb{U}$ is fixed, then Loewner's equation (11.4) is an ordinary differential equation for $g_t(z)$ with respect to the variable t (as long as $z \notin \gamma[0, t]$). Loewner's equation can also be considered an ordinary differential equation for g_t in the space of conformal maps with image in $\mathbb{U}$, but variable domain.

Just as before, the knowledge of the function $t \mapsto W_t$ allows to reconstruct the curve γ. The Markovian-like condition of loop-erased random walk leads to the idea that the process $t \to W_t$ has stationary and independent increments. Recall also that it is continuous and that the law of Brownian motion is symmetric (i.e. this implies that W has no bias). All this suggests that $\{W_t : t \geq 0\}$ must be a Brownian motion on the unit circle.

11.2.4 Definition of SLE

Now suppose that instead of starting with a curve γ, we start with a one dimensional continuous path $W \colon [0, \infty) \to \partial \mathbb{U}$. If $z \in \overline{\mathbb{U}} \setminus \{W_0\}$ is fixed, then we may consider the solution $g_t(z)$ of the ordinary differential equation (11.4) started at $g_0(z) = z$. There exists a unique solution to this initial value problem as long as $g_t(z) - W_t$ is bounded away from zero. Thus, there is some $\tau_z \in (0, \infty]$ such that the solution $g_t(z)$ is defined for all $t \in (0, \tau_z)$ and if $\tau_z < \infty$, then $\liminf_{t \uparrow \tau_z} |g_t(z) - W_t| = 0$. (In fact, it is easy to see that the lim inf may be replaced by a lim.) Set $K_t := \{z \in \overline{\mathbb{U}} \colon \tau_z \leqslant t\}$ (we take $\tau_{W_0} = 0$). It is immediate to verify that $g_t \colon \mathbb{U} \setminus K_t \to \mathbb{U}$ is a conformal homeomorphism satisfying $g_t(0) = 0$ and $g_t'(0) = e^t$. The one parameter family of maps g_t is called the **Loewner evolution** driven by W_t. The set K_t is often called the **hull** of the Loewner evolution at time t. At this point we should point out that the set K_t constructed in this way does not have to be a simple path. This brings us to the definition of SLE:

Definition 11.6. Fix some $\kappa \geqslant 0$, and set $W_t = \exp(iB(\kappa t))$, where $\{B(t) \colon t \geqslant 0\}$ is Brownian motion. Then the Loewner evolution driven by W_t is called **radial stochastic Loewner evolution** with parameter κ in $\mathbb{U}$, or just **radial SLE(κ)**, from W_0 to 0.   ◇

To define radial SLE in another simply-connected domain $D \subsetneqq \mathbb{C}$, we may start with a conformal homeomorphism $G \colon D \to \mathbb{U}$, and solve (11.4) with $g_0 = G$. Set $K_t^D := \{z \in D \colon g_t(z) \text{ is undefined}\}$. Of course, the resulting process $K_{\cdot}^D$ will depend on G. The point $G^{-1}(0)$ is referred to as the **target** of the SLE.

If G_1 and G_2 are two conformal homeomorphisms from D to $\mathbb{U}$ such that $G_1^{-1}(0) = G_2^{-1}(0)$, then $G_2 = \lambda G_1$ for some $\lambda \in \partial \mathbb{U}$. Since the law of W_t is invariant under rotations, it follows that the law of the evolution $g_{\cdot}$ starting at G_2 is obtained from the law of the evolution starting at G_1 by appropriately rotating the maps g_t by λ. Consequently, the law of $K_{\cdot}^D$ is the same for G_1 as for G_2. This is also the same as the law of $G_1^{-1}(K_{\cdot} \cap \mathbb{U})$, where K_t are the hulls of radial SLE in $\mathbb{U}$.

Our argument based on the assumptions of conformal invariance and the analogue of Lemma 11.5 for the loop-erased walk scaling limit γ shows that for some choice of the constant κ, the law of the process $K_{\cdot}^D$ is the same as the law of $\gamma(0, \cdot)$ (where the starting point of the SLE is started uniformly on the unit circle). It turns out that the correct κ for loop-erased random walk is 2. This is explained in [Sc00, LSW04].

11.2.5 Critical percolation and SLE(6)

It will turn out that SLE(6) is a useful SLE in order to study planar Brownian motion. To better understand why this is the case, we first turn to a model of percolation in the plane. Let D be some simply connected domain in the plane whose boundary is a simple closed curve. Fix two points $a, b \in \partial D$, and let A denote the counterclockwise arc from a to b along D (not including a and b). Fix some small $\delta > 0$, and let $\mathcal{H}_\delta$ denote the planar hexagonal grid of hexagons with edge length δ as in Figure 11.6. If H is a connected component of the intersection of D with a hexagon in $\mathcal{H}_\delta$, we colour H white if its boundary meets A, and colour H black if its boundary meets ∂D but does not meet A. If H is a hexagon of $\mathcal{H}_\delta$ that lies entirely inside D, we colour H white or black with probability $1/2$, independently. Let $\mathcal{W}$ denote the closure of the union of the white coloured tiles in D and let $\mathcal{B}$ denote the closure of the union of the black coloured tiles. We assume that δ is sufficiently small so that $\mathcal{B} \cap \partial D \neq \emptyset$. It is then easy to see that there is a unique path γ, which is the connected component of $\partial \mathcal{B} \cap \partial \mathcal{W}$ that meets ∂D. (See Figure 11.6.)

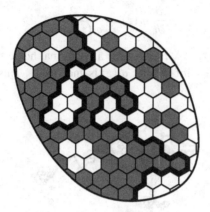

Fig. 11.6. The percolation interface

Smirnov [Sm01, Sm07] has shown that the limit as $\delta \to 0$ of the law of this interface γ exists, and is conformally invariant, in the following sense. If $D' \subset \mathbb{C}$ is a simply connected domain whose boundary is a simple closed curve and $G : D \to D'$ is a conformal homeomorphism, then the image of the limit law in D is the limit law in D', provided that in D' we take the points $a' := G(a)$ and $b' := G(b)$ as the two special boundary points. (It is known that G extends to a homeomorphism from ∂D to $\partial D'$.) We have chosen to discuss domains whose boundary is a simple closed curve for the sake of simplicity, but this is by no means necessary.

Next, we consider the analogue of Lemma 11.5 in this setting. If we condition on the first k steps of the discrete curve γ from its (deterministic) endpoint near a, the conditioning involves only the colours of those tiles which meet this initial segment β. Moreover, on the right hand side of β we find white tiles while on the left hand side we find black tiles. Consequently, conditioned on β, the law of $\gamma \setminus \beta$ is just the law of the interface in the domain $D \setminus \beta$, where the special points are chosen as the terminal point of β and b. (If we are to be entirely precise, we should replace the domain $D \setminus \beta$ with $D \setminus \hat{\beta}$, where $\hat{\beta}$ is an appropriate small neighbourhood of $\beta \setminus \{$a small piece of its last segment$\}$, so that the resulting domain is a simple closed path that does not intersect a hexagon whose colour has not been determined.) This is indeed analogous to Lemma 11.5.

Since we have conformal invariance for the scaling limit of the percolation interface and the analogue of Lemma 11.5, we would expect the scaling limit of the interface to be given by an SLE curve. However, the setting is different, since the percolation interface connects two boundary points of the domain (which are fixed), while the loop-erased random walk connects a fixed interior point with a random boundary point. Indeed, the percolation interface scaling is described not by a radial SLE, but by a different version of SLE called **chordal** SLE.

In the chordal setting, the base domain is normally chosen to be the upper half plane $\mathbb{H}$. We let $W_t := B(\kappa t)$, where $\{B(t) : t \geqslant 0\}$ is a standard one dimensional Brownian motion, and let $g_t(z)$ denote the solution of the differential equation

$$\frac{\partial}{\partial t} g_t(z) = \frac{2}{g_t(z) - W_t} \tag{11.5}$$

with $g_0(z) = z$. Then $g_t : \mathbb{H} \setminus K_t \to \mathbb{H}$ is a conformal homeomorphism, where $K_t := \{z \in \overline{\mathbb{H}} : \tau_z \leqslant t\}$ and $\tau_z := \inf\{t > 0 : g_t(z) \text{ is undefined}\}$. This defines the chordal SLE from 0 to infinity in $\mathbb{H}$.

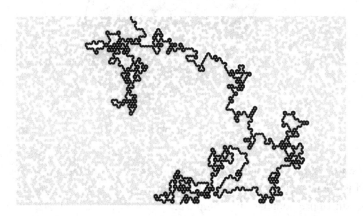

Fig. 11.7. Beginning of the percolation interface in the upper half-plane

Chordal SLEs are often more natural than radial SLEs in the context of models from statistical physics, and both variants are very closely related to each other. In particular, see [LSW01b], if one looks at the beginning of a radial SLE in the unit disc, and the image of the beginning of a chordal SLE under the conformal map from the unit disc on the upper half-plane that sends the points 0 and i onto 1 and 0, for the same parameter κ, then the two laws are absolutely continuous with respect to each other (the fractal dimensions of the curves are therefore the same). In view of applications to Brownian motion (and more precisely to those questions that we raised at the beginning of this appendix), we can however mostly restrict ourselves to the study of radial SLE.

It turns out that the value $\kappa = 6$ is the one that corresponds to the scaling limit of percolation interfaces [Sm01] (as conjectured in [Sc00]). At this point, it is worth stressing the following subtle point. In our discussion, we have described the construction of SLE as if it would anyway define a simple curve. This is indeed the fact when $\kappa \leqslant 4$, see [RS05]. But, in the case of the scaling limit of percolation, one expects the scaling limit of the discrete interfaces to have double points. Indeed, on whatever scale, the discrete interface will 'bounce' on its remote past, and this will produce (in the limit when the mesh of the lattice goes to zero) double points. Hence, one has to change the construction of the SLE as follows (we describe it in the radial case – the chordal case is treated in a similar way):

- Start with the Brownian motion $\{W_t : t \geqslant 0\}$ on the unit circle.
- For each z in the unit disc, solve the ordinary differential equation (11.4) up to the (random and possibly infinite) time τ_z.
- At each time t, denote $K_t = \{z \in \overline{\mathbb{U}} : \tau_z \leqslant t\}$ and $U_t = \mathbb{U} \setminus K_t$. Then, g_t is the normalised conformal map from U_t onto $\mathbb{U}$.
- Call the increasing family $\{K_t : t \geqslant 0\}$ the **SLE Loewner chain**.

Then, with some substantial work when $\kappa > 4$, it is in fact possible [RS05] to prove that there exists almost surely a continuous curve $\{\gamma_t : t \geqslant 0\}$ such that at each $t \geqslant 0$, the domain U_t is the connected component containing the origin of $\mathbb{U} \setminus \gamma[0, t]$ and that this curve γ is determined by the SLE Loewner chain. We call this curve γ **the SLE curve**.

In order to discuss the consequences for Brownian motion, it is in fact not necessary to know that SLE chains are 'generated' by curves γ. One can just work with the chain instead of the path, but it is helpful to have this in mind in order to guide our intuition about what goes on. In the case $\kappa = 6$, the convergence of critical percolation interfaces to SLE(6), see [Sm07], provides a rather direct alternative proof of the fact that SLE(6) chains are generated by paths (see also [We07]).

11.3 Special properties of SLE(6)

It is possible to prove directly via stochastic calculus methods [LSW01b, We04, La05] that the law of the beginning of radial SLE(6) and chordal SLE(6) curves are the same. Here is a precise statement:

Proposition 11.7 *Consider a chordal SLE(6) process γ^1 from 1 to -1 in the unit disc $\mathbb{U}$, and a radial SLE(6) process γ^2 from 1 to 0 in $\mathbb{U}$. Define*

$$T^l = \inf\{t > 0 : |\gamma^l[0, t] - 1| > 1/2\}$$

for $l = 1, 2$. Then, the two paths $\gamma^1[0, T^1]$ and $\gamma^2[0, T^2]$ defined modulo time-reparametrisation have the same law.

We omit the proof here. The main idea is basically to express the radial Loewner evolution as a chordal Loewner chain, and to compute how the time-parametrisations and driving functions are transformed. It turns out that a seemingly miraculous cancellation occurs when κ is equal to 6, that leads to this result. In fact, it is also possible to derive this relation between radial SLE(6) and chordal SLE(6) using the relation with critical percolation (see e.g. [We07]); this provides a transparent justification of this 'miraculous result'.

It may be useful at this point to have a picture of the radial exploration process for percolation in the discrete setting. We start with a fine-mesh δ approximation of the unit disc. Our goal is to define a path from the boundary point 1 to the origin. We are going to define this path dynamically. We start with the same rule as the exploration process from 1 in the chordal case, except that we do not fix a priori the colours of the sites on the $\partial \mathbb{U}$. Note that as long as the discrete exploration path does not disconnect the origin from infinity, there is some arc I of points on $\partial \mathbb{U}$ that are connected to the origin without intersecting the exploration path. We then use the same boundary conditions to define the exploration process as if we would do the exploration process from 1 to one of the points in I. Note that I is non-increasing in time, and the rule that we just described indeed determines the exploration path up to the first time at which it disconnects the origin from $\partial \mathbb{U}$. In this case, note that the connected component of the complement of the path that contains the origin is simply connected, and that it has a boundary point at distance δ of the tip of the exploration process. We now force the exploration process to move to this point. Then, the exploration process is at a boundary point of the connected component that contains the origin. Now, we start again, as if the colours of the boundary of this domain would not have been known, and we start exploring interfaces in this domain using the same algorithm (replacing 1 by the end-point of the exploration).

Theorem 11.8 *When the mesh of the lattice goes to zero, then the law of the radial discrete exploration process converges to that of radial SLE(6).*

We will really not use this result here, so we will not discuss its detailed proof. We refer to [We07] for a self-contained proof in the spirit of Smirnov's paper [Sm07], see also [CN07].

We can also use a similar construction to define the continuous analogue of our 'discrete' curve that is growing towards infinity from a given point. The idea is to use exactly the same definition, except that this time the initial domain is the complement of the disc of radius r, and the target point is infinity. Then, when the mesh of the lattice goes to 0 and $r \to 0$, this exploration process converges to a random curve η started at the origin that possesses the following two properties:

- For any simply connected domain U' that contains the origin, the exit point of U' by η is distributed according to harmonic measure from the origin in U' (this follows either from the locality properties of SLE(6), or alternatively, from the conformal invariance properties of percolation).
- For any t, the conditional law of $\eta[t, \infty)$ given η up to time t, is that of radial SLE(6) from η_t to infinity in the unbounded connected component of $\mathbb{R}^2 \setminus \eta[0, t]$.

We can therefore conclude (either by using the relation to critical percolation or via direct derivations of the special properties of SLE(6)) that the dynamics of an 'outwards growing' radial SLE(6) provide a way to construct a path η that satisfies our 'harmonic measure condition'. It looks indeed as if computations for SLE(6) will provide useful information for Brownian hulls.

11.4 Exponents of stochastic Loewner evolution

11.4.1 A radial computation

We now briefly browse through the computations that lead to the determination of the exponents that we are looking for. This section will certainly seem quick to the first-time reader. The goal is not to give a complete proof, but rather to give a flavour of the type of stochastic calculus arguments that are used in this derivation.

Define, for $z = \exp(ix)$ on the unit circle, the event $\mathcal{H}(x,t)$ that one radial SLE(6) (in the usual parametrisation) started from 1 did not disconnect the point z from the origin in $\mathbb{U}$ before time t. For reasons that we will explain in a moment, we will focus on the moments of the derivative of g_t at $\exp(ix)$ on the event $\mathcal{H}(x,t)$. Note already that on a heuristic level, $|g_t'(e^{ix})|$ measures how 'far' e^{ix} is from the origin in $\mathbb{U} \setminus \gamma[0,t]$.

More precisely, we define

$$f(x,t) := \mathbb{E}\Big[|g_t'(\exp(ix))|\, 1_{\mathcal{H}(x,t)}\Big].$$

The main result of this section is the following estimate:

Proposition 11.9 *There is a constant $c > 0$ such that for all $t \geqslant 1$, for all $x \in (0, 2\pi)$,*

$$e^{-5t/4}\big(\sin(x/2)\big)^{1/3} \leqslant f(x,t) \leqslant c\, e^{-5t/4}\big(\sin(x/2)\big)^{1/3}$$

Proof. Let $W_t = \exp(i\sqrt{6}B(t))$ be the driving process of the radial SLE(6), with $B(0) = 0$. For all $x \in (0, 2\pi)$, we define Y_t^x the continuous function (with respect to t) such that

$$g_t(e^{ix}) = W_t \exp(iY_t^x)$$

and $Y_0^x = x$. The function Y_t^x is defined as long as $\mathcal{H}(x,t)$ holds. Since g_t satisfies Loewner's differential equation, we get immediately that

$$d(Y_t^x - B(6t)) = \cot(Y_t^x/2)\, dt. \tag{11.6}$$

Let

$$\tau^x := \inf\{t \geqslant 0 : Y_t^x \in \{0, 2\pi\}\}$$

denote the time at which $\exp(ix)$ is absorbed by K_t, so that

$$\mathbb{P}\big(\mathcal{H}(x,t)\big) = \mathbb{P}\{\tau^x > t\}.$$

We therefore want to estimate the probability, weighted by some power of $|g_t'(\exp(e^{ix}))|$ that the diffusion Y^x (started from x) has not hit $\{0, 2\pi\}$ before time t as $t \to \infty$. This turns out to be a rather standard problem that can be treated via the general theory of diffusion processes: Define, for all $t < \tau^x$,

$$\Phi_t^x := |g_t'(\exp(ix))| \,.$$

On $t \geqslant \tau^x$ set $\Phi_t^x := 0$. Note that on $t < \tau^x$ we have $\Phi_t^x = \partial_x Y_t^x$ and

$$Y_t^x = B(6t) + \int_0^t \cot(Y_s^x/2)\, ds.$$

Hence, we have that, for $t < \tau^x$,

$$\partial_t \log \Phi_t^x = -\frac{1}{2\sin^2(Y_t^x/2)}, \tag{11.7}$$

so that, for $t < \tau^x$,

$$\Phi_t^x = \exp\left(-\frac{1}{2}\int_0^t \frac{ds}{\sin^2(Y_s^x/2)}\right). \tag{11.8}$$

Hence,

$$f(x,t) = \mathbb{E}\left[1_{\mathcal{H}(x,t)} \exp\left(-\frac{1}{2}\int_0^t \frac{ds}{\sin^2(Y_s^x/2)}\right)\right].$$

Hence, the weighting by Φ_t^x can be interpreted as a (space-dependent) killing rate for the process Y, and $f(x,t)$ is just the probability that a given Markov process (the process Y with the given killing rate and additional killing when it exits $(0, 2\pi)$) survives up to time t when it starts at x. In order to estimate such probabilities, one has to look for the first eigenfunction of the generator of this process.

It is, for instance, not difficult to see that the right hand side of (11.8) is 0 when $t = \tau^x$ and that

$$\lim_{x \to 0} f(x,t) = \lim_{x \to 2\pi} f(x,t) = 0 \qquad (11.9)$$

holds for all fixed $t > 0$. Let $F \colon [0, 2\pi] \to \mathbb{R}$ be a continuous function with $F(0) = F(2\pi) = 0$, which is smooth in $(0, 2\pi)$, and set

$$h(x,t) = h_F(x,t) := \mathbb{E}\left[\Phi_t^x F(Y_t^x)\right].$$

By (11.8) and the general theory of diffusion Markov processes, we know that h is smooth in $(0, 2\pi) \times \mathbb{R}_+$. The Markov property for Y_t^x and (11.8) show that $h(Y_t^x, t' - t) \times \Phi_t^x$ is a local martingale on $t < \min\{\tau^x, t'\}$. Hence, the drift term of the stochastic differential $d(h(Y_t^x, t' - t)\Phi_t^x)$ is zero at $t = 0$. By Itô's formula, this means that

$$\partial_t h = \frac{6}{2} \partial_x^2 h + \cot(x/2)\, \partial_x h - \frac{1}{2\sin^2(x/2)}\, h. \qquad (11.10)$$

The corresponding positive eigenfunction is $\left(\sin(x/2)\right)^{1/3}$. We therefore define F to be this function, so that $F(x)e^{-5t/4} = h_F$ because both satisfy (11.10) on $(0, 2\pi) \times [0, \infty)$ and have the same boundary values. The proposition then follows easily.

11.4.2 Consequences

Let us now explain some steps that enable us to transform the previous considerations and computations into an actual proof of the fact that $\xi = 5/8$. Consider a radial SLE(6) path in the unit disc, parametrised by capacity. Define, for each $r < 1$, its hitting time τ_r of the disc $\mathcal{B}(0, r)$ of radius r around the origin. A standard result from complex analysis, Koebe's $1/4$ theorem, shows that the path can not reach distance r from the origin before time $(\log(1/r))/4$ and that it has to do so before time $\log(1/r)$. In other words, $\log(1/r) \leqslant 4\tau_r \leqslant 4\log(1/r)$. Furthermore, the map $t \mapsto |g_t'(e^{ix})|$ is decreasing with t (this can for instance be seen from its expression as a killing probability). The previous estimate can therefore be transformed into an estimate of

$$\mathbb{E}\left[\left|g_{\tau_r}'\left(\exp(ix)\right)\right| 1_{\mathcal{H}(x, \tau_r)}\right] \qquad \text{as } r \to 0.$$

We can also integrate this quantity when x spans $[0, 2\pi]$. In fact, the integral expression $\int_0^{2\pi} |g_{\tau_r}'(\exp(ix))|1_{\mathcal{H}(x, \tau_r)}\, dx$ is the harmonic measure (from the origin) of $\partial \mathbb{U}$ in the domain $\mathbb{U} \setminus \gamma[0, \tau_r]$. In other words, if we start a planar Brownian motion Z from the origin, and stop it at the first time σ at which it hits the unit circle, we get that, for some absolute constants $c_1, c_2, \ldots,$

$$c_1\, r^{5/4} \leqslant \mathbb{P}\{Z[0,\sigma] \cap \gamma[0,\tau_r] = \emptyset\} \leqslant c_2\, r^{5/4}.$$

Modulo some additional arguments, this can be reformulated also in terms of a planar Brownian motion Y started uniformly on the unit circle and stopped at its first hitting time ρ_r of the circle of radius r (roughly speaking, we time-reverse Z):

$$c_3 r^{5/4} \leqslant \mathbb{P}\{Y[0,\rho_r] \cap \gamma[0,\tau_r] = \emptyset\} \leqslant c_4 r^{5/4}.$$

We can then use the transformation $z \mapsto 1/z$ to be back in our original setting. This last result (after a couple more uses of monotonicity and of Koebe's $1/4$ Theorem) can then be reformulated in this setting, and it shows that if η is our curve that is growing 'from the origin to infinity' in the plane, and if Z is a Brownian motion started uniformly on the circle of radius r,

$$c_5 r^{5/4} \leqslant \mathbb{P}\{Z[0,\sigma] \cap \eta[0,\tau] = \emptyset\} \leqslant c_6\, r^{5/4}.$$

Hence, we indeed get a precise answer to Question 11.1 with $\xi = 5/8$ (the factor 2 comes in because of the scaling relation between time and space).

Arguments of a similar type can be used to prove that if $\{Z^1(t): 0 \leqslant t \leqslant \tau_1\}$ and $\{Z^2(t): 0 \leqslant t \leqslant \tau_2\}$ are two planar Brownian paths started at the origin and stopped at their hitting times of the unit circle, then the probability that $Z^1[0,\tau_1] \cup Z^2[0,\tau_2]$ does not disconnect the point r from infinity, does decay like $r^{2/3+o(1)}$ as $r \to 0$. This exponent $\alpha = 2/3$ comes in fact from a similar radial SLE computation. This time, one has to consider the moment of order $1/3$ of $|g_t(e^{ix})|$ as $t \to \infty$ (we do however not explain here why this $1/3$ moment comes in, this has in fact to do with a chordal SLE computation, the interested reader might consult [LSW01a, LSW01b, We04, La05]).

11.4.3 From exponents to dimensions

In papers [La96a, La96b] (before the mathematical determination of the values of the exponents in [LSW01a, LSW01b]), Greg Lawler showed how to derive and use moment bounds in order to express the Hausdorff dimension of special random subsets of the planar Brownian curve in terms of the corresponding exponents.

More precisely, let $\{Z(t): t \geqslant 0\}$ denote a planar Brownian motion. Recall that $p = Z(t)$ is a cut point if $Z[0,t] \cap Z(t,1] = \emptyset$. Note that, loosely speaking, near p, there are two independent Brownian paths starting at p: The future $\{Z^1(s): s \in [0,1-t]\}$, given by $Z^1(s) = Z(t+s)$, and the past $\{Z^1(s): s \in [0,t]\}$, given by $Z^2(s) = Z(t-s)$. Furthermore, p is a cut point if $Z^1[0,1-t] \cap Z^2[0,t] = \{p\}$. Similarly, $p = Z(t)$ is a boundary point if $Z^1[0,1-t] \cup Z^2[0,t]$ does not disconnect p from infinity.

Hence, the previous estimates enable us to control the probability that a given point $x \in \mathbb{C}$ is in the ε-neighbourhood of a cut point (resp. boundary point). Independence properties of planar Brownian paths then make it also possible to derive second moment estimates (i.e. the probability that two given points x and x' are both in the ε-neighbourhood of such points) and to show that the Hausdorff dimension of the set of cut times is almost surely $1 - \xi$, and that the Hausdorff dimension of the set of boundary points is almost surely equal to $1 - \alpha/2$.

The proofs use (just as in the case of cone points described in Section 10.4, see in particular Theorem 10.43) first and second moment estimates. In fact, if one uses the relation with critical percolation, it turns out that life can be somewhat simplified in the derivation of the second moment estimates (see for instance [Be04]).

Recall that on the other hand, we know from SLE calculations that $2 - 2\xi = 3/4$, $2 - \alpha = 4/3$. In view of Kaufman's dimension doubling theorem, see Theorem 9.28, we can therefore answer our three initial questions:

Theorem 11.10

(1) The exponent ξ is equal to $5/8$.

(2) The Hausdorff dimension of the set of cut points is almost surely equal to $3/4$.

(3) The Hausdorff dimension of the outer boundary is almost surely equal to $4/3$.

Notes and comments

The idea to use Loewner's equation to study random growth models probably first appeared in the works of Carleson and Makarov in the context of diffusion limited aggregation (DLA), see [CM01, CM02].

Conformal invariance of lattice models has now been established in various cases. Aizenman [Ai96] was probably the first one to emphasise that the conformal invariance conjectures that were present in various forms in the physics literature could be expressed in terms of conformally invariant laws on curves. Kenyon used determinant computations and estimates in order to prove several conformal invariance properties of the loop-erased random walk (and its companion model called the uniform spanning tree), see [Ke00a, Ke00b]. Later [LSW04] showed stronger conformal invariance properties and the convergence of the loop-erased random walk to SLE(2) in the fine-mesh limit. Smirnov [Sm01, Sm07, Sm08] proved conformal invariance for the particular critical percolation model that we presented here, and also for the Ising model on the square lattice.

The idea that one probably had to compute the value of the Brownian exponents using another model (that should be closely related to critical percolation scaling limits) appeared in [LW00]. The mathematical derivation of the value of the exponents was performed in the series of papers [LSW01a, LSW01b, LSW02]. The properties of SLE that were later derived in [LSW03] enable us to shorten some parts of some proofs and to derive various direct identities in law between SLE(6) boundaries and Brownian boundaries, see [We05, We08a].

A good reference for the relation between Brownian exponents and Hausdorff dimensions is Lawler's review paper [La99]. See also Beffara [Be04, Be08]. Determining the Hausdorff dimensions of the SLE curves is a rather difficult question. It turns out to be $1 + \kappa/8$ when $\kappa \leqslant 8$. This has been proved by Rohde and Schramm [RS05] for the upper bound and Beffara [Be08] for the tricky lower bound.

The value of the Brownian intersection exponents had been predicted/conjectured before: Duplantier–Kwon [DK88] had for instance predicted the values of ξ using numerics and non-rigorous conformal field theory considerations. Later, Duplantier [Du04] used also 'quantum gravity techniques' to produce the values of all exponents. The fact that planar Brownian motion contains cut points had first been proved by Burdzy ([Bu89] and [Bu95]). A different shorter proof was given by Lawler in [La96a].

The fact that the dimension of the Brownian boundary is $4/3$ was first observed visually and conjectured by Mandelbrot [Ma82]. Before the proof of this conjecture, some rigorous bounds had been derived, for instance that the dimension of the Brownian boundary is strictly larger than 1 and strictly smaller than $3/2$ (see [BJPP97, BL90, We96]). The two exponents that we have chosen to focus on are just two examples from a continuous family of intersection exponents, that can all be derived using these SLE methods.

Appendix B: Background and prerequisites

12.1 Convergence of distributions

In this section we collect the basic facts about convergence in distribution, see for example the books of Billingsley [Bi95, Bi99] for more extensive treatment. While this is a familiar concept for real valued random variables, for example in the central limit theorem, we need a more abstract viewpoint, which allows to study convergence in distribution for random variables with values in metric spaces, like for example function spaces.

If random variables $\{X_n : n \geqslant 0\}$ converge in distribution, strictly speaking it is their *distributions* and not the *random variables* themselves which converge. This just means that the shape of the distributions of X_n for large n is like the shape of the distribution of X: Sample values from X_n allow no inference towards sample values from X and, indeed, there is no need to define X_n and X on the same probability space.

Definition 12.1. Suppose (E, ρ) is a metric space and $\mathcal{A}$ the Borel-σ-algebra on E. Suppose that X_n and X are E-valued random variables. Then we say that X_n **converges in distribution** to X, if, for every bounded continuous $g \colon E \to \mathbb{R}$,

$$\lim_{n \to \infty} \mathbb{E}[g(X_n)] = \mathbb{E}[g(X)].$$

We write $X_n \xrightarrow{\mathrm{d}} X$ for convergence in distribution. ◇

Remark 12.2 $X_n \xrightarrow{\mathrm{d}} X$ is equivalent to *weak convergence* of the distributions. ◇

Remark 12.3 If $X_n \xrightarrow{\mathrm{d}} X$ and $g \colon E \to \mathbb{R}$ is continuous, then $g(X_n) \xrightarrow{\mathrm{d}} g(X)$. But note that, if $E = \mathbb{R}$ and $X_n \xrightarrow{\mathrm{d}} X$, this does not imply that $\mathbb{E}[X_n]$ converges to $\mathbb{E}[X]$, as $g(x) = x$ is not a bounded function on $\mathbb{R}$. ◇

Example 12.4

- Suppose $E = \{1, \ldots, m\}$ is finite and $\rho(x, y) = 1 - 1_{\{x=y\}}$. Then $X_n \xrightarrow{\mathrm{d}} X$ if and only if $\lim_{n \to \infty} \mathbb{P}\{X_n = k\} = \mathbb{P}\{X = k\}$ for all $k \in E$.

- Let $E = [0, 1]$ and $X_n = 1/n$ almost surely. Then $X_n \xrightarrow{\mathrm{d}} X$, where $X = 0$ almost surely. However, note that $\lim_{n \to \infty} \mathbb{P}\{X_n = 0\} = 0 \neq \mathbb{P}\{X = 0\} = 1$. ◇

Theorem 12.5 *Suppose a sequence $\{X_n : n \geqslant 0\}$ of random variables converges almost surely to a random variable X (of course, all on the same probability space). Then X_n converges in distribution to X.*

Proof. Suppose g is bounded and continuous. The $g(X_n)$ converges almost surely to $g(X)$. As the sequence is bounded it is also uniformly integrable, hence convergence holds also in the $\mathbf{L}^1$-sense and this implies convergence of the expectations, i.e. $\mathbb{E}[g(X_n)] \to \mathbb{E}[g(X)]$. ∎

Theorem 12.6 (Portmanteau theorem) *The following statements are equivalent*

(i) $X_n \xrightarrow{\mathrm{d}} X$.

(ii) *For all closed sets* $K \subset E$, $\limsup_{n\to\infty} \mathbb{P}\{X_n \in K\} \leqslant \mathbb{P}\{X \in K\}$.

(iii) *For all open sets* $G \subset E$, $\liminf_{n\to\infty} \mathbb{P}\{X_n \in G\} \geqslant \mathbb{P}\{X \in G\}$.

(iv) *For all Borel sets* $A \subset E$ *with* $\mathbb{P}\{X \in \partial A\} = 0$, *we have*

$$\lim_{n\to\infty} \mathbb{P}\{X_n \in A\} = \mathbb{P}\{X \in A\}.$$

(v) *For all bounded measurable functions* $g \colon E \to \mathbb{R}$ *with*

$$\mathbb{P}\{g \text{ is discontinuous at } X\} = 0$$

we have $\mathbb{E}[g(X_n)] \to \mathbb{E}[g(X)]$.

Proof. **(i)⇒(ii)** Let $g_n(x) = 1 - (n\rho(x, K) \wedge 1)$, which is continuous and bounded, is 1 on K and converges pointwise to 1_K. Then, for every n,

$$\limsup_{k\to\infty} \mathbb{P}\{X_k \in K\} \leqslant \limsup_{k\to\infty} \mathbb{E}[g_n(X_k)] = \mathbb{E}[g_n(X)].$$

Let $n \to \infty$. The integrand on the right hand side is bounded by 1 and converges pointwise and hence in the $\mathbf{L}^1$-sense to $1_K(X)$.

(ii)⇒(iii) Follows from $1_G = 1 - 1_K$ for the closed set $K = G^c$.

(iii)⇒(iv) Let G be the interior and K the closure of A. Then, by assumption, $\mathbb{P}\{X \in G\} = \mathbb{P}\{X \in K\} = \mathbb{P}\{X \in A\}$ and we may use (iii) and (ii) (which follows immediately from (iii)) to get

$$\limsup_{n\to\infty} \mathbb{P}\{X_n \in A\} \leqslant \limsup_{n\to\infty} \mathbb{P}\{X_n \in K\} \leqslant \mathbb{P}\{X \in K\} = \mathbb{P}\{X \in A\},$$

$$\liminf_{n\to\infty} \mathbb{P}\{X_n \in A\} \geqslant \liminf_{n\to\infty} \mathbb{P}\{X_n \in G\} \geqslant \mathbb{P}\{X \in G\} = \mathbb{P}\{X \in A\}.$$

(iv)⇒(v) From (iv) we infer that the convergence holds for g of the form $g(x) = \sum_{n=1}^{N} a_n 1_{A_n}$, where A_n satisfies $\mathbb{P}\{X \in \partial A_n\} = 0$. Let us call such functions elementary. Given g as in (v) we observe that for every $a < b$ with possibly a countable set of exceptions

$$\mathbb{P}\{X \in \partial\{x \colon g(x) \in (a, b]\}\} = 0.$$

Indeed, if $X \in \partial\{x \colon g(x) \in (a, b]\}$ then either g is discontinuous in X or $g(X) = a$ or $g(X) = b$. The first event has probability zero and so have the last two except possibly for a countable set of values of a, b. By decomposing the real axis in small suitable intervals we thus obtain an increasing sequence g_n and a decreasing sequence h_n of elementary functions both converging pointwise to g. Now, for all k,

$$\limsup_{n\to\infty} \mathbb{E}[g(X_n)] \leqslant \limsup_{n\to\infty} \mathbb{E}[h_k(X_n)] = \mathbb{E}[h_k(X)],$$

348

and

$$\liminf_{n\to\infty} \mathbb{E}[g(X_n)] \geqslant \liminf_{n\to\infty} \mathbb{E}[g_k(X_n)] = \mathbb{E}[g_k(X)].$$

and the right sides converge, as $k \to \infty$, by bounded convergence, to $\mathbb{E}[g(X)]$.
(v)⇒(i) This is obvious. ∎

To remember the directions of the inequalities in the Portmanteau theorem it is useful to recall the last example $X_n = 1/n \to 0$ and choose $G = (0,1)$ and $K = \{0\}$ to obtain cases where the opposite inequalities fail. We now show that the convergence of distribution as defined here agrees with the familiar concept in the case of real random variables.

Theorem 12.7 (Helly-Bray theorem) *Let X_n and X be real valued random variables and define the associated distribution functions $F_n(x) = \mathbb{P}\{X_n \leqslant x\}$ and $F(x) = \mathbb{P}\{X \leqslant x\}$. Then the following assertions are equivalent.*

(a) X_n *converges in distribution to X,*
(b) $\lim\limits_{n\to\infty} F_n(x) = F(x)$ *for all x such that F is continuous in x.*

Proof. **(a)⇒(b)** Use property (iv) for the set $A = (-\infty, x]$.
(b)⇒(a) We choose a dense sequence $\{x_n\}$ with $\mathbb{P}\{X = x_n\} = 0$ and note that every open set $G \subset \mathbb{R}$ can be written as the countable union of disjoint intervals $I_k = (a_k, b_k]$ with a_k, b_k chosen from the sequence. We have

$$\lim_{n\to\infty} \mathbb{P}\{X_n \in I_k\} = \lim_{n\to\infty} F_n(b_k) - F_n(a_k) = F(b_k) - F(a_k) = \mathbb{P}\{X \in I_k\}.$$

Hence, for all N,

$$\liminf_{n\to\infty} \mathbb{P}\{X_n \in G\} \geqslant \sum_{k=1}^{N} \liminf_{n\to\infty} \mathbb{P}\{X_n \in I_k\} = \sum_{k=1}^{N} \mathbb{P}\{X \in I_k\},$$

and as $N \to \infty$ the last term converges to $\mathbb{P}\{X \in G\}$. ∎

Finally, we note the useful fact that for nonnegative random variables X_n, rather then testing convergence of $\mathbb{E}[g(X_n)]$ for *all* continuous bounded functions g, it suffices to consider functions of a rather simple form.

Proposition 12.8 *Suppose $(X_1^{(n)}, \ldots, X_m^{(n)})$ are random vectors with nonnegative entries, then*

$$(X_1^{(n)}, \ldots, X_m^{(n)}) \xrightarrow{\mathrm{d}} (X_1, \ldots, X_m),$$

if and only if, for any $\lambda_1, \ldots, \lambda_m \geqslant 0$,

$$\lim_{n\uparrow\infty} \mathbb{E}\Big[\exp\big\{-\sum_{j=1}^{m}\lambda_j X_j^{(n)}\big\}\Big] = \mathbb{E}\Big[\exp\big\{-\sum_{j=1}^{m}\lambda_j X_j\big\}\Big].$$

The function $\phi(\lambda_1, \ldots, \lambda_m) = \mathbb{E}[\exp\{-\sum_{j=1}^{m}\lambda_j X_j\}]$ is called the **Laplace transform** of $(X_1, \ldots, X_m)$ and thus the proposition states in other words that the convergence of nonnegative random vectors is equivalent to convergence of their Laplace transforms. The proof, usually done by approximation, can be found as Theorem 5.3 in [Ka02].

12.2 Gaussian random variables

In this section we have collected the facts about Gaussian random vectors, which are used in this book. We start with a useful estimate for standard normal random variables, which is quite precise for large x.

Lemma 12.9 *Suppose X is standard normally distributed. Then, for all $x > 0$,*

$$\frac{x}{x^2 + 1} \frac{1}{\sqrt{2\pi}} e^{-x^2/2} \leqslant \mathbb{P}\{X > x\} \leqslant \frac{1}{x} \frac{1}{\sqrt{2\pi}} e^{-x^2/2}.$$

Proof. The right inequality is obtained by the estimate

$$\mathbb{P}\{X > x\} \leqslant \frac{1}{\sqrt{2\pi}} \int_x^\infty \frac{u}{x} e^{-u^2/2} \, du = \frac{1}{x} \frac{1}{\sqrt{2\pi}} e^{-x^2/2}.$$

For the left inequality we define

$$f(x) = x e^{-x^2/2} - (x^2 + 1) \int_x^\infty e^{-u^2/2} \, du.$$

Observe that $f(0) < 0$ and $\lim_{x\to\infty} f(x) = 0$. Moreover,

$$f'(x) = (1 - x^2 + x^2 + 1) e^{-x^2/2} - 2x \int_x^\infty e^{-u^2/2} \, du = -2x \left(\int_x^\infty e^{-u^2/2} \, du - \frac{e^{-x^2/2}}{x} \right),$$

which is positive for $x > 0$, by the first part. Hence $f(x) \leqslant 0$, proving the lemma. ∎

We now look more closely at random vectors with normally distributed components. Our motivation is that they arise, for example, as vectors consisting of the increments of a Brownian motion. Let us clarify some terminology.

Definition 12.10. A random variable $X = (X_1, \ldots, X_d)^{\mathrm{T}}$ with values in $\mathbb{R}^d$ has the *d-dimensional standard Gaussian distribution* if its d coordinates are standard normally distributed and independent. ◇

More general Gaussian distributions can be derived as linear images of standard Gaussians. Recall, e.g. from Definition 1.5, that a random variable Y with values in $\mathbb{R}^d$ is called *Gaussian* if there exists an m-dimensional standard Gaussian X, a $d \times m$ matrix A, and a d dimensional vector b such that $Y^{\mathrm{T}} = AX + b$. The *covariance matrix* of the (column) vector Y is then given by

$$\mathrm{Cov}(Y) = \mathbb{E}\big[(Y - \mathbb{E}Y)(Y - \mathbb{E}Y)^{\mathrm{T}}\big] = AA^{\mathrm{T}},$$

where the expectations are defined componentwise.

Our next lemma shows that applying an orthogonal $d \times d$ matrix does not change the distribution of a standard Gaussian random vector, and in particular that the standard Gaussian distribution is rotationally invariant. We write I_d for the $d \times d$ identity matrix.

Lemma 12.11 *If A is an orthogonal $d \times d$ matrix, i.e. $AA^{\mathrm{T}} = I_d$, and X is a d-dimensional standard Gaussian vector, then AX is also a d-dimensional standard Gaussian vector.*

Proof. As the coordinates of X are independent, standard normally distributed, X has a density

$$f(x_1, \ldots, x_d) = \prod_{i=1}^{d} \frac{1}{\sqrt{2\pi}} \, e^{-x_i^2/2} = \frac{1}{(2\pi)^{d/2}} \, e^{-|x|^2/2} \, ,$$

where $| \cdot |$ is the Euclidean norm. The density of AX is (by the transformation rule) $f(A^{-1}x) \, |\det(A^{-1})|$. The determinant is 1 and, since orthogonal matrices preserve the Euclidean norm, the density of X is invariant under A. ∎

Corollary 12.12 *Let X_1 and X_2 be independent and normally distributed with zero expectation and variance $\sigma^2 > 0$. Then $X_1 + X_2$ and $X_1 - X_2$ are independent and normally distributed with expectation 0 and variance $2\sigma^2$.*

Proof. The vector $(X_1/\sigma, X_2/\sigma)^{\mathrm{T}}$ is standard Gaussian by assumption. Look at

$$A = \begin{pmatrix} \frac{1}{\sqrt{2}} & \frac{1}{\sqrt{2}} \\ \frac{1}{\sqrt{2}} & -\frac{1}{\sqrt{2}} \end{pmatrix}.$$

This is an orthogonal matrix and applying it to our vector yields $((X_1+X_2)/(\sqrt{2}\sigma), (X_1-X_2)/(\sqrt{2}\sigma))$, which thus must have independent standard normal coordinates. ∎

The next proposition shows that the distribution of a Gaussian random vector is determined by its expectation and covariance matrix.

Proposition 12.13 *If X and Y are d-dimensional Gaussian vectors with $\mathbb{E}X = \mathbb{E}Y$ and $\mathrm{Cov}(X) = \mathrm{Cov}(Y)$, then X and Y have the same distribution.*

Proof. It is sufficient to consider the case $\mathbb{E}X = \mathbb{E}Y = 0$. By definition, there are standard Gaussian random vectors X_1 and X_2 and matrices A and B with $X = AX_1$ and $Y = BX_2$. By adding columns of zeros to A or B, if necessary, we can assume that X_1 and X_2 are both k-vectors, for some k, and A, B are both $d \times k$ matrices. Let $\mathcal{A}$ and $\mathcal{B}$ be the vector subspaces of $\mathbb{R}^k$ generated by the row vectors of A and B, respectively. To simplify notation assume that the first $l \leqslant d$ row vectors of A form a basis of $\mathcal{A}$. Define the linear map $L: \mathcal{A} \to \mathcal{B}$ by

$$L(A_i) = B_i \text{ for } i = 1, \ldots, l.$$

Here A_i is the ith row vector of A, and B_i is the ith row vector of B. Our aim is to show that L is an orthogonal isomorphism and then use the previous proposition. Let us first show that L is an isomorphism. Our covariance assumption gives that $AA^{\mathrm{T}} = BB^{\mathrm{T}}$. Assume there is a vector $v_1 A_1 + \ldots v_l A_l$ whose image is 0. Then the d-vector

$$v = (v_1, \ldots, v_l, 0, \ldots, 0)$$

satisfies $vB = 0$. Hence

$$\|vA\|^2 = vAA^{\mathrm{T}}v^{\mathrm{T}} = vBB^{\mathrm{T}}v^{\mathrm{T}} = 0 \, .$$

We conclude that $vA = 0$. Hence L is injective and $\dim \mathcal{A} \leqslant \dim \mathcal{B}$. Interchanging the rôle of A and B gives that L is an isomorphism. As the entry (i, j) of $AA^{\mathrm{T}} = BB^{\mathrm{T}}$ is the scalar product of A_i and A_j as well as B_i and B_j, the mapping L is orthogonal. We can extend it on the orthocomplement of $\mathcal{A}$ to an orthogonal map $L \colon \mathbb{R}^k \to \mathbb{R}^k$ (or an orthogonal $k \times k$-matrix). Then $X = AX_1$ and $Y = BX_2 = AL^{\mathrm{T}}X_2$. As $L^{\mathrm{T}}X_2$ is standard Gaussian, by Lemma 12.11, X and Y have the same distribution. ∎

In particular, comparing a d-dimensional Gaussian vector with $\mathrm{Cov}(X) = I_d$ with a Gaussian vector with d independent entries and the same expectation, we obtain the following fact.

Corollary 12.14 *A Gaussian random vector X has independent entries if and only if its covariance matrix is diagonal. In other words, the entries in a Gaussian vector are uncorrelated if and only if they are independent.*

We now show that the Gaussian nature of a random vector is preserved under taking limits.

Proposition 12.15 *Suppose $\{X_n : n \in \mathbb{N}\}$ is a sequence of Gaussian random vectors and $\lim_n X_n = X$, almost surely. If $b := \lim_{n \to \infty} \mathbb{E}X_n$ and $C := \lim_{n \to \infty} \mathrm{Cov}\, X_n$ exist, then X is Gaussian with mean b and covariance matrix C.*

Proof. A variant of the argument in Proposition 12.13 shows that X_n converges in law to a Gaussian random vector with mean b and covariance matrix C. As almost sure convergence implies convergence of the associated distributions, this must be the law of X. ∎

Lemma 12.16 *Suppose X, Y are independent and normally distributed with mean zero and variance σ^2, then $X^2 + Y^2$ is exponentially distributed with mean $2\sigma^2$.*

Proof. For any bounded, measurable $f \colon \mathbb{R} \to \mathbb{R}$ we have, using polar coordinates,

$$
\begin{aligned}
\mathbb{E}f(X^2 + Y^2) &= \frac{1}{2\pi\sigma^2} \int f(x^2 + y^2) \exp\left\{ -\tfrac{x^2 + y^2}{2\sigma^2} \right\} dx\, dy \\
&= \frac{1}{\sigma^2} \int_0^\infty f(r^2) \exp\left\{ -\tfrac{r^2}{2\sigma^2} \right\} r\, dr \\
&= \frac{1}{2\sigma^2} \int_0^\infty f(a) \exp\left\{ -\tfrac{a}{2\sigma^2} \right\} da = \mathbb{E}f(Z),
\end{aligned}
$$

where Z is exponential with mean $2\sigma^2$. ∎

12.3 Martingales in discrete time

In this section we recall the essentials from the theory of martingales in discrete time. A more thorough introduction to this subject is Williams [Wi91].

Definition 12.17. A *filtration* $(\mathcal{F}_n : n \geqslant 0)$ is an increasing sequence

$$
\mathcal{F}_0 \subset \mathcal{F}_1 \subset \cdots \subset \mathcal{F}_n \subset \cdots
$$

of σ-algebras.

Let $\{X_n : n \geqslant 0\}$ be a stochastic process in discrete time and $(\mathcal{F}_n : n \geqslant 0)$ be a filtration. The process is a *martingale* relative to the filtration if, for all $n \geqslant 0$,

- X_n is measurable with respect to $\mathcal{F}_n$,

- $\mathbb{E}|X_n| < \infty$, and

- $\mathbb{E}[X_{n+1} \mid \mathcal{F}_n] = X_n$, almost surely.

If we have '$\geqslant$' in the last condition, then $\{X_n : n \geqslant 0\}$ is called a *submartingale*, if '$\leqslant$' holds it is called a *supermartingale*. ◇

Remark 12.18 Note that for a submartingale $\mathbb{E}[X_{n+1}] \geqslant \mathbb{E}[X_n]$, for a supermartingale $\mathbb{E}[X_{n+1}] \leqslant \mathbb{E}[X_n]$, and hence for a martingale we have $\mathbb{E}[X_{n+1}] = \mathbb{E}[X_n]$. ◇

Loosely speaking, a stopping time is a random time such that the knowledge about a random process at time n suffices to determine whether the stopping time has happened at time n or not. Here is a formal definition.

Definition 12.19. A random variable T with values in $\{0, 1, 2, \ldots\} \cup \{\infty\}$ is called a *stopping time* if $\{T \leqslant n\} = \{\omega : T(\omega) \leqslant n\} \in \mathcal{F}_n$ for all $n \geqslant 0$. ◇

If $\{X_n : n \geqslant 0\}$ is a supermartingale and T a stopping time, then it is easy to check that the process

$$\{X_n^T : n \geqslant 0\} \qquad \text{defined by } X_n^T = X_{T \wedge n}$$

is a supermartingale. If $\{X_n : n \geqslant 0\}$ is a martingale, then both $\{X_n : n \geqslant 0\}$ and $\{-X_n : n \geqslant 0\}$ are supermartingales and, hence, we have,

$$\mathbb{E}[X_{T \wedge n}] = \mathbb{E}[X_0], \qquad \text{for all } n \geqslant 0.$$

Doob's optional stopping theorem gives criteria when, letting $n \uparrow \infty$, we obtain $\mathbb{E}[X_T] = \mathbb{E}[X_0]$.

Theorem 12.20 (Doob's optional stopping theorem) *Let T be a stopping time and X a martingale. Then X_T is integrable and $\mathbb{E}[X_T] = \mathbb{E}[X_0]$, if one of the following conditions hold:*

(1) *T is bounded, i.e. there is N such that $T < N$ almost surely;*

(2) *$\{X_n^T : n \geqslant 0\}$ is dominated by an integrable random variable Z, i.e. $|X_{n \wedge T}| \leqslant Z$ for all $n \geqslant 0$ almost surely;*

(3) *$\mathbb{E}[T] < \infty$ and there is $K > 0$ such that $\sup_n |X_n - X_{n-1}| \leqslant K$.*

Proof. Recall that $\mathbb{E}[X_{T \wedge n} - X_0] = 0$. The result follows in case (1) by choosing $n = N$. In case (2) let $n \to \infty$ and use dominated convergence. In case (3) observe that $|X_{T \wedge n} - X_0| = |\sum_{k=1}^{T \wedge n}(X_k - X_{k-1})| \leqslant KT$. By assumption KT is an integrable function and dominated convergence can be used again. ∎

Doob's famous forward convergence theorem gives a sufficient condition for the almost sure convergence of supermartingales to a limiting random variable. See 11.5 in [Wi91] for the proof.

Theorem 12.21 (Doob's supermartingale convergence theorem) *Let $\{X_n : n \geqslant 0\}$ be a supermartingale, which is bounded in $\mathbf{L}^1$, i.e. there is $K > 0$ such that $\mathbb{E}|X_n| \leqslant K$ for all n. Then there exists an integrable random variable X on the same probability space such that*

$$\lim_{n \to \infty} X_n = X \text{ almost surely.}$$

Remark 12.22 Note that if $\{X_n : n \geqslant 0\}$ is nonnegative, we have $\mathbb{E}[|X_n|] = \mathbb{E}[X_n] \leqslant \mathbb{E}[X_0] := K$ and thus X_n is bounded in $\mathbf{L}^1$ and $\lim_{n \to \infty} X_n = X$ exists. ◇

A key question is when the almost sure convergence in the supermartingale convergence theorem can be replaced by $\mathbf{L}^1$-convergence (which in contrast to almost sure convergence implies convergence of expectations). A necessary and sufficient criterion for this is *uniform integrability*. A stochastic process $\{X_n : n \geqslant 0\}$ is called *uniformly integrable* if, for every $\varepsilon > 0$, there exists $K > 0$ such that

$$\mathbb{E}\big[|X_n| 1\{|X_n| \geqslant K\}\big] < \varepsilon \qquad \text{for all } n \geqslant 0.$$

Sufficient criteria for uniform integrability are

- $\{X_n : n \geqslant 0\}$ is dominated by an integrable random variable,
- $\{X_n : n \geqslant 0\}$ is $\mathbf{L}^p$-bounded for some $p > 1$,
- $\{X_n : n \geqslant 0\}$ is $\mathbf{L}^1$-convergent.

The following lemma is proved in Section 13.1 of [Wi91].

Lemma 12.23 *Any stochastic process $\{X_n : n \geqslant 0\}$, which is uniformly integrable and almost surely convergent, converges also in the $\mathbf{L}^1$-sense.*

The next result is one of the highlights of martingale theory.

Theorem 12.24 (Martingale closure theorem) *Suppose that the martingale $\{X_n : n \geqslant 0\}$ is uniformly integrable. Then there is an integrable random variable X such that*

$$\lim_{n \to \infty} X_n = X \text{ almost surely and in } \mathbf{L}^1.$$

Moreover, $X_n = \mathbb{E}[X \mid \mathcal{F}_n]$ for every $n \geqslant 0$.

Proof. Uniform integrability implies that $\{X_n : n \geqslant 0\}$ is $\mathbf{L}^1$-bounded and thus, by the martingale convergence theorem, almost surely convergent to an integrable random variable X. Convergence in the $\mathbf{L}^1$-sense follows from Lemma 12.23. To check the last assertion, we note that X_n is $\mathcal{F}_n$-measurable and let $F \in \mathcal{F}_n$. For all $m \geqslant n$ we have, by the martingale property, $\int_F X_m \, d\mathbb{P} = \int_F X_n \, d\mathbb{P}$. We let $m \to \infty$. Then $|\int_F X_m \, d\mathbb{P} - \int_F X \, d\mathbb{P}| \leqslant \int |X_m - X| \, d\mathbb{P} \to 0$, hence we obtain $\int_F X \, d\mathbb{P} = \int_F X_n \, d\mathbb{P}$, as required. ∎

There is a natural converse to the martingale closure theorem, see Section 14.2 in [Wi91] for the proof.

Theorem 12.25 (Lévy's upward theorem) *Suppose that X is an integrable random variable and $X_n = \mathbb{E}[X \mid \mathcal{F}_n]$. Then $\{X_n : n \geqslant 0\}$ is a uniformly integrable martingale and*

$$\lim_{n \to \infty} X_n = \mathbb{E}[X \mid \mathcal{F}_\infty] \text{ almost surely and in } \mathbf{L}^1,$$

where $\mathcal{F}_\infty = \left(\bigcup_{n=1}^\infty \mathcal{F}_n \right)$ is the smallest σ-algebra containing the entire filtration.

There is also a convergence theorem for 'reverse' martingales, which is called Lévy's downward theorem and is a natural partner to the upward theorem, see Section 14.4 in [Wi91] for the proof.

Theorem 12.26 (Lévy's downward theorem) *Suppose that $(\mathcal{G}_n : n \in \mathbb{N})$ is a collection of σ-algebras such that*

$$\mathcal{G}_\infty := \bigcap_{k=1}^\infty \mathcal{G}_k \subset \cdots \subset \mathcal{G}_{n+1} \subset \mathcal{G}_n \subset \cdots \subset \mathcal{G}_1.$$

*An integrable process $\{X_n : n \in \mathbb{N}\}$ is a **reverse martingale** if almost surely,*

$$X_n = \mathbb{E}[X_{n-1} \mid \mathcal{G}_n] \qquad \text{for all } n \geqslant 2.$$

Then

$$\lim_{n \uparrow \infty} X_n = \mathbb{E}[X_1 \mid \mathcal{G}_\infty] \qquad \text{almost surely.}$$

An important consequence of Theorems 12.20 and 12.24 is that the martingale property holds for well-behaved stopping times. For a stopping time T define $\mathcal{F}_T$ to be the σ-algebra of events A with $A \cap \{T \leqslant n\} \in \mathcal{F}_n$. Observe that X_T is $\mathcal{F}_T$-measurable.

Theorem 12.27 (Optional sampling theorem) *If the martingale $\{X_n : n = 1, 2, \ldots\}$ is uniformly integrable, then for all stopping times $0 \leqslant S \leqslant T$ we have $\mathbb{E}[X_T \mid \mathcal{F}_S] = X_S$ almost surely.*

Proof. By the martingale closure theorem, X_n^T converges to X_T in $\mathbf{L}^1$ and $\mathbb{E}[X_T \mid \mathcal{F}_n] = X_{T \wedge n} = X_n^T$. Dividing X_T in its positive and its nonpositive part if necessary, we may assume that $X_T \geqslant 0$ and therefore $X_n^T \geqslant 0$ almost surely. Taking conditional expectation with respect to $\mathcal{F}_{S \wedge n}$ gives $\mathbb{E}[X_T \mid \mathcal{F}_{S \wedge n}] = X_{S \wedge n}$. Now let $A \in \mathcal{F}_S$. We have to show that $\mathbb{E}[X_T 1_A] = \mathbb{E}[X_S 1_A]$. Note first that $A \cap \{S \leqslant n\} \in \mathcal{F}_{S \wedge n}$. Hence, we get $\mathbb{E}[X_T 1\{A \cap \{S \leqslant n\}\}] = \mathbb{E}[X_{S \wedge n} 1\{A \cap \{S \leqslant n\}\}] = \mathbb{E}[X_S 1\{A \cap \{S \leqslant n\}\}]$. Letting $n \uparrow \infty$ and using monotone convergence gives the required result. ∎

Of considerable practical importance are martingales $\{X_n : n \geqslant 0\}$, which are *square integrable*. Note that in this case we can calculate, for $m \geqslant n$,

$$\begin{aligned}
\mathbb{E}[X_m^2 \mid \mathcal{F}_n] &= \mathbb{E}[(X_m - X_n)^2 \mid \mathcal{F}_n] + 2\mathbb{E}[X_m \mid \mathcal{F}_n] X_n - X_n^2 \\
&= \mathbb{E}[(X_m - X_n)^2 \mid \mathcal{F}_n] + X_n^2 \geqslant X_n^2,
\end{aligned} \tag{12.1}$$

so that $\{X_n^2 : t \geqslant 0\}$ is a submartingale.

Theorem 12.28 (Convergence theorem for L²-bounded martingales) *Suppose that the martingale $\{X_n : t \geqslant 0\}$ is $\mathbf{L}^2$-bounded. Then there is a random variable X such that*

$$\lim_{n \to \infty} X_n = X \text{ almost surely and in } \mathbf{L}^2.$$

Proof. From (12.1) and $\mathbf{L}^2$-boundedness of $\{X_n : t \geqslant 0\}$ it is easy to see that, for $m \geqslant n$,

$$\mathbb{E}\big[(X_m - X_n)^2\big] = \sum_{k=n+1}^{m} \mathbb{E}\big[(X_k - X_{k-1})^2\big] \leqslant \sum_{k=1}^{\infty} \mathbb{E}\big[(X_k - X_{k-1})^2\big] < \infty.$$

Recall that $\mathbf{L}^2$-boundedness implies $\mathbf{L}^1$-boundedness, and hence, by the martingale convergence theorem, X_n converges almost surely to an integrable random variable X. Letting $m \uparrow \infty$ and using Fatou's lemma in the last display, gives $\mathbf{L}^2$-convergence. ∎

We now discuss two *martingale inequalities* that have important counterparts in the continuous setting. The first one is Doob's weak maximal inequality.

Theorem 12.29 (Doob's weak maximal inequality) *Let $\{X_j : j \geqslant 0\}$ be a submartingale and denote $M_n := \max_{1 \leqslant j \leqslant n} X_j$. Then, for all $\lambda > 0$,*

$$\lambda \mathbb{P}\{M_n \geqslant \lambda\} \leqslant \mathbb{E}\big[X_n 1\{M_n \geqslant \lambda\}\big].$$

Proof. Define the stopping time

$$\tau := \begin{cases} \min\{k : X_k \geqslant \lambda\} & \text{if } M_n \geqslant \lambda \\ n & \text{if } M_n < \lambda \end{cases}$$

Note that $\{M_n \geqslant \lambda\} = \{X_\tau \geqslant \lambda\}$. This implies

$$\lambda \mathbb{P}\{M_n \geqslant \lambda\} = \lambda \mathbb{P}\{X_\tau \geqslant \lambda\} = \mathbb{E}\lambda 1\{X_\tau \geqslant \lambda\}$$
$$\leqslant \mathbb{E}X_\tau 1\{X_\tau \geqslant \lambda\} = \mathbb{E}X_\tau 1\{M_n \geqslant \lambda\},$$

and the result follows once we demonstrate $\mathbb{E}X_\tau 1\{M_n \geqslant \lambda\} \leqslant \mathbb{E}X_n 1\{M_n \geqslant \lambda\}$. But, as τ is bounded by n and X^τ is a submartingale, we have $\mathbb{E}[X_\tau] \leqslant \mathbb{E}[X_n]$, which implies

$$\mathbb{E}\big[X_\tau 1\{M_n < \lambda\}\big] + \mathbb{E}\big[X_\tau 1\{M_n \geqslant \lambda\}\big]$$
$$\leqslant \mathbb{E}\big[X_n 1\{M_n < \lambda\}\big] + \mathbb{E}\big[X_n 1\{M_n \geqslant \lambda\}\big].$$

Because, by definition of τ, we have $X_\tau 1\{M_n < \lambda\} = X_n 1\{M_n < \lambda\}$, this reduces to

$$\mathbb{E}\big[X_\tau 1\{M_n \geqslant \lambda\}\big] \leqslant \mathbb{E}\big[X_n 1\{M_n \geqslant \lambda\}\big],$$

and this concludes the proof. ∎

The most useful martingale inequality for us is Doob's L^p-maximal inequality.

Theorem 12.30 (Doob's $\mathbf{L}^p$ maximal inequality) *Suppose $\{X_n : n \geqslant 0\}$ is a martingale or nonnegative submartingale. Let $M_n = \max_{1 \leqslant k \leqslant n} X_k$ and $p > 1$. Then*

$$\mathbb{E}\big[M_n^p\big] \leqslant \big(\tfrac{p}{p-1}\big)^p \mathbb{E}\big[|X_n|^p\big].$$

We make use of the following lemma, which allows us to compare the $\mathbf{L}^p$-norms of two nonnegative random variables.

Lemma 12.31 *Suppose nonnegative random variables X and Y satisfy, for all $\lambda > 0$,*

$$\lambda \mathbb{P}\{Y \geqslant \lambda\} \leqslant \mathbb{E}[X 1\{Y \geqslant \lambda\}].$$

Then, for all $p > 1$,

$$\mathbb{E}[Y^p] \leqslant \left(\tfrac{p}{p-1}\right)^p \mathbb{E}[X^p].$$

Proof.   Using the fact that $X \geqslant 0$ and $x^p = \int_0^x p\lambda^{p-1} d\lambda$, we can express $\mathbb{E}[X^p]$ as a double integral and apply Fubini's theorem,

$$\mathbb{E}[X^p] = \mathbb{E}\int_0^\infty 1\{X \geqslant \lambda\} p\lambda^{p-1} \, d\lambda = \int_0^\infty p\lambda^{p-1} \, \mathbb{P}\{X \geqslant \lambda\} \, d\lambda.$$

Similarly, using the hypothesis,

$$\mathbb{E}[Y^p] = \int_0^\infty p\lambda^{p-1} \, \mathbb{P}\{Y \geqslant \lambda\} \, d\lambda \leqslant \int_0^\infty p\lambda^{p-2} \mathbb{E}[X 1\{Y \geqslant \lambda\}] \, d\lambda.$$

We can rewrite the right hand side, using Fubini's theorem again, and then integrating $p\lambda^{p-2}$ and using Hölder's inequality with $q = p/(p-1)$,

$$\int_0^\infty p\lambda^{p-2} \, \mathbb{E}[X 1\{Y \geqslant \lambda\}] \, d\lambda = \mathbb{E}\Big[X \int_0^Y p\lambda^{p-2} \, d\lambda\Big]$$
$$= q \, \mathbb{E}[XY^{p-1}] \leqslant q \, \|X\|_p \|Y^{p-1}\|_q \, .$$

Altogether, this gives $\mathbb{E}[Y^p] \leqslant q(\mathbb{E}[X^p])^{1/p} (\mathbb{E}[Y^p])^{1/q}$ So, assuming $\mathbb{E}[Y^p] < \infty$, the above inequality gives,

$$\left(\mathbb{E}[Y^p]\right)^{1/p} \leqslant q\left(\mathbb{E}[X^p]\right)^{1/p},$$

from which the result follows by raising both sides to the pth power. In general, if $\mathbb{E}[Y^p] = \infty$, then for any $n \in \mathbb{N}$, the random variable $Y_n = Y \wedge n$ satisfies the hypothesis of the lemma, and the result follows by letting $n \uparrow \infty$ and applying the monotone convergence theorem.   ∎

Proof of Theorem 12.30.   If $\{X_n : n \geqslant 0\}$ is a martingale, then $\{|X_n| : n \geqslant 0\}$ is a nonnegative submartingale. Hence it suffices to prove the result for nonnegative submartingales. By Doob's weak maximal inequality,

$$\lambda \mathbb{P}\{M_n \geqslant \lambda\} \leqslant \mathbb{E}[X_n 1\{M_n \geqslant \lambda\}],$$

and applying Lemma 12.31 with $X = X_n$ and $Y = M_n$ gives the result.   ∎

We end this section with a useful version of the Radon-Nikodým theorem, which can be proved using martingale arguments, cf. [Du95], Chapter 4, Theorem 3.3.

Theorem 12.32 *Let μ, ν be two probability measures on a space with σ-algebra $\mathcal{F}$. Assume that $(\mathcal{F}_n : n = 1, 2, \ldots)$ is a filtration such that $\mathcal{F}_n \nearrow \mathcal{F}$ (i.e. the union of all $\mathcal{F}_n$ generates $\mathcal{F}$) and denote $\mu_n = \mu_{|\mathcal{F}_n}$ and $\nu_n = \nu_{|\mathcal{F}_n}$. Suppose $\mu_n \ll \nu_n$ for all n and let*

$$X_n = \frac{d\mu_n}{d\nu_n}.$$

(a) $\{X_n : n \geqslant 0\}$ is a nonnegative martingale and therefore ν-almost surely convergent. We denote

$$X = \limsup_{n \to \infty} X_n.$$

(b) For any $A \in \mathcal{F}$ we have

$$\mu(A) = \int_A X \, d\nu + \mu\big(A \cap \{X = \infty\}\big). \tag{12.2}$$

In particular,

    (i) If $\nu\{X = 0\} = 1$, then $\mu \perp \nu$.

    (ii) If $\mu\{X = \infty\} = 0$, then $\mu \ll \nu$.

    (iii) If $\nu\{X > 0\} = 1$, then $\nu \ll \mu$.

Proof. Note that, for any $A \in \mathcal{F}_n$, we have

$$\int_A X_{n+1} \, d\nu = \int_A \frac{d\mu_{n+1}}{d\nu_{n+1}} \, d\nu_{n+1} = \mu_{n+1}(A) = \mu_n(A) = \int_A \frac{d\mu_n}{d\nu_n} \, d\nu_n = \int_A X_n \, d\nu,$$

and hence $\{X_n : n \geqslant 0\}$ is a martingale. Moreover $X_n \geqslant 0$ and hence, by Remark 12.22, it is convergent, which proves (a). Claims (i), (ii) and (iii) follow easily from (12.2), so it suffices to establish the latter. Rewrite (12.2) in the equivalent form

$$\mu\big(A \cap \{X < \infty\}\big) = \int_A X \, d\nu \quad \text{for all } A \in \mathcal{F}. \tag{12.3}$$

For $A \in \mathcal{F}_k$ and $n > k$ we have $\mu(A) = \int_A X_n \, d\nu$ whence $\mu(A) \geqslant \int_A X \, d\nu$ by Fatou's lemma. It follows that the last inequality holds for all $A \in \mathcal{F}$, whence for all $A \in \mathcal{F}$ we have

$$\mu\big(A \cap \{X < \infty\}\big) \geqslant \int_{A \cap \{X < \infty\}} X \, d\nu = \int_A X \, d\nu. \tag{12.4}$$

On the other hand, for $A \in \mathcal{F}_k$ and $n > k$ we also have

$$\mu\big(A \cap \{X_n < M\}\big) = \int_{A \cap \{X_n < M\}} X_n \, d\nu \leqslant \int_A X_n \wedge M \, d\nu$$

whence $\mu(A \cap \{\sup_{\ell \geqslant n} X_\ell < M\}) \leqslant \int_A X_n \wedge M \, d\nu$. Taking $n \to \infty$, bounded convergence yields $\mu(A \cap \{X < M\}) \leqslant \int_A X \wedge M \, d\nu$ so that letting $M \to \infty$ gives $\mu(A \cap \{X < \infty\}) \leqslant \int_A X \, d\nu$. Thus (12.3) holds for all $A \in \mathcal{F}$. ∎

12.4 Trees and flows on trees

In this section we provide the notation for the discussion of trees, and the basic facts about trees, which we use in this book.

Definition 12.33. A **tree** $T = (V, E)$ is a connected graph described by a finite or countable set V of **vertices**, which includes a distinguished vertex ϱ designated as the root, and a set $E \subset V \times V$ of ordered **edges**, such that

- for every vertex $v \in V$ the set $\{w \in V : (w, v) \in E\}$ consists of exactly one element $\overline{v}$, the **parent**, except for the **root** $\varrho \in V$, which has no parent;

- for every vertex v there is a unique self-avoiding path from the root to v and the number of edges in this path is the **order** or **generation** $|v|$ of the vertex $v \in V$;

- for every $v \in V$, the set of **offspring** or **children** of $\{w \in V : (v, w) \in E\}$ is finite. ◇

Remark 12.34 Sometimes, the notation is slightly abused and the tree T is identified with its vertex set. This should not cause any confusion. ◇

We introduce some further notation. For any $v, w \in V$ we denote by $v \wedge w$ the element on the intersection of the paths from the root to v, respectively w with maximal order, i.e. the last common ancestor of v and w. We write $v \leqslant w$ if v is an ancestor of w, which is equivalent to $v = v \wedge w$.

The order $|e|$ of an edge $e = (u, v)$ is the order of its end-vertex v. Every infinite self-avoiding path started in the root is called a **ray**. The set of rays is denoted ∂T, the **boundary** of T. For any two rays ξ and η we define $\xi \wedge \eta$ the vertex in the intersection of the rays, which maximises the order. Note that $|\xi \wedge \eta|$ is the number of edges that two rays ξ and η have in common. The distance between two rays ξ and η is defined to be $|\xi - \eta| := 2^{-|\xi \wedge \eta|}$, and this definition makes the boundary ∂T a compact metric space.

Remark 12.35 The boundary ∂T of a tree is an interesting fractal in its own right. Its Hausdorff dimension is $\log_2 (\text{br } T)$ where br T is a suitably defined average offspring number. This, together with other interesting aspects of trees, is discussed in depth in [LP05]. ◇

For infinite trees, we are interested in flows on the tree. We suppose that **capacities** are assigned to the edges of a tree T, i.e. there is a mapping $C \colon E \to [0, \infty)$. A **flow** of strength $c > 0$ through a tree with capacities C is a mapping $\theta \colon E \to [0, c]$ such that

- for the root we have $\sum_{\overline{w} = \varrho} \theta(\varrho, w) = c$, for every other vertex $v \neq \varrho$ we have

$$\theta(\overline{v}, v) = \sum_{w \,:\, \overline{w} = v} \theta(v, w),$$

  i.e. the flow into and out of each vertex other than the root is conserved.

- $\theta(e) \leqslant C(e)$, i.e. the flow through the edge e is bounded by its capacity.

A set Π of edges is called a **cutset** if every ray includes an edge from Π.

We now give a short proof of a famous result of graph theory, the max-flow min-cut theorem of Ford and Fulkerson [FF56] in the special case of infinite trees.

Theorem 12.36 (Max-flow min-cut theorem)

$$\max\left\{\text{strength}\,(\theta)\colon \theta \text{ a flow with capacities } C\right\} = \inf\left\{\sum_{e\in\Pi} C(e)\colon \Pi \text{ a cutset}\right\}.$$

Proof. The proof is a festival of compactness arguments.

First observe that on the left hand side the infimum is indeed a maximum, because if $\{\theta_n\}$ is a sequence of flows with capacities C, then at every edge we have a bounded sequence $\{\theta_n(e)\}$ and by the diagonal argument we may pass to a subsequence such that $\lim \theta_n(e)$ exists simultaneously for all $e \in E$. This limit is obviously again a flow with capacities C. Secondly observe that every cutset Π contains a finite subset $\Pi' \subset \Pi$, which is still a cutset. Indeed, if this was not the case, we had for every positive integer j a ray $e_1^j, e_2^j, e_3^j, \ldots$ with $e_i^j \notin \Pi$ for all $i \leqslant j$. By the diagonal argument we find a sequence j_k and edges e_l of order l such that $e_l^{j_k} = e_l$ for all $k \geqslant l$. Then $e_1, e_2, \ldots$ is a ray and $e_l \notin \Pi$ for all l, which is a contradiction.

Now let θ be a flow with capacities C and Π an arbitrary cutset. We let A be the set of vertices v such that there is a sequence of edges $e_1, \ldots, e_n \notin \Pi$ with $e_1 = (\rho, v_1)$, $e_n = (v_{n-1}, v)$ and $e_j = (v_{j-1}, v_j)$. By our previous observation this set is finite. Let

$$\phi(v, e) := \begin{cases} 1 & \text{if } e = (v, w) \text{ for some } w \in V, \\ -1 & \text{if } e = (w, v) \text{ for some } w \in V. \end{cases}$$

Then, using the definition of a flow and finiteness of all sums,

$$\begin{aligned} \text{strength}\,(\theta) &= \sum_{e\in E} \phi(\rho, e)\theta(e) = \sum_{v\in A}\sum_{e\in E} \phi(v, e)\theta(e) \\ &= \sum_{e\in E} \theta(e) \sum_{v\in A} \phi(v, e) \leqslant \sum_{e\in\Pi} \theta(e) \leqslant \sum_{e\in\Pi} C(e). \end{aligned}$$

This proves the first inequality.

For the reverse inequality we restrict attention to finite trees. Let T_n be the tree consisting of all vertices V_n and edges E_n of order $\leqslant n$ and look at cutsets Π consisting of edges in E_n. A flow θ of strength $c > 0$ through the finite tree T_n with capacities C is defined as in the case of infinite trees, except that the main condition

$$\theta(\overline{v}, v) = \sum_{w\,:\,\overline{w}=v} \theta(v, w),$$

is only required for vertices $v \neq \rho$ with $|v| < n$. We shall show that

$$\begin{aligned} \max\Big\{\text{strength}\,(\theta)\colon &\theta \text{ a flow in } T_n \text{ with capacities } C\Big\} \\ &\geqslant \min\Big\{\sum_{e\in\Pi} C(e)\colon \Pi \text{ a cutset in } T_n\Big\}. \end{aligned} \tag{12.5}$$

Once we have this, we get a sequence (θ_n) of flows in T_n with capacities C and strength at least $c = \min\{\sum_{e \in \Pi} C(e) : \Pi \text{ a cutset in } T\}$. By using the diagonal argument once more we can get a subsequence such that the limits of $\theta_n(e)$ exist for every edge, and the result is a flow θ with capacities C and strength at least c, as required.

To prove (12.5) let θ be a flow of maximal strength c with capacities C in T_n and call a sequence $\rho = v_0, v_1, \ldots, v_n$ with $(v_i, v_{i+1}) \in E_n$ an *augmenting sequence* if $\theta(v_i, v_{i+1}) < C(v_i, v_{i+1})$. If there are augmenting sequences, we can construct a flow $\tilde{\theta}$ of strength $> c$ by just increasing the flow through every edge of the augmenting sequence by a sufficiently small $\varepsilon > 0$. As θ was maximal this is a contradiction. Hence there is a minimal cutset Π consisting entirely of edges in E_n with $\theta(e) \geqslant C(e)$. Let A, as above, be the collection of all vertices which are connected to the root by edges not in Π. As before, we have

$$\text{strength}\,(\theta) = \sum_{e \in E} \theta(e) \sum_{v \in A} \phi(v, e) = \sum_{e \in \Pi} \theta(e) \geqslant \sum_{e \in \Pi} C(e),$$

where in the penultimate step we use minimality. This proves (12.5). ∎

Finally, we discuss the most important class of random trees, the **Galton–Watson trees**. For their construction we pick an **offspring distribution**, given as the law of a random variable N with values in the nonnegative integers. To initiate the recursive construction of the tree, we sample from this distribution to determine the number of offspring of the root. Having constructed the tree up to the nth generation and supposing this generation is nonempty, we sample an independent copy of N for each vertex in this generation and attach the corresponding number of offspring to it. If this procedure is infinite, i.e. if it produces an infinite tree, we say that the Galton–Watson tree *survives* otherwise that it becomes *extinct*. The sharp criterion below is at least as old as the work of Galton and Watson in the middle of the nineteenth century.

Proposition 12.37 *If $N \neq 1$ with positive probability, a Galton–Watson tree survives with positive probability if and only if $\mathbb{E}N > 1$. Moreover, the extinction probability is the smallest nonnegative fixed point of the generating function $f \colon [0,1] \to [0,1]$ given by $f(z) = \mathbb{E}z^N$.*

Proof. Note that the generating function of the number Z_n of vertices in the nth generation is the iterate $f_n = f \circ \overset{n}{\cdots} \circ f$. Elementary analysis shows that $f_n(0)$ converges increasingly to the smallest nonnegative fixed point of f. At the same time

$$\lim_{n \to \infty} f_n(0) = \lim_{n \to \infty} \mathbb{P}\{Z_n = 0\} = \lim_{n \to \infty} \mathbb{P}\{Z_i = 0 \text{ for some } 1 \leqslant i \leqslant n\}$$
$$= \mathbb{P}\{Z_i = 0 \text{ for some } i \geqslant 1\} = \mathbb{P}\{\text{ extinction }\}.$$

It is again an exercise in elementary analysis to see that, unless f is the identity, the smallest nonnegative fixed point of f is one if and only if $\mathbb{E}N = f'(1) \leqslant 1$. ∎

Hints and solutions for selected exercises

Here we give hints, solutions or additional references for the exercises marked with the symbol Ⓢ in the main body of the text.

Exercise 1.2. Using the notation from Theorem 1.3, the Brownian motion is defined on a probability space $(\Omega, \mathcal{A}, \mathbb{P})$ on which a collection $\{Z_t : t \in \mathcal{D}\}$ of independent, standard normally distributed random variables are defined. It is easy to see from the construction that, for any $n \in \mathbb{N}$, the functions F_n are jointly measurable as a function of $Z_d, d \in \mathcal{D}_n$ and $t \in [0, 1]$. Therefore it is also jointly measurable as a function of $\omega \in \Omega$ and $t \in [0, 1]$, and this carries over to $(\omega, t) \mapsto B(\omega, t)$ by summation and taking a limit.

Exercise 1.3. Fix times $0 < t_1 < \ldots < t_n$. Let

$$
M := \begin{pmatrix} 1 & 0 & \cdots & 0 \\ -1 & \ddots & \ddots & \vdots \\ 0 & \ddots & \ddots & 0 \\ 0 & 0 & -1 & 1 \end{pmatrix}, \qquad
D := \begin{pmatrix} \frac{1}{\sqrt{t_1}} & 0 & \cdots & 0 \\ 0 & \frac{1}{\sqrt{t_2 - t_1}} & \ddots & \vdots \\ \vdots & & \ddots & 0 \\ 0 & \cdots & 0 & \frac{1}{\sqrt{t_n - t_{n-1}}} \end{pmatrix}.
$$

Then, for a Brownian motion $\{B(t) : t \geq 0\}$ with start in x, by definition, the vector

$$
X := D M \left(B(t_1) - x, \ldots, B(t_n) - x \right)^{\mathsf{T}}
$$

has independent standard normal entries. As both D and M are nonsingular, the matrix $A := M^{-1} D^{-1}$ is well-defined and, denoting also $b = (x, \ldots, x)$, we have that

$$
\left(B(t_1), \ldots, B(t_n) \right)^{\mathsf{T}} = AX + b.
$$

By definition, this means that $(B(t_1), \ldots, B(t_n))$ is a Gaussian random vector.

Exercise 1.5. Note that $\{X(t) : 0 \leq t \leq 1\}$ is a Gaussian process, while the distributions given in (a) determine Gaussian random vectors. Hence it suffices to identify the means and covariances of $(X(t_1), \ldots, X(t_n))$ and compare them with those given in (a). Starting with the mean, on the one hand we obviously have $\mathbb{E}X(t) = x(1 - t) + ty$, on the other

361

hand

$$\int z \, \frac{\mathfrak{p}(t, x, z) \mathfrak{p}(1 - t, z, y)}{\mathfrak{p}(1, x, y)}$$

$$= \frac{1}{\mathfrak{p}(1, x, y)} \int \left(z - x(1 - t) - ty\right) \mathfrak{p}(t, x, z) \mathfrak{p}(1 - t, z, y) \, dz + x(1 - t) + ty,$$

and the integral can be seen to vanish by completing the square in the exponent of the integrand. To perform the covariance calculation one may assume that $x = y = 0$, which reduces the complexity of expressions significantly, see (8.5) in Chapter 7 of [Du96] for more details.

Exercise 1.6. $B(t)$ does not oscillate too much between n and $n + 1$ if

$$\limsup_{n \to \infty} \frac{1}{n} \Big[\max_{n \leqslant t \leqslant n+1} B(t) - B(n) \Big] = 0 \,.$$

Estimate $\mathbb{P}\{\max_{0 \leqslant t \leqslant 1} B(t) \geqslant \varepsilon n\}$ and use the Borel–Cantelli lemma.

Exercise 1.7. One has to improve the lower bound, and show that, for every constant $c < \sqrt{2}$, almost surely, there exists $\varepsilon > 0$ such that, for all $0 < h < \varepsilon$, there exists $t \in [0, 1 - h]$ with

$$\left| B(t + h) - B(t) \right| \geqslant c \sqrt{h \log(1/h)} \,.$$

To this end, given $\delta > 0$, let $c < \sqrt{2} - \delta$ and define, for integers $k, n \geqslant 0$, the events

$$A_{k,n} = \left\{ B\big((k + 1)e^{-\sqrt{n}}\big) - B\big(ke^{-\sqrt{n}}\big) > c \big(\sqrt{n}\, e^{-\sqrt{n}}\big)^{\frac{1}{2}} \right\} \,.$$

Then, using Lemma 12.9, for any $k \geqslant 0$,

$$\mathbb{P}(A_{k,n}) = \mathbb{P}\left\{ B\big(e^{-\sqrt{n}}\big) > c \big(\sqrt{n}\, e^{-\sqrt{n}}\big)^{\frac{1}{2}} \right\} = \mathbb{P}\left\{ B(1) > c \, n^{\frac{1}{4}} \right\} \geqslant \frac{c \, n^{\frac{1}{4}}}{c^2 \sqrt{n} + 1} \, e^{-c^2 \sqrt{n}/2} \,.$$

Therefore, by our assumption on c, and using that $1 - x \leqslant e^{-x}$ for all $x \geqslant 0$,

$$\sum_{n=0}^{\infty} \mathbb{P}\left(\bigcap_{k=0}^{\lfloor e^{\sqrt{n}} - 1 \rfloor} A_{k,n}^c \right) \leqslant \sum_{n=0}^{\infty} \big(1 - \mathbb{P}(A_{0,n})\big)^{e^{\sqrt{n}} - 1} \leqslant \sum_{n=0}^{\infty} \exp\big(-(e^{\sqrt{n}} - 1) \mathbb{P}(A_{0,n})\big) < \infty \,.$$

From the Borel–Cantelli lemma we thus obtain that, almost surely, there exists $n_0 \in \mathbb{N}$ such that, for all $n \geqslant n_0$, there exists $t \in [0, 1 - e^{-\sqrt{n}}]$ of the form $t = ke^{-\sqrt{n}}$ such that

$$\left| B\big(t + e^{-\sqrt{n}}\big) - B(t) \right| > c \big(\sqrt{n}\, e^{-\sqrt{n}}\big)^{\frac{1}{2}} \,.$$

In addition, we may choose n_0 big enough to ensure that $e^{-\sqrt{n_0}}$ is sufficiently small in the sense of Theorem 1.12. Then we pick $\varepsilon = e^{-\sqrt{n_0}}$ and, given $0 < h < \varepsilon$, choose n such that $e^{-\sqrt{n+1}} < h \leqslant e^{-\sqrt{n}}$. Then, for t as above,

$$\left| B(t + h) - B(t) \right| \geqslant \left| B\big(t + e^{-\sqrt{n}}\big) - B(t) \right| - \left| B(t + h) - B\big(t + e^{-\sqrt{n}}\big) \right|$$

$$> c \big(\sqrt{n}\, e^{-\sqrt{n}}\big)^{\frac{1}{2}} - C \sqrt{\big(e^{-\sqrt{n}} - e^{-\sqrt{n+1}}\big) \log\big(1/(e^{-\sqrt{n}} - e^{-\sqrt{n+1}})\big)} \,.$$

It is not hard to see that the second (subtracted) term decays much more rapidly than the first, so that modifying n_0 to ensure that it is below $\delta \big(\sqrt{n}\, e^{-\sqrt{n}}\big)^{\frac{1}{2}}$ gives the result.

Exercise 1.8. Given $f \in \mathbf{C}[0,1]$ and $\varepsilon > 0$ there exists n such that the function $g \in \mathbf{C}[0,1]$, which agrees with f on the dyadic points in $\mathcal{D}_n$ and is linearly interpolated inbetween, satisfies $\sup |f(t) - g(t)| < \varepsilon$. Then use Lévy's construction of Brownian motion and the fact that normal distributions have full support to complete the proof.

Exercise 1.9. It suffices to show that, for fixed $\varepsilon > 0$ and $c > 0$, almost surely, for all $t \geqslant 0$, there exists $0 < h < \varepsilon$ with $|B(t+h) - B(t)| > ch^\alpha$. By Brownian scaling we may further assume $\varepsilon = 1$. Note that, after this simplification, the complementary event means that there is a $t_0 \geqslant 0$ such that

$$\sup_{h \in (0,1)} \frac{B(t_0+h) - B(t_0)}{h^\alpha} \leqslant c \quad \text{or} \quad \inf_{h \in (0,1)} \frac{B(t_0+h) - B(t_0)}{h^\alpha} \geqslant -c.$$

We may assume that $t_0 \in [0,1)$. Fix $l \geqslant 1/(\alpha - \frac{1}{2})$. Then $t_0 \in \left[\frac{k-1}{2^n}, \frac{k}{2^n}\right)$ for any large n and some $0 \leqslant k < 2^n - l$. Then, by the triangle inequality, for all $j \in \{1, \ldots, 2^n - k\}$,

$$\left| B\left(\tfrac{k+j}{2^n}\right) - B\left(\tfrac{k+j-1}{2^n}\right) \right| \leqslant 2c \left(\tfrac{j+1}{2^n}\right)^\alpha.$$

Now, for any $0 \leqslant k < 2^n - l$, let $\Omega_{n,k}$ be the event

$$\left\{ \left| B\left(\tfrac{k+j}{2^n}\right) - B\left(\tfrac{k+j-1}{2^n}\right) \right| \leqslant 2c \left(\tfrac{j+1}{2^n}\right)^\alpha \text{ for } j = 1, 2, \ldots, l \right\}.$$

It suffices to show that, almost surely for all sufficiently large n and all $k \in \{0, \ldots, 2^n - l\}$ the event $\Omega_{n,k}$ does not occur. Observe that

$$\mathbb{P}(\Omega_{n,k}) \leqslant \left[\mathbb{P}\left\{ |B(1)| \leqslant 2^{n/2} 2c \left(\tfrac{l+1}{2^n}\right)^\alpha \right\} \right]^l \leqslant \left[2^{n/2} 2c \left(\tfrac{l+1}{2^n}\right)^\alpha \right]^l,$$

since the normal density is bounded by $1/2$. Hence, for a suitable constant C,

$$\mathbb{P}\left(\bigcup_{k=0}^{2^n - l} \Omega_{n,k} \right) \leqslant 2^n \left[2^{n/2} 2c \left(\tfrac{l+1}{2^n}\right)^\alpha \right]^l = C \left[2^{(1 - l(\alpha - 1/2))} \right]^n,$$

which is summable. Thus

$$\mathbb{P}\left(\limsup_{n \to \infty} \bigcup_{k=0}^{2^n - l} \Omega_{n,k} \right) = 0.$$

This is the required statement and hence the proof is complete.

Exercise 1.10. The proof can be found in Chapter 3 of [Du95], or Theorem 3.15 of [Ka02].

Exercise 1.12. Argue as in the proof of Theorem 1.30 with B replaced by $B + f$. The resulting term

$$\mathbb{P}\left\{ |B(1) + \sqrt{2^n} f((k+j)/2^n) - \sqrt{2^n} f((k+j-1)/2^n)| \leqslant 7M/\sqrt{2^n} \right\}$$

can be estimated in exactly the same manner as for the unshifted Brownian motion.

Exercise 1.13. This can be found, together with stronger and more general results, in [BP84]. Put $I = \left[B(1), \sup_{0 \leqslant s \leqslant 1} B(s)\right]$, and define a function $g \colon I \to [0,1]$ by setting

$$g(x) = \sup\{s \in [0,1] \colon B(s) = x\}.$$

First check that almost surely the interval I is nondegenerate, g is strictly decreasing, left continuous and satisfies $B(g(x)) = x$. Then show that almost surely the set of discontinuities of g is dense in I. We restrict our attention to the event of probability 1 on which these assertions hold. Let

$$V_n = \left\{x \in I \colon g(x-h) - g(x) > nh \text{ for some } h \in (0, n^{-1})\right\}.$$

Now show that V_n is open and dense in I. By the Baire category theorem, $V := \bigcap_n V_n$ is uncountable and dense in I. Now if $x \in V$ then there is a sequence $x_n \uparrow x$ such that $g(x_n) - g(x) > n(x - x_n)$. Setting $t = g(x)$ and $t_n = g(x_n)$ we have $t_n \downarrow t$ and $t_n - t > n(B(t) - B(t_n))$, from which it follows that $D^*B(t) \geqslant 0$. On the other hand $D_*B(t) \leqslant 0$ since $B(s) \leqslant B(t)$ for all $s \in (t,1)$, by definition of $t = g(x)$.

Exercise 1.14. We first fix some positive ε and positive a. For some small h and an interval $I \subset [\varepsilon, 1-\varepsilon]$ with length h, we consider the event A that $t_0 \in I$ and we have

$$B(t_0 + \tilde{h}) - B(t_0) > -2ah^{1/4} \qquad \text{for some } h^{1/4} < \tilde{h} \leqslant 2h^{1/4}.$$

We now denote by t_L the left endpoint of I. Using Theorem 1.12 we see there exists some positive C so that

$$B(t_0) - B(t_L) \leqslant C\sqrt{h\log(1/h)}.$$

Hence the event A implies the following events

$$A_1 = \left\{B(t_L - s) - B(t_L) \leqslant C\sqrt{h\log(1/h)} \text{ for all } s \in [0, \varepsilon]\right\},$$

$$A_2 = \left\{B(t_L + s) - B(t_L) \leqslant C\sqrt{h\log(1/h)} \text{ for all } s \in [0, h^{1/4}]\right\}.$$

We now define $T := \inf(s > t_L + h^{1/4} : B(s) > B(t_L) - 2ah^{1/4})$. Then by definition we have that $T \leqslant t_L + 2h^{1/4}$ and this implies the event

$$A_3 = \left\{B(T + s) - B(T) \leqslant 2ah^{1/4} + C\sqrt{h\log(1/h)} \text{ for all } s \in [0, \varepsilon]\right\}.$$

Now by the strong Markov property, these three events are independent and we obtain

$$\mathbb{P}(A) \leqslant \mathbb{P}(A_1)\,\mathbb{P}(A_2)\,\mathbb{P}(A_3).$$

We estimate the probabilities of these three events and obtain

$$\mathbb{P}(A_1) = \mathbb{P}\left\{B(\varepsilon) \leqslant C\sqrt{h\log(1/h)}\right\} \leqslant \frac{1}{\sqrt{2\pi\varepsilon}}\,2C\sqrt{h\log(1/h)},$$

$$\mathbb{P}(A_2) = \mathbb{P}\left\{B(h^{1/4}) \leqslant C\sqrt{h\log(1/h)}\right\} \leqslant \frac{1}{\sqrt{2\pi h^{1/4}}}\,2C\sqrt{h\log(1/h)},$$

$$\mathbb{P}(A_3) = \mathbb{P}\left\{B(\varepsilon) \leqslant 2ah^{1/4} + C\sqrt{h\log(1/h)}\right\} \leqslant \frac{1}{\sqrt{2\pi\varepsilon}}\,2\left(C\,h^{1/4} + 2ah^{1/4}\right).$$

Hence we obtain, for a suitable constant $K > 0$, depending on a and ε, that

$$\mathbb{P}(A) \leqslant K\,h^{9/8}\log(1/h).$$

Summing over a covering collection of $1/h$ intervals of length h gives the bound

$$\mathbb{P}\{t_0 \in [\varepsilon, 1 - \varepsilon] \text{ and } B(t_0 + \tilde{h}) - B(t_0) > -2ah^{1/4} \text{ for some } h^{1/4} < \tilde{h} \leqslant 2h^{1/4}\}$$
$$\leqslant K \log(1/h)h^{1/8}.$$

Taking $h = 2^{-4n-4}$ in this bound and summing over n, we see that

$$\sum_{n=1}^{\infty} \mathbb{P}\left\{t_0 \in [\varepsilon, 1 - \varepsilon] \text{ and } \sup_{2^{-n-1} < h \leqslant 2^{-n}} \frac{B(t_0 + h) - B(t_0)}{h} > -a\right\} < \infty,$$

and from the Borel–Cantelli lemma we obtain that, almost surely, either $t_0 \notin [\varepsilon, 1 - \varepsilon]$, or

$$\limsup_{h \downarrow 0} \frac{B(t_0 + h) - B(t_0)}{h} \leqslant -a.$$

Now recall that a and ε are arbitrary positive numbers, so taking a countable union over a and ε gives that, almost surely, $D^* B(t_0) = -\infty$, as required.

Exercise 1.15. By Brownian scaling it suffices to consider the case $t = 1$.
(a) We first show that, given $M > 0$ large, for any fixed point $s \in [0, 1]$, almost surely there exists $n \in \mathbb{N}$ such that the dyadic interval $I(n, s) := [k2^{-n}, (k+1)2^{-n}]$ containing s satisfies

$$\left|B\big((k+1)2^{-n}\big) - B\big(k2^{-n}\big)\right| \geqslant M\, 2^{-n/2}. \tag{13.1}$$

To see this, it is best to consider the construction of Brownian motion, see Theorem 1.3. Using the notation of that proof, let $d_0 = 1$ and $d_{n+1} \in \mathcal{D}_{n+1} \setminus \mathcal{D}_n$ be the dyadic point that splits the interval $[k2^{-n}, (k+1)2^{-n})$ containing s. This defines a sequence $Z_{d_n}, n = 0, 1, \ldots$ of independent, normally distributed random variables. Now let

$$n = \min\{k \in \{0, 1, \ldots\}: |Z_{d_k}| \geqslant 3M\},$$

which is almost surely well-defined. Moreover,

$$3M \leqslant |Z_{d_n}| = 2^{\frac{n-1}{2}} \left|2B(d_n) - B(d_n - 2^{-n}) - B(d_n + 2^{-n})\right|$$
$$\leqslant 2^{\frac{n+1}{2}} \left|B(d_n) - B(d_n \pm 2^{-n})\right| + 2^{\frac{n-1}{2}} \left|B(d_n + 2^{-n}) - B(d_n - 2^{-n})\right|,$$

where $\pm$ indicates that the inequality holds with either choice of sign. This implies that either $I(n, s)$ or $I(n - 1, s)$ satisfies (13.1). We denote by $N(s)$ the smallest nonnegative integer n, for which (13.1) holds.
By Fubini's theorem, almost surely, we have $N(s) < \infty$ for almost every $s \in [0, 1]$. On this event, we can pick a finite collection of disjoint dyadic intervals $[t_{2j}, t_{2j+1}], j = 0, \ldots, k-1$, with summed lengths exceeding $1/2$, say, such that the partition $0 = t_0 < \cdots < t_{2k} = 1$ given by their endpoints satisfies

$$\sum_{j=1}^{2k} \big(B(t_j) - B(t_{j-1})\big)^2 \geqslant M^2 \sum_{j=1}^{k}(t_{2j+1} - t_{2j}) \geqslant \frac{M}{2},$$

from which (a) follows, as M was arbitrary.

(b) Note that the number of (finite) partitions of $[0, 1]$ consisting of dyadic points is countable. Hence, by (a), given $n \in \mathbb{N}$, we can find a finite set P_n of partitions such that the probability that there exists a partition $0 = t_0 < \cdots < t_k = 1$ in P_n with the property that

$$\sum_{j=1}^{k} \left(B(t_j) - B(t_{j-1})\right)^2 \geqslant n$$

is bigger than $1 - \frac{1}{n}$. Successively enumerating the partitions in $P_1, P_2, \ldots$ yields a sequence satisfying the requirement of (b).

Exercise 1.16. To see convergence in the $\mathbf{L}^2$-sense one can use the independence of the increments of a Brownian motion,

$$\mathbb{E}\left[\sum_{j=1}^{k(n)} \left(B(t_{j+1}^{(n)}) - B(t_j^{(n)})\right)^2 - t\right]^2 = \sum_{j=1}^{k(n)} \mathbb{E}\left[\left(B(t_{j+1}^{(n)}) - B(t_j^{(n)})\right)^2 - (t_{j+1}^{(n)} - t_j^{(n)})\right]^2$$

$$\leqslant \sum_{j=1}^{k(n)} \mathbb{E}\left[\left(B(t_{j+1}^{(n)}) - B(t_j^{(n)})\right)^4 + (t_{j+1}^{(n)} - t_j^{(n)})^2\right].$$

Now, using that the fourth moment of a centred normal distribution with variance σ^2 is $3\sigma^4$, this can be estimated by a constant multiple of

$$\sum_{j=1}^{k(n)} (t_{j+1}^{(n)} - t_j^{(n)})^2,$$

which goes to zero. Moreover, by the Markov inequality

$$\mathbb{P}\left\{\left|\sum_{j=1}^{k(n)} \left(B(t_{j+1}^{(n)}) - B(t_j^{(n)})\right)^2 - t\right| > \varepsilon\right\} \leqslant \varepsilon^{-2} \mathbb{E}\left[\left(\sum_{j=1}^{n} \left(B(t_{j+1}^{(n)}) - B(t_j^{(n)})\right)^2 - t\right)^2\right],$$

and summability of the right hand side together with the Borel–Cantelli lemma ensures almost sure convergence.

Exercise 1.17. Recall (1.5) from Lemma 1.41 and note that it implies

$$\nabla_{2j-1}^{(n)} B = \tfrac{1}{2} \nabla_j^{(n-1)} B + \sigma_n Z\left(\tfrac{2j-1}{2^n}\right), \quad \nabla_{2j}^{(n)} B = \tfrac{1}{2} \nabla_j^{(n-1)} B - \sigma_n Z\left(\tfrac{2j-1}{2^n}\right),$$

where $\sigma_n = 2^{-(n+1)/2}$ and $Z(t)$ for $t \in \mathcal{D}_n \setminus \mathcal{D}_{n-1}$ are i.i.d. standard normal random variables independent of $\mathcal{F}_{n-1}$. Hence

$$\mathbb{E}\left[\exp\left\{-2^n \left(\nabla_{2j-1}^{(n)} B\right)\left(\nabla_{2j-1}^{(n)} F\right) - 2^n \left(\nabla_{2j}^{(n)} B\right)\left(\nabla_{2j}^{(n)} F\right)\right\} \,\big|\, \mathcal{F}_{n-1}\right]$$

$$= \exp\left\{-2^{n-1} \left(\nabla_j^{(n-1)} B\right)\left(\nabla_j^{(n-1)} F\right)\right\}$$

$$\times \mathbb{E}\left[\exp\left\{-2^n \sigma_n Z\left(\tfrac{2j-1}{2^n}\right)\left(\nabla_{2j-1}^{(n)} F - \nabla_{2j}^{(n)} F\right)\right\}\right].$$

The expectation equals

$$\exp\left\{2^{n-2} \left(\nabla_{2j-1}^{(n)} F - \nabla_{2j}^{(n)} F\right)^2\right\}$$

$$= \exp\left\{2^{n-1} \left(\nabla_{2j-1}^{(n)} F\right)^2 + 2^{n-1} \left(\nabla_{2j}^{(n)} F\right)^2 - 2^{n-2} \left(\nabla_j^{(n-1)} F\right)^2\right\}.$$

Rearranging the terms completes the proof.

Exercise 1.19. Write $F(a+h) - F(a)$ as an integral and apply Cauchy-Schwarz.

Exercise 2.3.
(i) If $A \in \mathcal{F}(S)$, then $A \cap \{T \leqslant t\} = (A \cap \{S \leqslant t\}) \cap \{T \leqslant t\} \in \mathcal{F}^+(t)$.
(ii) By (i), $\mathcal{F}(T) \subset \mathcal{F}(T_n)$ for all n, which proves $\subset$. On the other hand, if $A \in \bigcap_{n=1}^{\infty} \mathcal{F}(T_n)$, then for all $t \geqslant 0$,

$$A \cap \{T < t\} = \bigcup_{k=1}^{\infty} \bigcap_{n=k}^{\infty} A \cap \{T_n < t\} \in \mathcal{F}^+(t).$$

(iii) Look at the discrete stopping times T_n defined in the previous example. We have, for any Borel set $A \subset \mathbb{R}^d$,

$$\{B(T_n) \in A\} \cap \{T_n \leqslant k2^{-n}\} = \bigcup_{m=0}^{k} \left(\{B(m2^{-n}) \in A\} \cap \{T_n = m2^{-n}\} \right) \in \mathcal{F}^+(k2^{-n}).$$

Hence $B(T_n)$ is $\mathcal{F}(T_n)$-measurable, and as $T_n \downarrow T$, we get that $B(T) = \lim B(T_n)$ is $\mathcal{F}(T_n)$-measurable for any n. Hence $B(T)$ is $\mathcal{F}(T)$-measurable by part (ii).

Exercise 2.7. If $T = 0$ almost surely there is nothing to show, hence assume $\mathbb{E}[T] > 0$.
(a) By construction, T_n is the sum of n independent random variables with the law of T, hence, by the law of large numbers, almost surely,

$$\lim_{n \to \infty} \frac{T_n}{n} = \mathbb{E}[T] > 0,$$

which, by assumption, is finite. This implies, in particular, that $T_n \to \infty$ almost surely, and together with the law of large numbers for Brownian motion, Corollary 1.11, we get almost surely, $\lim_{n\to\infty} B(T_n)/T_n = 0$. The two limit statements together show that, almost surely,

$$\lim_{n \to \infty} \frac{B(T_n)}{n} = \lim_{n \to \infty} \frac{B(T_n)}{T_n} \lim_{n \to \infty} \frac{T_n}{n} = 0.$$

(b) Again by construction, $B(T_n)$ is the sum of n independent random variables with the law of $B(T)$, which we conveniently denote $X_1, X_2, \ldots$. As

$$\lim_{n \to \infty} \frac{X_n}{n} = \lim_{n \to \infty} \frac{B(T_n)}{n} - \lim_{n \to \infty} \frac{B(T_{n-1})}{n} = 0,$$

the event $\{|X_n| \geqslant n\}$ occurs only finitely often, so that the Borel–Cantelli lemma implies

$$\sum_{n=0}^{\infty} \mathbb{P}\{|X_n| \geqslant n\} < \infty.$$

Hence we have that

$$\mathbb{E}[B(T)] = \mathbb{E}|X_n| \leqslant \sum_{n=0}^{\infty} \mathbb{P}\{|X_n| \geqslant n\} < \infty.$$

(c) By the law of large numbers, almost surely,

$$\lim_{n \to \infty} \frac{B(T_n)}{n} = \lim_{n \to \infty} \frac{1}{n} \sum_{j=1}^{n} X_j = \mathbb{E}[B(T)].$$

Exercise 2.9. Let S be a nonempty, closed set S with no isolated points. To see that it is uncountable, we construct a subset with the cardinality of $\{1, 2\}^{\mathbb{N}}$. Start by choosing a point $x_1 \in S$. As this point is not isolated there exists a further, different point $x_2 \in S$. Now pick two disjoint closed balls B_1, B_2 around these points. Again, as x_1 is not isolated, we can find two points in $B_1 \cap S$, around which we can put disjoint balls contained in $B_1 \cap S$, similarly for $B_2 \cap S$, and so on. Now there is a bijection between $\{1, 2\}^{\mathbb{N}}$ and the decreasing sequences of balls in our construction. The intersection of the balls in each such sequence contains, as S is closed, at least one point of S, and two points belonging to two different sequences are clearly different. This completes the proof.

Exercise 2.13. By Fubini's theorem,

$$\mathbb{E}[T^\alpha] = \int_0^\infty \mathbb{P}\{T > x^{1/\alpha}\}\, dx \leqslant 1 + \int_1^\infty \mathbb{P}\{M(x^{1/\alpha}) < 1\}\, dx.$$

Note that, by Brownian scaling, $\mathbb{P}\{M(x^{1/\alpha}) < 1\} \leqslant C\, x^{-\frac{1}{2\alpha}}$ for a suitable constant $C > 0$, which implies that $\mathbb{E}[T^\alpha] < \infty$, as required.

Exercise 2.16. By Exercise 2.15 the process $\{X(t) \colon t \geqslant 0\}$ defined by

$$X(t) = \exp\left\{2bB(t) - 2b^2 t\right\} \qquad \text{for } t \geqslant 0,$$

defines a martingale. Observe that $T = \inf\{t > 0 \colon B(t) = a + bt\}$ is a stopping time for the natural filtration, which is finite exactly if $B(t) = a + bt$ for some $t > 0$. Then

$$\mathbb{P}\{T < \infty\} = e^{-2ab}\, \mathbb{E}\big[X(T)\, 1\{T < \infty\}\big],$$

and because $\{X^T(t) \colon t \geqslant 0\}$ is bounded, the right hand side equals e^{-2ab}.

Exercise 2.17. Use the binomial expansion of $(B(t) + (B(t+h) - B(t)))^3$ to deduce that $X(t) = B(t)^3 - 3tB(t)$ defines a martingale. We know that $\mathbb{P}_x\{T_R < T_0\} = x/R$. Write $\tau_* = \tau(\{0, R\})$. Then

$$\begin{aligned} x^3 &= \mathbb{E}_x[X(0)] = \mathbb{E}_x[X(\tau_*)] = \mathbb{P}_x\{T_R < T_0\}\, \mathbb{E}_x\big[X(\tau_*) \,|\, T_R < T_0\big] \\ &= \mathbb{P}_x\{T_R < T_0\}\, \mathbb{E}_x\big[R^3 - 3\tau_* R \,|\, T_R < T_0\big] = (x/R)(R^3 - 3\gamma R) = x\,(R^2 - 3\gamma). \end{aligned}$$

Solving the last equation for γ gives the claim.

Exercise 2.20. Part (a) can be proved similarly to Theorem 2.51, which in fact is the special case $\lambda = 0$ of this exercise. For part (b) choose $u \colon U \to \mathbb{R}$ as a bounded solution of

$$\tfrac{1}{2}\Delta u(x) = \lambda\, u(x), \qquad \text{for } x \in U,$$

with $\lim_{x \to x_0} u(x) = f(x_0)$ for all $x_0 \in \partial U$. Then

$$X(t) = e^{-\lambda t} u(B(t)) - \int_0^t e^{-\lambda s} \big(\tfrac{1}{2}\Delta u(B(s)) - \lambda u(B(s))\big)\, ds$$

defines a martingale. For any compact $K \subset U$ we can pick a twice continuously differentiable function $v \colon \mathbb{R}^d \to \mathbb{R}$ with $v = u$ on K and $v = 0$ on U^c. Apply the optional stopping theorem to stopping times $S = 0$, $T = \inf\{t \geqslant 0\colon B(T) \notin K\}$ to get, for every $x \in K$,

$$u(x) = \mathbb{E}[X(0)] = \mathbb{E}[X(T)] = \mathbb{E}_x\left[e^{-\lambda T} f(B(T))\right].$$

Now choose a sequence $K_n \uparrow U$ of compacts and pass to the limit on the right hand side of the equation.

Exercise 3.3. To prove the result for $k = 1$ estimate $|u(x) - u(y)|$ in terms of $|x - y|$ using the mean value formula for harmonic functions and the fact that, if x and y are close, the volume of the symmetric difference of $\mathcal{B}(x,r)$ and $\mathcal{B}(y,r)$ is bounded by a constant multiple of $r^{d-1}|x - y|$. For general k note that the partial derivatives of a harmonic function are themselves harmonic, and iterate the estimate.

Exercise 3.5. Define a random variable Y by $Y := X$, if $X > \lambda\mathbb{E}[X]$, and $Y := 0$, otherwise. Applying the Cauchy–Schwarz inequality to $\mathbb{E}[Y] = \mathbb{E}[Y1\{Y > 0\}]$ gives

$$\mathbb{E}[Y\,1\{Y > 0\}] \leqslant \mathbb{E}[Y^2]^{1/2}\left(\mathbb{P}\{Y > 0\}\right)^{1/2},$$

hence, as $X \geqslant Y \geqslant X - \lambda\mathbb{E}[X]$, we get

$$\mathbb{P}\{X > \lambda\mathbb{E}[X]\} = \mathbb{P}\{Y > 0\} \geqslant \frac{\mathbb{E}[Y]^2}{\mathbb{E}[Y^2]} \geqslant (1 - \lambda)^2\,\frac{\mathbb{E}[X]^2}{\mathbb{E}[X^2]}.$$

Exercise 3.7. For $d \geqslant 3$, choose a and b such that $a + br^{2-d} = \tilde{u}(r)$, and $a + bR^{2-d} = \tilde{u}(R)$. Notice that the harmonic functions given by $u(x) = \tilde{u}(|x|)$ and $v(x) = a + b|x|^{2-d}$ agree on ∂D. They also agree on D by Corollary 3.7. So $u(x) = a + b|x|^{2-d}$. By similar consideration we can show that $u(x) = a + b\log|x|$ in the case $d = 2$.

Exercise 3.8. Let $x, y \in \mathbb{R}^d$, $a = |x - y|$. Suppose u is a positive harmonic function. Then

$$u(x) = \frac{1}{\mathcal{L}\mathcal{B}(x,R)} \int_{\mathcal{B}(x,R)} u(z)\,dz$$

$$\leqslant \frac{\mathcal{L}\mathcal{B}(y,R+a)}{\mathcal{L}\mathcal{B}(x,R)}\,\frac{1}{\mathcal{L}\mathcal{B}(y,R+a)} \int_{\mathcal{B}(y,R+a)} u(z)\,dz = \frac{(R+a)^d}{R^d}\,u(y).$$

This converges to $u(y)$ as $R \to \infty$, so $u(x) \leqslant u(y)$, and by symmetry, $u(x) = u(y)$ for all x, y. Hence u is constant.

Exercise 3.11. Uniqueness is clear, because there is at most one *continuous* extension of u. Let $D_0 \subset D$ be a ball whose closure is contained in D, which contains x. u is bounded and harmonic on $D_1 = D_0 \setminus \{x\}$ and continuous on $\overline{D}_1 \setminus \{x\}$. Show that this already implies that $u(z) = \mathbb{E}_z[u(\tau(D_1))]$ on D_1 and that the right hand side has an obvious harmonic extension to $D_1 \cup \{x\}$, which defines the global extension.

Exercise 3.14. To obtain joint continuity one can show equicontinuity of $G(x, \cdot)$ and $G(\cdot, x)$ in $D \setminus \mathcal{B}(x,\varepsilon)$ for any $\varepsilon > 0$. This follows from the fact that these functions are harmonic, by Theorem 3.35, and the estimates of Exercise 3.3.

Exercise 3.15. Recall that

$$G(x,y) = -\tfrac{1}{\pi} \log|x-y| + \tfrac{1}{\pi} \mathbb{E}_x [\log|B(\tau)-y|].$$

The expectation can be evaluated (one can see how in the proof of Theorem 3.44). The final answer is

$$G(x,y) = \begin{cases} -\tfrac{1}{\pi} \log|x/R - y/R| + \tfrac{1}{\pi} \log\left|\tfrac{x}{|x|} - |x|yR^{-2}\right|, & \text{if } x \neq 0, \, x,y \in \mathcal{B}(0,R), \\ -\tfrac{1}{\pi} \log|y/R| & \text{if } x = 0, \, y \in \mathcal{B}(0,R). \end{cases}$$

Exercise 3.16. Suppose $x,y \notin \overline{\mathcal{B}(0,r)}$ and $A \subset \mathcal{B}(0,r)$ compact. Then, by the strong Markov property applied to the first hitting time of $\partial\mathcal{B}(0,r)$,

$$\mu_A(x,\,\cdot\,) = \int_{\partial\mathcal{B}(0,r)} \mu_A(z,\,\cdot\,)\, d\mu_{\partial\mathcal{B}(0,r)}(x,dz).$$

Use Theorem 3.44 to show that, for $B \subset A$ Borel, $\mu_{\partial\mathcal{B}(0,r)}(x,B) \leqslant C\mu_{\partial\mathcal{B}(0,r)}(y,B)$ for a constant C not depending on B. Complete the argument from there.

Exercise 4.1. Let $\alpha = \log 2/\log 3$. For the upper bound it suffices to find an efficient covering of C by intervals of diameter ε. If $\varepsilon \in (0,1)$ is given, let n be the integer such that $1/3^n < 2\varepsilon \leqslant 1/3^{n-1}$ and look at the sets

$$\left[\sum_{i=1}^n \frac{x_i}{3^i}, \sum_{i=1}^n \frac{x_i}{3^i} + \varepsilon \right] \text{ for } (x_1,\ldots,x_n) \in \{0,2\}^n.$$

These sets obviously cover C and each of them is contained in an open ball centred in an interval of diameter 2ε. Hence

$$M(C,\varepsilon) \leqslant 2^n = 3^{\alpha n} = 3^\alpha \left(3^{n-1} \right)^\alpha \leqslant 3^\alpha (1/\varepsilon)^\alpha.$$

This implies $\overline{\dim}_M C \leqslant \alpha$.

For the lower bound we may assume we have a covering by intervals $(x_k - \varepsilon, x_k + \varepsilon)$, with $x_k \in C$, and let n be the integer such that $1/3^{n+1} \leqslant 2\varepsilon < 1/3^n$. Let $x_k = \sum_{i=1}^\infty x_{i,k} 3^{-i}$. Then

$$B(x_k - \varepsilon, x_k + \varepsilon) \cap C \subset \left\{ \sum_{i=1}^\infty \frac{y_i}{3^i} : y_1 = x_{1,k}, \ldots, y_n = x_{n,k} \right\},$$

and we need at least 2^n sets of the latter type to cover C. Hence,

$$M(C,\varepsilon) \geqslant 2^n = 3^{\alpha n} = (1/3)^\alpha \left(3^{n+1} \right)^\alpha \geqslant (1/3)^\alpha (1/\varepsilon)^\alpha.$$

This implies $\underline{\dim}_M C \geqslant \alpha$.

Exercise 4.2. Given $\varepsilon \in (0,1)$ find the integer n such that $1/(n+1)^2 \leqslant \varepsilon < 1/n^2$. Then the points in $\{1/k : k > n\} \cup \{0\}$ can be covered by $n+1$ intervals of diameter ε, and n further balls suffice to cover the remaining n points. Hence

$$M(E,\varepsilon) \leqslant 2n+1 \leqslant \tfrac{2n+1}{n} \left(1/\varepsilon\right)^{1/2},$$

implying $\overline{\dim}_M (E) \leqslant 1/2$. On the other hand, as the distance between neighbouring points is

$$\frac{1}{k} - \frac{1}{k+1} = \frac{1}{k(k+1)} \geqslant \frac{1}{(k+1)^2},$$

we always need at least $n-1$ sets of diameter ε to cover E, which implies

$$M(E,\varepsilon) \geqslant n-1 \geqslant \tfrac{n-1}{n+1} \left(1/\varepsilon\right)^{1/2},$$

hence $\underline{\dim}_M (E) \geqslant 1/2$.

Exercise 4.3. Suppose E is a bounded metric space with $\underline{\dim}_M E < \alpha$. Choose $\varepsilon > 0$ such that $\underline{\dim}_M E < \alpha - \varepsilon$. Then, for every k there exists $0 < \delta < \frac{1}{k}$ and a covering $E_1, \ldots, E_n$ of E by sets of diameter at most δ with $n \leqslant \delta^{-\alpha+\varepsilon}$. The α-value of this covering is at most $n\delta^\alpha \leqslant \delta^\varepsilon$, which tends to zero for large k. Hence $\mathcal{H}_\infty^\alpha(E) = 0$, and $\dim E \leqslant \alpha$.

Exercise 4.4. Indeed, as $E \subset F$ implies $\dim E \leqslant \dim F$, it is obvious that

$$\dim \bigcup_{k=1}^\infty E_k \geqslant \sup \{ \dim E_k : k \geqslant 1 \}.$$

To see the converse, we use

$$\mathcal{H}_\infty^\alpha \left(\bigcup_{k=1}^\infty E_k \right) \leqslant \inf \left\{ \sum_{k=1}^\infty \sum_{j=1}^\infty |E_{j,k}|^\alpha : E_{1,k}, E_{2,k}, \ldots \text{ covers } E_k \right\}$$

$$= \sum_{k=1}^\infty \inf \left\{ \sum_{j=1}^\infty |E_{j,k}|^\alpha : E_{1,k}, E_{2,k}, \ldots \text{ covers } E_k \right\} = \sum_{k=1}^\infty \mathcal{H}_\infty^\alpha(E_k).$$

Hence,

$$\dim \bigcup_{k=1}^\infty E_k \leqslant \sup \left\{ \alpha \geqslant 0 : \mathcal{H}_\infty^\alpha \left(\bigcup_{k=1}^\infty E_k \right) > 0 \right\} \leqslant \sup \left\{ \alpha \geqslant 0 : \sum_{k=1}^\infty \mathcal{H}_\infty^\alpha(E_k) > 0 \right\}$$

$$\leqslant \sup_{k=1}^\infty \sup \left\{ \alpha \geqslant 0 : \mathcal{H}_\infty^\alpha(E_k) > 0 \right\}.$$

This proves the converse inequality.

Exercise 4.6. Suppose that f is surjective and α-Hölder continuous with Hölder constant $C > 0$, and assume that $\mathcal{H}^{\alpha\beta}(E_1) < \infty$. Given $\varepsilon, \delta > 0$ we can cover E_1 with sets $B_1, B_2, \ldots$ of diameter at most δ such that

$$\sum_{i=1}^{\infty} |B_i|^{\alpha\beta} \leqslant \mathcal{H}^{\alpha\beta}(E_1) + \varepsilon.$$

Note that the sets $f(B_1), f(B_2), \ldots$ cover E_2 and that $|f(B_i)| \leqslant C\,|B_i|^{\alpha} \leqslant C\,\delta^{\alpha}$. Hence

$$\sum_{i=1}^{\infty} |f(B_i)|^{\beta} \leqslant C^{\beta} \sum_{i=1}^{\infty} |B_i|^{\alpha\beta} \leqslant C^{\beta}\,\mathcal{H}^{\alpha\beta}(E_1) + C^{\beta}\,\varepsilon,$$

from which the claimed result for the Hausdorff measure readily follows.

Exercise 4.8. Start with $d = 1$. For any $0 < a < 1/2$ let $C(a)$ be the Cantor set obtained by iteratively removing from each construction interval a central interval of $1 - 2a$ of its length. Note that at the nth level of the construction we have 2^n intervals each of length a^n. It is not hard to show that $C(a)$ has Hausdorff dimension $\log 2 / \log(1/a)$, which solves the problem for the case $d = 1$.

For arbitrary dimension d and given α we find a such that $\dim C(a) = \alpha/d$. Then the Cartesian product $C(a) \times \overset{d}{\ldots} \times C(a)$ has dimension α. The upper bound is straightforward, and the lower bound can be verified, for example, from the mass distribution principle, by considering the natural measure that places mass $1/2^{dn}$ to each of the 2^{dn} cubes of side length a^n at the nth construction level.

Exercise 4.14. Recall that it suffices to show that $\mathcal{H}^{1/2}(\mathrm{Rec}) = 0$ almost surely. In the proof of Lemma 4.21 the maximum process was used to define a measure on the set of record points: this measure can be used to define 'big intervals' analogous to the 'big cubes' in the proof of Theorem 4.18. A similar covering strategy as in this proof yields the result.

Exercise 5.1. Use the Borel–Cantelli lemma for the events

$$E_n = \left\{ \sup_{n \leqslant t < n+1} B(t) - B(n) \geqslant \sqrt{a \log n} \right\}$$

and test for which values of a the series $\mathbb{P}(E_n)$ converges. To estimate the probabilities, the reflection principle and Lemma 12.9 will be useful.

Exercise 5.2. The lower bound is immediate from the one-dimensional statement. For the upper bound pick a finite subset $S \subset \partial \mathcal{B}(0,1)$ of directions such that, for every $x \in \partial \mathcal{B}(0,1)$ there exists $\tilde{x} \in S$ with $|x - \tilde{x}| < \varepsilon$. Almost surely, all Brownian motions in dimension one obtained by projecting $\{B(t) : t \geqslant 0\}$ on the line determined by the vectors in S satisfy the statement. From this one can infer that the limsup under consideration is bounded from above by $1 + \varepsilon$.

Exercise 5.3. Let $T_a = \inf\{t > 0\colon B(t) = a\}$. The proof of the upper bound can be based on the fact that, for $A < 1$ and $q > 1$,

$$\sum_{n=1}^{\infty} \mathbb{P}\{\psi(T_1 - T_{1-q^n}) < \tfrac{1}{A} 2^{-n}\} < \infty.$$

Exercise 5.4. Define the stopping time $\tau_{-1} = \min\{k\colon S_k = -1\}$ and recall the definition of p_n from (5.4). Then

$$p_n = \mathbb{P}\{S_n \geqslant 0\} - \mathbb{P}\{S_n \geqslant 0, \tau_{-1} < n\}.$$

Let $\{S_j^*\colon j \geqslant 0\}$ denote the random walk reflected at time τ_{-1}, that is

$$\begin{aligned}
S_j^* &= S_j & \text{for } j \leqslant \tau_{-1}, \\
S_j^* &= (-1) - (S_j + 1) & \text{for } j > \tau_{-1}.
\end{aligned}$$

Note that if $\tau_{-1} < n$ then $S_n \geqslant 0$ if and only if $S_n^* \leqslant -2$, so

$$p_n = \mathbb{P}\{S_n \geqslant 0\} - \mathbb{P}\{S_n^* \leqslant -2\}.$$

Using symmetry and the reflection principle, we have

$$p_n = \mathbb{P}\{S_n \geqslant 0\} - \mathbb{P}\{S_n \geqslant 2\} = \mathbb{P}\{S_n \in \{0, 1\}\},$$

which means that

$$\begin{aligned}
p_n &= \mathbb{P}\{S_n = 0\} &= \binom{n}{n/2} 2^{-n} & \qquad \text{for } n \text{ even,} \\
p_n &= \mathbb{P}\{S_n = 1\} &= \binom{n}{(n-1)/2} 2^{-n} & \qquad \text{for } n \text{ odd.}
\end{aligned}$$

Recall that Stirling's Formula gives $m! \sim \sqrt{2\pi} m^{m+1/2} e^{-m}$, where the symbol $\sim$ means that the ratio of the two sides approaches 1 as $m \to \infty$. One can deduce from Stirling's Formula that $p_n \sim \sqrt{2/\pi n}$, which proves the result.

Exercise 5.5. Denote by $I_n(k)$ the event that k is a point of increase for $S_0, S_1, \ldots, S_n$ and by $F_n(k) = I_n(k) \setminus \bigcup_{i=0}^{k-1} I_n(i)$ the event that k is the first such point. The events that $\{S_k$ is largest among $S_0, S_1, \ldots S_k\}$ and that $\{S_k$ is smallest among $S_k, S_{k+1}, \ldots S_n\}$ are independent, and therefore $\mathbb{P}(I_n(k)) = p_k p_{n-k}$.
Observe that if S_j is minimal among $S_j, \ldots, S_n$, then any point of increase for $S_0, \ldots, S_j$ is automatically a point of increase for $S_0, \ldots, S_n$. Therefore for $j \leqslant k$ we can write

$$F_n(j) \cap I_n(k) =$$

$$F_j(j) \cap \{S_j \leqslant S_i \leqslant S_k \text{ for all } i \in [j, k]\} \cap \{S_k \text{ is minimal among } S_k, \ldots, S_n\}.$$

The three events on the right hand side are independent, as they involve disjoint sets of summands; the second of these events is of the type considered in Lemma 5.9. Thus,

$$\begin{aligned}
\mathbb{P}(F_n(j) \cap I_n(k)) &\geqslant \mathbb{P}(F_j(j)) p_{k-j}^2 p_{n-k} \\
&\geqslant p_{k-j}^2 \mathbb{P}(F_j(j)) \mathbb{P}\{S_j \text{ is minimal among } S_j, \ldots, S_n\},
\end{aligned}$$

since $p_{n-k} \geqslant p_{n-j}$. Here the two events on the right are independent, and their intersection is precisely $F_n(j)$. Consequently $\mathbb{P}(F_n(j) \cap I_n(k)) \geqslant p_{k-j}^2 \mathbb{P}(F_n(j))$.

Decomposing the event $I_n(k)$ according to the first point of increase gives

$$\sum_{k=0}^{n} p_k p_{n-k} = \sum_{k=0}^{n} \mathbb{P}(I_n(k)) = \sum_{k=0}^{n} \sum_{j=0}^{k} \mathbb{P}(F_n(j) \cap I_n(k))$$

$$\geqslant \sum_{j=0}^{\lfloor n/2 \rfloor} \sum_{k=j}^{j+\lfloor n/2 \rfloor} p_{k-j}^2 \mathbb{P}(F_n(j)) \geqslant \sum_{j=0}^{\lfloor n/2 \rfloor} \mathbb{P}(F_n(j)) \sum_{i=0}^{\lfloor n/2 \rfloor} p_i^2 . \tag{13.2}$$

This yields an upper bound on the probability that $\{S_j : j = 0, \ldots, n\}$ has a point of increase by time $n/2$; but this random walk has a point of increase at time k if and only if the reversed walk $\{S_n - S_{n-i} : i = 0, \ldots, n\}$ has a point of increase at time $n - k$. Thus, doubling the upper bound given by (13.2) proves the statement.

Exercise 5.7. In the proof of Exercise 5.5 we have seen that,

$$\sum_{k=0}^{n} p_k p_{n-k} = \sum_{k=0}^{n} \mathbb{P}(I_n(k)) = \sum_{k=0}^{n} \sum_{j=0}^{k} \mathbb{P}(F_n(j) \cap I_n(k)) .$$

By Lemma 5.9, we have, for $j \leqslant k \leqslant n$,

$$\mathbb{P}(F_n(j) \cap I_n(k)) \leqslant \mathbb{P}(F_n(j) \cap \{S_j \leqslant S_i \leqslant S_k \text{ for } j \leqslant i \leqslant k\})$$

$$\leqslant \mathbb{P}(F_n(j)) p_{\lfloor (k-j)/2 \rfloor}^2 .$$

Thus,

$$\sum_{k=0}^{n} p_k p_{n-k} \leqslant \sum_{k=0}^{n} \sum_{j=0}^{k} \mathbb{P}(F_n(j)) p_{\lfloor (k-j)/2 \rfloor}^2 \leqslant \sum_{j=0}^{n} \mathbb{P}(F_n(j)) \sum_{i=0}^{n} p_{\lfloor i/2 \rfloor}^2 .$$

This implies the statement.

Exercise 5.10. Suppose that X is an arbitrary random variable with vanishing expectation and finite variance. For each $n \in \mathbb{N}$ divide the intersection of the support of X with the interval $[-n, n]$ into finitely intervals with mesh $< \frac{1}{n}$. If $x_1 < \cdots < x_m$ are the partition points, construct the law of X_n by placing, for any $j \in \{0, \ldots, m\}$, atoms of size $P\{X \in [x_j, x_{j+1})\}$ in position $E[X \mid x_j \leqslant X < x_{j+1}]$, using the convention $x_0 = -\infty$ and $x_{m+1} = \infty$. By construction, X_n takes only finitely many values.

Observe that $E[X_n] = 0$ and X_n converges to X in distribution. Moreover, one can show that $\tau_n \to \tau$ almost surely. This implies that $B(\tau_n) \to B(\tau)$ almost surely, and therefore also in distribution, which implies that X has the same law as $B(\tau)$. Fatou's lemma implies that

$$\mathbb{E}[\tau] \leqslant \liminf_{n \uparrow \infty} \mathbb{E}[\tau_n] = \liminf_{n \uparrow \infty} E[X_n^2] < \infty .$$

Hence, by Wald's second lemma, $E[X^2] = \mathbb{E}[B(\tau)^2] = \mathbb{E}[\tau]$.

Exercise 5.11. Note that

$$\left| n\, \mathcal{L}\{t \in [0,1]: S_n^*(t) > 0\} - \#\{k \in \{1,\ldots,n\}: S_k > 0\} \right|$$

is bounded by $\#\{k \in \{1,\ldots,n\}: S_k S_{k-1} \leqslant 0\}$. Hence it suffices to show that

$$\frac{1}{n} \sum_{k=1}^{n} \mathbb{P}\{S_k S_{k-1} \leqslant 0\} \longrightarrow 0.$$

Note that, for any $M > 0$, we have $\{S_k S_{k-1} \leqslant 0\} \subset \{|S_k - S_{k-1}| > M\} \cup \{|S_{k-1}| < M\}$. One can now choose $M > 0$ so large that the probability of the first event on the right, which does not depend on k, is arbitrarily close to zero. Donsker's invariance principle implies that, for any $M > 0$, one has $\mathbb{P}\{|S_{k-1}| < M\} \to 0$, as $k \to \infty$.

Exercise 5.12 (b). For a continuous function $f: [0,\infty) \to \mathbb{R}$ and any $a > 0$ define $\tau_a^f = \inf\{t \geqslant 0: f(t) = a\}$, $\tau_{a,0}^f = \inf\{t \geqslant \tau_a^f: f(t) = 0\}$ and

$$\sigma_{0,a}^f = \sup\{0 \leqslant t \leqslant \tau_{a,0}^f: f(t) = 0\}.$$

Define a mapping Φ^a on the set of continuous functions by letting $\Phi^a f = f$ if $\tau_{a,0}^f = \infty$ and otherwise

$$\Phi^a f(t) = \begin{cases} f(t) & \text{if } t \leqslant \sigma_{0,a}^f \text{ or } t \geqslant \tau_{a,0}^f, \\ f(\tau_{a,0}^f + \sigma_{a,0}^f - t) & \text{if } \sigma_{a,0}^f \leqslant t \leqslant \tau_{a,0}^f. \end{cases}$$

For fixed $n \in \mathbb{N}$, we look at the functions $S_n^*: [0,\infty) \to \mathbb{R}$ associated to a simple random walk as in Donsker's invariance principle. It is easy to see that the laws of S_n^* and $\Phi^a S_n^*$ coincide.

The function Φ^a is continuous on the set of all continuous functions taking positive and negative values in every neighbourhood of every zero. By Theorem 2.28, Brownian motion is almost surely in this set. Hence, by property (v) in the Portmanteau theorem, see Theorem 12.6 in the appendix, and Donsker's invariance principle, the laws of $\{B(t): t \geqslant 0\}$ and $\{\Phi^a B(t): t \geqslant 0\}$ coincide, which readily implies our claim.

Exercise 6.6. From Exercise 2.17 we get, for any $x \in (0,1)$ that

$$\mathbb{E}_x\big[T_1 \,\big|\, T_1 < T_0\big] = \frac{1-x^2}{3}, \qquad \mathbb{E}_x\big[T_0 \,\big|\, T_0 < T_1\big] = \frac{2x-x^2}{3},$$

where T_0, T_1 are the first hitting times of the points 0, resp. 1.
Define stopping times $\tau_0^{(x)} = 0$ and, for $j \geqslant 1$,

$$\sigma_j^{(x)} = \inf\{t > \tau_{j-1}^{(x)}: B(t) = x\}, \qquad \tau_j^{(x)} = \inf\{t > \sigma_j^{(x)}: B(t) \in \{0,1\}\}.$$

Let $N^{(x)} = \min\{j \geqslant 1: B(\tau_j^{(x)}) = 1\}$. Then $N^{(x)}$ is geometric with parameter x. We have

$$\int_0^{T_1} \mathbf{1}\{0 \leqslant B(s) \leqslant 1\}\, ds = \lim_{x \downarrow 0} \sum_{j=1}^{N^{(x)}} (\tau_j^{(x)} - \sigma_j^{(x)}).$$

and this limit is increasing. Hence

$$
\mathbb{E} \int_0^{T_1} 1\{0 \leqslant B(s) \leqslant 1\}\, ds
$$
$$
= \lim_{x \downarrow 0} \mathbb{E}\big[N^{(x)} - 1\big] \mathbb{E}\big[\tau_1^{(x)} - \sigma_1^{(x)} \mid B(\tau_1^{(x)}) = 0\big] + \lim_{x \downarrow 0} \mathbb{E}\big[\tau_1^{(x)} - \sigma_1^{(x)} \mid B(\tau_1^{(x)}) = 1\big]
$$
$$
= \lim_{x \downarrow 0} \Big(\frac{1}{x} - 1\Big)\frac{2x - x^2}{3} + \lim_{x \downarrow 0} \frac{1 - x^2}{3} = 1\,.
$$

Exercise 6.7. Observe that $\mathbb{E}\exp\{\lambda X_j\} = e^\lambda/(2 - e^\lambda)$ for all $\lambda < \log 2$, and hence, for a suitable constant C and all small $\lambda > 0$,

$$
\mathbb{E}\exp\big\{\lambda(X_j - 2)\big\} \leqslant \exp\{\lambda^2 + C\lambda^3\},
$$

by a Taylor expansion. Using this for $\lambda = \frac{\varepsilon}{2}$ we get from Chebyshev's inequality,

$$
\mathbb{P}\Big\{\sum_{j=1}^{k}(X_j - 2) > m\varepsilon\Big\} \leqslant \exp\{-m\tfrac{\varepsilon^2}{2}\}\Big(\mathbb{E}\exp\{\tfrac{\varepsilon}{2}(X_j - 2)\}\Big)^k
$$
$$
\leqslant \exp\big\{-m\tfrac{\varepsilon^2}{2}\big\}\exp\big\{m\big(\tfrac{\varepsilon^2}{4} + C\tfrac{\varepsilon^3}{8}\big)\big\},
$$

which proves the more difficult half of the claim. The inequality for the lower tail is obvious.

Exercise 6.8. We have that

$$
\mathbb{P}\Big\{\frac{(X + \ell)^2}{2} \leqslant t\Big\} = \mathbb{P}\big\{-\sqrt{2t} - \ell \leqslant X \leqslant \sqrt{2t} - \ell\big\}.
$$

So the density of the left hand side is

$$
\frac{1}{2\sqrt{\pi t}}\, e^{-(2t + \ell^2)/2}\Big[e^{\ell\sqrt{2t}} + e^{-\ell\sqrt{2t}}\Big],
$$

which by Taylor expansion is

$$
\frac{1}{\sqrt{\pi t}}\, e^{-(2t + \ell^2)/2}\sum_{k=0}^{\infty}\frac{(\ell\sqrt{2t})^{2k}}{(2k)!}\,.
$$

Recall that $X^2/2$ is distributed as Gamma$(\frac{1}{2})$, and given N the sum $\sum_{i=1}^{N} Z_i$ is distributed as Gamma(N). By conditioning on N, we get that the density of the right hand side is

$$
\sum_{k=0}^{\infty}\frac{\ell^{2k} e^{-\ell^2/2} t^{k-1/2} e^{-t}}{2^k k!\,\Gamma(k + \frac{1}{2})}\,.
$$

Recall that

$$
\Gamma\Big(k + \frac{1}{2}\Big) = \frac{\sqrt{\pi}\,(2k)!}{2^{2k} k!}\,,
$$

and so the densities of both sides are equal.

Exercise 7.1. Given $F \in \mathbf{D}[0,1]$ approximate $f = F'$ by the deterministic step process

$$f_n = \sum_{i=1}^{2^n} 1_{((i-1)2^{-n},i2^{-n}]} 2^n \left[F\left(i2^{-n}\right) - F\left((i-1)2^{-n}\right) \right].$$

Exercise 7.2. Use that $\int_0^T H(s)\, dB(s) = \int_0^\infty H^T(s)\, dB(s)$.

Exercise 7.4. First establish a Taylor formula of the form

$$\left| f(x,y) - f(x_0, y_0) - \nabla_y f(x_0, y_0) \cdot (y - y_0) \right.$$
$$\left. - \nabla_x f(x_0, y_0) \cdot (x - x_0) - \tfrac{1}{2}(x - x_0)^T \operatorname{Hes}_x f(x_0, y_0)(x - x_0) \right|$$
$$\leqslant \omega_1(\delta, M) |y - y_0| + \omega_2(\delta, M)|x - x_0|^2,$$

where $\operatorname{Hes}_x f = (\partial_{ij} f)$ is the $d \times d$-Hessian matrix of second derivatives in the directions of x, and

$$\omega_1(\delta, M) = \sup_{\substack{x_1,x_2 \in [-M,M]^d, y_1,y_2 \in [-M,M]^m \\ |x_1-x_2| \wedge |y_1-y_2| < \delta}} \left| \nabla_y f(x_1,y_1) - \nabla_y f(x_2,y_2) \right|,$$

and the modulus of continuity of $\operatorname{Hes}_x f$ by

$$\omega_2(\delta, M) = \sup_{\substack{x_1,x_2 \in [-M,M]^d, y_1,y_2 \in [-M,M]^m \\ |x_1-x_2| \wedge |y_1-y_2| < \delta}} \left\| \operatorname{Hes}_x f(x_1,y_1) - \operatorname{Hes}_x f(x_2,y_2) \right\|,$$

where $\| \cdot \|$ is the operator norm of a matrix. Then argue as in the proof of Theorem 7.14.

Exercise 7.5. First use Brownian scaling and the Markov property, as in the original proof of Theorem 2.37 to reduce the problem to showing that the distribution of $B(T(1))$ (using the notation of Theorem 2.37) is the Cauchy distribution.
The map defined by $f(z) = \frac{z}{2-z}$, for $z \in \mathbb{C}$, takes the half-plane $\{(x,y)\colon x < 1\}$ onto the unit disk and $f(0) = 0$. The image measure of harmonic measure on $V(1)$ from 0 is the harmonic measure on the unit sphere from the origin, which is uniform. Hence the harmonic measure $\mu_{V(1)}(0, \cdot)$ is the image measure of the uniform distribution ϖ on the unit sphere under f^{-1}, which can be calculated using the derivative of f.

Exercise 7.6. Use that $\theta(t) = W_2(H(t))$ and $\lim_{t \uparrow \infty} H(t) = \infty$.

Exercise 7.9. Suppose h is supported by $[0,b]$ and look at the partitions given by $t_k^{(n)} = bk2^{-n}$, for $k = 0, \ldots, 2^n$. By Theorem 7.33 and Theorem 6.19 we can choose a continuous modification of the process $\{ \int_0^t \operatorname{sign}(B(s) - a)\, dB(s)\colon a \in \mathbb{R} \}$. Hence the Lebesgue integral on the left hand side is also a Riemann integral and can be approximated by the sum

$$\sum_{k=0}^{2^n-1} b2^{-n} h(t_k^{(n)}) \left(\int_0^t \operatorname{sign}(B(s) - t_k^{(n)})\, dB(s) \right) = \int_0^t F_n(B(s))\, dB(s),$$

where

$$F_n(x) = \sum_{k=0}^{2^n-1} b2^{-n} h(t_k^{(n)}) \operatorname{sign}(x - t_k^{(n)}), \qquad \text{for } n \in \mathbb{N}.$$

This is a uniformly bounded sequence, which is uniformly convergent to the Lebesgue integral

$$F(x) = \int_{-\infty}^{\infty} h(a)\,\text{sign}(x-a)\,da.$$

Therefore the sequence of stochastic integrals converges in $\mathbf{L}^2$ to the stochastic integral $\int_0^{\infty} F(B(s))\,dB(s)$, which is the right hand side of our formula.

Exercise 7.12. By Theorem 5.35 we may replace T by the first exit time τ from the interval $(-1,1)$ by a linear Brownian motion.
For statement (a) we use that

$$\mathbb{P}\{\tau < x\} = 2\mathbb{P}\Big\{\max_{0 \leqslant t \leqslant x} B(t) > 1\Big\} - \mathbb{P}\Big\{\max_{0 \leqslant t \leqslant x} B(t) > 1, \min_{0 \leqslant t \leqslant x} B(t) < -1\Big\}.$$

The subtracted term is easily seen to be of smaller order. For the first term we can use the reflection principle and Lemma 12.9 to see that

$$\mathbb{P}\Big\{\max_{0 \leqslant t \leqslant x} B(t) > 1\Big\} = 2\,\mathbb{P}\{B(t) > 1\} \sim 2\sqrt{\tfrac{x}{2\pi}}\,e^{-\frac{1}{2x}}.$$

Combining these results leads to the given asymptotics.

Statement (b) can be inferred from the equation

$$\mathbb{P}\{\tau > x\} = \mathbb{P}_1\big\{B(s) \in (0,2) \text{ for all } 0 \leqslant s \leqslant x\big\}$$

and the representation of the latter probability in (7.15).

Exercise 8.1. Suppose that u is subharmonic and $\mathcal{B}(x,r) \subset U$. Let τ be the first exit time from $\mathcal{B}(x,r)$, which is a stopping time. As $\Delta u(z) \geqslant 0$ for all $z \in U$ we see from the multidimensional version of Itô's formula that

$$u(B(t \wedge \tau)) \leqslant u(B(0)) + \sum_{i=1}^{d} \int_0^{t \wedge \tau} \frac{\partial u}{\partial x_i}(B(s))\,dB_i(s).$$

Note that $\partial u/\partial x_i$ is bounded on the closure of $\mathcal{B}(x,r)$, and thus everything is well-defined. We can now take expectations, and use Exercise 7.2 to see that

$$\mathbb{E}_x\big[u(B(t \wedge \tau))\big] \leqslant \mathbb{E}_x\big[u(B(0))\big] = u(x).$$

Now let $t \uparrow \infty$, so that the left hand side converges to $\mathbb{E}_x[u(B(\tau))]$ and note that this gives the mean value property for spheres. The result follows by integrating over r.

Exercise 8.3. Let u be a solution of the Poisson problem on U. Define open sets $U_n \uparrow U$ by

$$U_n = \big\{x \in U : |x - y| > \tfrac{1}{n} \text{ for all } y \in \partial U\big\}.$$

Let τ_n be the first exit time of U_n, which is a stopping time. As $\frac{1}{2}\Delta u(x) = -g(x)$ for all $x \in U$ we see from the multidimensional version of Itô's formula that

$$u(B(t \wedge \tau_n)) = u(B(0)) + \sum_{i=1}^{d} \int_0^{t \wedge \tau_n} \frac{\partial u}{\partial x_i}(B(s))\,dB_i(s) - \int_0^{t \wedge \tau_n} g(B(s))\,ds.$$

Note that $\partial u/\partial x_i$ is bounded on the closure of U_n, and thus everything is well-defined. We can now take expectations, and use Exercise 7.2 to see that

$$\mathbb{E}_x\big[u(B(t \wedge \tau_n))\big] = u(x) - \mathbb{E}_x \int_0^{t \wedge \tau_n} g(B(s))\, ds.$$

Note that both integrands are bounded. Hence, as $t \uparrow \infty$ and $n \to \infty$, bounded convergence yields that

$$u(x) = \mathbb{E}_x \int_0^\tau g(B(s))\, ds,$$

where we have used the boundary condition to eliminate the left hand side.

Exercise 8.5. First note that the lower bound is elementary, because $\tau > 1$ with positive probability. For the upper bound we proceed in three steps. In the first step, we prove an inequality based on Harris' inequality, see Theorem 5.7.

Let f_1, f_2 be densities on $[0, \infty)$. Suppose that the likelihood ratio $\psi(r) = \frac{f_2(r)}{f_1(r)}$ is increasing, and $h : [0, \infty) \to [0, \infty)$ is decreasing on $[a, \infty)$. Then

$$\frac{\int_0^\infty h(r) f_2(r)\, dr}{\int_0^\infty h(r) f_1(r)\, dr} \leqslant \psi(a) + \frac{\int_a^\infty f_2(r)\, dr}{\int_a^\infty f_1(r)\, dr}. \tag{13.3}$$

To see this, observe first that $\int_0^a h(r) f_2(r)\, dr \leqslant \psi(a) \int_0^a h(r) f_1(r)\, dr$. Write $T_a = \int_a^\infty f_1(r)\, dr$. Using Harris' inequality, we get

$$\int_a^\infty h(r) f_2(r)\, dr = T_a \int_a^\infty h(r) \psi(r) \frac{f_1(r)}{T_a}\, dr$$

$$\leqslant T_a \int_a^\infty h(r) \frac{f_1(r)}{T_a}\, dr \int_a^\infty \psi(r) \frac{f_1(r)}{T_a}\, dr$$

$$= \frac{1}{T_a} \int_a^\infty h(r) f_1(r)\, dr \int_a^\infty f_2(r)\, dr,$$

Combining the two inequalities proves (13.3).
As a second step, we show that, for $t_1 \leqslant t_2$,

$$\mathbb{P}_0\big\{B[t_2, t_2 + s] \cap A \neq \emptyset\big\} \leqslant C_a\, \mathbb{P}_0\big\{B[t_1, t_1 + s] \cap A \neq \emptyset\big\},$$

where

$$C_a = \frac{f_2(a)}{f_1(a)} + \frac{1}{\mathbb{P}_0\{|B(t_1)| > a\}} \leqslant e^{\frac{|a|^2}{2t_1}} + \frac{1}{\mathbb{P}_0\{|B(t_1)| > a\}}$$

and f_j is the density of $|B(t_j)|$. This follows by applying (13.3) with

$$h(r) = \int \mathbb{P}_y\{B[0, s] \cap A \neq \emptyset\}\, d\varpi_{0,r}(y).$$

Finally, to complete the proof, we show that

$$\mathbb{P}_0\big\{B(0, \tau) \cap A \neq \emptyset\big\} \leqslant \frac{C_a}{1 - e^{-1/2}}\, \mathbb{P}\{B[0, 1] \cap A \neq \emptyset\},$$

where $C_a \leqslant e^{|a|^2} + \mathbb{P}_0\{|B(\tfrac{1}{2})| > a\}^{-1}$. To this end, let $H(I) = \mathbb{P}_0\{B(I) \cap A \neq \emptyset\}$, where I is an interval. Then H satisfies $H[t, t + \frac{1}{2}] \leqslant C_a H[\frac{1}{2}, 1]$ for $t \geqslant \frac{1}{2}$. Hence, we can conclude that

$$\mathbb{E}H[0, \tau] \leqslant H[0, 1] + \sum_{j=2}^{\infty} e^{-j/2} H[\tfrac{j}{2}, \tfrac{j+1}{2}] \leqslant C_a \sum_{j=0}^{\infty} e^{-j/2} H[0, 1],$$

which is the required statement.

Exercise 8.10. Note that $X \times X$ is itself a compact metric space. Then, by the Stone–Weierstrass theorem, the vector space spanned by the functions of the form $f(x, y) = g(x)h(y)$, where g, h are continuous functions on X, is dense in the space $\mathbf{C}(X \times X)$ of continuous functions on $X \times X$. Hence weak convergence is implied by the fact that,

$$\lim_{n \to \infty} \int f \, d\mu_n \otimes \mu_n = \lim_{n \to \infty} \int g \, d\mu_n \int h \, d\mu_n = \int g \, d\mu \int h \, d\mu = \int f \, d\mu \otimes \mu.$$

Exercise 8.11. For the proof of the upper bound, choose $M > 0$ such that

$$\inf_{x \in \mathcal{B}(0,1)} \mathbb{P}_x\{B(t) \in \mathcal{B}(0, M) \text{ for all } 0 \leqslant t \leqslant 1\} \geqslant \tfrac{1}{2}.$$

Then, for all $t \geqslant 1$,

$$\int \mathbb{P}_x\{\tau_{\mathcal{B}(0,1)} < t\} \, dx \leqslant \sum_{j=1}^{\lceil t \rceil} \int \mathbb{P}_x\{B[j-1, j] \cap \mathcal{B}(0, 1) \neq \emptyset\} \, dx$$

$$\leqslant 2 \sum_{j=1}^{\lceil t \rceil} \int \mathbb{P}_x\{B(j) \in \mathcal{B}(0, M)\} \, dx$$

$$= 2 \sum_{j=1}^{\lceil t \rceil} \int_{\mathcal{B}(0,M)} \int \mathsf{p}_j(x, y) \, dx \, dy \leqslant \left(4\mathcal{L}(\mathcal{B}(0, 1))M^d\right) t.$$

For the lower bound, we argue that

$$\int \mathbb{P}_x\{\tau_{\mathcal{B}(0,1)} < t\} \, dx$$

$$\geqslant \sum_{j=1}^{\lfloor t \rfloor} \int \mathbb{P}_x\{B[0, j-1] \cap \mathcal{B}(0, 1) = \emptyset, \ B(j) \in \mathcal{B}(0, 1)\} \, dx$$

$$\geqslant \sum_{j=1}^{\lfloor t \rfloor} \int \mathbb{P}_0\{B[1, j] \cap \mathcal{B}(0, 2) = \emptyset, \ B(j) \in \mathcal{B}(x, 1)\} \, dx,$$

reversing time in the last step. Using Fubini's theorem, we rewrite the right hand side as

$$\sum_{j=1}^{\lfloor t \rfloor} \mathbb{E}_0\left[\mathbb{1}\{B[1, j] \cap \mathcal{B}(0, 2) = \emptyset\} \int_{\mathcal{B}(B(j), 1)} dx\right]$$

$$\geqslant \left(\tfrac{1}{2}\mathcal{L}(\mathcal{B}(0, 1))\mathbb{P}_0\{B[1, \infty) \cap \mathcal{B}(0, 2) = \emptyset\}\right) t.$$

Exercise 9.3. Use arguments as in the proof of Theorem 9.22 to transfer the results of Theorem 9.8 from intersections of independent Brownian motions to self-intersections of one Brownian motion.

Exercise 9.8. An example can be constructed as follows: Let A_1 and A_2 be two disjoint closed sets on the line such that the Cartesian squares A_i^2 have Hausdorff dimension less than $1/2$ yet the Cartesian product $A_1 \times A_2$ has dimension strictly greater than $1/2$. Let A be the union of A_1 and A_2. Then Brownian motion $\{B(t): t \geqslant 0\}$ on A is 1-1 with positive probability (if $B(A_1)$ is disjoint from $B(A_2)$) yet with positive probability $B(A_1)$ intersects $B(A_2)$.

For instance let A_1 consist of points in $[0, 1]$ where the binary n^{th} digit vanishes whenever $(2k)! \leqslant n < (2k + 1)!$ for some k. Let A_2 consist of points in $[2, 3]$ where the binary n^{th} digit vanishes whenever $(2k - 1)! \leqslant n < (2k)!$ for some k. Then $\dim(A_i^2) = 0$ for $i = 1, 2$ yet $\dim(A_1 \times A_2) \geqslant \dim(A_1 + A_2) = 1$, in fact $\dim(A_1 \times A_2) = 1$.

Exercise 9.10. Let $\{B_1(t): 0 \leqslant t \leqslant 1\}$ be the first component of the planar motion. By Kaufman's theorem, almost surely,

$$\dim S(a) = 2 \dim\{t \in [0, 1]: B_1(t) = a\}$$

and, as in Corollary 9.30, the dimension on the right equals $1/2$ for every $a \in (\min\{x: (x, y) \in B[0, t]\}, \max\{x: (x, y) \in B[0, t]\})$.

Exercise 10.2. For every decomposition $E = \bigcup_{i=1}^{\infty} E_i$ of E into bounded sets, we have, using countable stability of Hausdorff dimension,

$$\sup_{i=1}^{\infty} \overline{\dim}_M E_i \geqslant \sup_{i=1}^{\infty} \dim E_i = \dim \bigcup_{i=1}^{\infty} E_i = \dim E,$$

and passing to the infimum yields the statement.

Exercise 10.7. The argument is sketched in [La99].

Exercise 10.9. For (a) note that Theorem 10.28 can be read as a criterion to determine the packing dimension of a set E by hitting it with a limsup random fractal. Hence $\dim_P(A \cap E)$ can be found by evaluating $\mathbb{P}\{A \cap A' \cap E = \emptyset\}$ for A' an independent copy of A. Now use that $A \cap A'$ is also a discrete limsup fractal.

Exercise 10.10. To apply Theorem 7.25 for the proof of Lemma 10.40 (a) we shift the cone by defining a new tip $\tilde{z}$ as follows:

- If $\alpha < \pi$ the intersection of the line through x parallel to the central axis of the cone with the boundary of the dual cone,
- if $\alpha > \pi$ the intersection of the line through x parallel to the central axis of the cone with the boundary of the cone.

Note that $z + W[\alpha, \xi] \subset \tilde{z} + W[\alpha, \xi]$ and there exists a constant $C > 1$ depending only on α such that $|z - \tilde{z}| < C\delta$. There is nothing to show if $C\delta > \varepsilon/2$ and otherwise

$$\mathbb{P}_x\big\{B(0, T_\varepsilon(z)) \subset z + W[\alpha, \xi]\big\} \leqslant \mathbb{P}_x\big\{B(0, T_{\varepsilon/2}(\tilde{z})) \subset \tilde{z} + W[\alpha, \xi]\big\}.$$

By shifting, rotating and scaling the Brownian motion and by Theorem 7.25 we obtain an upper bound for the right hand side of

$$\mathbb{P}_1\big\{B(0, T_{\frac{\varepsilon}{\delta}(C+\frac{1}{2})^{-1}}(0)) \subset W[\alpha, 0]\big\} = \tfrac{2}{\pi} \arctan\big(C_0 \big(\tfrac{\delta}{\varepsilon}\big)^{\frac{\pi}{\alpha}}\big) \leqslant C_1 \big(\tfrac{\delta}{\varepsilon}\big)^{\frac{\pi}{\alpha}},$$

where $C_0, C_1 > 0$ are suitable constants.

Selected open problems

In this section we give a personal selection of problems related to the material of this book, which are still open.

(1) *Given an almost sure property of Brownian paths, characterise those continuous functions f such that $B + f$ also has this property almost surely.*

Recall that by the Cameron–Martin theorem, Theorem 1.38, for the functions $f \in D[0,1]$ all almost sure properties of B carry over to $B + f$. Hence only functions $f \in C[0,1] \setminus D[0,1]$ are of interest.

The answer to this problem depends on the property one is looking at. Some problems are easy (and fully resolved) and others are very tricky. Here are some examples:

(a) *Nowhere differentiable.* We have seen in Exercise 1.12 that for *all* continuous functions $f \colon [0,1] \to \mathbb{R}$, the function $B + f$ is nowhere differentiable.

(b) *Not hitting points.* Taking $d \geqslant 2$ the problem is to characterise the functions $f \colon [0,1] \to \mathbb{R}^d$ with the property

$$\mathbb{P}\{\exists t \in (0,1) \text{ such that } B(t) + f(t) = 0\} = 0. \tag{13.1}$$

Recall that there are continuous space-filling curves f, so that it is plausible that some continuous f violate the statement in the display. For $d = 2$ Graversen [Gr82] shows that, for any $\alpha < 1/2$, there exist α-Hölder continuous functions f violating (13.1), and Le Gall [LG88a] shows that any α-Hölder continuous f with $\alpha \geqslant 1/2$ satisfies (13.1). The latter paper also contains finer results near the critical case $\alpha = 1/2$ and results for dimensions $d \geqslant 3$. An extension to Lévy processes is given by Mountford [Mo89].

(c) *No isolated zeros.* Recall from Theorem 2.28 that, for a linear Brownian motion $\{B(t) \colon t \in [0,1]\}$, the set Zeros $= \{t \in [0,1] \colon B(t) = 0\}$ has no isolated points. Using the law of the iterated logarithm in the form of Corollary 5.3, one can easily construct functions $f \in C[0,1]$ such that $\{t \in [0,1] \colon B(t) + f(t) = 0\}$ has an isolated point in the origin. The problem is therefore to characterise those $f \in C[0,1]$ such that the process $\{B(t) + f(t) \colon 0 < t \leqslant 1\}$ has no isolated zeros.

(d) *No double points.* Take Brownian motion $\{B(t) \colon 0 \leqslant t \leqslant 1\}$ in dimension $d = 4$. Characterise those functions $f \in C([0,1], \mathbb{R}^4)$ such that the process $\{B(t) + f(t) \colon 0 \leqslant t \leqslant 1\}$ has no double points.

(2) *What is the minimal Hausdorff dimension of a curve contained in the path of planar Brownian motion?*

We have seen in Theorem 11.10 that the outer boundary is a curve contained in planar Brownian motion, which has Hausdorff dimension $4/3$. It is not known whether this is the curve of minimal dimension. The best known lower bound stems from Pemantle [Pe97], where it is shown that the planar Brownian path does not contain a line segment. It is also unknown whether there exists a Lipschitz curve intersecting the range of planar Brownian motion in a set of positive length.

(3) *Is the set of double points of planar Brownian motion totally disconnected?*

It is natural to conjecture that, almost surely, all connected components of the set of double points of a planar Brownian motion are singletons, but no proof is known. For Brownian motion in $\mathbb{R}^3$ this follows from the fact that the set of double points has $\mathcal{H}^1$-measure zero, together with a general fact from geometric measure theory, see e.g. [Fa97a].

(4) *Can one move between any two domains of the complement of the range of a planar Brownian motion by passing through only a finite number of points of the range?*

This question is due to Wendelin Werner. To put it more formally let $\{B(t): t \geqslant 0\}$ be a planar Brownian motion. We ask whether, almost surely, for any $x, y \in \mathbb{R}^2 \setminus B[0,1]$ there exists a curve $\gamma \colon [0,1] \to \mathbb{R}^2$ with $\gamma(0) = x, \gamma(1) = y$ such that

$$\gamma[0,1] \cap B[0,1]$$

is a finite set.

(5) *For which gauge functions ϕ does a planar Brownian motion visit some (random) point $z \in \mathbb{R}^2$ in a set of positive ϕ-Hausdorff measure?*

This problem is related to finding the 'maximal multiplicity' of points on a planar Brownian curve $\{B(t): t \geqslant 0\}$. We know from Corollary 9.29 that, almost surely,

$$\dim \{t > 0 \colon B(t) = z\} = 0 \quad \text{for all } z \in \mathbb{R}^2.$$

It is however unknown for which gauge functions ϕ we can find an (exceptional) point z such that $\mathcal{H}^\phi \{t > 0 \colon B(t) = z\} > 0$.

(6) *What is the Hausdorff dimension of the set of points where the 'local time' of planar Brownian motion takes a particular value?*

This problem, which is due to Bass, Burdzy and Khoshnevisan [BBK94], requires some background from that paper. Recall from Theorem 9.24 that planar Brownian motion has points of infinite multiplicity. Similar arguments can also be used to show that the Hausdorff dimension of the set of points of infinite multiplicity is still two. How far can we go before we see a reduction in the dimension?

A natural way is to count the number of excursions from a point. To be explicit, let $\{B(s)\colon s \geqslant 0\}$ be a planar Brownian motion and fix $x \in \mathbb{R}^2$ and $\epsilon > 0$. Let $S_{-1} = 0$ and, for any integer $j \geqslant 0$, let $T_j = \inf\{s > S_{j-1}\colon B(s) = x\}$ and $S_j = \inf\{s > T_j\colon |B(s) - x| \geqslant \epsilon\}$. Then define

$$N_\epsilon^x = \max\{j \geqslant 0\colon T_j < \infty\},$$

which is the number of completed excursions from x reaching $\partial\mathcal{B}(x, \epsilon)$. Observe that $\lim_{\epsilon \downarrow 0} N_\epsilon^x = \infty$ if and only if x has infinite multiplicity. It is therefore a natural question to ask how rapidly N_ϵ^x can go to infinity when $\epsilon \downarrow 0$. Bass, Burdzy and Khoshnevisan [BBK94] show that, almost surely,

$$\frac{1}{2} \leqslant \sup_{x \in \mathbb{R}^2} \limsup_{\epsilon \downarrow 0} \frac{N_\epsilon^x}{\log(1/\epsilon)} \leqslant 2e,$$

where the limsup represents a 'local time' of planar Brownian motion in x. It is an open problem to find the value of the supremum and to identify, for any $0 < a < 2$, the value of

$$\dim\left\{x \in \mathbb{R}^2\colon \lim_{\epsilon \downarrow 0} \frac{N_\epsilon^x}{\log(1/\epsilon)} = a\right\}.$$

Partial progress on this problem was made by Bass, Burdzy and Khoshnevisan [BBK94], who show a lower bound of $2 - a$ for the Hausdorff dimension for all $0 < a < \frac{1}{2}$, and an upper bound of $2 - \frac{a}{e}$ for all $0 < a < 2e$.

(7) *Does planar Brownian motion have triple points which are also pioneer points?*

Let $\{B(t)\colon 0 \leqslant t \leqslant 1\}$ be a planar Brownian motion. A point $x \in \mathbb{R}^2$ is called a *pioneer point* if there exists $0 < t \leqslant 1$ such that $x = B(t)$ and x lies on the outer boundary of $B[0, t]$, i.e. on the boundary of the unbounded component of $\mathbb{R}^2 \setminus B[0, t]$. Note that all points on the outer boundary of $B[0, 1]$ itself are pioneer points, but not vice versa. Indeed, Lawler, Schramm and Werner [LSW02] show, using arguments like in Chapter 11, that the Hausdorff dimension of the set of pioneer points is $\frac{7}{4}$.

Burdzy and Werner in [BW96] show that, almost surely, there are no triple points of planar Brownian motion on the outer boundary and conjecture that there are also no triple points which are pioneer points. It is not hard to see (using nontrivial knowledge about intersection exponents) that the set of triple points which are also pioneer points has Hausdorff dimension zero, but it is open whether this set is empty.

Bibliography

[dA83]   A. DE ACOSTA. A new proof of the Hartman–Wintner law of the iterated logarithm. *Ann. Probab.* **11**, 270–276 (1983).

[Ad85]   O. ADELMAN. Brownian motion never increases: a new proof of a theorem of Dvoretzky, Erdős and Kakutani. *Israel J. of Math.* **50**, 189–192 (1985).

[ABP98]   O. ADELMAN, K. BURDZY and R. PEMANTLE. Sets avoided by Brownian motion. *Ann. Probab.* **26**, 429–464 (1998).

[AD85]   O. ADELMAN and A. DVORETZKY. Plane Brownian motion has strictly n-multiple points. *Israel J. of Math.* **52**, 361–364 (1985).

[Ad90]   R. ADLER. *An introduction to continuity, extrema and related topics for general Gaussian processes.* Institute Math. Statistics, Hayward, CA (1990).

[Ah78]   L.V. AHLFORS. *Complex Analysis.* McGraw-Hill, New York (1978).

[Ai96]   M. AIZENMAN. The geometry of critical percolation and conformal invariance. In: *Statphys19 (Xiamen, 1995)*, 104-120, World Scientific, River Edge, NJ (1996).

[An87]   D. ANDRÉ. Solution directe du problème résolu par M. Bertrand. *C. R. Acad. Sci. Paris* **105**, 436–437 (1887).

[AN04]   K. B. ATHREYA and P.E. NEY. *Branching processes.* Dover, Mineola, NY (2004).

[AY79]   J. AZÉMA and M. YOR. Une solution simple au problème de Skorokhod. In: *Séminaire de probabilités XIII*, 90–115, Lecture Notes in Mathematics 721. Springer, Berlin (1979).

[Ba00]   L. BACHELIER. Théorie de la speculation. *Ann. Sci. Ecole Norm. Sup.* **17**, 21–86 (1900).

[Ba01]   L. BACHELIER. Théorie mathématique du jeu. *Ann. Sci. Ecole Norm. Sup.* **18**, 143–210 (1901).

[BS97]   R. BAÑUELOS and R. SMITS. Brownian motion in cones. *Probab. Theory Related Fields* **108**, 299–319 (1997).

[BP84]   M. T. BARLOW and E. PERKINS. Levels at which every Brownian excursion is exceptional. In: *Séminaire de probabilités XVIII*, 1–28, *Lecture Notes in Mathematics* 1059. Springer, Berlin (1984).

[Ba95]   R. BASS. *Probabilistic Techniques in Analysis.* Springer, New York (1995).

[Ba98]   R. BASS. *Diffusions and elliptic operators.* Springer, New York (1998).

[BB97]   R. BASS and K. BURDZY. Cutting Brownian paths. *Memoir Amer. Math. Soc.* **137** (1997).

[BBK94]   R.F. BASS, K. BURDZY and D. KHOSHNEVISAN. Intersection local time for points of infinite multiplicity. *Ann. Probab.* **22**, 566–625 (1994).

[Ba62]   G. BAXTER. Combinatorial methods in fluctuation theory. *Z. Wahrscheinlichkeitstheorie verw. Gebiete* **1**, 263–270 (1962).

[Be04]   V. BEFFARA. Hausdorff dimensions for SLE$_6$. *Ann. Probab.* **32**, 2606-2629 (2004).

[Be08]   V. BEFFARA. The dimension of the SLE curves. *Ann. Probab.* **36**, 1421-1452 (2008).

[BPP95]   I. BENJAMINI, R. PEMANTLE and Y. PERES. Martin capacity for Markov chains. *Ann. Probab.* **23**, 1332–1346 (1995).

[BP94]   I. BENJAMINI and Y. PERES. Tree-indexed random walks on groups and first passage percolation. *Probab. Theory Related Fields* **98**, 91–112 (1994).

[Be83]   S. M. BERMAN. Nonincrease almost everywhere of certain measurable functions with applications to stochastic processes. *Proc. Amer. Math. Soc.* **88**, 141–144 (1983).

[Be91]   J. BERTOIN. Increase of a Lévy process with no positive jumps. *Stochastics* **37**, 247–251 (1991).

[Be96]   J. BERTOIN. *Lévy processes*. Cambridge University Press, Cambridge (1996).

[Bi67]   P. BICKEL. Some contributions to the theory of order statistics. *Proc. 5th Berkeley Symp. Math. Statist. Probab.* **2**, 575–591 (1967).

[Bi99]   P. BILLINGSLEY. *Convergence of probability measures.* Second edition. Wiley, New York (1999).

[Bi95]   P. BILLINGSLEY. *Probability and measure.* Third edition. Wiley, New York (1995).

[Bi82]   P. BILLINGSLEY. Van der Waerden's continuous nowhere differentiable function. *Amer Math. Monthly* **89**, 691 (1982).

[Bi86]   N. H. BINGHAM. Variants on the law of the iterated logarithm. *Bull. London Math. Soc.* **18**, 433–467 (1986).

[BP96]   C. J. BISHOP and Y. PERES. Packing dimension and Cartesian products. *Trans. Amer. Math. Soc.* **348**, 4433–4445 (1996).

[BJPP97]   C. J. BISHOP, P. W. JONES, R. PEMANTLE and Y. PERES. The outer boundary of Brownian motion has dimension greater than one. *J. Funct. Analysis* **143**, 309-336 (1997).

[Bl57]   R.M. BLUMENTHAL. An extended Markov property. *Trans. Amer. Math. Soc.* **82**, 52–72 (1957).

[BG68]   R.M. BLUMENTHAL and R. GETOOR. *Markov processes and potential theory.* Academic Press, New York (1968).

[Bo89]   A.N. BORODIN. Brownian local time. *Russian Math. Surveys* **44**, 1–51 (1989).

[BS02]   A.N. BORODIN and P. SALMINEN. *Handbook of Brownian motion. Facts and formulae.* Second edition. Birkhäuser, Basel (2002).

[Br76]   G.R. BROSAMLER. A probabilistic version of the Neumann problem. *Math. Scand.* **38**, 137–147 (1976).

[Br28]   R. BROWN. A brief description of microscopical observations made in the months of June, July and August 1827, on the particles contained in the pollen of plants; and on the general existence of active molecules in organic and inorganic bodies. *Ann. Phys.* **14**, 294–313 (1828).

[Bu89]   K. BURDZY. Cut points on Brownian paths. *Ann. Probab.* **17**, 1012–1036 (1989).

[Bu90]   K. BURDZY. On nonincrease of Brownian motion. *Ann. Probab.* **18**, 978–980 (1990).

[Bu95]   K. BURDZY. Labyrinth dimension of Brownian trace. Dedicated to the memory of Jerzy Neyman. *Probability and Mathematical Statistics* **15**, 165–193 (1995).

[BL90]   K. BURDZY and G. F. LAWLER. Non-intersection exponents for Brownian paths. Part II: Estimates and applications to a random fractal. *Ann. Probab.* **18**, 981–1009 (1990).

[BSM89]   K. BURDZY and J. SAN MARTÍN. Curvature of the convex hull of planar Brownian motion near its minimum point. *Stoch. Proc. Appl.* **33**, 89–103 (1989).

[BW96]   K. BURDZY and W. WERNER. No triple point of planar Brownian motion is accessible. *Ann. Probab.* **24**, 125–147 (1996).

[Bu77] D. L. BURKHOLDER. Exit times of Brownian motion, harmonic majorization and Hardy spaces. *Adv. Math.* **26**, 182–205 (1977).

[BDG72] D. BURKHOLDER, B. DAVIS and R. GUNDY. Integral inequalities for convex functions of operators on martingales. *Proc. 6th Berkeley Symp. Math. Statist. Probab.* **2**, 223–240 (1972).

[CN07] F. CAMIA and C. NEWMAN. Critical percolation exploration path and SLE(6): a proof of convergence. *Probab. Theory Related Fields* **139**, 473-520 (2007).

[Ca14] C. CARATHÉODORY. Über das lineare Maß von Punktmengen, eine Verallgemeinerung des Längenbegriffs. *Nachrichten Ges. Wiss. Göttingen* 406–426 (1914).

[Ca84] J.L. CARDY. Conformal invariance and surface critical behavior. *Nucl. Phys.* **B 240**, 514-532 (1984).

[Ca67] L. CARLESON. *Selected Problems on Exceptional Sets.* Van Nostrand, Princeton, NJ (1967).

[CM01] L. CARLESON and N.G. MAKAROV. Aggregation in the plane and Loewner's equation. *Comm. Math. Phys.* **216**, 583-607 (2001).

[CM02] L. CARLESON and N.G. MAKAROV. Laplacian path models. Dedicated to the memory of Thomas H. Wolff. *J. Anal. Math.* **87**, 103–150 (2002).

[Ch07] S. CHATTERJEE. A new approach to strong embeddings. *ArXiv math* **0711.0501** (2007).

[Ch72] J.P.R. CHRISTENSEN. On sets of Haar measure zero in abelian Polish groups. *Israel J. of Math.* **13**, 255–260 (1972).

[Ch73] K.L. CHUNG. Probabilistic approach in potential theory to the equilibrium problem. *Ann. Inst. Fourier, Grenoble* **23**, 313–322 (1973).

[Ch82] K.L. CHUNG. *Lectures from Markov processes to Brownian motion.* Springer, New York (1982).

[Ch02] K.L. CHUNG. *Green, Brown and Probability & Brownian motion on the line.* World Scientific, River Edge, NJ (2002).

[CW90] K.L. CHUNG and R. J. WILLIAMS. *Introduction to stochastic integration.* Second edition. Birkhäuser, Boston, MA (1990).

[CT62] Z. CIESIELSKI and S. J. TAYLOR. First passage times and sojourn times for Brownian motion in space and the exact Hausdorff measure of the sample path. *Trans. Amer. Math. Soc.* **103**, 434–450 (1962).

[CFL28] R. COURANT, K. FRIEDRICHS and H. LEWY. Über die partiellen Differentialgleichungen der mathematischen Physik. *Math. Annalen* **100**, 32–74 (1928).

[CHM89] M. CRANSTON, P. HSU and P. MARCH. Smoothness of the convex hull of planar Brownian motion. *Ann. Probab.* **17**, 144–150 (1989).

[CHu07] E. CSÁKI and Y. HU. Strong approximations of three-dimensional Wiener sausages. *Acta Math. Hungar.* **114**, 205–226 (2007).

[CR81] M. CSÖRGÖ and P. REVESZ. *Strong Approximations in Probability and Statistics.* Academic Press, New York (1981).

[Da65] K.E. DAMBIS. On the decomposition of continuous martingales. *Theor. Prob. Appl.* **10**, 401–410 (1965).

[DK57] D.A. DARLING and M. KAC. On occupation times of Markoff processes. *Trans. Amer. Math. Soc.* **84**, 444–458 (1957).

[Da75] B. DAVIS. Picard's theorem and Brownian motion. *Trans. Amer. Math. Soc.* **213**, 353–362 (1975).

[Da83] B. DAVIS. On Brownian slow points. *Z. Wahrscheinlichkeitstheorie verw. Gebiete* **64**, 359–367 (1983).

[DE06]   M. DAVIS and A. ETHERIDGE. Louis Bachelier's theory of speculation: The origins of modern finance. Princeton University Press, Princeton, NJ (2006).

[DM98]   P. DEHEUVELS and D.M. MASON. Random fractal functional laws of the iterated logarithm. *Studia Sci. Math.Hungar.* **34**, 89–106 (1998).

[DM04]   P. DEL MORAL. *Feynman–Kac formulae. Genealogical and interacting particle systems with applications.* Springer, New York (2004).

[DPRZ00a]   A. DEMBO, Y. PERES, J. ROSEN and O. ZEITOUNI. Thick points for spatial Brownian motion: multifractal analysis of occupation measure. *Ann. Probab.* **28(1)**, 1–35 (2000).

[DPRZ00b]   A. DEMBO, Y. PERES, J. ROSEN and O. ZEITOUNI. Thin points for Brownian motion. *Ann. Inst. H. Poincaré Probab. Statist.* **36**, 749–774 (2000).

[DPRZ01]   A. DEMBO, Y. PERES, J. ROSEN and O. ZEITOUNI. Thick points for planar Brownian motion and the Erdős–Taylor conjecture on random walk. *Acta Math.* **186**, 239–270 (2001).

[DPRZ02]   A. DEMBO, Y. PERES, J. ROSEN and O. ZEITOUNI. Thick points for intersections of planar sample paths. *Trans. Amer. Math. Soc.* **354**, 4969–5003 (2002).

[Do96]   R. A. DONEY. Increase of Lévy processes. *Ann. Probab.* **24**, 961–970 (1996).

[Do51]   M. D. DONSKER. An invariance principle for certain probability limit theorems. *Mem. Amer. Math. Soc.* **6** (1951–52).

[Do49]   J. L. DOOB. Heuristic approach to the Kolmogorov–Smirnov theorems. *Ann. Math. Stat.* **20**, 393–403 (1949).

[Do53]   J.L. DOOB. *Stochastic Processes*. Wiley, New York (1953).

[Do84]   J.L. DOOB *Classical potential theory and its probabilistic counterpart.* Grundlehren der Mathematischen Wissenschaften, Springer, New York (1984).

[Du68]   L. E. DUBINS. On a theorem of Skorokhod. *Ann. Math. Statist.* **39**, 2094–2097 (1968).

[DS65]   L. E. DUBINS and G. SCHWARZ. On continuous martingales. *Proc. Nat. Acad. Sci. USA* **53**, 913–916 (1965).

[Du02]   R. M. DUDLEY. *Real Analysis and Probability.* Cambridge University Press, Cambridge (2002).

[Du04]   B. DUPLANTIER. Conformal fractal geometry and boundary quantum gravity. In: *Fractal Geometry and Applications: A Jubilee of Benoît Mandelbrot.* Proc. Symposia Pure Math. **72**, 365–482, American Mathematical Society, Providence, RI (2004).

[DK88]   B. DUPLANTIER and K.-H. KWON. Conformal invariance and intersection of random walks. *Phys. Rev. Let.* **61**, 2514–2517 (1988).

[DL02]   T. DUQUESNE and J.-F. LE GALL. Random trees, Lévy processes and spatial branching processes. *Astérisque* **281**, 1–147 (2002).

[Du83]   P.L. DUREN. *Univalent functions*, Grundlehren der mathematischen Wissenschaften 259. Springer, New York (1983).

[Du84]   R. DURRETT. *Brownian motion and martingales in analysis.* Wadsworth, Belmont, CA (1984).

[Du95]   R. DURRETT. *Probability: Theory and examples.* Duxbury Press, Belmont, CA (1995).

[Du96]   R. DURRETT. *Stochastic calculus: a practical introduction.* CRC Press, Boca Raton, FL (1996).

[Dv63]   A. DVORETZKY. On the oscillation of the Brownian motion process. *Israel J. of Math.* **1**, 212–214 (1963).

[DE51]   A. DVORETZKY and P. ERDŐS. Some problems on random walk in space. *Proc. 2nd Berkeley Symp. Math. Statist. Probab.* 353–367 (1951).

[DEK50]   A. DVORETZKY, P. ERDŐS and S. KAKUTANI Double points of paths of Brownian motion in n-space. *Acta Sci. Math. Szeged* **12**, 75–81 (1950).

[DEK54]   A. DVORETZKY, P. ERDŐS and S. KAKUTANI Multiple points of paths of Brownian motion in the plane. *Bull. Res. Council Israel* **3**, 364–371 (1954).

[DEK58]   A. DVORETZKY, P. ERDŐS and S. KAKUTANI Points of multiplicity c of planar Brownian motion. *Bull. Res. Council Israel* **7**, 157–180 (1958).

[DEK61]   A. DVORETZKY, P. ERDŐS and S. KAKUTANI. Nonincrease everywhere of the Brownian motion process. *Proc. 4th Berkeley Symp. Math. Statist. Probab.* **2**, 103–116 (1961).

[DEKT57]   A. DVORETZKY, P. ERDŐS, S. KAKUTANI and S. J. TAYLOR Triple points of Brownian paths in 3-space. *Proc. Cambridge Philos. Soc.* **53**, 856–862 (1957).

[Dy57]   E. B. DYNKIN. Inhomogeneous strong Markov processes. *Dokl. Akad. Nauk. SSSR* **113**, 261–263 (1957).

[Ei05]   A. EINSTEIN. Über die von der molekularkinetischen Theorie der Wärme geforderte Bewegung von in ruhenden Flüssigkeiten suspendierten Teilchen. *Ann. Physik* **17**, 549–560 (1905).

[Ei56]   A. EINSTEIN. *Investigations on the theory of the Brownian movement.* Dover, New York (1956).

[Ei94]   N. EISENBAUM. Dynkin's isomorphism theorem and the Ray–Knight theorems. *Probab. Theory Related Fields* **99**, 321–335 (1994).

[EK46]   P. ERDŐS and M. KAC. On certain limit theorems in the theory of probability. *Bull. Amer. Math. Soc.* **52**, 292–302 (1946).

[EK47]   P. ERDŐS and M. KAC. On the number of positive sums of independent random variables. *Bull. Amer. Math. Soc.* **53**, 1011–1020 (1947).

[Ev85]   S. EVANS. On the Hausdorff dimension of Brownian cone points. *Math. Proc. Cambridge Philoph. Soc.* **98**, 343–353 (1985).

[Ev87a]   S. EVANS. Potential theory for a family of several Markov processes. *Ann. Inst. H. Poincaré Probab. Statist.* **23**, 499–530 (1987).

[Ev87b]   S. EVANS. Multiple points in the sample paths of a Lévy process. *Probab. Theory Related Fields* **76**, 359–367 (1987).

[Fa97a]   K.J. FALCONER. *Fractal geometry: mathematical foundations and applications.* Wiley, Chichester (1997).

[Fa97b]   K.J. FALCONER. *Techniques in fractal geometry.* Wiley, Chichester (1997).

[FH96]   K.J. FALCONER and J. D. HOWROYD. Projection theorems for box and packing dimension. *Math. Proc. Camb. Phil. Soc.* **119**, 287–295 (1996).

[Fe68]   W. FELLER. *An introduction to probability theory and its applications. Volume I.* Third edition. Wiley, New York (1968).

[Fe66]   W. FELLER. *An introduction to probability theory and its applications. Volume II.* Second edition. Wiley, New York (1966).

[FS89]   P.J. FITZSIMMONS and T. SALISBURY. Capacity and energy for multiparameter Markov processes. *Ann. Inst. Henri Poincaré, Probab.* **25**, 325–350 (1989).

[FF56]   L. R. FORD JR. and D. R. FULKERSON. Maximal flow through a network. *Canad. J. Math.* **8**, 399–404 (1956).

[FKG71]   C. M. FORTUIN, P. N. KASTELEYN and J. GINIBRE. Correlational inequalities for partially ordered sets. *Comm. Math. Phys.* **22**, 89–103 (1971).

[Fr83]   D. FREEDMAN. *Brownian motion and diffusion.* Second edition. Springer, New York (1983).

[Fr67]   B. FRISTEDT An extension of a theorem of S. J. Taylor concerning the multiple points of the symmetric stable process. *Z. Wahrscheinlichkeitstheorie verw. Gebiete* **9**, 62–64 (1967).

[Fr35]   O. FROSTMAN. Potential d'équilibre et capacité des ensembles avec quelques applications a la théorie des fonctions. *Meddel. Lunds Univ. Math. Sem.* **3**, 1–118 (1935).

[Fu80]   M. FUKUSHIMA.   *Dirichlet forms and Markov processes.* North-Holland, Amsterdam (1980).

[Ga40]   C. F. GAUSS. Allgemeine Lehrsätze in Beziehung auf die im verkehrten Verhältnisse des Quadrats der Entfernung wirkenden Anziehungs- und Abstoßungskräfte. *Gauss Werke* **5**, 197–242 (1840).

[GH80]   D. GEMAN and J. HOROWITZ. Occupation densities. *Ann. Probab.* **8**, 1–67 (1980).

[GHR84]   D. GEMAN, J. HOROWITZ and J. ROSEN. A local time analysis of intersections of Brownian motion in the plane. *Ann. Probab.* **12**, 86–107 (1984).

[Gr99]   D. GRABINER. Brownian motion in a Weyl chamber, non-colliding particles and random matrices. *Ann. Inst. Henri Poincaré, Probab. et Stat.* **35**, 177–204 (1999).

[Gr82]   S.E. GRAVERSEN. "Polar"-functions for Brownian motion. *Z. Wahrscheinlichkeitstheorie verw. Gebiete* **61**, 261–270 (1982).

[Gr28]   G. GREEN. *An essay on the application of mathematical analysis to the theories of electricity and magnetism.* Printed for the author by T. Wheelhouse, Nottingham (1828).

[GP83]   P. GREENWOOD and E. PERKINS. A conditioned limit theorem for random walk and Brownian local time on square root boundaries. *Ann. Probab.* **11**, 227–261 (1983).

[GP80]   P. GREENWOOD and J. PITMAN. Construction of local time and Poisson point processes from nested arrays *Journal London Math. Soc.* **22**, 182–192 (1980).

[Gr99]   G.R. GRIMMETT. *Percolation.* Second edition. Springer, Berlin (1999).

[HH80]   P. HALL and C.C. HEYDE. *Martingale limit theory and its applications.* Academic Press, New York (1980).

[HMO01]   B. HAMBLY, J. MARTIN and N. O'CONNELL. Pitman's $2M - X$ theorem for skip-free random walks with Markovian increments. *Electron. Comm. Probab.* **6**, 73–77 (2001).

[Ha19]   F. HAUSDORFF. Dimension und äußeres Maß. *Math. Ann.* **79**, 157–179 (1919).

[Ha60]   T. E. HARRIS. A lower bound for the critical probability in a certain percolation process. *Proc. Camb. Phil. Soc.* **56**, 13–20 (1960).

[HW41]   P. HARTMAN and A. WINTNER. On the law of the iterated logarithm. *J. Math.* **63**, 169–176 (1941).

[Ha81]   J. HAWKES. Trees generated by a simple branching process. *Journal London Math. Soc.*, **24**, 373–384 (1981).

[Ho95]   J.D. HOWROYD. On dimension and on the existence of sets of finite positive Hausdorff measure. *Proc. Lond. Math. Soc.*, III. **70**, 581–604 (1995).

[HT97]   X. HU and S. J. TAYLOR. The multifractal structure of stable occupation measure. *Stoch. Proc. Appl.* **66**, 283–299 (1997).

[HSY92]   B. R. HUNT, T. SAUER and J. A. YORKE. Prevalence: a translation-invariant "almost every" on infinite-dimensional spaces. *Bull. Amer. Math. Soc.* **27** (2), 217–238 (1992).

[Hu56]   G. A. HUNT. Some theorems concerning Brownian motion. *Trans. Amer. Math. Soc.* **81**, 294–319 (1956).

[IW89]   N. IKEDA and S. WATANABE. *Stochastic differential equations and diffusion processes.* Second edition. North-Holland, Amsterdam (1989).

[It44]   K. ITÔ. Stochastic integral. *Proc. Imp. Acad. Tokyo* **20**, 519–524 (1944).

[It51]   K. ITÔ. On a formula concerning stochastic differentials. *Nagoya Math. Journal* **3**, 55–55 (1951).

[IM74]   K. ITÔ and H.P. MCKEAN. *Diffusion processes and their sample paths.* Springer, Berlin (1974).

[Ja97]   S. JANSON. *Gaussian Hilbert spaces.* Cambridge University Press, Cambridge (1997).

[Ka51]   M. KAC. On some connections between probability theory and differential and integral equations. *Proc. 2nd Berkeley Symp. Math. Statist. Probab.* 189–215 (1951).

[Ka76]   J.-P. KAHANE. Sur les zéros et les instants de ralentissement du mouvement brownien. *C. R. Acad. Sci. Paris* **282**, 431–433 (1976).

[Ka85]   J.-P. KAHANE. *Some random series of functions.* Cambridge University Press, Cambridge (1985).

[Ka86]   J.-P. KAHANE. Sur la dimension des intersections. In: *Aspects of Mathematics and its Applications*, 419–430, North Holland, Amsterdam (1986).

[Ka44a]   S. KAKUTANI. On Brownian motions in n-space. *Proc. Imp. Acad. Tokyo* **20**, 648–652 (1944).

[Ka44b]   S. KAKUTANI. Two dimensional Brownian motion and harmonic functions. *Proc. Imp. Acad. Tokyo* **20**, 706–714 (1944).

[Ka45]   S. KAKUTANI. Markoff process and the Dirichlet problem. *Proc. Japan Acad.* **21**, 227–233 (1945).

[Ka02]   O. KALLENBERG. *Foundations of modern probability.* Second edition. Springer, New York (2002).

[Ka71]   G. KALLIANPUR. Abstract Wiener processes and their reproducing kernel Hilbert spaces. *Z. Wahrscheinlichkeitstheorie verw. Gebiete* **17**, 113–123 (1971).

[KR53]   G. KALLIANPUR and H. ROBBINS. Ergodic property of the Brownian motion process. *Proc. Nat. Acad. Sci. U.S.A.* **39**, 525–533 (1953).

[KS91]   I. KARATZAS and S. E. SHREVE. *Brownian motion and stochastic calculus.* Second edition. Springer, New York (1991).

[Ka69]   R. KAUFMAN. Une propriété métrique du mouvement brownien. *C. R. Acad. Sci. Paris* **268**, 727–728 (1969).

[Ka75]   R. KAUFMAN. Large increments of Brownian motion. *Nagoya Math. Journal* **56**, 139–145 (1975).

[KM09]   W. KENDALL and I. MOLCHANOV. *Perspectives in stochastic geometry.* Oxford University Press, Oxford (2009).

[Ke00a]   R. KENYON. Conformal invariance of domino tiling. *Ann. Probab.* **28**, 759–785 (2000).

[Ke00b]   R. KENYON. The asymptotic determinant of the discrete Laplacian. *Acta Math.* **185**, 239–286 (2000).

[Kh23]   A.Y. KHINCHIN. Über dyadische Brüche. *Math. Zeitschrift* **18**, 109–116 (1923).

[Kh24]   A.Y. KHINCHIN. Über einen Satz der Wahrscheinlichkeitsrechnung. *Fund. Math.* **6**, 9–20 (1924).

[Kh33]   A.Y. KHINCHIN. *Asymptotische Gesetze der Wahrscheinlichkeitsrechnung.* Springer Verlag, Berlin (1933).

[Kh02]   D. KHOSHNEVISAN. *Multiparameter processes.* Springer, New York (2002).

[Kh99]   D. KHOSHNEVISAN. Brownian sheet images and Bessel–Riesz capacity. *Trans. Amer. Math. Soc.* **351**, 2607–2622 (1999)

[KPX00]   D. KHOSHNEVISAN, Y. PERES and Y. XIAO. Limsup random fractals. *El. J. Probab.* **5**, Paper 4, pp.1–24 (2000).

[KS00]   D. KHOSHNEVISAN and Z. SHI. Fast sets and points for fractional Brownian motion. In: *Séminaire de Probabilités, XXXIV*, 393–416, Lecture Notes in Mathematics 1729. Springer, Berlin (2000).

[KX05]   D. KHOSHNEVISAN and Y. XIAO. Lévy processes: Capacity and Hausdorff dimension. *Ann. Probab.* **33**, 841–878 (2005).

[KM05]   A. KLENKE and P. MÖRTERS. The multifractal spectrum of Brownian intersection local times. *Ann. Probab.* **33**, 1255–1301 (2005).

[Kn63]   F. B. KNIGHT. Random walks and a sojourn density process of Brownian motion. *Trans. Amer. Math. Soc.* **109**, 56–86 (1963).

[Kn81]   F. B. KNIGHT. *Essentials of Brownian Motion and Diffusion.* American Mathematical Society, Providence, RI (1981).

[KS64]   S. KOCHEN and C. STONE. A note on the Borel-Cantelli lemma. *Illinois J. Math.* **8**, 248-251 (1964).

[Ko29]   A.N. KOLMOGOROV. Über das Gesetz des iterierten Logarithmus. *Math. Ann.* **101**, 126–135, (1929).

[KMT75]   J. KOMLÓS, P. MAJOR and G. TUSNÁDY. An approximation of partial sums of independent random variables and the sample distribution function. *Z. Wahrscheinlichkeitstheorie verw. Gebiete* **32**, 111–131 (1975).

[KM02]   W. KÖNIG and P. MÖRTERS. Brownian intersection local times: upper tails and thick points. *Ann. Probab.* **30**, 1605–1656 (2002).

[Ky06]   A.E. KYPRIANOU. *Introductory lectures on fluctuations of Lévy processes with applications.* Springer, Berlin (2006).

[La98]   M. LACZKOVICH. *Conjecture and proof.* TypoTeX, Budapest (1998).

[La09]   E. LANDAU. *Handbuch der Lehre von der Verteilung der Primzahlen.* Teubner, Leipzig (1909).

[LPS94]   R. LANGLANDS, Y. POULIOT and Y. SAINT–AUBIN. Conformal invariance in two-dimensional percolation. *Bull. Amer. Math. Soc.* **30**, 1–61 (1994).

[La96a]   G.F. LAWLER. Hausdorff dimension of cut points for Brownian motion. *El. Journal Probab.* **1** Paper 2, (1996).

[La96b]   G.F. LAWLER. The dimension of the frontier of planar Brownian motion. *El. Comm. Probab.* **1**, 29–47 (1996).

[La99]   G.F. LAWLER. Geometric and fractal properties of Brownian motion and random walk paths in two and three dimensions. In: *Random walks (Budapest 1998)* 219–258, Bolyaj Society Math. Stud., Volume 9, Budapest (1999).

[La05]   G.F. LAWLER. *Conformally invariant processes in the plane.* American Mathematical Society, Providence, RI (2005).

[La09]   G.F. LAWLER. Schramm-Loewner Evolutions (SLE). In : *Statistical mechanics.* IAS/Park City Mathematics Series 16. 231–295. American Mathematical Society, Providence, RI (2009).

[LSW01a]   G. F. LAWLER, O. SCHRAMM and W. WERNER. Values of Brownian intersection exponents I: Half-plane exponents. *Acta Mathematica* **187**, 237–273 (2001).

[LSW01b]   G. F. LAWLER, O. SCHRAMM and W. WERNER. Values of Brownian intersection exponents II: Plane exponents. *Acta Mathematica* **187**, 275–308 (2001).

[LSW01c]   G. F. LAWLER, O. SCHRAMM and W. WERNER. The dimension of the Brownian frontier is $4/3$. *Math. Res. Lett.* **8**, 401–411 (2001).

[LSW02]   G. F. LAWLER, O. SCHRAMM and W. WERNER. Analyticity of planar Brownian intersection exponents. *Acta Mathematica* **189**, 179–201 (2002).

[LSW03]   G. F. LAWLER, O. SCHRAMM and W. WERNER. Conformal restriction properties. The chordal case. *J. Amer. Math. Soc.* **16**, 917–955 (2003).

[LSW04]   G. F. LAWLER, O. SCHRAMM and W. WERNER. Conformal invariance of planar loop-erased random walks and uniform spanning trees. *Ann. Probab.* **32**, 939–996 (2004).

[LW00]   G. F. LAWLER and W. WERNER. Universality for conformally invariant intersection exponents. *J. Europ. Math. Soc.* **2**, 291–328 (2000).

[Le24]   H. LEBESQUE. Conditions de régularité, conditions d' irrégularité, conditions de impossibilité dans le problème de Dirichlet. *Comp. Rendu Acad. Sci.* **178**, 349–354 (1924).

[LG85]   J.-F. LE GALL. Sur la measure de Hausdorff de la courbe brownienne. In: *Séminaire de probabilités XIX*, 297–313, Lecture Notes in Mathematics 1123. Springer, Berlin (1985).

[LG86a]   J.-F. LE GALL. The exact Hausdorff measure of Brownian multiple points. In: *Seminar on Stochastic Processes* (1986), Birkhäuser, Basel, pp. 107–137.

[LG86b]   J.-F. LE GALL. Sur la saucisse de Wiener et les points multiples du mouvement brownien. *Ann. Probab.* **14**, 1219–1244 (1986).

[LG86c]   J.-F. LE GALL. Une approche élémentaire des théorèmes de décomposition de Williams. In: *Séminaire de probabilités XX*, 447–464, Lecture Notes in Mathematics 1204. Springer, Berlin (1986).

[LG87a]   J.-F. LE GALL. Le comportement du mouvement brownien entre les deux instants où il passe par un point double. *J. Funct. Anal.* **71**, 246–262 (1987).

[LG87b]   J.-F. LE GALL. The exact Hausdorff measure of Brownian multiple points. In: *Seimnar on Stochastic Processes 1986*, 107–137, Progress in Probability 13. Birkhäuser, Boston (1987).

[LG87c]   J.-F. LE GALL. Temps locaux d'intersection et points multiples des processus de Lévy. In: *Séminaire de probabilités XXI*, 341–374, Lecture Notes in Mathematics 1247. Springer, Berlin (1987).

[LG88a]   J.-F. LE GALL. Sur les fonctions polaires pour le mouvement brownien. In: *Séminaire de probabilités XXII*, 186–189, Lecture Notes in Mathematics 1321. Springer, Berlin (1988).

[LG88b]   J.-F. LE GALL. Fluctuation results for the Wiener sausage. *Ann. Probab.* **16**, 991–1018 (1988).

[LG92]   J.-F. LE GALL. Some properties of planar Brownian motion. In: *Lectures on probability theory and statistics*, 111–235, Lecture Notes in Mathematics 1527. Springer, Berlin (1992).

[LG99]   J.-F. LE GALL. *Spatial branching processes, random snakes and partial differential equations*. Lecture Notes in Mathematics (ETH Zürich). Birkhäuser, Basel (1999).

[LL98]   J.-F. LE GALL and Y. LE JAN. Branching processes in Lévy processes: the exploration process. *Ann. Probab.* **26**, 213–252 (1998).

[LRS89]   J.-F. LE GALL, J.S. ROSEN, and N.-R. SHIEH. Multiple points of Lévy processes. *Ann. Probab.* **17**, 503–515 (1989).

[Le66]   E. LEHMANN. Some concepts of dependence. *Ann. Math. Statist.* **37**, 1137–1153 (1966).

[Le37]   P. LÉVY. *Théorie de l'addition des variables aléatoires.* Gauthier-Villars, Paris (1937).

[Le39]   P. LÉVY. Sur certain processus stochastiques homogénes. *Comp. Math.* **7**, 283–339 (1939).

[Le40]   P. LÉVY. Le mouvement brownien plan. *Amer. J. Math.* **62**, 487–550 (1940).

[Le48]   P. LÉVY. *Processus stochastiques et mouvement Brownien.* Suivi d'une note de M. Loève. Gauthier-Villars, Paris (1948).

[Le51]   P. LÉVY. La mesure de Hausdorff de la courbe du mouvement brownien à n dimensions. *C. R. Acad. Sci. Paris* **233**, 600–602 (1951).

[Le54]   P. LÉVY. La mesure de Hausdorff de la courbe du mouvement brownien. *Giorn. Ist. Ital. Attuari* **16**, 1–37 (1954).

[Li95]   M.A. LIFSHITS. *Gaussian random functions.* Kluwer, Dordrecht (1995).

[Ly90]   R. LYONS. Random walks and percolation on trees. *Ann. Probab.* **18**, 931–958 (1990).

[Ly92]   R. LYONS. Random walks, capacity and percolation on trees. *Ann. Probab.* **20**, 2043–2088 (1992).

[Ly96]   R. LYONS. Probabilistic aspects of infinite trees and some applications. In: Trees, pp. 81–94, Birkhäuser, Basel (1996).

[LP05] R. LYONS with Y. PERES. *Probability on trees and networks*. In preparation. Preliminary version currently available at *http://php.indiana.edu/ rdlyons/prbtree/prbtree.html*

[Ly86] T.J. LYONS. The critical dimension at which quasi-every Brownian path is self-avoiding. *Adv. in Appl. Probab.* Suppl., 87–99 (1986).

[Ma85] N.G. MAKAROV. On the distortion of boundary sets under conformal mappings. *Proc. London Math. Soc.* **51**, 369–384 (1985).

[Ma82] B.B. MANDELBROT. *The Fractal Geometry of Nature*. Freeman, San Francisco, CA (1982).

[MR06] M.B. MARCUS and J. ROSEN. *Markov processes, Gaussian processes, and local times*. Cambridge University Press, Cambridge (2006).

[Ma06] A.A. MARKOV. Extension of the law of large numbers to dependent events. [Russian] *Bull. Soc. Phys. Math. Kazan* **15**, 135–156 (1906).

[Ma95] P. MATTILA. *Geometry of sets and measures in Euclidean spaces. Fractals and rectifiability*. Cambridge University Press, Cambridge (1999).

[MM97] P. MATTILA and R.D. MAULDIN. Measure and dimension functions: measurability and densities. *Math. Proc. Camb. Philos. Soc.* **121**, 81–100 (1997).

[McK55] H.P. MCKEAN JR. Hausdorff-Besicovitch dimension of Brownian motion paths. *Duke Math. J.* **22**, 229–234 (1955).

[McK69] H.P. MCKEAN. *Stochastic Integrals*. Academic Press, New York (1969).

[Me83] I. MEILIJSON. On the Azéma–Yor stopping time. In: *Séminaire de probabilités XVII*, 225–226, Lecture Notes in Mathematics 986. Springer, Berlin (1983).

[Me76] P. A. MEYER. Un cours sur les intégrales stochastiques. In: *Séminaire de probabilités X*, 246–400, Lecture Notes in Mathematics 511. Springer, Berlin (1976).

[MG84] D. P. MINASSIAN and J. W. GAISSER. A simple Weierstrass function. *Amer. Math. Monthly* **91**, 254–256 (1984).

[Mö00] P. MÖRTERS. Almost sure Kallianpur-Robbins laws for Brownian motion in the plane. *Probab. Theory Related Fields* **118**, 49–64 (2000).

[Mö02] P. MÖRTERS. A pathwise version of Spitzer's law. In: *Limit theorems in Probability and Statistics II*, János Bolyai Mathematical Society 427–436 (2002).

[MS09] P. MÖRTERS and N.-R. SHIEH. The exact packing measure of Brownian double points. *Probab. Theory Related Fields* **143**, 113–136 (2009).

[Mo89] T.S. MOUNTFORD. Time inhomogeneous Markov processes and the polarity of single points. *Ann. Probab.* **17** 573–585 (1989).

[Ne67] E. NELSON. *Dynamical theories of Brownian motion*. Princeton University Press, Princeton, NJ (1967).

[Ne70] R. NEVANLINNA. *Analytic functions*. Springer, Berlin (1970).

[Ne75] J. NEVEU. *Discrete-parameter martingales*. Revised edition. Translated from the French by T. P. Speed. North Holland, Amsterdam (1975).

[NP89] J. NEVEU and J.W. PITMAN. The branching process in a Brownian excursion. In: *Séminaire de probabilités XXIII*, 248–257, Lecture Notes in Mathematics 1372. Springer, Berlin (1989).

[Ob04] I. OBLÓJ The Skorokhod embedding problem and its offspring. *Probability Surveys* **1**, 321–390 (2004).

[Ok03] B. ØKSENDAL. *Stochastic differential equations*. Sixth edition. Springer, Berlin (2003).

[ON07] T.C. O'NEIL. The Hausdorff dimension of the visible sets of connected compact sets *Trans. Amer. Math. Soc.* **359**, 5141–5170 (2007).

[OT74] S. OREY and S. J. TAYLOR. How often on a Brownian path does the law of the iterated logarithm fail? *Proc. London Math. Soc.* **28**, 174–192 (1974).

[PWZ33]   R. E. A. C. PALEY, N. WIENER and A. ZYGMUND. Notes on random functions. *Math. Zeitschrift* **37**, 647–668 (1933).

[Pe95]   R. PEMANTLE. Tree-indexed processes. *Stat. Sci.* **10**, 200–213 (1995).

[Pe97]   R. PEMANTLE. The probability that Brownian motion almost contains a line. *Ann. Inst. H. Poincaré, Probab. Stat.* **33**, 147–165 (1997).

[PP07]   R. PEMANTLE and Y. PERES. What is the probability of intersecting the set of Brownian double points? *Ann. Probab.* **35**, 2044–2062 (2007).

[Pe89]   M.D. PENROSE. On the existence of self-intersections for quasi-every Brownian path in space. *Ann. Probab.* **17**, 482–502 (1986).

[Pe96a]   Y. PERES. Remarks on intersection-equivalence and capacity equivalence. *Ann. Inst. H. Poincaré (Physique Théorique)* **64**, 339–347 (1996).

[Pe96b]   Y. PERES. Intersection-equivalence of Brownian paths and certain branching processes. *Commun. Math. Phys.* **177**, 417–434 (1996).

[Pe96c]   Y. PERES. Points of increase for random walks. *Israel J. of Math.* **95**, 341–347 (1996).

[Pe99]   Y. PERES. Probability on trees: An introductory climb. In: *Lectures on probability theory and statistics*, 193–280, Lecture Notes in Mathematics 1717. Springer, Berlin (1999).

[Pe81]   E. PERKINS. The exact Hausdorff measure of the level sets of Brownian motion. *Z. Wahrscheinlichkeitstheorie verw. Gebiete* **58**, 373–388 (1981).

[Pe83]   E. PERKINS. On the Hausdorff dimension of Brownian slow points. *Z. Wahrscheinlichkeitstheorie verw. Gebiete* **64**, 369–399 (1983).

[PT87]   E. PERKINS and S.J. TAYLOR. Uniform measure results for the image of subsets under Brownian motion. *Probab. Theory Related Fields* **76**, 257–289 (1987).

[PT88]   E. PERKINS and S.J. TAYLOR. Measuring close approaches on a Brownian path. *Ann. Probab.* **16**, 1458–1480 (1988).

[PW23]   H. B. PHILLIPS and N. WIENER. Nets and Dirichlet problem. *J. Math. Phys.* **2**, 105–124 (1923).

[Pi75]   J.W. PITMAN. One-dimensional Brownian motion and the three-dimensional Bessel process. *Advances in Appl. Probability.* **7**, 511–526 (1975).

[PY86]   J. PITMAN and M. YOR. Asymptotic laws of planar Brownian motion. *Ann. Probab.* **14**, 733–779 (1986).

[PY92]   J. PITMAN and M. YOR. Arcsine laws and interval partitions derived from a stable subordinator. *Proc. London Math. Soc.* **65**, 326–356 (1992).

[PY03]   J. PITMAN and M. YOR. Hitting, occupation, and inverse local times of one-dimensional diffusions: martingale and excursion approaches. *Bernoulli* **9**, 1–24 (2003).

[PY07]   J. PITMAN and M. YOR. Itô's excursion theory and its applications. *Japanese Journal of Mathematics* **2**, 83–96 (2007).

[Po90]   H. POINCARÉ. Sur les équations aux derivées partielles de la physique mathématique. *Amer. J. Math.* **12**, 211–779 (1986).

[Po21]   G. PÓLYA. Über eine Aufgabe der Wahrscheinlichkeitsrechnung betreffend die Irrfahrt im Strassennetz. *Math. Ann.* **84**, 149–160 (1921).

[PS78]   S.C. PORT and C.J. STONE Brownian motion and classical potential theory. Academic Press, New York (1978).

[Pr56]   Y.V. PROHOROV. Convergence of random processes and limit theorems in probability. *Th. Probab. Appl.* **1**, 157–214 (1956).

[Pr90]   W.E. PRUITT. The rate of escape of random walk. *Ann. Probab.* **18**, 1417–1461 (1990).

[PU89]   F. PRZYTYCKI and M. URBAŃSKI. On the Hausdorff dimension of some fractal sets. *Studia Math.* **93**, 155–186 (1989).

[Ra77]   M. RAO *Brownian motion and classical potential theory.* Matematisk Institut, Aarhus University, Aarhus (1977).

[Ra63a]   D. RAY. Sojourn times and the exact Hausdorff measure of the sample path for planar Brownian motion. *Trans. Amer. Math. Soc.* **106**, 436–444 (1963).

[Ra63b]   D. RAY. Sojourn times of diffusion processes. *Illinois J. Math.* **7**, 615–630 (1963).

[RY94]   D. REVUZ and M. YOR. *Continuous martingales and Brownian motion.* Second edition. Springer, Berlin (1994).

[RP49]   H. ROBBINS and E.J.G. PITMAN. Application of the method of mixtures to quadratic forms. *Ann. Math. Statist.* **20**, 552–560 (1949).

[Ro99]   C.A. ROGERS. *Hausdorff measures.* Cambridge University Press, Cambridge (1999).

[RT61]   C.A. ROGERS and S.J. TAYLOR. Functions continuous and singular with respect to a Hausdorff measure. *Mathematika* **8**, 1–31 (1961).

[RW00a]   L.C.G. ROGERS and D. WILLIAMS. *Diffusions, Markov processes, and martingales. Vol. 1: Foundations* Cambridge University Press, Cambridge (2000).

[RW00b]   L.C.G. ROGERS and D. WILLIAMS. *Diffusions, Markov processes, and martingales. Vol. 2: Itô Calculus* Cambridge University Press, Cambridge (2000).

[RS05]   S. ROHDE and O. SCHRAMM. Basic properties of SLE. *Ann. Math.* **161**, 879–920 (2005)

[Ro69]   D. H. ROOT. The existence of certain stopping times on Brownian motion. *Ann. Math. Statist.* **40**, 715–718 (1969).

[Ro83]   J. ROSEN. A local time approach to the self-intersections of Brownian paths in space. *Comm. Math. Phys.* **88**, 327–338 (1983).

[Ru87]   W. RUDIN. *Real and complex analysis.* Third edition. McGraw-Hill, New York (1987).

[Sc00]   O. SCHRAMM. Scaling limits of loop-erased random walks and uniform spanning trees. *Israel J. of Math.* **118**, 221–288 (2000).

[Se95]   L. SERLET. Some dimension results for super-Brownian motion. *Probab. Theory Related Fields* **101**, 371–391 (1995).

[Sh67]   L.A. SHEPP. A first passage problem for the Wiener process. *Ann. Math. Statist.* **38**, 1912–1914 (1967).

[Sh98]   Z. SHI. Windings of Brownian motion and random walks in the plane. *Ann. Probab.* **26**, 112–131 (1998).

[ST99]   N.-R. SHIEH and S.J. TAYLOR. Logarithmic multifractal spectrum of stable occupation measure. *Stochastic Process Appl.* **79**, 249–261 (1998).

[Sk65]   A. SKOROKHOD. *Studies in the theory of random processes.* Addison-Wesley, Reading, MA (1965).

[Sm01]   S. SMIRNOV. Critical percolation in the plane: conformal invariance, Cardy's formula, scaling limits. *C. R. Acad. Sci. Paris Ser. I Math.* **333**, 239–244 (2001).

[Sm07]   S. SMIRNOV. Towards conformal invariance of 2D lattice models. *International Congress of Mathematicians. Vol. II,* 1421–1451, European Mathematical Society, Zürich (2007).

[Sm08]   S. SMIRNOV. Conformal invariance in random cluster models. I. Holomorphic fermions in the Ising model. *Ann. Math.*, to appear (2008).

[Sp58]   F. SPITZER. Some theorems concerning 2-dimensional Brownian motion. *Trans. Amer. Math. Soc.* **87**, 187–197 (1958).

[Sp64]   F. SPITZER. Electrostatic capacity, heat flow, and Brownian motion. *Z. Wahrscheinlichkeitstheorie verw. Gebiete* **3**, 110–121 (1964).

[Sp76]   F. SPITZER. *Principles of random walk.* Second edition. Springer, New York (1976).

[St75]   F. STERN. Conditional expectation of the duration in the classical ruin problem. *Mathematics magazine* **48**, 200–203 (1975).

[St64]   V. STRASSEN. An invariance principle for the law of the iterated logarithm. *Z. Wahrscheinlichkeitstheorie verw. Gebiete* **3**, 211–226 (1964).

[Ta63]   H. TANAKA Note on continuous additive functionals of the one-dimensional Brownian path. *Z. Wahrscheinlichkeitstheorie verw. Gebiete* **1**, 251–157 (1963).

[Ta53]   S. J. TAYLOR. The Hausdorff α-dimensional measure of Brownian paths in n-space. *Proc. Cambridge Philos. Soc.* **49**, 31–39 (1953).

[Ta55]   S. J. TAYLOR. The α-dimensional measure of the graph and set of zeros of a Brownian path. *Proc. Cambridge Philos. Soc.* **51**, 265–274 (1955).

[Ta64]   S. J. TAYLOR. The exact Hausdorff measure of the sample path for planar Brownian motion. *Proc. Cambridge Philos. Soc.* **60**, 253–258 (1964).

[Ta66]   S. J. TAYLOR. Multiple points for the sample paths of the symmetric stable process. *Zeitschr. Wahrscheinlichkeitstheorie verw. Gebiete* **5**, 247–264 (1966).

[Ta72]   S. J. TAYLOR. Exact asymptotic estimates of Brownian path variation. *Duke Math. J.* **39**, 219–241 (1972).

[Ta86]   S. J. TAYLOR. The measure theory of random fractals. *Proc. Cambridge Philos. Soc.* **100**, 383–406 (1986).

[TT85]   S. J. TAYLOR and C. TRICOT. Packing measure and its evaluation for a Brownian path. *Trans. Amer. Math. Soc.* **288**, 679–699 (1985).

[TW66]   S. J. TAYLOR and J. G. WENDEL. The exact Hausdorff measure of the zero set of a stable process. *Zeitschr. Wahrscheinlichkeitstheorie verw. Gebiete* **6**, 170–180 (1966).

[To88]   N. TONGRING. Which sets contain multiple points of Brownian motion? *Math. Proc. Camb. Phil. Soc.* **103**, 181–187 (1988).

[Tr82]   C. TRICOT. Two definitions of fractional dimension. *Math. Proc. Cambridge Philos. Soc.* **91**, 57–74 (1982).

[Tr58]   H. TROTTER. A property of Brownian motion paths. *Illinois J. Math.* **2**, 425–433 (1958).

[Wa78]   J.B. WALSH. Temps locaux. *Astérisque* 52-53, 159-192 (1978).

[We96]   W. WERNER. Bounds for disconnection exponents. *Electr. Comm. Probab.* **1**, 19–28 (1996).

[We04]   W. WERNER. Random planar curves and Schramm-Loewner evolutions. In: *Lectures on probability theory and statistics*, 107–195, Lecture Notes in Mathematics 1840. Springer, Berlin (2004).

[We05]   W. WERNER. Conformal restriction and related questions. *Probability Surveys* **2**, 145–190 (2005).

[We07]   W. WERNER Lectures on two-dimensional critical percolation. In : *Statistical mechanics.* IAS/Park City Mathematics Series 16. 297–360. American Mathematical Society, Providence, RI (2009).

[We08a]   W. WERNER. The conformal invariant measure on self-avoiding loops. *J. Amer. Math.Soc.* **21**, 137–169 (2008).

[Wi23]   N. WIENER. Differential space. *Journal Math. Phys.* **2**, 131–174 (1923).

[Wi70]   D. WILLIAMS. Decomposing the Brownian path. *Bull. Amer. Math. Soc.* **76**, 871–873 (1970).

[Wi74]   D. WILLIAMS. Path decomposition and continuity of local time for one-dimensional diffusions. I *Proc. London Math. Soc.* **28**, 738–768 (1974).

[Wi77]   D. WILLIAMS. Lévy's downcrossing theorem. *Z. Wahrscheinlichkeitstheorie verw. Gebiete* **40**, 157–158 (1977).

[Wi91] D. WILLIAMS. *Probability with martingales.* Cambridge University Press, Cambridge (1991).

[Xi04] Y. XIAO. Random fractals and Markov processes. In: *Fractal Geometry and Applications: A Jubilee of Benoît Mandelbrot.* Proc. Symposia Pure Math. **72**, 261–338, American Mathematical Society, Providence, RI (2004).

[Yo92] M. YOR. *Some aspects of Brownian motion. Part I: Some special functionals.* Birkhäuser, Basel (1992).

[Za11] S. ZAREMBA. Sur le principe de Dirichlet. *Acta Math.* **34**, 293–316 (1911).

Index